HANDBOOK OF
WATER SENSITIVE PLANNING and DESIGN

HANDBOOK OF WATER SENSITIVE PLANNING and DESIGN

Edited by
Robert L. France, B.Sc., M.Sc., Ph.D.
Associate Professor of Landscape Ecology
Science Director of the
Center for Technology and Environment (CTE)
Graduate School of Design
Harvard University
Cambridge, Massachusetts

LEWIS PUBLISHERS

A CRC Press Company
Boca Raton   London   New York   Washington, D.C.

TD
657
.H36
2002
c. 2

Library of Congress Cataloging-in-Publication Data

Handbook of water sensitive planning and design / edited by Robert L. France.
    p. cm. — (Integrative studies in water management and land development)
    Includes bibliographical references and index
    ISBN 1-56670-562-2 (alk. paper)
    1. Runoff—Management—Handbooks, manuals, etc. 2. Landscape ecology—Handbooks, manuals, etc. 3. Ecological landscape design—Handbooks, manuals, etc. 4. Water in landscape architecture—Handbooks, manuals, etc. 5. Buffer zones (Ecosystem management)—Handbooks, manuals, etc. 6. Watershed management—Handbooks, manuals, etc. I. France, R. L. (Robert Lawrence) II. Series.

TD657 .H36 2002
333.91—dc21

2002016075

This book contains information obtained from authentic and highly regarded sources. Reprinted material is quoted with permission, and sources are indicated. A wide variety of references are listed. Reasonable efforts have been made to publish reliable data and information, but the authors and the publisher cannot assume responsibility for the validity of all materials or for the consequences of their use.

Neither this book nor any part may be reproduced or transmitted in any form or by any means, electronic or mechanical, including photocopying, microfilming, and recording, or by any information storage or retrieval system, without prior permission in writing from the publisher.

All rights reserved. Authorization to photocopy items for internal or personal use, or the personal or internal use of specific clients, may be granted by CRC Press LLC, provided that $1.50 per page photocopied is paid directly to Copyright clearance Center, 222 Rosewood Drive, Danvers, MA 01923 USA The fee code for users of the Transactional Reporting Service is ISBN 1-56670-562-2/02/$0.00+$1.50. The fee is subject to change without notice. For organizations that have been granted a photocopy license by the CCC, a separate system of payment has been arranged.

The consent of CRC Press LLC does not extend to copying for general distribution, for promotion, for creating new works, or for resale. Specific permission must be obtained in writing from CRC Press LLC for such copying.

Direct all inquiries to CRC Press LLC, 2000 N.W. Corporate Blvd., Boca Raton, Florida 33431.

Trademark Notice: Product or corporate names may be trademarks or registered trademarks, and are used only for identification and explanation, without intent to infringe.

Visit the CRC Press Web site at www.crcpress.com

© 2002 by CRC Press LLC
Lewis Publishers is an imprint of CRC Press LLC

No claim to original U.S. Government works
International Standard Book Number 1-56670-562-2
Library of Congress Card Number 2002016075
Printed in the United States of America   3 4 5 6 7 8 9 0
Printed on acid-free paper

Series Statement

Integrative studies in water management and land development

Ecological issues and environmental problems have become exceedingly complex. Today, it is hubris to suppose that any single discipline can provide *all* the solutions for protecting and restoring ecological integrity. We have entered an age where professional humility is the only operational means for approaching environmental understanding and prediction. As a result, socially acceptable and sustainable solutions must be both imaginative and integrative in scope; in other words, garnered through combining insights gleaned from various specialized disciplines, expressed and examined together.

The purpose of the series *Integrative Studies in Water Management and Land Development* by Lewis Publishers, part of CRC Press, is to produce a set of books that transcend the disciplines of science and engineering alone. Instead, these efforts will be truly integrative in their incorporation of additional elements from landscape architecture, land-use planning, economics, education, and environmental management, history, and art. The emphasis of the series will be on the breadth of study approach, coupled with the depth of intellectual vigor required for the investigations undertaken.

Dr. Robert L. France
Series Editor
Harvard University

Foreword

"One can first of all simply wonder at the fact that it is only recently that humanity has begun to ponder the evolution and fate of water in the world when the very survival of our species depends on it." So begins the report from the UNESCO World Commission on the Ethics of Scientific Knowledge and Technology (*The Ethics of Freshwater Use: A Survey*, Lord Selborne, 2000). The report goes on to challenge our current culture in which water, in public policy agendas or in the news, is most frequently represented as either a hapless victim of pollution, as a malevolent foe causing flooding or drought, or as a guilty instigator of regional conflict. In contrast, the report stresses that "we need to take a constructive approach to water," where it is treated as "a foremost priority in every community from the local to the global."

Water infiltrates every aspect of our biological, cultural, and spiritual lives (*Deep Immersion: The Experience of Water*, Robert Lawrence France, 2002), yet we continue to regard it with derision if we even think about it at all. We imagine our lives as if somehow they were uncoupled from the hydrological cycle, giving little thought to the ramifications of our actions on those living downstream from our drains and streets. For many urban dwellers, the experience of water remains restricted to the distance it falls from the tap to the sink or that which carries away our waste upon flushing. Those championing the beauty and emotional benefits of water are, for the most part, little-heard voices in a deadening silence (Water-logged-in.com, Robert Lawrence France, 2001). The time is ripe for a major paradigm shift in how we regard water on both the exterior landscapes of our environments as well as the interior landscapes of our cultural sociology.

We urgently need to move toward achieving ecological integrity through water sensitive planning and design. The earth's hydrological cycle has been altered through centuries of human activity. Due to its corporal form, water is an important integrator of environmental disturbance across spatial scales from small subdivisions to regional drainage basins. As a result, local site-specific actions are cumulative on larger landscapes. This behooves us to exercise caution and to employ wisdom in the planning and design of individual development projects. Given the demonstrable decreases in environmental integrity already manifest, proactive planning and design can provide positive environmental benefits.

Landscape architects have become active participants and leaders in developing low-impact stormwater management, wetland park creation, riparian buffer-corridor planning, and watershed management. It is hoped that this book will come to be regarded as a landmark effort in which to shape and advance these emerging fields of truly integrative water management and land development. In that light, this book has three goals:

1. To present promising procedures, options, and limitations in water management across the spatial scale from parking lots to bioregions
2. To introduce novel approaches and explore future directions in water sensitive site-specific design, watershed planning, and waterside development policy
3. To examine case studies employing innovative techniques used by a variety of disciplines representing those concerned with water sensitive development

Koichiro Matsuura, the Director-General of UNESCO, concluded his message on the occasion of World Water Day 2000 with the following apt statement: "There is a fundamental truth which I would like to emphasize...the water supply does not run dry when it is drawn from the well of human wisdom." Readers will find a wellspring of such wisdom contained within these pages, arising from the contributors whose work is documented here. Therefore, drink deep, think hard, and act imaginatively.

Robert L. France, B.Sc., M.Sc., Ph.D.
Associate Professor of Landscape Ecology
Science Director of the Center for Technology and Environment
Graduate School of Design, Harvard University,
Founder and Principal
W.D.N.R.G. Limnetics

Preface

Origin of the book

In Spring 2000, a symposium was convened at the Graduate School of Design at Harvard University. Titled "Water Sensitive Ecological Planning and Design," this gathering was sponsored by the Department of Landscape Architecture and brought together more than 50 presenters from North America and Europe (see www.gsd.harvard.edu/watersymp for further details). The symposium was perhaps most successful and instrumental in demonstrating the extremely broad range of interests and professional approaches encompassed within water sensitive planning and design. In addition to landscape architecture, the following disciplines were represented by the presenters: governance, history, fisheries resource management, environmental economics, ecotoxicology, limnology, urban planning, hydrology, land-use planning, civil engineering, environmental conflict resolution, soil science, citizen activism, ornithology, environmental arts, cultural geography, extension school outreach, conservation biology, education, performance arts, and watershed policy management. The symposium was attended by over 300 practitioners, academics, students, and interested private citizens. Based on the formal written responses and informal comments received in the time since, many attendees felt the meeting to be one of the most significant in which they had ever participated. The present book was developed from a subset of 27 of the oral and poster presentations at the Harvard Design School symposium, in addition to another eight works that were independently solicited.

Purpose of the book

The book is designed to introduce, describe, and demonstrate new interpretations to water management in a form to engage the broadest audience possible. A few readers will find familiar elements here, but even the most educated will be surprised at the variety of novel approaches explored and developed within these pages. This collection of papers pushes the frontiers of standard water management toward new directions, challenging readers into abandoning the comfortable safety of conducting business-as-usual within narrow disciplinary confines, and instead directing views outward to the exciting and incompletely mapped regions of true interdisciplinary water sensitive planning and design. Represented in these pages are the pioneering efforts of those individuals working at the intersections of paradigms, those very regions where — as Thomas Kuhn informs us — visionary insight and conceptual evolution are most likely to occur.

Structure of the book

The book is arranged in two parts that loosely pertain to the physical scale of research and study scope. In Part I, 17 chapters address the subject of site-specific water sensitive design, and in Part II, another 17 chapters focus on issues relating to the water sensitive planning of riparian buffers and watersheds. Even a brief perusal of these chapters, however, will strike perceptive readers that this structuring is somewhat artificial in that one person's concept of "design" might very well be another's idea of "planning." In other words, the thematic structuring of chapters in the book really has as much to do about convenience as it does about trying to conceptually confine an element such as water, which by its very nature is known for its ability to defy rigid compartmentalization.

The "Overview" papers leading each of the "design" and "planning" sections of the book are intended to provide a brief roadmap to navigate through the 17 subsequent chapters. At the end of each of these sections are two point-form summaries arising from taped discussions with presenters and a few invited participants attending the Harvard Design School symposium. These discussions provided opportunities for leaders to elaborate on past problems, current challenges, and future directions in their respective fields. Following each chapter is an accompanying "Response" whose purpose is to succinctly tease out several salient features from the chapter in more detail, as well as to emphasize cross-linking to other chapters in the book.

And finally, many of the figures in the book are available in color on an accompanying Web page: **www.gsd.harvard.edu/watercolors.** Use the password **lentic-lotic.** Readers are advised that the color versions of images may be particularly helpful in appreciating subtleties in the design projects and in understanding some of the planning maps.

Acknowledgments

No conference at the Harvard Design School is ever the product of the efforts of a sole individual. George Hargreaves, Chairman of the Department of Landscape Architecture, is to be credited for having the wisdom to insist that both design and planning be represented here together. His continued support, as well as that from the Dean of the Design School, Peter Rowe, were instrumental to bringing about the success that the symposium has since come to be regarded in professional circles. In terms of logistics, it is no exaggeration to state that the entire undertaking would have been impossible without the dedicated "above-and-beyond" efforts of a few individuals: Doug Cogger; Trevor O'Brien; and most especially, Lora Nielson, to whom I am ever indebted. I also wish to acknowledge the help of Francois de Kock in designing the Web page, Markley Bavinger in preparing the slides for the introductory lecture, and that of the moderators at the symposium: Nicholas Pouder, Francois de Kock, John Felkner, Todd Gilens, Mike Flaxman, and Markley Bavinger, in addition to all participants in the taped discussion workshops.

Transcription of a symposium to a final book is never as easy a process as one might imagine. I wish to thank the acquisition and production staff at Lewis Publishers, in particular Randi Gonzalez for her help in the "cat herding" process, Brian Kenet for his enthusiasm in initiating this new series, and especially the late Arline Massey, to whose judgment this present product owes its existence and to whom I would like to dedicate this book.

Finally, I would like to express my sincere thanks to the contributors to this publication from whom I have learned so much: Glenn Allen, Jim Bays, Diana Balmori, Judi Barnes, Brian Bear, Tom Benjamin, Catherine Berris, Mike Binford, David Blau, Margot Cantwell, Kelly Cave, Mike Clar, Larry Coffman, Tim Collins, Phil Craul, Jay Dorsey, John Felkner, Bruce Ferguson, Mike Flaxman, Chuck Flink, Wendi Goldsmith, Dennis Haag, Stephen Hust, Neil Hutchinson, Gail Krantzberg, Tom Liptan, Jim MacBroom, Kaki Martin, Frank Mitchell, Amir Mueller, Bob Murase, Richard Pinkham, Nick Pouder, Rob Rempel, Clarissa Rowe, Jeff Schloss, Larry Schwartz, Carl Steinitz, Mark Vian, Desheng Wang, Anne Weekes, Dan Williams, Jennifer Zielinski, and Leslie Zucker. It is the insightful vision and dedicated efforts of these individuals to whom we are all indebted for their instructional offerings about how to live sustainably and more sanely in a healthy world where water is regarded, protected, and cherished with the respect that it so deserves.

Robert L. France
Cambridge, Massachusetts
September 2001

About the Editor

Robert France has published more than 100 papers on the ecology and conservation biology of organisms from bacteria and algae to birds and whales, on research topics from environmental pollution to theoretical biodiversity, and in locations ranging from the High Arctic to the tropics (www.gsd.harvard.edu/info/directory/faculty/france/cv.htm).

He currently teaches courses at the Harvard Design School on the influence of landscape processes and development on aquatic systems, and on the ecopsychology of human–nature relationships. He is founder of the firm W.D.N.R.G. Limnetics, which specializes in the restoration of degraded waterways.

Dr. France is senior editor at Green Frigate Books and also serves as series editor at Lewis/CRC Press for Integrative Studies in Water Management and Land Development. His books include: *Designing Wetlands: Principles and Practices for Landscape Architects and Land-use Planners, Landscape Architecture of Created Wetlands: CD-ROM of 17 Virtual Visual Tours, Reflecting Heaven: Thoreau on Water, Profitably Soaked: Thoreau's Engagement with Water,* and *Deep Immersion: The Experience of Water.*

Contributors

Glenn Allen
Hargreaves Associates
Cambridge, Massachusetts

Diana Balmori
Balmori Associates
New York, New York

Judi Barnes
Ontario Ministry of Environment
Toronto, Ontario, Canada

James S. Bays
CH2M Hill
Tampa, Florida

Bryan J. Bear
City of Overland Park
Overland Park, Kansas

Thomas S. Benjamin
Rizzo Associates
Framingham, Massachusetts

Catherine Berris
Catherine Berris Associates Inc.
Vancouver, British Columbia, Canada

Michael W. Binford
Department of Geography
University of Florida
Gainesville, Florida

David Blau
EDAW, Inc.
San Francisco, California

Margot Young Cantwell
Environmental Design and Management, Ltd.
Halifax, Nova Scotia, Canada

Kelly A. Cave
Wayne County Department
 of Environment
Watershed Management Division
Detroit, Michigan

Michael L. Clar
Ecosite, Inc.
Laurel, Maryland

Larry S. Coffman
Prince George's County
Department of Environmental Resources
Largo, Maryland

Timothy Collins
Carnegie Mellon University
College of Fine Arts
Pittsburgh, Pennsylvania

Philip Craul
Harvard Design School
Cambridge, Massachusetts

Jay D. Dorsey
Oxbow River and Stream Restoration
Delaware, Ohio

John S. Felkner
University of Chicago
Department of Economics
Chicago, Illinois

Bruce K. Ferguson
University of Georgia
School of Environmental Design
Athens, Georgia

Michael Flaxman
Harvard Design School
Cambridge, Massachusetts

Charles A. Flink
Greenways, Inc.
Durham, North Carolina

Robert L. France
WDNRG Limnetics and Harvard Design
 School
Cambridge, Massachusetts

Wendi Goldsmith
The Bioengineering Group, Inc.
Salem, Massachusetts

Dennis A. Haag
Tetra Tech EM, Inc.
Lenexa, Kansas

Stephen A. Hurst
Kansas Water Office
Topeka, Kansas

Neil J. Hutchinson
Gartner Lee Ltd.
Markham, Ontario, Canada

Gail Krantzberg
International Joint Commission
Great Lakes Regional Office
Windsor, Ontario, Canada

Thomas Liptan
Bureau of Environmental Services
Portland, Oregon

James G. MacBroom
Milone & MacBroom, Inc.
Cheshire, Connecticut

Kaki Martin
Wallace Floyd Design Group
Boston, Massachusetts

Frank Mitchell
University of New Hampshire
 Cooperative Extension
Durham, New Hampshire

Amir Mueller
ADM Landscape Architecture
 Environmental Design
Raanana, Israel

Robert K. Murase
Murase Associates
Portland, Oregon

W. Erik Olson
Indian River County
Vero Beach, Florida

Richard D. Pinkham
Rocky Mountain Institute
Snowmass, Colorado

Nicholas Pouder
Pouder Design Group
Bedford, New York

Robert Rempel
Ontario Ministry
 of Natural Resources
Centre for Northern Ecosystem
 Research
Thunder Bay, Ontario, Canada

Clarissa Rowe
Brown, Richardson and Rowe, Inc.
Boston, Massachusetts

Jeffrey A. Schloss
University of New Hampshire
 Cooperative Extension
Durham, New Hampshire

Larry N. Schwartz
Camp Dresser and McKee, Inc.
Maitland, Florida

Carl Steinitz
Harvard Design School
Cambridge, Massachusetts

Mark A. Vian
DeepStreams Consulting
Shokan, New York

Desheng Wang
Carr Research Laboratory, Inc.
Natick, Massachusetts

Anne Weekes
University of Washington
Seattle, Washington

Daniel Williams
Daniel Williams Architect
Seattle, Washington

Lee P. Wiseman
Camp Dresser and McKee, Inc.
Maitland, Florida

Jennifer A. Zielinski
Center for Watershed Protection
Ellicott City, Maryland

Leslie Zucker
Department of Natural Resources
Cornell University
Ithaca, New York

Contents

Series statement
Foreword
Preface

Background — Perspectives of water management: Representative examples from the recent literature 1
Robert France

PART I
WATER SENSITIVE DESIGN
Responses by Robert France

Overview: New interpretations in stormwater management and wetland park creation 9

I.1 **Stormwater management and stormwater restoration** 11
Bruce K. Ferguson

Response — Stormwater infiltration: Curing the disease rather than treating the symptoms

I.2 **Successful stormwater management ponds (Massachusetts)** 31
Desheng Wang

Response — Centralized stormwater treatment: Improving performance through engineering design

I.3 **Open spaces and impervious surfaces: Model development principles and benefits** 49
Jennifer A. Zielinski

Response — Using computer scenarios to improve site design

I.4 **Post-industrial watersheds: Retrofits and restorative redevelopment (Pittsburgh, Pennsylvania)** 67
Richard D. Pinkham and Timothy Collins

Response — Raising consciousness through interdisciplinary design workshops

I.5 **Low-impact development: An alternative stormwater management technology** 97
Larry S. Coffman

Response — Thinking big, acting small: Multi-tasking and the benefits of dispersed micromanagement

I.6 **Water gardens as stormwater infrastructure (Portland, Oregon)** 125
Thomas Liptan and Robert K. Murase

Response — Letting it soak in

I.7 **Retaining water: Technical support for capturing parking lot runoff (Ithaca, New York)** 155
Robert France and Philip Craul

Response — To build an oxymoron: A green parking lot

I.8 **A productive stormwater park (Farmington, Minnesota)** 175
Diana Balmori

Response — Successfully marrying form and function in stormwater management

I.9 **A stormwater wetland becomes a nature park (British Columbia, Canada)** 193
Catherine Berris

Response — Naturalized design: Triumph of imagination and innovation

I.10 **Wetlands-based indirect potable reuse project (West Palm Beach, Florida)** 205
Larry N. Schwartz, Lee P. Wiseman, and W. Erik Olson

Response — Treating wastewater with innovative technology

I.11 **Restoring urban wetland — pond systems (Boston, Massachusetts)** 215
Clarissa Rowe

Response — Project development through concerned citizenry

I.12 **Water connections: Wetlands for science instruction (Wichita, Kansas)** 235
Robert France and Kaki Martin

Response — Project development of interpretive wetlands

I.13 **Constructed wetlands and stormwater management at the Northern Water Feature (Sydney Olympic Park)** 247
Glenn Allen

Response — Highly visible water: Recreating a landscape for public use

I.14 **Principles and applications of wetland park creation** 263
James S. Bays

Response — Designing wetlands for multiple benefits

I.15 Applications of low-impact development techniques (Maryland)................297
Michael L. Clar

Response — Values of demonstration projects and case studies
of stormwater source management

I.16 Restoring and protecting a small, urban lake (Boston, Massachusetts)...............317
Nicholas Pouder and Robert France

Response — Buying time by bioengineering

**I.17 Integrated ecology, geomorphology, and bioengineering
for watershed-friendly design**................341
Wendi Goldsmith

Response — Sustainability through interdisciplinarity

**Discussion summary: Constraints, challenges, and opportunities
in implementing innovative stormwater management techniques**................355

**Discussion summary: Moving from single-purpose treatment wetlands
toward multifunction designed wetland parks**................357

PART II
WATER SENSITIVE PLANNING
Responses by Robert France

**Overview: New interpretations in the management of watersheds and
riparian buffers and corridors**................359

**II.1 Shoreline buffers: Protecting water quality and biological diversity
(New Hampshire)**................361
Frank Mitchell

Response — Buffer strips: More than green eyelashes?

II.2 River restoration planning (Connecticut)................379
James G. MacBroom

Response — Water quality improvements are not enough

II.3 Greenways as green infrastructure in the new millennium................395
Charles A. Flink

Response — Corridors that integrate natural, societal, and social elements

**II.4 Natural resource stewardship planning and design: Fresh Pond Reservation
(Massachusetts)**................407
Thomas S. Benjamin

Response — Protecting and restoring treasured landscapes:
Complexity and integration

II.5 Treating rivers as systems to meet multiple objectives 431
Leslie Zucker, Anne Weekes, Mark Vian, and Jay D. Dorsey

Response — Beyond the banks: Holistic planning of rivers as more than the sum of their parts

II.6 What progress has been made in the Remedial Action Plan program after ten years of effort? (Ontario, Canada) 445
Gail Krantzberg and Judi Barnes

Response — Measuring recovery of impaired waters

II.7 Watershed management plans: Bridging from science to policy to operations (San Francisco, California) 459
David Blau

Response — Sociology of implementing adaptive management

II.8 Watershed assessment planning process assessment (Johnson County, Kansas) 477
Dennis A. Haag, Stephen A. Hurst, and Bryan J. Bear

Response — Managing suburban watersheds for multiple objectives

II.9 Urban watershed management (Detroit, Michigan) 491
Kelly A. Cave

Response — Looking beyond the end of the pipe

II.10 Modeling a soil moisture index using geographic information systems in a developing country context (Thailand) 513
John S. Felkner and Michael W. Binford

Response — Incorporating scientific information into land-use planning

II.11 The design of regions: A watershed planning approach to sustainability 541
Daniel Williams

Response — Expanding planning vision in space and time

II.12 GIS watershed mapping: Developing and implementing a watershed natural resources inventory (New Hampshire) 557
Jeffrey A. Schloss

Response — Janus planning: Using computer tools to look backward and forward simultaneously

II.13 The effect of spatial location in land–water interactions: A comparison of two modeling approaches to support watershed planning (Newfoundland, Canada) 577
Margot Young Cantwell

Response — Linking land use to landscapes for water quality protection

II.14 **Spatial investigation of applying Ontario's timber management guidelines: GIS analysis for riparian areas of concern**..................601
Robert France, John S. Felkner, Michael Flaxman, and Robert Rempel
Response — Size matters

II.15 **Aquifer recharge management model: Evaluating the impacts of urban development on groundwater resources (Galilee, Israel)**..................615
Amir Mueller, Robert France, and Carl Steinitz
Response — Planning by examining alternatives

II.16 **Factors influencing sediment transport from logging roads near boreal trout lakes (Ontario, Canada)**..................635
Robert France
Response — Empirically testing planning assumptions

II.17 **Limnology, plumbing and planning: Evaluation of nutrient-based limits to shoreline development in Precambrian Shield watersheds**..................647
Neil J. Hutchinson
Response — Land–lake linkages and land-use limits

Discussion summary: Social and political issues in managing riparian buffers and corridors..................683

Discussion summary: Multiple objectives in watershed management through use of GIS analysis..................685

Postscript: Implementing water sensitive planning and design..................687

Index..................689

Background — Perspectives of water management: Representative examples from the recent literature

Robert France

Water sensitive planning and design is not new. It lies at the historic base of civilization from whence we developed and can be expected to play a profound role in shaping our common future. As elaborated on in subsequent pages, the chapters comprising the present book put forward the very best of modern water sensitive planning and design. It is important to realize that though this new work offers imaginative and novel approaches and solutions to water management, it very much stands on the collective efforts of many involved in centuries of closely fostered and carefully studied water-culture relationships. For example, the Harvard Library system lists over 2000 documents going back more than 400 years under searches for "water planning" and "water design."

Giving credence to historic precedents in no way diminishes the contribution of those whose work is included in these pages. Instead, by understanding the cumulative wisdom and experience encapsulated within the state of the art of current water sensitive planning and design, readers can have a frame of reference from which to appreciate the additional benefits accruing from implementation of ideas advanced in the present compendium of work. To this end, eight representative publications from the recent literature were selected that demonstrate a snapshot of innovative thinking important for establishing what is in many respects a new paradigm of water sensitive planning and design. Salient features extracted from these publications concern the four major subject areas targeted in the present book — stormwater, wetlands, riparian buffers, and watersheds. This point-form synopsis therefore outlines the foundation upon which the subsequent chapters in the present book are further developed toward new, exciting, and profitable directions.

Stormwater

American Association of Landscape Architects. 1996. *Integrating Stormwater into the Urban Fabric.* Portland, Oregon Conference Proceedings. 21 multi-authored chapters. ASLA Publishing. 99 pp.

- Three-quarters of the job of water sensitive planning and design entails communicating and dealing with other people rather than producing construction drawings.
- By focusing on end-of-the-pipe controls for stormwater management, we fail to realize that it is the pipe itself that may be part of the problem.
- A critical need exists for the utilization of natural land forms such as trees and grasses to retain and filter stormwater.
- The publication emphasizes the truism that we can shrink our streets and parking lots (and therefore reduce impervious surfaces) without shrinking our economies.
- Sprawling suburbs and vacant city cores represent slash-and-burn urbanism.
- We must realize that floods are essential for healthy stream ecosystems.

- It is important to look at the site history for understanding the effects of urbanization on hydrology.
- In order to effectively manage river systems, we must realize that fish really live in the trees; i.e., they are part of the riparian system.
- Steam degradation can begin at levels of imperviousness as low as 10% watershed cover.
- Goals of urban water management should be balanced in respect to realistic expectations that are likely to occur with implementation of design improvements.
- Landscape architects, urban planners and designers, and civil engineers need to accept the precept that ecological functions are equally as important as other elements of design such as aesthetics and hydrology.
- Given that often the best designs are those that closely mimic nature, background data are needed from natural sites.
- Streams are most effectively managed from up in their watersheds than from within their restricted reaches.
- Effective stormwater management need not take up much space and can therefore be retrofitted into dense urban fabrics.
- "Daylighting" is increasingly becoming an important process in urban water restoration and urban neighborhood renewal; i.e., not only have we lost our buried streams, we have often lost the very memory of those streams.
- Development planning should consider concepts of watershed carrying capacity.
- One of the most important reasons for protecting urban streams and wetlands is for the education values they bring to city dwellers, whose concepts of water have become much reduced.
- Site planning measures need to be preventative instead of reactive.
- Land-use planners need to understand that water management is as critical as transportation issues in the urban environment.
- Stormwater must come to be looked upon as a resource for, not an impediment against, urban design, such as, for example, in the creation of "rain gardens."
- Development of green "eco-roofs" provide treatment and flood mitigation at the source and are thus cost-effective.

Bay Area Stormwater Management Agencies Association. 1999. *Start at the Source. Design Guidance Manual for Stormwater Quality Protection.* Tom Richman Associates. 165 pp.

- The best opportunity to reduce urban runoff occurs during the planning and design stage of project development.
- Given site idiosyncrasies and local regulation vagaries, it is impossible to produce a single set of specific design guidelines to be rigidly adhered to in all situations.
- Maintenance and operation of control measures are as critical to stormwater management as are their proper selection and design.
- Extent of impervious coverage provides a convenient barometer of environmental disturbance.
- A desperate need exists for a major paradigm shift to occur in urban stormwater management from an end-of-the-pipe, conveyance mentality to a dispersed, "start-at-the-source" form of micromanagement based on infiltration.
- We must stress the concept that every site is in a watershed.
- We need to think big in terms of beneficial effects while acting small in terms of establishing site solutions and in accommodating runoff from storms.
- Keep the design solutions as simple as possible.

- Every opportunity should be taken to integrate stormwater solutions into the overall site plan in order to provide recreational, aesthetic, and wildlife habitat benefits; i.e., stormwater needs to be looked upon as a design resource, not a planning hazard.
- Starting at the source for managing stormwater makes demonstrable economic sense.
- Conventional zoning practices typically do not address issues of water sensitive planning.
- Clustered development preserves open space in addition to providing room for inserting stormwater management practices.
- Streets and parking lots are often the major development on the land and must be carefully addressed in terms of minimizing runoff.
- Generation of public awareness is integral to implementing nonconventional water sensitive designs such as the use of porous pavement.
- Attention to site design is critical to improving stormwater management, in particular the following elements: define development envelope and protected areas; minimize directly connected impervious surfaces; maximize permeability wherever possible, and use drainage as a design element rather than hiding it away.
- Design for the entire drainage system to cost-effectively manage flooding, control streambank erosion, and protect water quality.
- Water sensitive design employs both structural and nonstructural elements and can be fit into all types of site conditions.

Wetlands

Azous, Amanda L. and Richard R. Horner (eds.), 2001. *Wetlands and Urbanization. Implications for the Future.* Lewis Publishers. 14 multi-authored chapters. 338 pp.

- Wetlands are very susceptible to the effects of urbanization, with these effects being felt on the scale of both the landscape as well as individual bodies of water.
- The need for a landscape approach to managing wetlands as parts of watersheds exists.
- It is useful to develop a comprehensive research program based on strong scientific acumen.
- It is important to place the study system within a framework of previous research conducted on other wetlands.
- Undertaking an inventory of land-use patterns through the use of GIS analysis produces notable benefits.
- We must emphasize the cardinal importance of imperviousness in affecting wetland health.
- We are required to understand wetland morphology and hydrology in order to assess the influences of urbanization.
- Comparisons of the functional ecology of wetlands in more pristine locations are critical toward gauging development effects upon urban wetlands.
- Consideration of soils and seasonality is important to assessment.
- Size alone is an inadequate criterion for judging the health of wetlands.
- Because wetlands are frequently characterized by relatively high biodiversity, even when located in urban areas, they represent natural arks.
- Cross-system empirical relationships, especially in relation to magnitude of deforestation, are useful for surveying the magnitude of urbanization effects.

- Detailed water balance models are essential for understanding how wetlands are affected by urbanization.
- Production of a list of stormwater management guidelines provides a useful communication tool to measure alterations in ecosystem integrity.

France, Robert Lawrence. 2002. *Designing Wetlands: Principles and Practices for Landscape Architects and Land-Use Planners.* W.W. Norton. In press.

- Technology transfer to aid communication among disciplines is needed.
- Assessments must be made of the cumulative impacts of many individual design projects from a larger watershed perspective.
- We must emphasize integrating concepts of wetlands as human amenities into the design process; i.e., the form of created wetlands may be just as important as their hydrological, chemical, or biological functions.
- Principles of wetland design should be approached as general concepts rather than construed as rigid instructions.
- Structuring and maintaining biological integrity may be the most important role of many wetlands.
- The retention and removal of contaminants in wetlands depends on a complex and interrelated system of chemical transformations.
- A strong, positive relationship exists between the percentage of upstream wetlands lost and the percentage increase in watershed peak flow discharges.
- Wetlands offer a myriad of opportunities for formalized natural history education.
- Careful attention to design, construction, and maintenance will ensure that created wetlands will operate effectively.
- Focusing on conceptual planning, site preparation and construction, planting, and long-term monitoring is essential to guaranteeing success in the creation of wetlands.
- On average, three-quarters of all analyses indicate that contaminant removal performances of 75% or greater are possible given adherence to appropriate treatment distances and transport times.
- Wetlands are created for a variety of different reasons, the choice of which will influence the initial screening and selection of suitable vegetation.
- The efficacy of treatment wetlands generally increases with their progressive size.
- A myriad of opportunities exist for exciting and imaginative landscape artistry that can augment wetland function.
- The interlinked networking of cells within an individual wetland or the specific associations of grouped wetland features within a larger combined water–wetland system will strongly influence resulting performance.
- Failed wetland creation/restoration projects abound due to opaque goal aspirations, poor engineering and ecological design, inadequate background data collection, absence of adaptive management, the erroneous assumption that project termination occurs with planning, and a cavalier process of critical evaluation.
- Recognition of the need for preserving undamaged wetlands as well as for restoring those that have already been degraded will be fostered by facilitating public access and easy interpretation at created wetland parks.
- An inventory of wetland resources is needed to assess their value to the community and the ecosystem as well as to help establish precise goals and recognizable steps for watershed management.
- Screening adjudication of potential mitigation sites involves assessment of wetland functions, project goals, and legal possibilities.

Riparian zones

Verry, Elon S., James W. Hornbeck, and C. Andrew Dolloff (eds.). 2000. *Riparian Management in Forests of the Continental Eastern United States.* Lewis Publishers. 20 multi-authored chapters. 402 pp.

- Riparian forests are in a constant state of dynamic equilibrium that integrates land-use alterations; in other words, they are defined as much by their functions as by their structures.
- It is important to consider development history on a site as a means for predicting future responses.
- Riparian forest management has a long-established tradition in Europe.
- Because the multifaceted roles of riparian areas are so closely dependent on watershed position, it may be necessary to manage them differently in relation to their steam order; i.e., no simple minimum buffer strip width is suitable for all situations.
- We need to consider floodplain geomorphology when managing riparian areas; i.e., a landscape perspective must be applied to riparian management.
- Riparian ecosystems are characterized by great biodiversity and high productivity far in excess relative to the physical area that they comprise.
- Riverine riparian areas operate as important wildlife corridors.
- Timber production should always be a secondary benefit in riparian management; instead, such areas should be managed for fish.
- Valley segments are convenient management units.
- Streams are closely linked to riparian forests through the role of the latter in water yields, erosion, nutrient transport, water temperature, and organic matter supply.
- Such ecotonal land–water linkages should form the basis for management practices.
- Stream health can be closely influenced by groundwater processes and soil regimes.
- Assessment of stream typology (based on morphology) is an important element in riparian management.
- Land clearance removes riparian sources of large woody debris needed to help structure stream morphology and sustain aquatic health.
- Road construction in riparian areas should be approached with caution and wisdom given justifiable concerns about erosion and sedimentation.
- Management of the aquatic side of riparian zones should not be ignored.
- Riparian zones are complex ecosystems necessitating complex (integrated) management solutions that focus on holistic options considering social, economic, and political interests at the earliest planning stage.
- Effective riparian management depends on closely considering human dimensions as much as it does on dealing with biogeochemical processes.
- The most successful management plans are those that engage the public in decision making.
- Planning is the most important best management practice (BMP).
- We should develop management guidelines in relation to science, practicality, and economics.
- Effective management is predicated on effective monitoring.

Palone, Roxane and Albert H. Todd. 1998. *Chesapeake Bay Riparian Handbook: A Guide for Establishing and Maintaining Riparian Forest Buffers.* USDA Forest Service. 890 pp.

- Riparian forest buffers perform a suite of water-protective functions in urban, suburban, agricultural, and forested landscapes.
- Management of riparian forest buffers can be implemented most effectively when approached on a spatial three-zone concept in relation to permitted development.
- Understanding the physiographic and hydrological attributes of riparian forests is necessary to effect wise management decisions.
- In addition to roles in influencing water quantity and quality, and in supporting shoreline and aquatic biodiversity, riparian forest buffer systems also offer recreational opportunities in an aesthetically pleasing environment.
- The essential need to understand soil characteristics and forest ecology as part of the process for planning the establishment of riparian buffer systems is essential.
- There is a need to relate upland contaminant loading before the removal capabilities of forest buffers can be assessed.
- Use of science-based criteria is essential for determining effective buffer strip widths in relation to site factors and predicted environmental threats.
- A whole suite of streamside restoration practices exist that can be implemented to return riparian forests to their protective functions.
- Stream systems rather than individual streams are the more appropriate scale for designing and managing riparian buffer networks.
- Forestry activities, road construction, and building development within riparian buffer systems must obviously be undertaken with caution and wisdom to limit deleterious aquatic effects.
- Riparian forest buffers offer a range of economic, in addition to ecologic, benefits.
- There is a need to educate the public and enforcement professionals about the important roles played by forest buffers in all types of landscapes.

Watersheds

Lal, Rattan (ed.) 2000. *Integrated Watershed Management in the Global Ecosystem.* CRC Press. 23 multi-authored chapters. 395 pp.

- There is need for a diverse interdisciplinary group considering social and economic, in addition to technical, issues.
- Focuses on the interface of land-use planning with watershed management.
- Interaction between water scarcity and food security in a global perspective, including tropical developing nations, exists.
- The watershed is the basic land management unit for agriculture.
- Soil erosion is the major threat to the quality of global water supplies as well as to cropland degradation.
- The health of nations is determined by the health of their watersheds is examined.
- Historical perspectives are important in order to help understand present problems.
- Social justice (equity, gender, etc.), community participation, and institutional infrastructure when managing watersheds is critical.
- We need good soil and hydrologic science to predict the extent and consequences of soil erosion.
- There are links among agricultural land-use, soil fertility, chemical nutrient applications, and aquatic eutrophication.

- Greater reliance on indigenous wisdom in addition to scientific knowledge is necessary.
- Experimental manipulations are required to identify the most useful best management practices (BMPs).
- Simulation models are an important tool for stimulating discussions through characterizing responses to land management interventions.
- An organizational structure that incorporates both physical and social elements for managing watershed agroecosystems must be developed.
- Initiation of local-level resource monitoring by watershed management organizations is needed.
- Holistic over reductionist approaches are required for effective management.
- Successful watershed management is built on the two pillars of practical technical innovation and participatory institutional innovation.
- Locals should be empowered in generating "bottom-up" solutions.
- Watersheds must be viewed from perspectives of both service and productivity functions.
- Concepts of sustainability must be infused into all stages of watershed management; i.e., the landscape–lifescape interaction.
- Scenario models linked to GIS analysis are a powerful means for evaluating actual and alternative management practices.
- Groundwater pollution in rural watersheds is significant.
- Downstream economic impacts of soil erosion on a lake-by-lake basis exist in terms of recreational boating and restorative dredging.
- Global climate change influences watershed functioning and management is examined.

EPA. 1997. *Top 10 Watershed Lessons Learned.* EPA. 59 pp.

- The best plans have clear visions, goals, and action items.
- Watershed approaches are developed over time and need dynamic leaders with strong interpersonal skills.
- Implementation of watershed plans benefit from the presence of a dedicated coordinator to help build relationships.
- Watershed management provides a means in which to integrate environmental, economic, and social aspirations under an umbrella of sustainability.
- Generating watershed management plans that stand a real likelihood of becoming implemented rather than just sitting on the shelf is necessary.
- It is important to engage all stakeholders, including industry, in a positive manner working toward a common goal of improving livability and environmental integrity in the watershed.
- Watershed groups should become familiar with a diversity of planning and communication tools such as guides, workshops, GIS mapping, etc.
- A procedure for monitoring progress in the implementation of watershed planning recommendations needs to be incorporated into management guidelines.
- Local empowerment through education provides tangible results in increasing awareness about watershed issues.
- Watershed plans work best when they are designed to advance incrementally forward in small, manageable components rather than as single, all-inclusive vague concepts.

part I

Water sensitive design

Overview: New interpretations in stormwater management and wetland park creation

Part I comprises 17 chapters and responses that deal with two of the most important issues associated with water sensitive design: stormwater management and wetland park creation. Together, these chapters go far toward demonstrating the precept that just as water is an integrator across physical landscapes, likewise it is an integrator across professional disciplines. For in addition to landscape architecture, elements of civil engineering, urban planning, environmental art, cultural geography, hydrology, soil science, biology, sociology, and geology are all present in these pages.

Seven chapters focus on managing stormwater on sites using a combination of decentralized, small-scale approaches that are representative of the emerging paradigm of "low-impact development." Landscape architect Bruce Ferguson provides an informative introduction to stormwater in urban and suburban settings and reviews concepts of porous pavement as an effective management technique. Jennifer Zielinski's chapter emphasizes the importance and wisdom of considering environmentally sensitive water management at the very earliest possible stage in the site design process. This topic is elaborated on in great detail by Larry Coffman, whose chapter summarizes his pioneering work in establishing low-impact development as a viable means for addressing stormwater management. Four of the chapters — those by Richard Pinkham and Timothy Collins, by Robert France and Philip Craul, by Michael Clar, and by landscape architects Thomas Liptan and Robert Murase — provide a diverse set of design examples of how stormwater can be managed and utilized in an environmentally sensitive and aesthetically pleasing fashion, true to the spirit of low-impact development.

Five chapters focus on managing stormwater through creating wetland parks. James Bays provides a comprehensive survey and discussion of the various attributes needed to be incorporated into designs in order to develop multifunctional wetland parks for regulating the quantity and treating the quality of stormwater at the same time as providing a whole suite of additional environmental and cultural benefits. Four of the chapters — those by landscape architect Diana Balmori, by landscape architect Catherine Berris, by Robert France and landscape architect Kaki Martin, and by landscape architect

Glenn Allen — show how these design ideas can be imaginatively adapted to varied situations of high visibility and ensuing public use.

Three chapters concern the restoration of degraded water bodies. Wendi Goldsmith provides a discussion of how the emerging techniques of shoreline bioengineering can be used in the broad context of both watersheds and disciplines. Landscape architect Clarissa Rowe demonstrates how the methods of wetland design can be implemented into larger management plans toward improving site conditions for urban ponds. In the study described by landscape architect Nicholas Pouder and Robert France, both of these restoration tools are integrated in a demonstration project designed to raise pubic consciousness about lake restoration.

The chapter by Desheng Wang describes methods needed to improve the design of stormwater detention basins, long the favored practice of many civil engineers, but unfortunately rarely approached with such environmental sensitivity as illustrated by the present examples. Finally, the chapter by Larry Schwartz presents an interesting example of utilizing wastewater as a functional amenity in the design of a treatment wetland park.

chapter I.1

Stormwater management and stormwater restoration

Bruce K. Ferguson

Abstract

A watershed maintains its natural health and its benefits to human beings by the accumulation, storage, and gradual flow of subsurface water. The fundamental disease of urban watersheds is sealing by impervious cover, which deflects runoff across the surface and carries pollutants into streams. Approaches that manage stormwater on the land surface treat only the downstream symptoms: conveyance, detention, and stormwater wetlands all fail to eliminate the fundamental urban problem of excess surface water volume. In contrast, stormwater infiltration forces surface runoff back into the underlying soil, curing the disease by restoring watershed process. Infiltration basins are well known in diverse geographic regions and have been integrated into diverse site circumstances. Porous pavements infiltrate urban runoff directly, where the rain falls; for many types of porous pavement the long-term infiltration rate is sufficient to absorb and treat the rain that falls during almost all storm events. Only a small capacity for daily infiltration is required to create a substantial cumulative effect over time. Unlike any surface management approach, infiltration is capable, within the limitations of specific sites, of solving all the problems of urban runoff, because it calls on the power of the underlying landscape.

Introduction

From the viewpoint of the health of the environment, the most important aspect of stormwater design is what the design causes water to do. Many other important aspects of stormwater design are involved: design can integrate water with all the other concerns of an urban site, and in this collection other fine practitioners discuss that kind of topic (for example, Pinkham and Collins, Liptan and Murase, Berris, and Flink). My subject here, however, is limited to the single question of where water goes in the environment, and the kinds of processes it experiences there.

The volumes of water involved in this question are vast, and what happens to it is a vital question in the welfare of people. In recent decades, designers have put quite a bit of ingenuity into various alternative answers to this question, with some degree of success if you look at them through sufficiently narrow glasses.

This chapter reviews some of those alternatives and then points to what we should be doing a lot more of, given our scientific understanding of what makes landscapes

Figure I.1.1 An apartment site in Long Beach, CA. (Color version available at www.gsd.harvard.edu/watercolors. Password: lentic-lotic.)

healthy. In the course of this, examples of installations from around the country will be presented, to illustrate that not only *should* we do these things, but we *can* do them.

The urban watershed problem

The discharge during a storm from typical urban culverts exemplifies the problem that urban watersheds present to us. Even the largest culverts flow nearly full. With every rain that falls on urban watersheds, floods bring oil, bacteria, and sediment from the watersheds' impervious pavements and eroding stream banks. In some urban watersheds, the floods get into sanitary sewers, adding overflows of raw sewage to the stream flow.

When the rain stops, little base flow remains in urban streams, because there is no water left in the watershed. Groundwater levels are low. Many cities are left, paradoxically, with local water shortages, and aquatic ecosystems are left without habitat.

The problem begins throughout urban watersheds, at sites similar to the one shown in Figure I.1.1. This site's dense impervious surfaces accumulate oils, bacteria, and metals. The same surfaces turn rainfall into surface runoff that carries the pollutants into streams. Here, at the site where the rain falls, the problem has been created: surface stormwater has been generated throughout the watershed, and pavements and channels are holding the concentrated water and its concentrated pollutants. Now we have to manage what we have created. This is the beginning of stormwater management.

It is possible to put the excess polluted water that comes off watersheds to constructive use. A successful example is at a horticultural nursery in Thomson, GA, which is fortuitously located on a small headwater watershed. The nursery has paved most of the watershed with plastic sheets and greenhouses. A pond captures the abundant runoff; from the pond, pumps recycle the water for the nursery's irrigation. This on-site "water harvesting" makes the use of water resources efficient and the future of this water-consuming industry sustainable, but implementing this approach requires special site conditions. Nationwide the occasions for doing it have been extremely limited. So we are left with the generic problem of urban watersheds, and we need to find an answer.

It is possible to eliminate excess runoff at the start. For example, at an office site in Atlanta, Robert Marvin dispersed 200 parking spaces across the wooded site, tucking a

Figure I.1.2 A detention basin at a commercial site in Georgia. (Color version available at www.gsd.harvard.edu/watercolors. Password: lentic-lotic.)

space or two at a time between the preexisting trees, and carefully preserving the forest floor adjacent to each pavement edge. The vegetated soil infiltrates and eliminates the runoff from each bit of pavement; but spreading out the parking here meant that the 200 parking spaces used up a greater amount of land, with a greater length of paved traveling lane connecting the dispersed spaces. The large amount of land this approach requires inhibits its widespread use. So, for most sites, we are still left with the generic urban problems of runoff and pollution, and we cannot escape the need to find an answer.

Some management alternatives that have been tried

Among the various approaches to stormwater management that have been developed over the years, the oldest is conveyance, which is the use of channels of various kinds, originally to move surface water away from urban sites to reduce on-site nuisance. Today, the ability to move surface water safely from place to place remains vital on every site: conveyance channels control the flow through any other type of system we put on a site, and they discharge the overflows from large, rare storms; however, conveyance, where it is used alone, passes all the urban problems on downstream, through successively larger channels, without the benefits of filtration or recharge. So designers have explored supplements and alternatives to it.

For the last several decades, designers have been supplementing urban conveyances with detention basins similar to the one shown in Figure I.1.2. The big culvert brings surface runoff from this highly impervious commercial site into a reservoir. During storms the water rises high in the basin. A relatively small orifice in the concrete structure limits the rate of outflow. Figure I.1.3 shows a typical result. A pulse of surface runoff comes in at a high rate. The basin stores the runoff for a while, while the outlet lets it out at a relatively low rate. Detention storage suppresses the peak rate of storm flow, as it has been intended to do since this practice started in the 1960s, but detention is only a relative, quantitative modification of conveyance; the total volume of runoff still continues downstream during the storm event, stretched out over time. The experience is that detention has failed to prevent urban flooding and erosion and has never done anything for water quality, groundwater replenishment, or urban water supplies (Ferguson, 1998, p. 164). So designers have considered further alternatives.

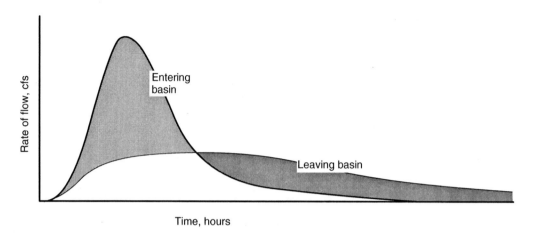

Figure I.1.3 Typical hydrologic outcome of detention during a storm event.

Figure I.1.4 A stormwater wetland at a convention center near Orlando, FL. (Color version available at www.gsd.harvard.edu/watercolors. Password: lentic-lotic.)

Relatively recently, designers have been supplementing conveyance more elaborately with extended detention and stormwater wetlands. The wetland pond shown in Figure I.1.4 is a particularly nice-looking example. It receives surface runoff from the roofs and parking lots of a large convention center. The still water of the pool helps water quality: suspended particles settle out, the wetland plants adsorb and biodegrade hydrocarbon pollutants, and the fountains in the background oxygenate the water. The design as a whole has multiple functions for aesthetics and wildlife habitat. On the other hand, the construction of the pool preempted the wetland and riparian functions that used to exist in this low-lying spot, barred fish migration through the natural stream system, raised the temperature of stream water, and shifted the stream system's trophic state. It also did not eliminate the fundamental problem of the unnatural volume of surface water. Wetlands such as this, as well as ordinary detention basins, modify the excess runoff before it passes on downstream, but do not change the fact that it has been created. They manage the problem of excess urban runoff, and perpetuate it. So we continue to consider still-further alternatives.

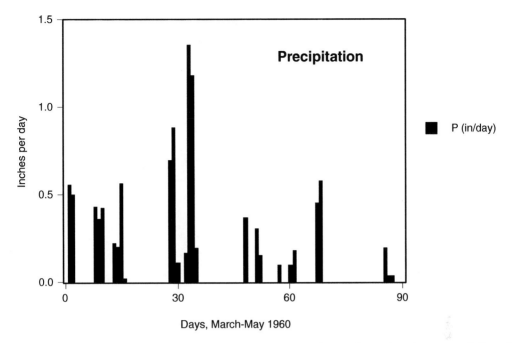

Figure I.1.5 Daily precipitation in Atlanta, GA, in the spring of 1960. (Data from Ferguson, B.K. and Suckling, P.W., Changing rainfall–runoff relationships in the urbanizing Peachtree Creek watershed, Atlanta, Georgia, *Water Resour. Bull.*, 26, 313, 1990.)

What makes landscapes healthy

As a basis for dealing completely and fundamentally with the problem of urban watersheds, we must first establish how landscapes fundamentally work. Landscapes are open, dynamic systems with inflows and outflows of resources (Ferguson, 1994, 1996).

Precipitation is the vast hydrologic inflow. Figure I.1.5 shows the daily precipitation in Atlanta for a 90-day period in the spring of 1960. Please note some of the specific numbers in this chart, because they will be referred to in the following figure. There is one precipitation event with almost 1.5 in./day, several others with about 0.3 to 0.5 in./day, and many very small events. This shows that rainfall comes to a landscape, over time, in discrete, isolated events, spaced apart from each other by a few days or a few weeks with no rain, and the events of different sizes come in quite random order.

Figure I.1.6 shows the stream flow coming out of Atlanta's Peachtree Creek watershed during the same 90-day period. This watershed is quite urban, with 30% impervious cover in 1960 (Ferguson and Suckling, 1990), so during storms the stream outflows shown in Figure I.1.6 are much bigger than they would have been without the impervious cover; however, this chart applies to the same time period as the previous one and is graphed in the same units, and you can see that the storm flows are smaller than the precipitation that created them, even in this urban watershed: the big event is reduced to 0.7 in. of outflow, and the smaller events are almost imperceptible. In place of the pattern of isolated precipitation events, the watershed has created a constant, moderate base flow, which is the base resource for perennial water supplies and aquatic ecosystems.

Figure I.1.7 again shows Peachtree Creek's stream flow, with daily evapotranspiration added in the light tone. In the Peachtree Creek watershed, as in most of the landscapes in the world, more than half of the annual rainfall discharges as evapotranspiration from soil and plants, without passing through a stream. Evapotranspiration supports the terrestrial

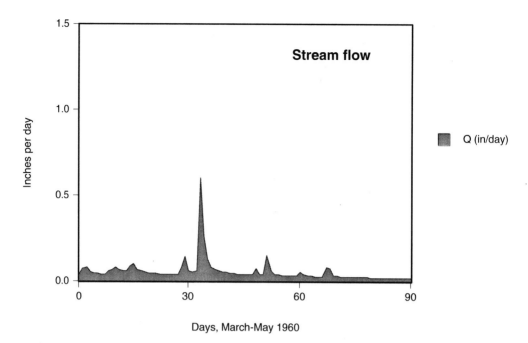

Figure I.1.6 Stream discharge in Peachtree Creek, Atlanta, during the same period shown in Figure I.1.5. (Data from Ferguson, B.K. and Suckling, P.W., Changing rainfall–runoff relationships in the urbanizing Peachtree Creek watershed, Atlanta, Georgia, *Water Resour. Bull.*, 26, 313, 1990.)

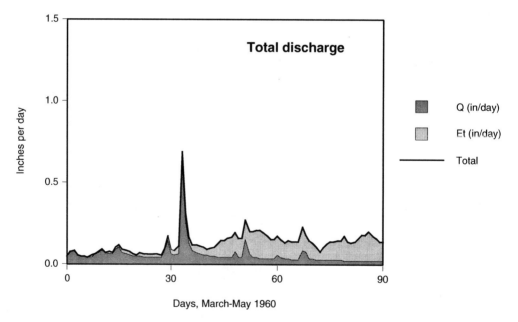

Figure I.1.7 Stream flow and evapotranspiration in the Peachtree Creek watershed, during the same period shown in Figures I.1.5 and I.1.6 (evapotranspiration calculated by the monthly Thornthwaite method and distributed daily in proportion to daily temperature).

ecosystem the same way stream base flow supports the aquatic ecosystem: all the terrestrial vegetation lives by the flow of water from the soil, through the roots, and out through the leaves. The heavy line at the top is the sum of stream flow and evapotranspiration; it is the

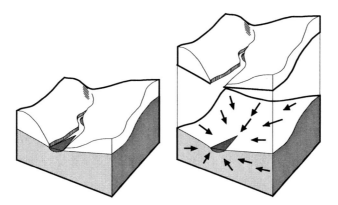

Figure I.1.8 Subsurface moisture in a landscape. (Adapted from Leopold, L.B., *Water, Rivers and Creeks*, University Science Books, Sausalito, CA, 1997.)

total discharge from the watershed. The continuity of that line shows that the watershed, obeying the simple laws of physics, transforms the random, isolated, unorganized, potentially destructive pulses of inflow into a perennial discharge that is, relatively, a resource to the ecosystem and to people.

Figure I.1.8 shows the underlying mechanism by which a landscape does that. The figure is a section through a landscape showing the moisture under the ground surface. On the left, the groundwater table can be seen gently reflecting the ground surface and discharging at low points as springs and streams. On the right, the overlying unsaturated soil has been lifted up to show that the water table is a continuous mass under the landscape, flowing slowly and continuously, under its own weight, from its stored mass in high areas toward the discharge points. From place to place in the earth, variations exist in the details of this diagram with the relative permeabilities of underground layers (Ferguson, 1985). But the accumulation, storage, and gradual flow of subsurface water happen in essentially all landscapes (Ferguson, 1992). The infiltration of rainfall into the subsurface of the Peachtree Creek watershed explains the reduction in that creek's storm flow volume compared with the precipitation that produced it. The storage of that water in subsurface pores explains the creation of stream base flow and continuous evapotranspiration.

The fact that landscapes evolve to work this way is not a matter of professional preference or bureaucratic convention. Infiltration and storage happen in nature, all over the world. A landscape, like any other open, dynamic system, absorbs its inflows, stores them inside itself, turns them into resources, and uses them to maintain itself (Ferguson, 1994, 1996). The result of a watershed's working this way is that, compared with the precipitation inflows it is given to work with, floods are moderate, erosion and sedimentation are in equilibrium, pollutants are degraded, the wetlands are sustained, and public water supplies are secure. So nature works. It evolves to work. It has a capacity to work, even in urban watersheds. It will work, greatly to our advantage, if we will only let it.

The understanding of this concept is a basis for us to move from treating the downstream symptoms of surface runoff to curing the disease of urban watersheds at the source. Figure I.1.9 shows the disease: the contrast between the hydrologic processes of open, vegetated soil and sealed, impervious cover. When rain falls on vegetated soil, it infiltrates. Evapotranspiration sustains the local ecosystem. Further infiltrating water removes pollutants if there are any present, recharges groundwater, and restores stream base flow. This process is the origin of a landscape's health and is of great benefit to people. In contrast, when an impervious cover seals the soil surface, rainwater is deflected across the surface. The surface runoff flushes pollutants directly into streams; flooding, erosion,

Handbook of water sensitive planning and design

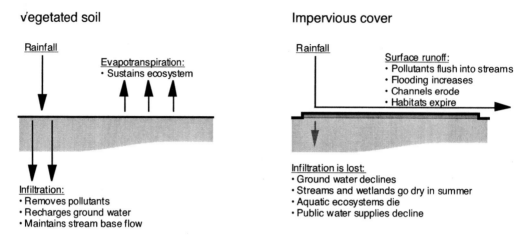

Figure I.1.9 The contrast between vegetated soil and urban impervious cover.

Figure I.1.10 A small infiltration basin near Orlando. (Color version available at www.gsd.harvard.edu/watercolors. Password: lentic-lotic.)

and habitat loss occur. Infiltration is lost, so groundwater declines; streams and wetlands go dry in summer; aquatic ecosystems die; and public water supplies decline. If we are going to cure the disease of urban watersheds, we have to minimize the process on the right and maximize that on the left.

Infiltration basins

Figure I.1.10 shows an approach that aids this process. This is an example of stormwater infiltration: the artificial forcing of surface water into the underlying soil. This infiltration basin is located in an office park near Orlando, FL. A miniature watershed on the parking lot pavement drains through a curb cut into a depression in the grassed "island." This is a dull-looking basin: you do not see the spectacular slash pine and palmetto vegetation that theoretically used to grow on this site. Nevertheless, this basin's process of infiltration

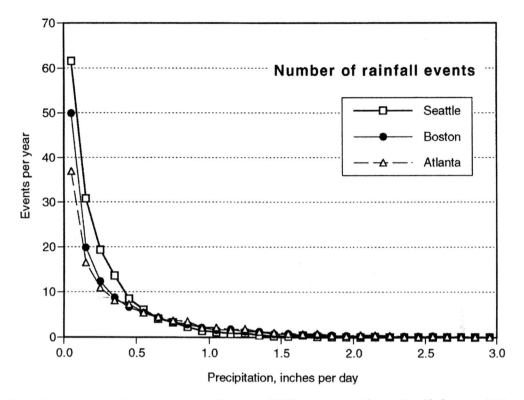

Figure I.1.11 Average frequency of rainfall events of different sizes, in three cities (daily precipitation data 1950–1989 from U.S. National Climatic Data Center, Asheville, NC).

is the restoration of hydrologic process in this landscape. Consequently, it is inherently the most complete possible solution to the environmental problems of urban stormwater.

Figure I.1.11 shows an important point about designing for infiltration, and the performance of which infiltration is capable. The chart shows the number of rainfall events that supply urban watersheds and that artificial infiltration would be given to exploit. On the horizontal axis is the size of individual rainfall events in inches per day; the size increases to the right. On the vertical axis is the number of times each event occurs in an average year. The chart shows data for three cities: Atlanta, Boston, and Seattle. They all tell us the same thing: the most common rainfall events — by far — are the smallest. Designers have been trained conventionally to design conveyances and detention systems for big flood events, such as the 10-year storm. Big floods are erosive and damaging when they occur, and we need to plan for them. From an ecological viewpoint, however, they have a problem: such events almost never happen. The 10-year, 24-hour storm is so rare that it is off Figure I.1.11's chart to the right; its contribution to average annual inflow is insignificant to the ecosystem and to people. The events that maintain the ecosystem are the small ones, because they are frequent: they replenish the system many times per year. Half of the average annual inflow comes in events of less than 1 in./day. Infiltrating only a small volume of water from each small runoff event can largely restore cumulative environmental function. The fact that only a small amount of water needs to be infiltrated each day makes effective stormwater restoration surprisingly feasible. Even basins on slowly permeable soils can have a substantial effect on water and pollution over the course of a year.

I estimate that 40,000 infiltration basins are in operation in the United States. They are well known in diverse geographic regions and have been integrated into diverse site circumstances.

Figure I.1.12 Infiltration swale at the Oregon Museum of Science and Industry in Portland. (Color version available at www.gsd.harvard.edu/watercolors. Password: lentic-lotic.)

Figure I.1.13 Swale at a Merrill Lynch training office near Princeton, NJ. (Color version available at www.gsd.harvard.edu/watercolors. Password: lentic-lotic.)

Figure I.1.12 shows one of Robert Murase's well-known projects in Portland, OR. This is hydrologically the same idea as the grass basin in Florida, but here the parking bays drain into swales thickly planted with native and adapted plants. Monitoring in the Portland area has shown that the roots of these native plants help maintain an open soil structure and cause more water to infiltrate than may be assumed in design calculations. Weirs like the one in the foreground hold small increments of water in contact with soil and vegetation to make sure they get treated and infiltrated, while the linear organization of swales allows overflows from large storms to move down the swale system.

The swale shown in Figure I.1.13 does the same thing hydrologically, although it looks more mechanical. This is an office site in New Jersey, designed by Roger Wells. Water from

Figure I.1.14 Infiltration basin doubling as a playground at Village Homes, Davis, CA. (Photograph by Roger D. Moore.) (Color version available at www.gsd.harvard.edu/watercolors. Password: lentic-lotic.)

the pavement surface flows across the grass slope, which acts as a kind of vegetated filter. At the bottom of the slope, in the swale, concentrated runoff flows on and through the stones. The stones slow the water down, prolonging its contact with soil and giving it a chance to be treated and to infiltrate. The inlet in the background carries the overflows from large storms.

The infiltration basin shown in Figure I.1.14 has been developed for multiple human use. This is in the Village Homes community in Davis, CA, the work of Robert Thayer and his colleagues. The sand-floored basin is the floor of a children's playground. The playground is located along a system of pedestrian ways that connect to the rest of the community. There is little annual rainfall here, but when there is a substantial storm, runoff comes down the system of drainage swales, paralleling the pedestrian ways, to points like this throughout the community. Village Homes' system of basins eliminates most of the community's surface runoff. Since it was installed in the late 1970s, it has very seldom overflowed into the City of Davis storm sewer (Thayer and Westbrook, 1989).

The large infiltration basin shown in Figure I.1.15 has been combined with water harvesting to support multiple ecological functions. This is on the campus of Hofstra University on Long Island, and it receives runoff from an adjacent highway. The university lined the floor of the basin with a plastic sheet to hold a permanent pool and planted it with native wetland vegetation. Biological surveys here have found extraordinarily diverse bird life, compared with the surrounding urban area. The university maintains the site as part of its arboretum and has designated it a wildlife sanctuary. When runoff enters the basin during storms, the water level rises; excess water infiltrates through the basin's unlined sandy sides. Due to the great capacity of this basin, there is no overflow swale; this is the absolute end of the local surface drainage system, even for the largest storms. Altogether, Long Island has more than 4,000 basins excavated into the sandy soil, recharging the underlying aquifer that is the exclusive water supply for 2.6 million persons (sum of the populations of Nassau and Suffolk Counties, 1990, listed in the Web site of the U.S. Census Bureau).

Figure I.1.16 shows stormwater infiltration lifted up to a massive scale of public works. This is the 570-acre Rio Hondo Spreading Grounds, operated by the Los Angeles County

Figure I.1.15 Infiltration basin at Hofstra University, Long Island, NY. (Color version available at www.gsd.harvard.edu/watercolors. Password: lentic-lotic.)

Figure I.1.16 The Rio Hondo Spreading Grounds near Los Angeles. (Color version available at www.gsd.harvard.edu/watercolors. Password: lentic-lotic.)

Figure I.1.17 Construction of a porous paver driveway in Athens, GA. (Color version available at www.gsd.harvard.edu/watercolors. Password: lentic-lotic.)

Department of Public Works. Throughout the Los Angeles region, for over 100 years, numerous public and private agencies have operated thousands of basins large and small, recharging water into aquifers in the region's alluvial fans and coastal plain. The Department of Public Works alone has 27 large facilities like this, through which it recharges each year 40,000 acre-ft of urban runoff, mixed with even huger quantities of mountain runoff, imported water, and reclaimed water, sustaining the coastal plain aquifer from which 3.5 million persons take their water supply (derived from data in the Web sites of the Los Angeles County Department of Public Works and the Water Replenishment District of Southern California).

Porous pavements

Porous pavements allow us to take care of a large part of the urban runoff problem at the source, where the rain falls, without having to get into downstream basins at all. For example, Figure I.1.17 shows concrete paver units being placed for a porous driveway in Georgia. The pavers rest on a sand setting bed with an aggregate base course. The pavers' configuration leaves drainage holes in 12% of the pavement area; the holes are filled with fine gravel. Figure I.1.18 shows the long-term infiltration rate for pavers of this kind. The rate declines in the first 4 to 6 years after installation, and then after 10 years it is nearly stable at a long-term level of about 4 in./hour. The light horizontal lines show, for comparison, the intensity of the 10-year, 1-hour storm in three different cities. You can see that the pavers' long-term infiltration rate is sufficient to absorb and treat the rain that falls even during intense storms, and even in Georgia where the 10-year storm approaches 3 in./hour.

In typical urban watersheds, pavements cover one third of the land; they produce two-thirds of the runoff and almost all the petroleum-based pollution. In the United States, we are paving or repaving over a quarter of a million acres per year. So paved areas that are reclaimed for infiltration take advantage of a vast and cost-free resource for watershed restoration at the seat of the problem of urban runoff and pollution.

Different types of porous materials are available to meet site-specific needs. For example, Figure I.1.19 shows porous concrete newly installed for a park road in Georgia. Porous

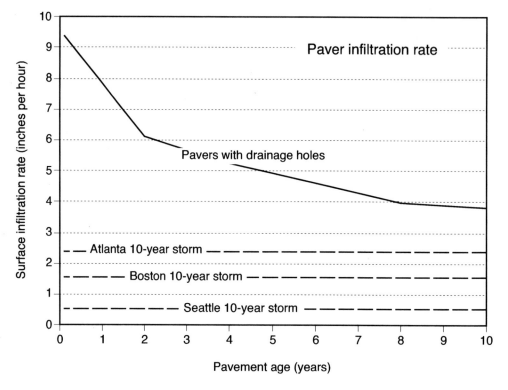

Figure I.1.18 Long-term infiltration rate in porous concrete pavers, in comparison with rainfall intensity in three cities. (Infiltration rate data from Borgwardt, 1997; 1-h rainfall data from U.S. Weather Bureau, 1955.)

Figure I.1.19 A porous concrete road in Webb Bridge Park, Alpharetta, GA. (Color version available at www.gsd.harvard.edu/watercolors. Password: lentic-lotic.)

Figure I.1.20 A parking lot in Medford, NJ, selectively combining impervious asphalt and porous aggregate. (Color version available at www.gsd.harvard.edu/watercolors. Password: lentic-lotic.)

concrete bears the high traffic load for the park's main entry with a durably high infiltration rate. This road's design illustrates an important detail for preserving the infiltration capacity of porous pavements. In the photograph, the recently graded clay soil is eroding freely; sediment is flowing visibly through the drainage swales. If that sediment were to get onto the pavement, it would enter the pores and greatly reduce the infiltration rate before the road has even been put into service. This site was properly configured by its designer, John Gnoffo, to protect the pavement from that. Drainage is away from the edges of the pavement in every direction, so sediment drains down eroding slopes without entering the pavement. Curbs are omitted, so overflow drainage from the pavement, when it happens, discharges freely into the swales, and potentially clogging surface debris is washed or blown off the pavement. This installation also exemplifies the cost saving that can come from the planned use of porous pavements. Because this pavement absorbs some of the runoff during storms, a reduction was allowed in the size of downstream conveyances and detention basins, producing a net cost saving when the porous road is properly counted as both a pavement structure and a functional part of the drainage system.

Figure I.1.20 illustrates the highly selective use of porous materials that is vital for maximizing porous pavements' cost advantages and for designating appropriate application of porous pavements anywhere. This is a parking lot in Medford, NJ. It was built with porous surfaces to satisfy the environmental protection objectives established in the famous study of the township by Ian McHarg and his colleagues (Juneja, 1974). Impermeable asphalt is used in the parking lot's traveling lane, where many vehicles travel and

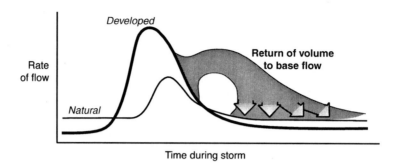

Figure I.1.21 The outcome of stormwater infiltration during an individual storm event.

turn each day. Impermeable asphalt is of course a well-known material and is a highly conservative response to the traffic load here. In contrast, simple stone aggregate is used in the individual parking stalls where there is only one car per day, standing still most of the time. The low traffic load freed the designers to select this material, which is highly permeable and cheaper per yd^2 than any other paving material, porous or nonporous.

The Medford designers' analysis of the detailed usage levels within this parking lot exemplifies the care, knowledge, and ingenuity that designers need to use in applying infiltration as they do in implementing any other feature of any project they have responsibility for. Stormwater infiltration should not be done in some places: for example, on steep unstable slopes, on grossly permeable gravel soils where groundwater pollution would be possible, and over toxic deposits in old industrial areas. Outside those local spots, carefully implemented stormwater infiltration succeeds environmentally and economically.

Infiltration outcomes in restoration

During an individual storm, infiltration returns runoff water from the surface to the soil like the flow shown in Figure I.1.21. The water is no longer part of the surface flow during the storm event. Instead it goes into the groundwater to discharge later as part of the long-term base flow.

Figure I.1.22 shows the outcome of infiltrating water from a series of storms, large and small, over the course of many years. This is the result of a modeling study for a high school site in Georgia (Ferguson et al., 1991). The average annual inflow of precipitation on this site amounts to 270 acre-ft. of water. Each column shows the disposition of that volume in the site's environment from some way of managing the stormwater. On the left is the disposition before development, when the site was entirely vegetated. Evapotranspiration was large. Of the stream flow, almost all was in steady, continuous base flow; direct runoff occurred only during large storms, and even then only in moderation. In the middle is the disposition from the developed site using surface stormwater discharge including culverts and detention basins. The impervious cover reduced evapotranspiration. Direct runoff dominated annual stream flow, claiming away water that used to go into both evapotranspiration and base flow.

On the right is the disposition from the developed site using infiltration basins. This plan captures the direct runoff from the impervious surfaces and transforms it, through infiltration, into base flow. Base flow is actually greater than it was before development, because the impervious surfaces divert water away from evapotranspiration. The residual direct runoff comes from peripheral impervious surfaces that could not direct their runoff into infiltration basins. According to studies like this, infiltration transforms, relatively

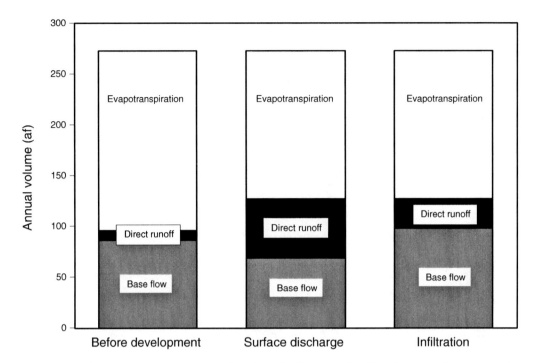

Figure I.1.22 The contrasting outcomes of surface discharge and infiltration over the course of an average year, for a high school site in Georgia.

speaking, the erosive filth of surface runoff into the resources of clean groundwater and steady base flow.

In conclusion, we can make a distinction between two fundamental parts of urban watersheds and the processes that go with them: the surface and the subsurface. A designer has the choice of directing urban stormwater toward one or the other. Design for stormwater on the surface of urban watersheds manages excess runoff by treating its downstream symptoms. Surface conveyance and detention ultimately continue the discharge of runoff to streams. Because they lie on the surface of landscapes, they are hydrologically isolated and artificially incomplete.

In contrast, infiltration into the subsurface of landscapes is restoration that cures the disease. Infiltration, unlike any surface approach to stormwater management, is capable, within the limitations of specific sites, of solving all the problems of urban runoff: peak flow, base flow, groundwater recharge and quality, because it calls on the power of the underlying landscape. The subsurface is a natural resource for filtration and storage waiting to be taken advantage of on essentially every site, as nature has always used it in the past.

Infiltration deserves to be the beginning of stormwater design, the fundamental tool for solving urban watershed problems, the approach that is implemented to its fullest feasible site-specific extent before anything else is attempted. Water belongs in the subsurface is a simple principle that is supported by the fundamental way landscapes maintain themselves, by the good it can do for human beings, and by decades of documented experience in thousands of field installations. The only part of stormwater we need to manage is the part we are not able, for some site-specific reason, to restore through infiltration.

Literature cited

Borgwardt, S., *Versickerungsfahige Pflastersysteme aus Beton: Voraussetzungen, Anforderungen, Einsatz (Infiltratable Concrete Block Pavement Systems: Prerequisites, Requirements, Applications)*, Bundesverband Deutsche Beton- und Fertigteilindustrie, Bonn, 1997.

Cahill, T., A second look at porous pavement/underground recharge, *Watershed Protection Tech.*, 1, 76, 1994.

Ferguson, B.K., Land environments of water resource management, *J. Environ. Syst.*, 14, 291, 1985.

Ferguson, B.K., Landscape hydrology, a component of landscape ecology, *J. Environ. Syst.*, 21, 3, 193, 1992.

Ferguson, B.K., *Stormwater Infiltration,* Boca Raton: Lewis Publishers, Boca Raton, FL, 1994a.

Ferguson, B.K., The concept of landscape health, *J. Environ. Manage.*, 40, 129, 1994b.

Ferguson, B.K., The maintenance of landscape health in the midst of land use change, *J. Environ. Manage.*, 48, 387, 1996.

Ferguson, B.K., *Introduction to Stormwater: Concept, Purpose, Design,* John Wiley & Sons, New York, 1998.

Ferguson, B.K. and Suckling, P.W., Changing rainfall–runoff relationships in the urbanizing Peachtree Creek watershed, Atlanta, Georgia, *Water Resour. Bull.*, 26, 313, 1990.

Ferguson, B.K., Ellington, M.M., and Gonnsen, P.R., Evaluation and control of the long-term water balance on an urban development site, p. 217 of *Proc. 1991 Georgia Water Resour. Conf.*, Kathryn J. Hatcher, ed., University of Georgia Institute of Natural Resources, Athens, GA, 1991.

Juneja, N., *Medford: Performance Requirements for the Maintenance of Social Values Represented by the Natural Environment of Medford Township, NJ,* Univ. Pennsylvania Dept. Landscape Architecture and Regional Planning, Philadelphia, 1974.

Leopold, L.B., *Water, Rivers and Creeks,* University Science Books, Sausalito, CA, 1997.

Los Angeles County Department of Public Works, *General information,* http://www.deltanet.com/sgvw/lacdpw/general/.

Thayer, R.L., Jr. and Westbrook, T., Open drainage systems for residential communities: Case studies from California's Central Valley, p. 152 in *Water, Proc. 1989 Annual Conf. Council of Educators in Landscape Architecture,* Landscape Architecture Foundation, Washington, 1989.

U.S. Census Bureau, *American Factfinder,* http://factfinder.census.gov.

U.S. Weather Bureau, *Rainfall Intensity-Duration-Frequency Curves,* Technical Paper No. 25, U.S. Weather Bureau, Washington, 1955.

Water Replenishment District of Southern California, *What Is the WRD?,* http://www.wrd.org/broch.htm.

Response

Stormwater infiltration:
Curing the disease rather than treating the symptoms

The title of one of the most useful design guidance manuals for stormwater quality protection is "Start at the Source." This is very much the objective expressed by Bruce Ferguson in this chapter. As Ferguson states, the fundamental disease of urbanized watersheds is the sealing of ground surfaces by impervious cover, thereby causing runoff to move across those surfaces and carry untreated contaminants directly into receiving waters. Other chapters in this book address means to ameliorate or negate the effects of impervious surfaces, and several chapters present approaches that move toward "starting at the source," including reducing impervious cover at the start as covered by Larry Coffman in Chapter I.5 and Michael Clar in Chapter I.15, setting limits to the amount of impervious cover as described by Jennifer Zielinski in Chapter I.3, and building bioretention swales or stormwater gardens as presented by Richard Pinkham and Timothy Collins in Chapter I.4, Thomas Liptan and Robert Murase in Chapter I.6, and Robert France and Philip Craul in Chapter I.7. Ferguson's chapter adds to the discussion of water sensitive design by proposing porous surfaces as viable alternatives to traditional impervious ones.

Ferguson reminds us that we need to regard landscapes as dynamic, evolving entities whose functions shift through time. What becomes very clear in this chapter is the cardinal requirement to understand processes in natural landscapes as the models upon which to pattern new water sensitive designs during development. Even simple "understanding" alone may be insufficient in this respect, for as Ferguson argues, it is imperative that we obtain advanced predictive knowledge about the patterns and dynamics of rainfall events for any local before we can generate truly effective, site-specific water sensitive designs. This is an extremely important lesson for design professionals and is a direction that should be embraced. For it is the shift from understanding to prediction that represents an intellectual maturity in any scientific investigation. This too underscores an important message that can be taken from Ferguson's chapter. If water sensitive design is to be fully integrated into landscape architecture, the profession needs to become much more informed about how the hydrological world works — in a predictive, *scientific* fashion.

Ferguson introduces a variety of alternative pavements in his chapter: interlinking blocks, concrete with pore spaces, stone aggregates, etc. Also touched upon are some concerns about site limitations for applying this technology, as for example, surface slopes and underlying soils. What is very much needed is a comprehensive examination of the relative strengths and weaknesses of these various approaches, followed through time. How well, for instance, does porous pavement perform in cold climates a decade after its installation?

One of the most fascinating aspects of this chapter is Ferguson's use of health metaphors in describing the "ailments" of urban watersheds. This is an exciting and rapidly expanding field of environmental scholarship and ecological ethics, and one to which those practicing water sensitive design could benefit from deeper exposure to. If, to run with Ferguson's analogy here, urban watersheds can be regarded as being diseased, can we then look upon the symptoms of that disease, i.e., surface runoff, as a form of hemorrhaging?

chapter I.2

Successful stormwater management ponds (Massachusetts)

Desheng Wang

Abstract

Stormwater management ponds have been one of the most widely used techniques in the history of flood control. Because of the growing knowledge of stormwater quality and its impact on natural resources such as water, wetlands, and wildlife habitat, federal, state, and local government agencies are tightening their stormwater management regulations. In this regulatory framework, stormwater management pond design naturally becomes more sophisticated, requiring a better understanding of the complicated soil and water conditions and different development needs of each project site. Therefore, designing a successful stormwater management pond becomes more challenging for every land development engineer. This chapter presents a framework extracted from design and review experience of over a hundred stormwater management ponds for industrial, commercial, and residential developments in the past five years. The discussions include (a) establishing design goals, (b) acquiring critical site information, (c) understanding the components of stormwater management ponds, and (d) proposing some thoughts for future design and study. Several projects are discussed to illustrate the scheme of designing successful stormwater management ponds, which includes

- A site where little space is left for a pond
- A 3:1 (H:V) slope of a pond that failed
- Stopping the discharge of muddy water from the pond

Introduction

Civil engineers and hydrologists would naturally think about detention/retention ponds or basins when they face stormwater problems. It is not surprising because stormwater management ponds are one of the most widely used techniques in the history of flood control. Stormwater pond design, however, is no longer only flood control oriented. In the past 20 years, it has been realized that water quality impacts of stormwater discharges on receiving waters are significant. The primary sources of area-wide (nonpoint source) pollutants are surface runoff from agricultural and urban watersheds (Loganathan et al., 1994). Many research and field studies were devoted to understand and enhance stormwater management systems, especially the pollution removal processes (Whipple et al.,

1980; EPA, 1986; Schueler, 1987, 1995a,b; Urbonas and Stahre, 1993). This resulted in guidelines for stormwater discharge permits in the National Pollutant Discharge Elimination System (NPDES) program (Code of Federal Regulations (CFR), 1994). A more stringent Stormwater Phase II rule of NPDES was published on December 8, 1999, and will become effective by early 2003. More information in this regard can be obtained at the Web site *http://www.epa.gov/owm/sw/phase2*. The stormwater quality control trend is also reflected in recent state and local environmental regulations and policies (MA DEP, 1997).

In current best management practices (BMPs), stormwater detention basins function as pollutant-trapping units by enhancing the detention time for pollutant settling. This is one of the most widely used methods due to pond reliability and effectiveness (Schueler, 1987). Whereas the method for designing detention ponds for flood control is relatively simple and well developed, the method for designing stormwater quality detention ponds is more complicated and relatively undeveloped (Wang and Carr, 1996). In addition, most stormwater ponds will finally end up as a seasonal or a perennial pond. Designers should know how the pond conditions will impact ecological systems, especially wildlife. These impacts are little known to the engineering community and are not adequately reflected in current stormwater management regulations.

This chapter will not address all stormwater pond design issues. Instead, it will try to present a systematic approach to the design of successful stormwater management ponds based on first-hand design experience and field observations of as-built ponds and basins. Then, a simple sequence of designing successful stormwater ponds used by the author will be discussed including planning, critical data acquisition, pond configuration, and outflow control structure. A few prototype stormwater management pond designs will be used to illustrate the discussion. More stormwater management design tips can be found in other chapters (I.1, I.6, I.7, I.8, I.11, and I.14). Further research needs to understand and enhance stormwater pond design are also discussed.

Functions of a stormwater pond

Humans create different kinds of ponds for different purposes: farm ponds, fish and wildlife ponds, water supply ponds, recreation/swimming ponds, landscape ponds, and stormwater/flood control ponds, which can be seasonal or perennial. Manmade pond history can be traced back thousands of years. For a stormwater management engineer facing this ancient topic in the new era, a question is constantly raised: how to design a successful stormwater management or stormwater BMP pond? As mentioned previously, stormwater management pond design has evolved from quantity oriented toward both quantity and quality oriented in the past 20 years (Whipple and Hunter, 1980; Schueler, 1987; Wang and Carr, 1996). Before the discussion gets into "how to," it will be helpful to look a little harder and deeper into what a stormwater pond can do for our environmental or ecological system.

Based on the aforementioned functions of manmade ponds, a stormwater pond may provide three basic functions: (1) flood control; (2) water quality enhancement; (3) ecological and aesthetic values. There will rarely be any disputes on the first two functions. The ecological and aesthetic functions of stormwater ponds can be controversial, however, and difficult to define due to different land uses and owners lacking quantitative evaluation methods. Today, this issue can no longer be overlooked because wildlife issues are receiving more attention in the permitting process. The pond aesthetics and recreational value are likely reflected in the sale and purchase process of a house that is located close to a stormwater pond. Based on the author's design and permitting experiences and field observations of as-built stormwater ponds, the following ecological and aesthetic functions are commonly observed:

1. Most stormwater basins with permanent pools attract wildlife and waterfowl including but not limited to wild ducks, frogs, turtles, spring peepers, and other macro-invertebrates.
2. Stormwater ponds with proper landscaping are usually attractive and may add value to commercial and industrial land.
3. The groundwater is recharged.
4. Stormwater ponds in residential subdivisions tend to trigger safety and mosquito concerns from neighbors.
5. Even stormwater ponds with temporary water provide some wildlife habitat value regardless of design purpose.

In general, therefore, a successful stormwater pond can provide optimal functions of flood control, stormwater quality enhancement, and ecological and aesthetic values for the environment. The following sections further discuss how to design a successful stormwater pond.

Designing a successful stormwater pond

Establishing design goals

As we know, there are different regulations from local, state, and federal governments for today's stormwater management system design. The most practical goal for designing successful stormwater ponds is to help a project meet these regulatory requirements. For example, when an engineer designs a stormwater management system in Massachusetts, he or she will have to meet nine stormwater management standards including (MA DEP, 1997):

1. No direct untreated discharge of stormwater
2. No increase in peak discharge rates (2-, 10-, and 100-year storms)
3. Compensation for loss of annual groundwater recharge due to development through infiltration measures to maximum extent practicable
4. Eighty percent removal of annual total suspended solids for new development
5. Special treatment of discharge from "hot spots" (auto service, commercial parking, etc.)
6. Special protection of critical resources (outstanding resource waters, shellfish beds, swimming beaches, coldwater fisheries, and recharge areas for public water supplies)
7. To the maximum extent practicable, application of the above standards for redevelopment of previously developed sites
8. Erosion maintenance and sediment control during construction and land disturbance
9. Providing an operation and maintenance plan to ensure that system function will be maintained as designed

To design a successful stormwater pond means meeting all regulatory requirements. Therefore, a successful stormwater pond cannot stand alone without other auxiliary components including planning, pretreatment, and operation and maintenance. In real life, lots of theoretically successfully designed ponds failed due to incorrect or missing site

Table I.2.1 Stormwater Pond Design Data

Design Goals	Design Data	Pond Factors
Flood control	Protected important resources in the watershed (wetlands, rivers, archaeological sites, wildlife habitat, water supply resources) Watershed topography Up-to-date precipitation data Existing and proposed land uses Groundwater conditions and soil infiltration rate	Location Volume Outflow control Emergency spillway Energy dissipation aprons at outfalls
Water quality	Pollutant sources in the watershed (sediment, hydrocarbons, metals, nutrients, bacteria, etc.) Characteristics of pollutants (concentration, particle size, and distribution in runoff)	Pretreatment (catchbasin, water quality inlets, and swales) In-pond features (sediment forebay, micropool, side slopes, planting) Flow distribution
Ecological impacts	Wildlife and public safety Recreation and landscape Water budget (precipitation, evapotranspiration, etc.)	Pondscape Landscape Food and water for wildlife

information. Some similar ponds in similarly sized watersheds can give quite different performance results (JSB and TRS, 1997).

The author inspected two ponds in northeast Massachusetts that failed during the early days of completion in the wintertime. The two ponds are located in different towns and were designed by different engineers. Both ponds did not provide a proper emergency spillway and were based on incorrect high groundwater and infiltration information. One pond ended up using significant storage volume below the groundwater table for flood control, which was overtopped during a 2-in. rain storm over frozen ground. The other pond was built above the actual high groundwater table, however, there was little depth below the outflow invert. Kids plugged the single-pipe outlet to create a skating rink, which also created an overtopping during a earlier spring melt. Both cases caused thousands of dollars in retrofitting and litigation enforcement fees. These case histories pointed to other important issues for successful stormwater ponds design: (1) design data acquisition and (2) maintenance-effective outflow control structures.

Critical data acquisition

As mentioned, a successful stormwater pond depends on a successful design after design goals are clearly defined. A successful design is the bridge between design goals and the materialized pond. On the other hand, the design data are the foundation of the design. A good design theory without complete and quality design data is foolish and greatly increases the chances of performance failing to meet design goals. When a project is undergoing review processes, incomplete design data not only delay the review process, but also make the project more expensive when actual data require changes in design. If a design with erroneous critical data "luckily" had passed review, the project would fail and be detrimental to downgradient resources. Therefore, design data acquisition is as important as design itself. In order to present the design data with a clear and concise relationship to design goals and pond features, the design data are divided into three groups: flood control, water quality, and ecological impacts, as shown in Table I.2.1. More detailed data information related to pond function design will be discussed.

Pond function design

Flood control

In general, in order to achieve flood control purpose, a stormwater pond shall be able to collect as much runoff as possible and as effectively as possible. The best place to locate a stormwater pond in a watershed is where surface runoff naturally goes. This will probably be the most effective way to collect runoff. In some cases, however, the best place may be occupied by other important resources: wetlands, endangered wildlife species habitat, or archaeological sites. Whenever a project comes across these issues, alternative locations or a mitigation plan would be required. Facing this situation, it is worth asking a question: Can a stormwater pond enhance the resource value (except for an archaeological site)? It is almost a "no" from a regulatory point of view at the current time in Massachusetts, but it is a "no" that lacks a good understanding and appreciation of what can be achieved in manmade systems. However, it is still a question for researchers to pursue and for a project team to evaluate, which will provide a significant benefit in the future design.

Land-use change is almost the most important factor triggering flood control in land development and urbanization. Different land uses generate different runoff volumes and rates. In the process of determining the total runoff, a representative runoff parameter is usually calculated for a combination of different land uses. When applying TR-55 or TR-20, the most popular methods in urban hydrology, the representative parameter is called the *curve number* (CN). A composite CN value is commonly calculated for different land use. When CN numbers for different land uses are significantly different, such as hydrological class A woods and sewered roadway, a simple weighted average CN number will significantly undervalue both runoff volume and peak rate, which can cause significant error in flood control calculation, even failure due to embankment overtopping. For such cases, the watershed should be divided into more subcatchments with relatively uniform CN numbers.

Groundwater conditions and the infiltration rate through pond side and bottom surface are critical in the determination of flood control volume and water budget. The latter can also impact water quality and ecological functions. Groundwater table fluctuations in the pond area can also cause slope failure, which will be further discussed in later case studies.

Water quality

The water quality function of stormwater ponds is commonly evaluated by removal efficiencies of pollutants. Different pollutants correspond to different removal rates or efficiencies. There are different methods to predict pollutant removal rates. Some methods are based on more sophisticated theory (Vittal and Raghav, 1997; Whipple and Hunter, 1980; Camp, 1964; Vanoni, 1975; Garde et al., 1990). Some are empirical (Wang and Carr, 1996; Meadows and Kollitz, 1995; Loganathan et al., 1994). Given so many parameters involved in real-life cases, it is not necessary that a theoretical method is better than an empirical method (JSB and TRS, 1997; Wang and Carr, 1996). More than a dozen parameters affect pollutant removal rates of a pond (Wang and Carr, 1996). Understanding what each parameter contributes to the pollutant removal can be helpful to simplify calculations and provide better guidelines in stormwater pond design. The main mechanism of pollutant removal in stormwater ponds is sedimentation, which has a more complete theoretical description. The three most important factors in sediment removal processes in a stormwater pond are (a) sediment size and particle distribution, which can be represented by settling velocity (ω); (b) pond detention time or residence time (t_d); (c) suspension or reentrainment factor (ω/u_*), where u_* is shear velocity of flow in the pond. As a rule of thumb, the removal rate of pollutant is proportional to ω and t_d, but inversely proportional to u_*. Then, the pond dimensions (width, depth, and length) can be determined based on these parameters for a required removal efficiency. As we can see, however, the sediment size and distribution would be very difficult data to collect for a designed site condition. Therefore, some theoretical statistical data will be valuable. Even though sediment size and distribution may change site to site, studies by different researchers revealed that particle size in stormwater mostly fall in the range of 10 to 35 µm (Urbonas and Stahre, 1993). If available, local data will help yield a more precise design. A significant effort is currently ongoing to build a nationwide database for different stormwater BMPs (ASCE et al., 1999). In real situations, the annual average pollutant removal is more important. It will be very tedious and unrealistic for a design engineer to calculate the annual average removal rate using theoretical methods and each rainfall amount, especially for small projects. Therefore, some simple methods — first flush (0.5" or 1") treatment volume, graphics method, and empirical formula — are more widely adopted in stormwater management policy and design guidelines (MA DEP, 1997; Schueler, 1987; Wang and Carr, 1996). Simple methods are usually accompanied by some general recommendations and requirements, which result from more detailed studies (MA DEP, 1997; Schueler, 1987). Some important considerations to enhance the pollutant removal rate include the following:

1. Detention time would be better between 24 h and 40 h. When in permeable soil, the time can be shorter depending on the infiltration rate. The larger the infiltrate rate, the smaller the detention time will be. Therefore, reliable infiltration rate calculation methods using accurate in-site permeability test data are necessary to achieve practically accurate groundwater recharge and water quality benefit due to infiltration in stormwater ponds or basins (Wang, 1999).
2. Flow distribution shall be made as uniform as possible across the pond section, which is perpendicular to the flow direction. This can be facilitated by shaping the pond as an ellipsis, and installing flow distribution berms at the inlet and outlet, a sediment forebay at the inlet, and a high marsh between the inlet and outlet.
3. Flow depth shall be no less than 1 ft for an annual mean storm event.
4. Maximize the distance between inlet(s) and outlet(s).
5. Minimize "clean water" through pond, such as roof runoff to maximize treatment for "dirty water," such as roadway runoff.

Ecological and safety impacts

As observed in the field, a stormwater pond will likely become some kind of habitat for wildlife and waterfowl. The water quality in the pond itself can be important to wildlife coming to use the pond, which somehow is contradictory to the stormwater quality treatment function of the pond design. To minimize this conflict, the following are recommended:

1. Stormwater runoff shall undergo pretreatment as much as possible before being discharged to the pond. Pretreatment includes deep sump catchbasins, oil/grit separators or water quality inlet, vegetation strips, and grass swales.
2. Pretreatment shall be installed offline unless detailed design requires online installation.
3. Pollutant tolerant vegetation shall be chosen for pondscape, such as cattail, sedges, and bulrushes, which can enhance in-pond water quality.
4. A safety leveled bench (5 to 10 ft) can be installed at a flow depth of 3 ft to minimize public hazard.
5. A minimum 3:1 slope shall be used.

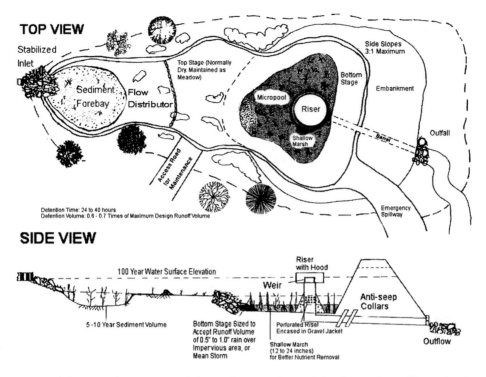

Figure I.2.1a Schematic detention pond design features. (From Schueler, T., *Controlling Urban Runoff: A Practical Manual for Planning and Designing Urban BMPs*, Dept. of Environmental Programs, Metropolitan Washington Council of Governments, Washington D.C., 1987. With permission.)

6. A 1- to 2-ft freeboard should be allowed for the maximum design storm (100-year) depending on the pond size. The larger the pond, the larger the freeboard should be.
7. Whenever the groundwater table will be penetrated, a slope stability analysis should be conducted to determine if the proposed slope is stable and slope stabilization is needed. For such a condition, a perimeter drain installed above the high groundwater table can increase slope stability.

A schematic stormwater pond is presented in Figure I.2.1a. Another public concern about stormwater ponds is mosquito breeding. In general, stormwater ponds do not have a high mosquito nuisance level compared with wastewater treatment ponds due to relatively low nutrient loads and more frequent replenishment. In case mosquitoes are breeding in the pond, many methods are available to control them, including biological control and chemical control. Biological controls are preferred because the biological controls specifically target mosquito larvae and are harmless to humans, unlike many chemicals even at standard doses (McLean, 1995).

In addition, a maintenance-effective outflow control structure is also important to assure the design function of the pond, especially the flood control function, which should have the following features:

- Multilevel discharge points, which can provide long enough detention time (>24 h) for storms less than a 2-year event
- A protection component, which can prevent debris and vandalism from interfering with the structure design function
- An overflow path for larger storm events (larger than a 50-year storm)

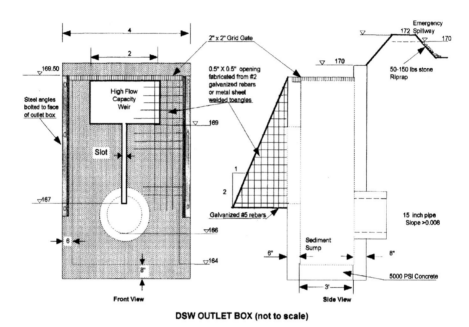

Figure I.2.1b A maintenance-effective outflow control for detention ponds.

Figure I.2.1b presents an outflow control structure based on the above criteria that has been proved effective in reality.

Roof detention

When space and soil are very limited for creating stormwater pond and subsurface infiltration trenches, the author has used a roof detention method for stormwater management to compensate for a smaller in-ground stormwater pond or subsurface detention. The mechanisms include ballast roof and water dams around roof drains, which can provide more storage and prolong the time of concentration for roof runoff and reduce the peak discharge rate. This method can only be applied to a relatively flat roof condition. The time of detention should be controlled within 6 h. The load of the water due to detention should be communicated to the architect and structure engineer for a corresponding roof supporting structure design. Based on design experience in Massachusetts, roof load due to detained water does not significantly increase cost in other components. One reason is that the roof has to be designed for significant snow load. The other advantage of roof detention is to reduce the number and pipe size of roof drains due to the reduced roof peak discharge rate. When applying this method to areas without snow load, costs may rise. Roof detention may be a new concept, however, in real situations, ballast roofs have

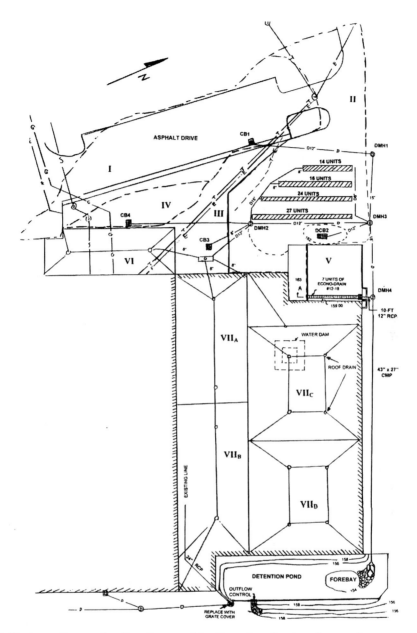

Figure I.2.2 Drainage System for Case I — combination of roof, subsurface, and open pond detention, Mansfield, MA.

been used to increase the roof stability for buildings along coasts subject to strong wind. At the same time, the ballast roof can slow down the roof runoff and provide flood control benefit. See Figure I.2.3a for a real-life ballast roof.

Project illustration

In order to demonstrate the importance of design data and design concept, three projects are briefly introduced and the highlights of each project are discussed.

Case I: A site where little space is left for a pond

A manufacturer needs to expand its factory on the same lot with limited space abutting bordering vegetated wetland in a industrial park in Mansfield, MA. The site conditions and project goals are as follows:

Total space available:	3.34 acres
Depth to high groundwater:	7 ft
On-site soil:	Silty with very low permeability (5.19×10^{-6} ft/s)
On-site utility lines:	Subsurface electric, phone, and gas lines
Building addition:	2.36 acres (70% of available space)
Space for pond:	190 ft × 47 ft
Effective depth for flood control:	1.5 ft due to the limits to inflow and outflow inverts
Effective flood storage:	6,000 ft^3
Flood storage need:	16,000 ft^3

As can be seen, 70% of the site will be building. Not enough space is available to create a large enough pond to achieve flood control. Low permeability does not allow the effective use of subsurface infiltration trenches. To overcome the difficulties, a small pond was created for stormwater quality and landscape purposes, which worked successfully together with roof detention and a small amount of subsurface detention to serve the overall stormwater management requirement. See Figure I.2.2 for details. Roof detention is achieved by a series of 4-in. PVC pipes fixed on the roof around each roof drain. One-in. orifices are drilled every 10 ft along the pipe bottom. The roof detention time was designed as 4 h, which compares to 5 min without the pipe check dams. See details of the water dam of roof detention in Figure I.2.3b.

Case II: A 3:1 (H:V) slope of a pond that failed

A detention pond (180 ft by 60 ft) in a residential subdivision in Marlboro, MA, seemed designed and constructed according to all MADEP criteria: 3:1 pond length-to-width ratio; 3:1 side slope; sediment forebay at inlet and micropool at outlet; high marsh (cattail) between sediment forebay and micropool; maintenance-effective outflow control structures; and energy dissipation aprons at inlet and outlet. Water coming out of the detention pond looked crystal clean (see pictures in Figure I.2.4a). The pond was built in the summer of 1997 by excavating in a 30-ft drop on a 10% slope. The maximum excavation was about 12 ft. The downgradient wetland is at an elevation similar to the bottom of the sediment forebay and micropool. In November 1997, part of the upgradient side of the pond slumped, which was simply fixed in the summer of 1998. By the late fall of 1998, the slope failed again. Then, the town engineer and conservation commission got worried and demanded a professional fix. The owner finally contacted the author in the spring of 1999. After a site inspection, the following was observed:

- The failure was a shallow, noncircular slide.
- Groundwater broke out at the upper limit of the failure.
- Soils in slope were a silty, sandy loam.
- A drainage channel was located on top of the slope running toward the corner with the failed slope.

Based on the information, it was concluded that the failure was caused by a large groundwater fluctuation (about 5 ft). When groundwater raises, it increases the pore

Figure I.2.3a Ballast roof, Milton, MA.

Figure I.2.3b Water dam roof detention, Mansfield, MA.

Figure I.2.4a Distant view of a detention pond with upper left slope slumped, Marlboro, MA.

pressure and reduces the soil strength against the gravity. The mitigation plan is as follows: a perimeter drain consisting of a 6-in. crushed stone blanket placed on the failed slope from the limit of failure to about 2 ft above the normal pond level and a 4-in. perforated pipe installed at the bottom of the trench. The crushed stone and drain pipe are underlain by filter fabric. Along the bottom of the drain, some 15-in. to 24-in. key stones are placed to increase the stability (see Figure I.2.4d).

Figure I.2.4b A close look of the outflow control structure in Figure I.2.4a.

Figure I.2.4c A close look of the slumped slope.

Figure I.2.4d Slope in Figure I.2.4c after restoration.

Case III: Stop the discharge of muddy water from the pond

On April 12, 1995, Carr Research Laboratory (CRL) received a call from a land developer in Merrimac, MA. A detention pond in his subdivision under construction was discharging muddy water to downgradient wetland and a neighbor's pond. The DEP issued an enforcement order to stop all construction activities unless the problem was solved. The CRL staff went to inspect the site the same afternoon:

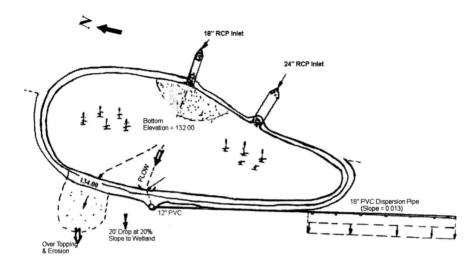

Figure I.2.5a A detention pond as observed, Merrimac, MA, April 1995.

- The soils in the watershed of the detention basin consisted of silty, sandy loam.
- The cause of muddy water was due to runoff from the disturbed lot.
- The detention basin is about 270 ft × 108 ft.
- Two inlets to the pond were observed: an 18-in. RCP and a 24-in. RCP.
- Only the 18-in. pipe was discharging water into the pond.
- Only a 12-in. PVC outlet with an invert at the pond bottom was observed, which was connected to an 18-in. level spreader covered with pea stone to the southwest of the pond.
- A sediment deposition delta was observed above pond level below the 18-in. inlet.
- Riprap aprons were at the inlets and outlet.
- No trash rack was observed at the outlet.
- No emergency spillway was observed for the pond.
- The inlets and outlet were installed across the width of the pond, which short-circuits the flow.
- Neither sediment forebay nor micropool was created in the pond bottom.
- The average water depth in the pond was less than 0.6 ft.
- Withered cattails spread in the bottom of the pond.
- An overtopping of the pond embankment had taken place and caused erosion and siltation in an area 80-ft long by 70-ft wide and about 0.5 ft deep in the downgradient wetland.
- Not enough sediment source control existed in the subdivision. All temporary disturbed lots were barren.

The overtopping was caused by a blockage of the outlet, which children had done to create a skating rink. Because there was no designed emergency spillway, the water rose to a low spot of the earth embankment. Figure I.2.5a shows the pond condition as observed. The pond was obviously not designed according to stormwater BMP criteria. Not enough stormwater treatment volume was designed to remove enough silt and clay sediment in the incoming water before discharging it to the downgradient. Based on observed information, a retrofit plan was created and approved by the MA DEP and the Town Conservation Commission. The highlights of the retrofit plan of the pond are as follows:

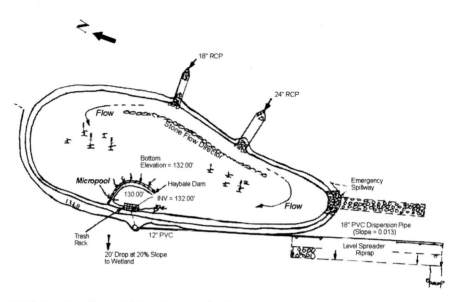

Figure I.2.5b Pond in Figure I.2.5a after retrofit, September 1995.

- Remove sediment deposition at the inlet.
- Install a 1-ft-tall stone flow distributor at the inlets of the pond.
- Excavate a micropool at the outlet and install a hay bale dam around it to increase the normal pool level for more water treatment volume.
- Install a trash rack at the outlet to prevent debris or humans from blocking it.
- Restore the eroded embankment and install an emergency spillway.
- Install a series of check dams along the slope at the end of the outfall of the outflow pipe.

See Figure I.2.5b for the retrofitted pond conditions. The results turned out to be satisfactory.

Summary and conclusions

This chapter illustrates a design philosophy for stormwater ponds. A stormwater pond is only a component in one stormwater management system. A successful pond design can only be achieved through successful stormwater management system planning, acquisition of important data, clear pond function definition, and a sound pond structure design, which provides the best pond function in the most cost-effective way.

The chapter also introduced a new concept: roof detention and its advantages and limits. The use of roof detention can be critical for some site conditions to make an effective stormwater management system practical.

Almost all stormwater ponds provide some wildlife habitat value to the ecological system. It is not clear how the pond water quality will impact the wildlife community using the pond, which is critical in evaluating future design functions of the stormwater ponds. More field data and studies would be needed in this area. In the meantime, it is recommended that stormwater from pollutant-laden areas should be pretreated as much as possible before being discharged to stormwater ponds, which is important to the wildlife community.

Acknowledgments

The author is thankful to Dr. Jerome Carr, Carr Research Laboratory, Inc. for his review and useful comments and for many opportunities provided to test design strategies, and Pamela Carlson for her meticulous help in the preparation of figures. Special thanks to my wife, Amy, who made this work possible, for her understanding and support during many nights and weekends.

Literature cited

ASCE and EPA, *User's Guide — National Stormwater Best Management Practices (BMP) Database*, Version 1.0, prepared by Urban Water Resources Research Council (UWRRC) of ASCE, Wright Water Engineering, Inc., Urban Drainage and Flood Control District, URS Griner Woodward Clyde, and Office of Water, U.S. Environmental Protection Agency, Washington, DC, 1999.

Camp, T.R., Sedimentation and design of settling tanks, *Trans. ASCE*, 111, 895, 1964.

Cheng, N.S., Simplified settling velocity formula for sediment, *J. Hydr. Eng.*, 123, 149, 1997.

Code of Federal Regulations (CFR), *The National Pollutant Discharge Elimination System*, 40 CFR 122, U.S. Government Printing Office, Washington, DC, 1994.

Garde, R.J., Ranga Raju, K.G., and Sujudi, A.W.R., Design of settling basins, *J. Hydr. Eng.*, 18, 81, 1990.

JSB and TRS, A tale of two regional wet extended detention ponds, Technical Note 98, *Watershed Protection Tech.*, 2, 529, 1997.

Loganathan, G.V., Watkins, E.W., and Kibler, D.F., Sizing stormwater detention basins for pollutant removal, *J. Environ. Eng.*, 120, 138, 1994.

MA DEP, *Stormwater Management — Stormwater Policy Handbook*, prepared by MA Department of Environmental Protection and MA Office of Coastal Zone Management, March 1997.

McLean, L., Mosquitoes in constructed wetlands — A management bugaboo, *Watershed Protection Tech.*, 1, Center for Watershed Protection, 203, 1995.

Meadows, M.E. and Kollitz, M.M., Sediment pond trapping efficiency curves, *Water Resources Engineering: Proceedings of the First International Conference*, San Antonio, Texas, Vol. 2, 1258–1262, August 14–18, 1995.

Reed, S.C., *Natural Systems for Wastewater Treatment — Manual of Practice FD-16*, Water Pollution Control Federation, 1990.

Schueler, T., *Controlling Urban Runoff: A Practical Manual for Planning and Designing Urban BMPs*, Dept. of Environmental Programs, Metropolitan Washington Council of Governments, Washington DC, 1987.

Schueler, T., *Design of Stormwater Wetland Systems: Guidelines for Creating Diverse and Effective Stormwater Wetland Systems in the Mid-Atlantic Region*, Anacostia Restoration Team, Dept. of Environmental Programs, Metropolitan Washington Council of Governments, Washington DC, 1992.

Schueler, T.R., Performance of a stormwater pond/wetland system in Colorado, *Watershed Protection Tech.*, 1, Center for Watershed Protection, 68, 1995a.

Schueler, T.R., Pollutant dynamics of pond muck, *Watershed Protection Tech.*, 1, Center for Watershed Protection, 39, 1995b.

UDFCD, *Douglas County Storm Drainage Design and Technical Criteria*, prepared by WRC Engineers, Inc., under a contract to Urban Drainage and Flood Control District, Denver, CO, Jan. 1986.

Urbonas, B. and Stahre, P., *Stormwater — Best Management Practices and Detention for Water Quality, Drainage, and CSO Management*, PTR Prentice Hall, Englewood Cliffs, NJ, 1993.

U.S. Environmental Protection Agency (EPA), *Methodology for Analysis of Detention Basins for Control of Urban Runoff Quality*, Nonpoint Source Branch, Office of Water, Washington, DC, EPA-440-5-87-001, 1986.

Vanoni, V.A. (ed.), *Sedimentation Engineering, Manuals and Rep. on Eng. Pract.*, ASCE, New York, 54, p. 582, 1975.

Vittal, N. and Raghav, M.S., Design of single-chamber settling basins, *J. Hyd. Eng.*, ASCE, 123, 469, 1997.

Wang, D.S., A simple mathematical model for infiltration BMP design, *J. Hydrol. Sci. Tech.*, November, 117, 1999.

Wang, D.S. and Carr, B.J., Pollutant removal rates for stormwater detention ponds, *Proc. of 1996 AIH Annual Meeting*, Boston, AMBP 12–21, 1996.

Whipple, William, Jr. and Hunter, Joseph V., Detention basin settleability of urban runoff pollution, *Final Technical Completion Report for Project A-058-NJ*, Water Resources Resource Research Institute, Rutgers University, New Brunswick, NJ, 1980.

Response

Centralized stormwater treatment:
Improving performance through engineering design

Detention ponds continue to be one of the most widely used techniques in stormwater management. As Desheng Wang describes, due to a tightening of stormwater regulations, detention pond design has become more sophisticated, requiring better understanding of complicated soil and water conditions and different development needs of particular sites. In particular, the most notable effect has been a shift from designing such ponds solely for purposes of flood control, to more sophisticated water sensitive designs that improve stormwater quality while simultaneously producing ecological benefits such as wildlife attraction and groundwater recharge.

In this chapter, Wang outlines a systematic approach to the design of successful stormwater ponds that includes planning, critical data acquisition, consideration of pond configuration, and attention to outflow control structures. He also stresses the need for auxiliary components such as watershed planning beyond the specifics of the site, pre-treatment to reduce the contamination of inflowing stormwater (and thereby reduce the opportunities for bioaccumulation by resident wildlife), and a strategic plan for operations and maintenance. By establishing clear design goals, and by fitting the pond into the topographic idiosyncrasies of the site rather than being imposed against them, it is possible for such stormwater management controls to have a strong positive role in boosting nearby property values while improving environmental conditions.

chapter I.3

Open spaces and impervious surfaces: Model development principles and benefits

Jennifer A. Zielinski

Abstract

Recent research has demonstrated that subwatershed impervious cover has a strong influence on the quality of streams and lakes, and that impairment occurs at levels as low as 5 to 15%. Most current suburban development patterns, or the "status quo," create impervious cover that equals or exceeds this level. This chapter reviews the potential of better site design techniques to minimize the impervious cover created by new development, conserve open space, and more effectively manage stormwater at development sites. A series of 22 model development principles developed by a consortium of planning, road, banking, engineering, development, and public safety organizations are described along with an overview of the research data on economic, market, legal, safety and social benefits of each. The environmental benefits of applying the principles will be demonstrated in a series of redesign comparisons of residential development projects. The redesign comparisons document changes in runoff rates and nutrient export between the "status quo" and subdivision designs incorporating the principles. The comparisons consistently indicate that application of the model development principles significantly reduces impervious cover, runoff, nutrient loads, and construction cost.

Introduction

Though they may not realize it, site planners have an excellent opportunity to reduce stormwater runoff and pollutant export simply by changing the way they lay out new residential subdivisions. Planners that employ open space design techniques can collectively reduce the amount of impervious cover, increase the amount of natural land conserved, and improve the performance of stormwater treatment practices at new residential developments.

Simply put, open space designs concentrate density on one portion of a site in order to conserve open space elsewhere by relaxing lot sizes, frontages, road sections, and other subdivision geometry. Although site designs that employ these techniques go by many different names, such as clustering or conservation design, they all incorporate some or all of the following better site design techniques:

- Using narrower, shorter streets and rights-of-way
- Applying smaller lots and setbacks, and narrow frontages to preserve significant open space
- Reducing the amount of site area devoted to residential lawns
- Spreading stormwater runoff over pervious surfaces
- Using open channels rather than curb and gutters
- Protecting stream buffers
- Enhancing the performance of septic systems, when applicable

In this chapter, I describe some of the benefits of employing better site design techniques as they apply to residential subdivisions. The analysis utilizes a simple spreadsheet computer model to compare actual residential sites constructed in the 1990s using conventional design techniques with the same sites "redesigned" utilizing better site design techniques. For each development scenario, site characteristics such as total impervious and vegetative cover, infrastructure quantities, and type of stormwater management practice are estimated.

The Simplified Urban Nutrient Output Model (SUNOM) was used to perform a comparative analysis for two subdivisions. The first is a large-lot subdivision known as Duck Crossing, and the second is a medium-density subdivision known as Stonehill Estates. In each case, the model was used to simulate five different development scenarios:

- Predeveloped conditions
- Conventional design without stormwater practices
- Conventional design with stormwater practices
- Open space design without stormwater practices
- Open space design with stormwater practices

This chapter compares the hydrology, nutrient export, and development cost for these sites under both conventional and open space design, and with and without stormwater treatment. The article also summarizes other research on the benefits of open space design and discusses the implications it can have for the watershed manager.

Duck Crossing — a low-density residential subdivision

Duck Crossing is a large-lot residential development located in Wicomico County on Maryland's Eastern Shore. Prior to development, the low-gradient coastal plain site contained a mix of tidal and nontidal wetlands, natural forest, and meadow (Figure I.3.1). Its sandy soils were highly permeable (hydrologic soil group A). Three existing homes were located on the parcel, which relied on septic systems for on-site sewage disposal. The existing septic systems discharged a considerable nutrient load to shallow groundwater.

A conventional large-lot subdivision of eight single-family homes was constructed on the 24-acre site in the early 1990s. The subdivision is reasonably typical of rural residential development along the Chesapeake Bay waterfront during this era (Figure I.3.2). Each new lot ranged from three to five acres in size and was set back several hundred feet from an access road. The access road was 30 ft wide and terminated in a large-diameter cul-de-sac. Sidewalks were located on both sides of the street. Each lot was served by a conventional septic system with a primary and reserve field of about 10,000 ft^2. Stormwater management consisted of curb and gutters that conveyed runoff into a storm drain system that, in turn, discharged to a small dry pond (designed for the water quality volume, only).

Chapter I.3: Open spaces and impervious surfaces 51

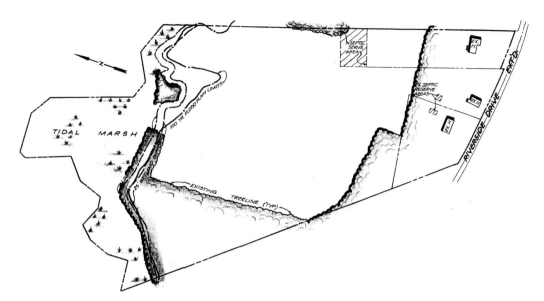

Figure I.3.1 Predevelopment conditions at the Duck Crossing site. (From Zielinski, J.A., The benefits of better site design in residential subdivisions, in *Watershed Protection Techniques*, Center for Watershed Protection, Ellicott City, MD, 2000. With permission.)

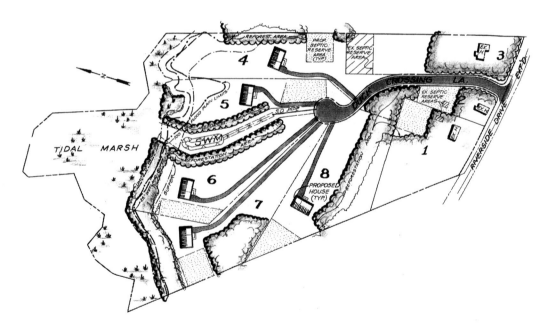

Figure I.3.2 The low-density conventional subdivision built at Duck Crossing (eight lots). (From Zielinski, J.A., The benefits of better site design in residential subdivisions, in *Watershed Protection Techniques*, Center for Watershed Protection, Ellicott City, MD, 2000. With permission.)

The entire site was privately owned, with the exception of the tidal marsh, which was protected under state and federal wetland laws and represented the only common open space on the site. As a result of construction, the existing meadow was entirely converted to lawn, and the impervious cover for the site increased to slightly over 8%.

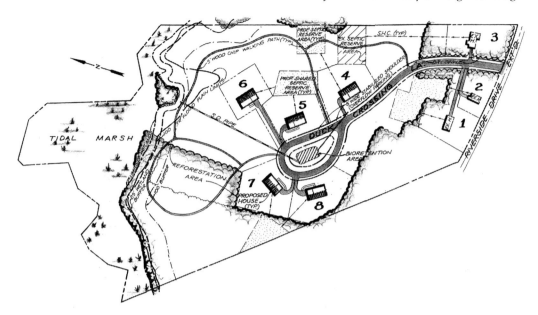

Figure I.3.3 The open space subdivision that could have been built at Duck Crossing (eight lots). (From Zielinski, J.A., The benefits of better site design in residential subdivisions, in *Watershed Protection Techniques*, Center for Watershed Protection, Ellicott City, MD, 2000. With permission.)

Open space design for Duck Crossing

The critical ingredient of the open space redesign was a reduction in lot size from several acres to about 30,000 ft^2. This enabled about 74% of the site to be protected and managed as common open space, which included most of the existing forest, wetlands and meadow (Figure I.3.3). Consequently, only 19% of the site was managed as turf, nearly all of which was located on the private lots.

The open space redesign at Duck Crossing also incorporated a narrower access road (20 ft wide) along with shorter, shared driveways that served six of the eight lots. The road turnaround was designed as a loop rather than a cul-de-sac bulb. Also, a wood chip trail system was provided through the open space instead of sidewalks along the road. Each home site was carefully located away from sensitive natural areas and the 100-year flood plain. Taken together, these better site design techniques reduced impervious cover for the site by about a third compared to the conventional design (from 8 to 5%).

The redesigned stormwater conveyance system utilized dry swales rather than a curb and gutter system and featured the use of bioretention areas in the roadway loop to treat stormwater quality. This combination of stormwater practices provided greater pollutant removal through filtering and infiltration.

One of the most important objectives in the redesign strategy was to improve the location and performance of the septic systems that dispose of wastewater at the site. Home sites were oriented to be near soils that were most suitable for septic system treatment. In addition, six homes shared three common septic fields located within open space rather than on individual private lots. Last, given the permeability of the soils, advanced recirculating sand filters were installed to provide better nutrient removal than could be achieved by conventional septic systems.

Comparative hydrology for Duck Crossing

Given its low impervious cover and permeable soils, the water balance at Duck Crossing was dominated by infiltration, even after development. The comparative hydrology under the

Chapter I.3: Open spaces and impervious surfaces

Table I.3.1 Annual Water Budget of Duck Crossing

		Predeveloped	Conventional Design	Open Space Design
Runoff (in./year)	No practices	2.3	4.8	3.9
	Practices	—	4.8	3.7
Infiltration (in./year)	No practices	18.2	15.3	17.0
	Practices	—	15.3	17.2

Source: Zielinski, J.A., The benefits of better site design in residential subdivisions, in *Watershed Protection Techniques,* Center for Watershed Protection, Ellicott City, MD, 2000. With permission.

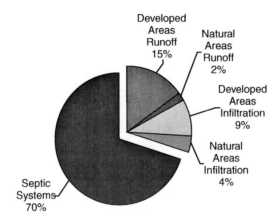

Figure I.3.4 Nitrogen load distribution from the conventional design of Duck Crossing, without stormwater practices. (From Zielinski, J.A., The benefits of better site design in residential subdivisions, in *Watershed Protection Techniques,* Center for Watershed Protection, Ellicott City, MD, 2000. With permission.)

five development scenarios is presented in Table I.3.1. As might be expected, the conventional design yielded the greatest volume of surface runoff and the least amount of infiltration. The open space design produced about 25% less annual surface runoff and 12% more infiltration than the conventional design but did not come close to replicating predevelopment conditions. The use of stormwater practices did not materially change the water balance under either the conventional or open space design at Duck Crossing (see Table I.3.1).

Comparative nutrient output at Duck Crossing

Nutrient export at Duck Crossing was dominated more by subsurface water movement than by surface runoff. Indeed, stormwater runoff seldom comprised more than 15% of the annual nitrogen or phosphorus load from this lightly developed site. The SUNOM model indicated that the major source of nutrients was subsurface discharges from septic systems, which typically accounted for 60 to 80% of the total load in every development scenario (see Figure I.3.4).

The open space design sharply reduced nutrient export, primarily because recirculating sand filters were used in the shared septic systems and helped to reduce (but not eliminate) subsurface nutrient discharge. The other elements of the open space design (reduced impervious cover, reduced lawn cover, and multiple stormwater practices) also helped to reduce nutrient export, but by a much smaller amount. The comparative nutrient export from each Duck Crossing development scenario is detailed in Figure I.3.5.

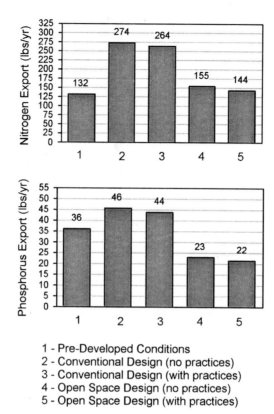

Figure I.3.5 Annual nitrogen and phosphorus loads for each development scenario at Duck Crossing. (From Zielinski, J.A., The benefits of better site design in residential subdivisions, in *Watershed Protection Techniques*, Center for Watershed Protection, Ellicott City, MD, 2000. With permission.)

Comparative cost of development

The cost to build infrastructure for the open space design was estimated to be 25% less than the conventional design at Duck Crossing, due primarily to the necessity for less road paving, sidewalks, and curbs and gutters. Even when higher costs were factored in for the more sophisticated stormwater and on-site wastewater treatment used in the open space design, the total cost was still 12% lower than the conventional design. In addition, the open space design had seven fewer acres that needed to be cleared and graded, or served by erosion and sediment controls, compared with the conventional design (these costs are not currently evaluated by the SUNOM model). Overall, the SUNOM model estimated that the conventional design at Duck Crossing had a total infrastructure cost of $143,600, compared with $126,400 for the open space design.

Summary

The comparative results for the Duck Crossing redesign analysis are summarized in Figure I.3.6. The open space design increased natural area conservation and reduced impervious cover, stormwater runoff, nutrient export, and infrastructure costs compared with the conventional subdivision design.

Stonehill Estates — a medium-density residential subdivision

Stonehill Estates, located near Fredericksburg, VA, is situated in the rolling terrain of the Piedmont. The undeveloped parcel was 45 acres in size, nearly all of which was mature

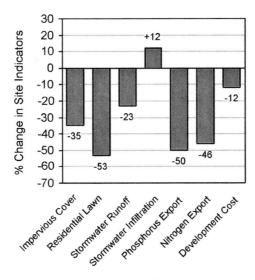

Figure I.3.6 Percentage change in key site conditions from a conventional design to an open space design, both with stormwater practices. (From Zielinski, J.A., The benefits of better site design in residential subdivisions, in *Watershed Protection Techniques*, Center for Watershed Protection, Ellicott City, MD, 2000. With permission.)

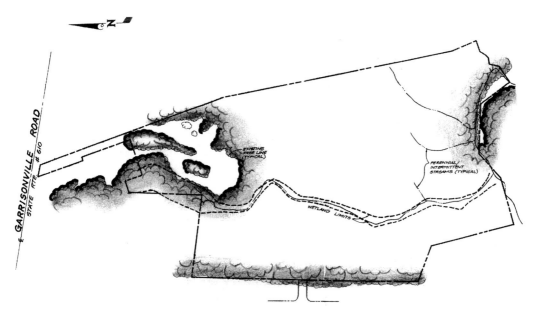

Figure I.3.7 Predevelopment conditions at the Stonehill Estates site. (From Zielinski, J.A., The benefits of better site design in residential subdivisions, in *Watershed Protection Techniques*, Center for Watershed Protection, Ellicott City, MD, 2000. With permission.)

hardwood forest (Figure I.3.7). An intermittent stream bisected the site, discharging into a perennial stream near the southern edge of the parcel. Roughly 3.6 acres of forested wetlands were found along the stream corridors, and an extensive floodplain was located along the perennial stream. Soils at the site were primarily silt loams and were moderately permeable (hydrologic soil groups C and D).

The site was highly attractive for development, given the excellent access provided by two existing roads, both of which had public water and sewer lines that could be easily tapped to serve the new subdivision. The conventional design was zoned for three dwelling units per acre. After unbuildable lands were excluded, the parcel yielded a total of 108 house lots, each of which was about 9,000 ft^2 in size (Figure I.3.8). The subdivision design typifies medium-density residential subdivisions developed in the last two decades in the Mid-Atlantic region, where lots sizes were uniform in size and shape and homes were set back a generous and fixed distance from the street. The design utilized a mix of wide and moderate street sections (34 ft and 26 ft), and included six large-diameter cul-de-sacs for turnarounds. Sidewalks were generally installed on both sides of the street.

The stormwater management system for the conventional design represents the typical "pipe and pond" approach utilized in many medium-density residential subdivisions. Street runoff was conveyed by curbs and gutters into a storm drain system that discharged into the intermittent stream channel and then traveled downstream to a dry extended detention pond. The pond was primarily designed to control flooding but also provided some limited removal of stormwater pollutants.

Interestingly, about 25% of the site was reserved as open space in the conventional design at Stonehill Estates. Nearly all of these lands were unbuildable because of environmental and site constraints (e.g., floodplains, steep slopes, wetlands, and stormwater facilities), and the resulting open space was highly fragmented. Even so, about a fourth of the forested wetlands were impacted by two roads crossing over the intermittent stream. Almost 90% of the original forest cover was cleared as a result of the conventional design and was replaced by lawns and impervious cover. Overall, about 60% of the site was converted to lawns, and another 27% was converted to impervious cover.

Open space design for Stonehill Estates

In the redesign analysis, Stonehill Estates was designed to incorporate many of the open space design techniques advocated by Arendt (1994). The resulting design retained the same number of lots as the conventional design but had a much different layout (Figure I.3.9). The average lot size declined from about 9,000 ft^2 in the conventional design to 6,300 ft^2 in the open space design. This reduced lot size allowed about 44% of the site to be protected as open space, most of which was managed as a single unit that included an extensive natural buffer along the perennial and intermittent stream corridor.

The basic open space layout was augmented by several other better site design practices, including narrower streets, shorter driveways, and fewer sidewalks. Loop roads were used as an alternative to cul-de-sacs. In some portions of the site, irregularly shaped lots and shared driveways were used to reduce overall road length. Each individual lot was located adjacent to open space, so that the more compact open space lots would not feel as crowded. As a result of these techniques, the open space design for Stonehill Estates reduced impervious cover from 27 to 20%. In addition, lawn cover declined from 60 to 30% of the total site area.

The innovative stormwater collection system utilized dry swales instead of storm drains in gently sloping portions of the site. The dry swales and several bioretention areas located in loop turnarounds were used to initially treat stormwater quality. Each of these practices then discharged to a small micropool detention pond, with an embankment created by the single road crossing over the intermittent stream.

Chapter I.3: Open spaces and impervious surfaces

Figure I.3.8 The conventional subdivision design that was built at Stonehill Estates (108 lots). (From Zielinski, J.A., The benefits of better site design in residential subdivisions, in *Watershed Protection Techniques*, Center for Watershed Protection, Ellicott City, MD, 2000. With permission.)

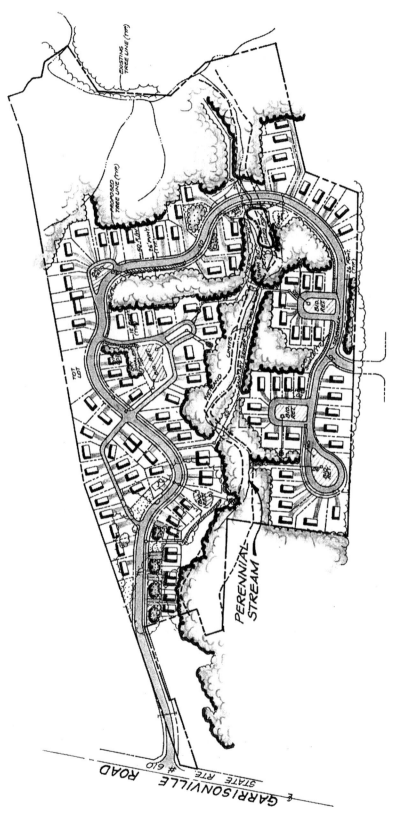

Figure 1.3.9 The open space subdivision that could have been built at Stonehill Estates (108 lots). (From Zielinski, J.A., The benefits of better site design in residential subdivisions, in *Watershed Protection Techniques*, Center for Watershed Protection, Ellicott City, MD, 2000. With permission.)

Chapter I.3: Open spaces and impervious surfaces

Table I.3.2 Comparative Hydrology of Stonehill Estates

		Predeveloped	Conventional Design	Open Space Design
Runoff (in./year)	No practices	2.1	10.6	8.8
	Practices	—	10.6	8.0
Infiltration (in./year)	No practices	4.9	3.1	4.0
	Practices	—	3.1	4.8

Source: Zielinski, J.A., The benefits of better site design in residential subdivisions, in *Watershed Protection Techniques*, Center for Watershed Protection, Ellicott City, MD, 2000. With permission.

Comparative hydrology

Prior to its development, the highly wooded site produced very little surface runoff but, because of relatively tight soils, generated only a modest amount of infiltration. After the site was converted into the conventional subdivision, however, surface runoff increased by a factor of 5, and infiltration was reduced by about 40% (Table I.3.2). In contrast, the open space design worked to reduce stormwater runoff and increase stormwater infiltration compared to the conventional design, although it did not come close to replicating the original hydrology of the forested site (Table I.3.2).

Comparative nutrient output

As might be expected, the conversion of the forest into a conventional subdivision greatly increased nutrient export from the site; the model indicated that annual phosphorus and nitrogen export would increase by a factor of 7 and 9, respectively, after development (see Figure I.3.10). Unlike Duck Crossing, nutrient export at Stonehill Estates was dominated by stormwater runoff after development. The SUNOM model indicated that stormwater runoff contributed about 94% of the annual nutrient export from the site, with subsurface water movement adding only 6% to the total export. Nutrient loads were not greatly reduced by the dry extended detention pond installed at the conventional subdivision; the model indicated that nutrient export from the conventional design would still be six to seven times greater than the predevelopment condition even with this stormwater treatment practice.

In contrast, the open space design resulted in greater nutrient reduction (Figure I.3.10). For example, the open space design scenario without stormwater practices produced a lower nutrient load than the conventional design scenario with stormwater practices. This was primarily due to lower impervious cover associated with the open space design. When the open space design was combined with more sophisticated stormwater practices (i.e., bioretention, dry swales, and wet ponds), nutrient export was half that of the conventional design. It is interesting to note, however, that even when the most innovative site design and stormwater techniques were applied to the site, nutrient export was still three to four times greater than that produced by the forest prior to development.

Infrastructure costs

The total cost to build infrastructure at Stonehill Estates was about 20% less for the open space design than for the conventional design. Considerable savings were realized in the form of less road paving and shorter lengths of sidewalks, water and sewer lines, and curbs and gutters. The cost difference between the open space and conventional designs would have been greater if not for the fact that higher costs were incurred for the more sophisticated stormwater practices used in the open space design. It was estimated that

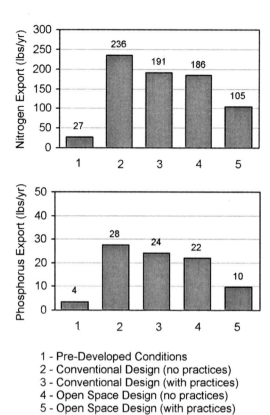

Figure I.3.10 Annual nitrogen and phosphorus loads for each Stonehill Estates development scenario. (From Zielinski, J.A., The benefits of better site design in residential subdivisions, in *Watershed Protection Techniques*, Center for Watershed Protection, Ellicott City, MD, 2000. With permission.)

the infrastructure cost for the conventional design was $1.54 million, compared with $1.24 million for the open space design.

Summary

The comparative results for the Stonehill Estates redesign analysis are summarized in Figure I.3.11. The open space design reduced impervious cover, natural area conversion, stormwater runoff, nutrient export, and development costs compared with the conventional subdivision design.

Other redesign research

Several other researchers have employed redesign comparisons to demonstrate the benefits of open space subdivisions, over a wide range of base lot sizes. The results are shown in Table I.3.3. It should be recognized that each study used slightly different models and assumptions, and as such, strict comparisons should be avoided. The redesign comparisons clearly show that open space designs can sharply reduce impervious cover and stormwater runoff while accommodating the same number of dwelling units, at least to base lot sizes of an eighth of an acre. The reductions in impervious cover and runoff range from 7 to 65%. The ability of open space design to reduce impervious cover starts to diminish for residential zones that exceed densities of four dwelling units per acre.

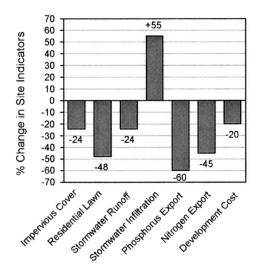

Figure I.3.11 Percentage change in key site conditions from a conventional design to an open space design, both with stormwater practices. (From Zielinski, J.A., The benefits of better site design in residential subdivisions, in *Watershed Protection Techniques*, Center for Watershed Protection, Ellicott City, MD, 2000. With permission.)

Table I.3.3 Redesign Analyses Comparing Impervious Cover and Stormwater Runoff from Conventional and Open Space Subdivisions

Residential Subdivision	Original Zoning for Subdivision	Impervious Cover at the Site			Reduction in Stormwater Runoff (%)
		Conventional Design (%)	Open Space Design (%)	Net Change (%)	
Remlik Hall[1]	5-acre lots	5.4	3.7	−31	20
Tharpe Knoll[2]	1-acre lots	13	7	−46	44
Chapel Run[2]	½-acre lots	29	17	−41	31
Pleasant Hill[2]	½-acre lots	26	11	−58	54
Prairie Crossing[3]	½- to ⅓-acre lots	20	18	−20	66
Buckingham Greene[2]	⅛-acre lots	23	21	−7	8
Belle-Hall[4]	High density	35	20	−43	31

[1]Maurer, 1996; [2]DE DNREC, 1997; [3]Dreher, 1994; [4]SCCCL, 1995

Source: Zielinski, J.A., The benefits of better site design in residential subdivisions, in *Watershed Protection Techniques*, Center for Watershed Protection, Ellicott City, MD, 2000. With permission.

These studies reinforce the conclusion that open space designs are usually less expensive to build than conventional subdivisions. The projected construction cost savings associated with open space designs ranged from 40 to 66% (Table I.3.4). Most of the cost savings were due to the reduced need for road building and stormwater conveyance. In another study, Liptan and Brown (1996) report that open space design produced infrastructure construction costs savings of $800 per home in a California subdivision.

Numerous economic studies have shown that well-designed and marketed open space designs are very desirable to homebuyers and very profitable for developers. Strong evidence indicates that open space subdivisions sell faster, produce better cash flow, yield a higher return on investment, and appreciate faster than their traditional counterparts

Table I.3.4 Projected Construction Cost Savings for Open Space Design from Redesign Analyses

Residential Subdivision	Construction Savings (%)	Notes
Remlik Hall[1]	52	Includes costs for engineering, road construction, and obtaining water and sewer permits
Tharpe Knoll[2]	56	Includes roads and stormwater management
Chapel Run[2]	64	Includes roads, stormwater management, and reforestation
Pleasant Hill[2]	43	Includes roads, stormwater management, and reforestation
Buckingham Greene[2]	63	Includes roads and stormwater management

[1]Maurer, 1996; [2]DE DNREC, 1997

Source: Zielinski, J.A., The benefits of better site design in residential subdivisions, in *Watershed Protection Techniques*, Center for Watershed Protection, Ellicott City, MD, 2000. With permission.

(Arendt, 1994; Ewing, 1996; NAHB, 1997; ULI, 1988; CWP, 1998a; and Porter et al., 1988). Although open space designs are often perceived as applying only to upscale and affluent consumers, several successful open space subdivisions have been built for moderate- to lower-income buyers. Both ULI (1988) and Ewing (1996) report that open space designs can be an effective tool to promote affordable housing within local communities.

The relatively high demand for open space designs reflects two important economic trends. The first trend is that the tastes and preferences of many new homebuyers are gradually changing. Recent market surveys indicate that homebuyers increasingly desire natural areas, smaller lawns, better pedestrian access, wildlife habitat, and open space in the communities in which they choose to live. The second trend is that open space developments that can provide these amenities seldom comprise more than 5% of the new housing offered in most communities. Consequently, there appears to be a large and relatively untapped potential demand for more open space developments. Other compelling benefits of open space design are detailed in CWP (1998a) and Schueler (1995).

Evaluating the quality of individual open space developments

In the real world, site designers must satisfy a wide range of economic objectives, and water quality or resource protection is usually not on the top of the list. It is certainly possible to design a lousy open space design, and communities should expect a wide range in the quality of open space designs they review. How can a community objectively evaluate the quality of individual open space design proposals, and differentiate poor or mediocre projects from the good and outstanding ones?

Nerenberg and Freil (1999) have recently developed a simple rating system to evaluate the quality of individual open space design proposals. The rating system, known as the Conservation Development Evaluation System (CeDES), was developed in consultation with a host of planning agencies and organizations. The CeDES employs 10 core criteria to test how well a proposed open space design reduces impervious cover, minimizes grading, prevents soil loss, reduces and treats stormwater, manages open space, protects sensitive areas, and conserves trees or native vegetation. Each of the 10 core criteria has a quantitative benchmark for comparison. An example of one benchmark that rates the quantity and quality of open space is provided in Table I.3.5. A full description of the CeDES rating can be found in Conservation Fund (1999).

Based on the total score achieved under the 10 core criteria, an open space design project can earn anywhere from zero "oak leaves" up to four "oak leaves." The more oak leaves earned, the better the quality of the proposed project. Based on initial testing, the CeDES seems to do a good job of sorting the poor projects from the outstanding ones.

Table I.3.5 Sample Evaluation Criteria for the Quantity and Quality of Open Space Development (Conservation Fund, 1999)

Points Achieved by the Development	Percent of Open Space Achieved for Different Residential Zones				
	More Than 4 Units per Acre (%)	From 2 to 4 Units per Acre (%)	From 1 to 2 Units per Acre (%)	From 0.5 to 1 Unit per Acre (%)	Less Than 0.5 Unit per Acre (%)
−2	0 to 9	Less than 15	15 to 24	25 to 34	Less than 40
−1	10 to 14	15 to 24	25 to 34	35 to 49	Less than 50
0	15 to 24	25 to 34	35 to 49	50 to 59	Less than 60
+1	25 to 30	35 to 40	50 to 55	60 to 70	Less than 70
+2	More than 30	More than 40	More than 55	More than 70	More than 80

Note: The total open space achieved by the site is computed using the following formula:
[A(0.2) + B(0.2) + C(0.5) + D] * 100/E
A = open space acres in managed landscape
B = open space acres in annual crops
C = open space acres in perennial crops
D = open space acres in native vegetation
E = total undeveloped acres in open space

Source: Zielinski, J.A., The benefits of better site design in residential subdivisions, in *Watershed Protection Techniques*, Center for Watershed Protection, Ellicott City, MD, 2000. With permission.

The CeDES is intended for use as a tool for local development review, but it can also be used as a marketing tool to let homebuyers know how green their new subdivision actually is.

Implications for the watershed manager

The redesign comparisons have several implications for the watershed manager. First, they offer compelling quantitative evidence that open space design can sharply reduce stormwater and nutrient export from new development and, as such, can serve as an effective tool for watershed protection. It is interesting to note that open space design alone produced nutrient reductions roughly equivalent to those achieved by structural stormwater practices. In other words, nutrient export from open space designs without stormwater treatment was comparable to the conventional designs with stormwater treatment. When open space designs were combined with effective stormwater treatment, nutrient loads were sharply reduced but were still greater than predevelopment conditions.

A second, more troubling implication is that it may well be impossible to achieve a strict goal of no increase in nutrient load for new development, even when the best site design and most sophisticated stormwater practices are applied. A handful of communities have adopted stormwater criteria that mandate that no net increase in phosphorus load occur as a result of development, but as the redesign comparisons in this article show, such criteria are not likely to be actually achieved. Thus, if nutrient loads are capped in a watershed, managers may need to remove pollutants at existing developments with stormwater retrofits in order to offset increases in nutrient loads produced by new development.

The redesign research also has some implications for watershed-based zoning. Quite simply, a shift from conventional to open space design can reduce the impervious cover of many residential zoning categories by as much as 30 to 40%. In some watersheds, an aggressive shift to open space design in new residential zones is an essential strategy to meet an impervious cover cap for protecting sensitive or impacted streams.

Another notable finding is that large-lot subdivisions have the potential to generate the same unit area nutrient export as higher-density subdivisions. The high nutrient loading from large-lot developments in unsewered areas is attributed to subsurface discharges from septic systems. From a nutrient management standpoint, it may be more cost-effective to regulate septic system performance than stormwater performance in very low-density residential subdivisions located on permeable soils.

Last, watershed managers have only a few tools at their disposal that offer developers a real chance to save money. The economic evidence clearly suggests that open space design is such a tool and has the potential to either reduce the cost of development or at least offset the cost of other watershed protection measures. Despite its economic and environmental benefits, however, open space design is neither a development option in many communities nor widely used by most developers even when available. Many communities will need to fundamentally change their local development rules in order to make open space design an attractive development option.

Site planning roundtables that involve the local players that shape new residential development, described later in this issue, are an effective way to bring this change about. The ultimate goal is to make open space design a "by-right" form of development, so that its design, review, and approval are just as easy and certain as a conventional subdivision. The day may come when a special exception or permit is needed to build a conventional subdivision.

Literature cited

Arendt, Randall, *Designing Open Space Subdivisions: A Practical Step-by-Step Approach,* Natural Lands Trust, Media, PA, 1994.

Center for Watershed Protection (CWP), *Better Site Design: A Handbook for Changing Development Rules in Your Community,* Ellicott City, MD, 1998a.

CWP, *Nutrient Loading from Conventional and Innovative Site Development,* prepared for the Chesapeake Research Consortium, Ellicott City, MD, 1998b.

Conservation Fund, Pilot Conservation Development Evaluation System, Great Lakes Office, available at www.conservationfund.org/conservation/sustain/gloindex.html, 1999.

Delaware Dept. of Natural Resources and Environmental Conservation (DE DNREC), *Conservation Design for Stormwater Management,* Dover, DE, 1997.

Dreher, D. and Price, T., *Reducing the Impact of Urban Runoff: The Advantages of Alternative Site Design Approaches,* Northeastern Illinois Planning Commission, Chicago, IL, 1994.

Ewing, R., *Best Development Practices: Doing the Right Thing and Making Money at the Same Time,* American Planning Association, Chicago, IL, 1996.

Liptan, T. and Brown, C., *A Cost Comparison of Conventional and Water Quality-based Stormwater Designs,* City of Portland, Portland, OR, 1996.

Maurer, G., *A Better Way to Grow: For More Livable Communities and a Healthier Chesapeake Bay,* Chesapeake Bay Foundation. Annapolis, MD, 1996.

National Association of Homebuilders, *Cost Effective Site Planning,* Washington, DC, 1986.

Nerenberg, S. and Freil, K., The conservation development evaluation system (CeDES): Evaluating environmentally friendly developments, *Land Development,* Fall 22, 1999.

Porter, D., Phillips, P., and Lassar, T., *Flexible Zoning: How It Works,* Urban Land Institute, Washington, DC, 1988.

Schueler, T., *Site Planning for Stream Protection,* Center for Watershed Protection. Ellicott City, MD, 1995.

Urban Land Institute (ULI). *Density by Design,* J. Wetling and L. Bookout, Eds., Urban Land Institute, Washington, DC, 1988.

Zielinski, J.A., The benefits of better site design in residential subdivisions, in *Watershed Protection Techniques,* Center for Watershed Protection, Ellicott City, MD, 2000.

Response

Using computer scenarios to improve site design

Studies have shown that water can be impaired at levels as low as 5 to 15% subwatershed impervious cover, amounts most suburban development patterns exceed. As a result, we need to explore better site designs to minimize pavement from new development, while at the same time conserving more open space and enabling more effective stormwater management. One way to further this exploration of viable development alternatives is to undertake iterative analyses of conceptual redesigns through computer simulations or models.

Jennifer Zielinski's method is to examine a series of model development principles developed by a consortium of planning, transportation, financial, engineering, real estate, wastewater, and public safety experts and to see what the effect of varying these principles might have on changes in runoff rates and consequent nutrient export. These redevelopment principles are similar to those discussed elsewhere in this book, as for example, low-impact development as in Chapter I.5 by Larry Coffman and various open space approaches as in Chapter II.3 by Charles Flink.

The lessons presented in this chapter for water sensitive designers are of major importance. We need to shift our focus from reacting to damages already produced, to instead proactively planning to circumvent those damages from occurring in the first place. The use of conceptual redesigns is an optimal way to generate such predictive understanding. In other words, as Zielinski states, improvements should begin right at the initial planning stage for new site developments.

The key to all this is a move toward embracing the principles of cluster development, what Zielinski refers to as "open space designs." By concentrating or "clustering" development to one portion of the site through relaxing lot sizes, frontages, road sections, and other subdivision geometry, open spaces are left with the positive benefits of reduced overall impervious surfaces at the same time as increased areas for recreation activities or wildlife habitat with a corresponding improvement in the quality of life for the residents. With the additions of other low-impact development techniques such as narrow setbacks, riparian buffers, and removal of curbs and gutters, an added benefit is a reduction in development costs by as much as one-half.

Zielinski reports that the benefits ensuing from undertaking such open space planning in terms of stormwater runoff reductions can also be substantial. Indeed, her modeling exercises raise the very interesting possibility that, through the inclusion of open spaces and clustered development alone, runoff can be reduced by 30 to 40%; in other words, roughly equivalent to that occurring through adoption of structural stormwater BMPs (best management practices). Thus, a very compelling case can be made that the earlier water sensitive design can be applied to newly planned developments, the more effective (and less expensive) can be our solutions for protecting water bodies from the consequences of those developments.

chapter I.4

Post-industrial watersheds: Retrofits and restorative redevelopment (Pittsburgh, Pennsylvania)

Richard D. Pinkham and Timothy Collins

Abstract

Although concepts and techniques to manage urban runoff in ways that maintain ecosystem health are being effectively demonstrated in many developing watersheds, efforts to change stormwater management and restore ecosystem function in older, highly developed watersheds are still in their infancy. This chapter highlights the "restorative redevelopment" concept first proposed in the Pittsburgh region. When development plans for a brownfield site in Pittsburgh called for filling in and culverting a local stream, an interdisciplinary, artist-led team at Carnegie Mellon University's STUDIO for Creative Inquiry developed and implemented a program of public education and dialogue, stream corridor evaluation, and planning for alternatives. A greenway concept for Nine Mile Run was developed with the support and participation of multiple agencies and local stakeholders. Greenway proponents also addressed watershed-scale problems. Nine Mile Run suffers from combined and sanitary sewer overflows, erosive pulses of stormwater, contributions of pollutants carried in urban runoff, and reduced base flow — problems traced to high levels of impervious surfaces in its watershed and crumbling sewer and stormwater infrastructure. The STUDIO and Rocky Mountain Institute hosted an interdisciplinary charrette to select and illustrate measures for retrofitting streetscapes, parks, and properties in the upper watershed to infiltrate, detain, and treat runoff at the site scale. The charrette also developed policy action plans to encourage the widespread micromeasure implementation necessary to produce cumulative benefit for Nine Mile Run. A key theme of the restorative redevelopment approach envisioned for Nine Mile Run is to integrate infrastructure, ecosystem, and community issues. The charrette results exemplify seven principles of restorative redevelopment.

Introduction: The opportunity of older urban watersheds

The watersheds of America's older urban areas suffer a multitude of problems: their high proportions of impervious surfaces produce polluted runoff, erosive storm flows, and reduced stream base flows. Frequently, crumbling or inadequate infrastructure results in sewer overflows that compromise ecosystem and human health. Financial resources are typically very limited, and the social fabric of these places is often frayed and faded. Traditional solutions such as stormwater detention basins, increased conveyance capacity, and wastewater plant expansions require space that is often not available, impose high monetary costs, and treat only the hydrologic results of urbanization rather than the diverse causes of the problems.

On the other hand, urban watersheds present remarkable opportunities if we look carefully and creatively for new solutions. This requires consideration of both the "green infrastructure" — the open spaces, vegetation, and soils remaining and recoverable even in highly urban areas — and the conventional pipes and plants of the "gray infrastructure" as an integrated system. Although we cannot expect to restore urban watersheds to predevelopment conditions, the rehabilitation of ecosystem services can assist in the resolution of wet weather problems and expose millions to the values and benefits of functional hydrological and biologic systems.

Many techniques have emerged in the new development context to prevent degradation of streams, lakes, and other water bodies. There are scores of ways to reduce impervious surfaces and incorporate measures that infiltrate, retain, and filter runoff. And in recent years the stormwater field has moved from isolated application of best management practices (BMPs) to comprehensive, integrative schemes for protecting watersheds as they develop. These include "model development principles" (see Chapter I.3 by Zielinski) and "low-impact development" (see Chapter I.5 by Coffman). These approaches focus on preservation of natural function, which requires going beyond the conventional stormwater detention emphasis on maintaining predevelopment peak flows, to include maintenance of groundwater recharge and the use of soil and vegetation to neutralize pollutants (see Chapter I.1 by Ferguson; also Chapter I.7 by France and Craul). The BMPs emphasized by these approaches are typically "micromeasures" — permeable pavements, dry wells, rain barrels, vegetated roof covers, subsurface recharge beds, bioretention cells, infiltration trenches, vegetated swales, and other techniques — that are installed near buildings and pavements, on individual lots, in public rights of way, and otherwise distributed throughout the built environment. They are implemented as close to where the rainfalls as is feasible, rather than some downstream point below conveyance systems that have concentrated runoff.

Can these concepts and techniques be applied to problems in already-built urban landscapes? Scattered applications are occurring. Toronto, Ontario, has encouraged disconnection of residential roof downspouts from combined sewers by doing providing free rain barrels or free labor to divert downspouts to vegetated areas ("Recycle Your Rain," 1998). Portland, OR, has trialed a variety of retrofits, including vegetated roof covers, vegetated swales, and water gardens (see Chapter I.6 by Liptan and Murase). The Low Impact Development Center is developing retrofits for the U.S. Navy Yard in Washington, D.C. and is preparing a U.S. EPA-funded guidance document on the use of low-impact development techniques to control wet weather flows and combined sewer overflows in urban areas (Low Impact Development Center homepage, 2001). In southern California, TreePeople developed a model retrofit program (Condon and Moriarty, 1999), has implemented a comprehensive residential retrofit as well as infiltration systems and cisterns at two schools, and is working with regional agencies on impervious surface reductions and an ambitious retrofit project for a 2,700-acre watershed as an alternative to a $42 million storm drain (T.R.E.E.S. Project Overview, 2001; Lipkis, 2001).

Further study will be necessary to demonstrate the effectiveness of microscale retrofit and redevelopment techniques in restoring more natural function to urban watersheds; but already, many cities and developers are motivated to try these measures by the staggering costs of conventional management and infrastructure investments. The U.S. Environmental Protection Agency (1997) estimates national combined sewer overflow remediation costs at $45 billion and sanitary sewer overflow remediation costs at up to $87 billion (Parsons Engineering Science et al., 2000). The agency estimates the costs for municipalities to come into compliance with the "Phase I" stormwater regulations (excluding, importantly, many O&M costs and all costs borne by private parties) to be $7.4 billion nationwide over 20 years (U.S. EPA, 1997). The EPA (1999) also estimates public and private costs for implementing "Phase II" stormwater regulations at $848 to $981 million annually.

What is also clear is that urban watersheds face a host of other problems, and their residents desire and deserve social, economic, and environmental progress. Yet most cities view wet weather management as a technical matter, best managed by specialists in technical disciplines concerned with conveyance infrastructure. The question we want to ask is: Cannot investments in sewer and stormwater infrastructure be made in ways that address additional urban goals? The opportunity in older urban watersheds is to expand the sewer overflow and stormwater management agendas — to redefine the wet weather problem and make infrastructure improvements that will provide multiple benefits and energize multiple constituencies. To do so, we must link infrastructure rehabilitation with the restoration of ecosystems and a broader cultural restoration. In turn, the resources for, and likely implementation rates of, sewer and stormwater measures will be increased. Exploring this uncharted territory is the story of this chapter.

We know that wet weather management takes place in the context of continual redevelopment and retrofitting of urban properties. Pavements are relaid; buildings are renovated and reconstructed; transportation systems are reorganized; and utilities are maintained and replaced. These changes are opportunities for reevaluating, correcting, healing, and educating. We can find ways to make redevelopment serve ecosystem restoration. We can also find ways to make infrastructure and ecosystem rehabilitation projects serve broader economic and social purposes. This give-and-take, this expanding of agendas, is the heart of what we call "restorative redevelopment," which we define as "redevelopment and retrofit projects that improve the value and livability of the city while effectively restoring natural processes and functions." The concept emerged from a watershed project in metropolitan Pittsburgh, Pennsylvania (Ferguson, Pinkham, and Collins, 1999).

Envisioning restorative redevelopment: Pittsburgh's Nine Mile Run

The watershed: Industrial legacies and a redevelopment opportunity

Nine Mile Run is a watershed and stream that joins the Monongahela River nine miles from "the Point," the spit by downtown Pittsburgh where the Monongahela and Allegheny rivers join to form the Ohio River (Figure I.4.1). The run (a regional term for creek or stream) drains five municipalities, flowing through wooded Frick Park and then into an urban brownfield that dominates the bottom of the watershed.

The brownfield is a mountain of slag — the detritus from steel making — as much as 20 stories high surrounding and overshadowing the creek. It covers Nine Mile Run's former flood plain, identified by Frederick Law Olmsted, Jr., for a new city park in 1910, and instead purchased by the steel industry. Roughly 240 acres were covered by slag dumped over a 50-year period. In 1993, Pittsburgh City Planning developed a conceptual plan for houses and open space on this mound of industrial by-product. An economic development team working with the city decided that the water in Nine Mile Run was dirty enough and the benefits to development attractive enough to propose that the stream be buried once and for all. Culverting would result in a 20% increase in housing on the site. Thus, the initial proposal called for a flat development site that would bury what was left of the stream under 150 ft of slag — a deed even the steel industry could not accomplish.

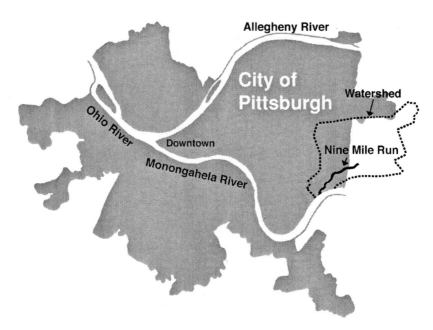

Figure I.4.1 The Nine Mile Run watershed in the Pittsburgh region. (From Ferguson, B.K., Pinkham, R., and Collins, T., *Re-Evaluating Stormwater: The Nine Mile Run Model for Restorative Redevelopment*, Rocky Mountain Institute, Snowmass, CO, 1999. With permission.)

In response to this unique brownfield site linking the Frick Park and Monongahela River ecosystems, and the proposal to develop it, a group of artists in the STUDIO for Creative Inquiry at Carnegie Mellon University began a community-based research program about post-industrial public space. The thesis was simple: instead of destroying the stream, why not let it define the development? The research agenda would focus on developing knowledge to inform public values, targeting the lack of care that had defined and denigrated this stream for over 100 years. Talking points included the potential for these lands to instigate a renewed civic dialogue, the role of artists and academia in this program, and the application of restoration ecology in the redevelopment of post-industrial public space. The contention from the beginning was that urban brownfields, the artificial great estates handed down from the industrial era, are an important public space opportunity, particularly for rust belt cities like Pittsburgh. The Nine Mile Run Greenway Project team, directed by three artists, included a diverse group of professionals from academia and industry. The project had a partnership with local government through Pittsburgh City Planning but retained its autonomy through external funding from the Heinz Endowments and others.

Arguably, the issues that drove the private development were different than the issues that might drive a public investment in a post-industrial open space. The first question to address was: Was the dump really a dump? Did the stream have any residual or emergent values? Was the riparian link between the Monongahela and Frick Park of any import? The project team developed a program to understand community interest, then followed that up with an ecological study to confirm the nature of the opportunities and constraints. The effort focused on:

- *History and public policy.* Purchased a year before the first zoning laws would have protected it, the Nine Mile Run dump site was a key component of the steel industry. For each ton of steel, two to three tons of slag were produced. Its transport

and disposal cut into profits, making disposal at nearby Nine Mile Run essential. A complicit city government looked away as the original grandfathered parcel was filled and slag began to creep toward communities and parklands.
- *The stream.* Erosive urban runoff has downcut the stream channel and revealed adjacent sewer systems. The aging, cracked, and disjointed sewers leak sewage into the stream in wet weather. In dry weather, the hydraulics reverse and much of the stream disappears into the sewers. The results are alarmingly high pathogen counts (in a city park) and a diminished aquatic ecology. Nonetheless, fish are found in a number of pools and beavers are regularly seen each spring.
- *The slag slopes.* The development interests wanted to bulldoze the slopes; the project team thought that they could be successfully vegetated in place. Porosity, alkalinity, fines on the surface, dark color, and south-facing slopes makes for a challenging environment for plants and seedlings. Yet the north-facing slopes and the oldest south-facing slopes are populated with Box Elder, Sycamore, Bigtooth Aspen, Black Locust and Slippery Elm, Pin Oak, Red Maple, and a curious find — the Hop Tree, a Pennsylvania endangered species not usually found in this area.
- *Sustainable open space.* Pittsburgh is only now planning to resuscitate its four largest city parks. The meaning, form, and function of post-industrial parklands were not on the primary agenda of the brownfield site developer. The challenge in Nine Mile Run was to see the post-industrial landscape as a contemporary mirror to Frick Park and as a demonstration site for ecological restoration. The goal was to seek a public–private model of development that would result in a new era of sustainable open space.

With an understanding of the existing development program, the team knew it had to develop specific programmatic goals to engage the public in an alternative development dialogue focusing on the potential form of this post-industrial public space and its function within a social and ecological context. The first year of the project, dubbed "Ample Opportunity: The Community Dialogue," was designed to create opportunities for on-site experience and community dialogue; expand the understanding of the issues and the discourse on public space; and increase understanding of the environmental and public space issues of a range of local constituencies (*Ample Opportunity: A Community Dialogue*, 1998).

The second year, *Ample Opportunity: The Ecology of a Brownfield*, built on the work of the first year and followed through with expert studies to define the range of opportunities. The team developed an exhaustive biological and landscape ecology study and a set of development alternatives (*Nine Mile Run Rivers Conservation Plan*, 1998). This study provided a baseline upon which further work on the ecosystems and the infrastructure that affects them could be judged, and it set out the content for design alternatives in year three. In addition to outlining alternatives, the goals of the second year included grounding the professional effort in a rigorous public discourse, addressing the watershed-wide issues impacting Nine Mile Run, and constraining the site development and greenway opportunity.

In the final year, in *Ample Opportunity: The Brownfield Transformation*, the experts' recommended alternatives were developed into a set of effective communication tools to inform community decision making. The team produced a series of images, texts, and experiences directed toward the goal of achieving a community consensus conceptual design for the site, the greenway, and Frick Park. Concurrently, the STUDIO and its advisory board developed an institutional and economic plan to place the alternatives and the final community consensus process in a realistic light, and provide the program with the inertia and support to move forward independently of the STUDIO's involvement. (Comprehensive materials for the Nine Mile Run Greenway Project and related, evolving activities are available at http://slaggarden.cfa.cmu.edu.)

The overarching goal of the Nine Mile Run Greenway Project was to enable an equitable public dialogue about brownfields, nature, and public space. The team saw its work in terms of a community consensus process and a public policy discussion about the form and function of post-industrial public space, a discourse that was missing and continues to be an anomaly in the current program of local brownfield development. To do this, the project leaders sought support from the specific municipal agencies managing the current development program. At the same time, to retain its objectivity as a separate autonomous entity, the STUDIO obtained independent funding from the Heinz Endowments. As the team began to clarify its role, members realized they were revealing a complex aesthetic based on discourse, restoration ecology, and sustainable landscape systems. It became clear that Nine Mile Run was a better site for experimentation and modeling new approaches to discourse and dialogue than originally expected. Nine Mile Run provided both the context and the subject for experimentation. Issues that arose from efforts to reclaim, restore, or heal the site enveloped diverse disciplines and areas of knowledge — engineering, ecology, social and political democracy, and the arts. The project showed that examining sites like this in the context of their watersheds and communities has a curious potential to transform culture in unexpected ways.

The ecosystems and infrastructure charrette

In the second year of the Nine Mile Run Greenway Project, it became clear to its organizers that long-term restoration of the aquatic and riparian ecosystem in the lower watershed (the brownfield site, the potential greenway, and Frick Park) would require actions to address the source of the excess stormwater flows: the highly built-up upper watershed (Figure I.4.2). Also at this time, regulators heightened concerns over sewer overflows in the Pittsburgh region by threatening fines and other legal actions if local communities did not move quickly toward compliance with relevant laws. In 1998, the STUDIO and Rocky Mountain Institute convened a three-day stormwater management charrette with 60 local and national landscape architects, engineers, architects, artists, planners, policy analysts, and local citizens.

The interdisciplinary teams at the charrette developed conceptual designs for sample sites to concretely illustrate the restorative development concept. They also produced recommendations for institutional arrangements and policy initiatives that could drive adoption of restorative redevelopment measures in the watershed. The charrette's results show how implementation of progressive stormwater management measures at the scale of individual properties and neighborhoods could benefit both the upper and lower watershed by linking infrastructure rehabilitation, ecosystem restoration, and community development.

Four sample designs

The sample designs for four sites in the upper portions of Nine Mile Run reuse, restore, and revitalize their sites by resolving existing site-specific issues, adapting techniques of construction and stormwater management to Pittsburgh's fine-textured soil, frequent frosts, steep hillsides, and unstable geology. They infiltrate or detain the runoff from a "2-year, 24-hour" storm on-site, within a construction budget of $2/gallon of hydraulic capacity. These parameters are consistent with standards and conventional project costs established in Allegheny County in recent years.

A variety of on-site measures is available for removing stormwater from sewers and restoring beneficial natural processes. The following general approaches were employed in many of the sample designs:

Figure I.4.2 The Nine Mile Run watershed, showing impervious streets and buildings, topography, and the stormwater charrette sites. (From Ferguson, B.K., Pinkham, R., and Collins, T., *Re-Evaluating Stormwater: The Nine Mile Run Model for Restorative Redevelopment*, Rocky Mountain Institute, Snowmass, CO, 1999. With permission.) (Color version available at www.gsd.harvard.edu/watercolors. Password: lentic-lotic.)

- *Capturing roof runoff* in tanks or cisterns for irrigation or indoor graywater use
- *Disconnecting pavement and roof drainage* from sewer lines and directing it to adjacent vegetated soil or to infiltration basins
- *Engineering infiltration basins* — "water gardens," dry wells, and subsurface recharge beds — to collect runoff and percolate it into the soil
- *Planting trees* to intercept a portion of rainwater
- *Rehabilitating soils* to increase infiltration rates and pollutant-neutralizing microbial activity
- *Reconfiguring driveways, parking lots, and streets* to turn more of a site over to pervious, vegetated soil
- *Using porous pavements* — special varieties of asphalt, concrete, masonry, and other materials with open pores that allow water to pass through
- *Routing runoff through vegetated surface channels* — "swales" — to slow its velocity, remove pollutants, and infiltrate it into the soil
- *Restoring ("daylighting") historic streams* by excavating culverts and creating naturalized open channels (Pinkham, 2000)

Each design integrates several stormwater management strategies into the built environment of its site. Additionally, each exemplifies restorative redevelopment by integrating the physical strategies into the social and economic life of the site and its neighborhood.

Hunter Park

Hunter Park is located near the headwaters of the Nine Mile Run watershed. It is a neighborhood park in the community of Wilkinsburg. This is a low-income area; neighborhood streets and sidewalks are in poor condition. The park's ball field, wading pool, basketball courts, and small playground are in disrepair, although all are heavily used in season (photographs of original conditions at the four sites are available in Ferguson, Pinkham, and Collins, 1999).

The bulk of the park is in pervious turf, although it hides the culverted remains of a natural stream. In contrast, the surrounding residential blocks are mostly impervious with densely built houses, streets, and sidewalks. Most of the runoff from the impervious surfaces drains into sewers, contributing to downstream flood pulses, sewer overflows, water pollution, and reduced base flow.

Despite the currently neglected condition of the Hunter Park area, it has a vigorous past that symbolizes the industrial development of the region and the character of its people. Before settlement of the area, the site was a V-shaped headwater stream valley. In the 19th century a coal mine filled and flattened the site with yards and spoil piles; the stream was diverted around the periphery. The mining industry brought in a working population and built company housing nearby. In the early part of the 20th century, the industrially created land forms served as a baseball field for the Negro League. Some of the best baseball players in the country played as semipros at "Hunter Field." Beginning in the 1950s, a series of developments gradually transformed the site into a general-purpose recreational park.

The site is in a valley with a drainage area of 59 acres, of which impervious roads and rooftops cover approximately 9 acres, or 16% (Figure I.4.3). (The percent impervious area figure for this and other catchments in this article does not include driveways, sidewalks, and most parking lots.) Most drainage inlets are clogged with sediment; some drainage pipes are broken. Some grass swales in the park improve water quality to a degree but are undersized even for the small amount of water they carry. Concentrated runoff from nearby impervious surfaces has eroded some of the park's drainage swales and steep side slopes.

The proposed design. The proposed design is a convergence of history, hydrology, recreation, and neighborhood revitalization, wedding the site's social history to its hydrologic future. Water is brought through a sequence of historical–recreational spaces and celebrated at the end. The hydrologic strategies are given form by the park's natural and cultural history; in turn, the forms illuminate the park's environmental and historic features.

The design uses complementary strategies for various portions of the catchment (Figure I.4.4). At the upper end of the park, a woodland "bioretention" area consists of sand and soil mixtures planted with native plants. It includes a pretreatment area to dissipate the energy of inflowing runoff and to collect coarse sediment. Then a constructed wetland treats water at the upstream end of the ball field. It is planted with emergent and scrub-shrub plants in a complex microtopography. It filters pollutants, reduces peak flow rates, and stabilizes the flow of water into the grass swales below.

Swales take overflow drainage from the wetlands, and runoff from the fields and surrounding slopes, around the ball field and through the lower part of the park. The swales have grass and other vegetation, which help remove pollutants from runoff. For further infiltration and filtering, they are enhanced with beds made of sand and topsoil 1 to 2 ft deep and 10 to 15 ft wide (Figure I.4.5).

At the bottom of the park, the area where coal mine shanty houses once stood is made into a public square for the neighborhood. The once-culverted stream is reopened ("daylighted") through the square to convey stormwater in restored stream habitat as an amenity and focal point for the park. The square includes a stage for public plays and festivals,

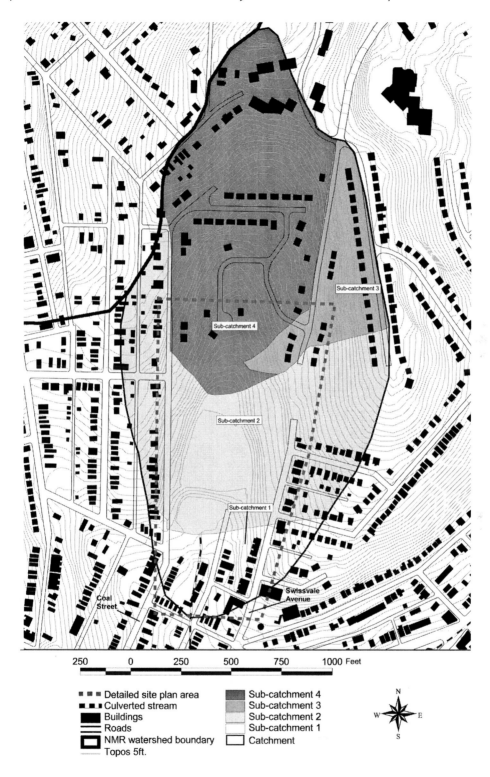

Figure I.4.3 The Hunter Park site, catchment, and surrounding area. (From Ferguson, B.K., Pinkham, R., and Collins, T., *Re-Evaluating Stormwater: The Nine Mile Run Model for Restorative Redevelopment*, Rocky Mountain Institute, Snowmass, CO, 1999. With permission.) (Color version available at www.gsd.harvard.edu/watercolors. Password: lentic-lotic.)

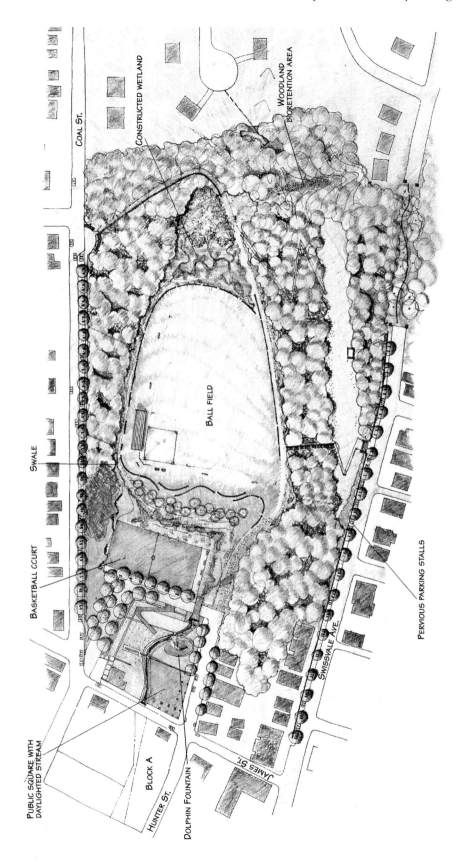

Figure I.4.4 The proposed plan for rehabilitation of Hunter Park. (From Ferguson, B.K., Pinkham, R., and Collins, T., *Re-Evaluating Stormwater: The Nine Mile Run Model for Restorative Redevelopment*, Rocky Mountain Institute, Snowmass, CO, 1999. With permission.) (Color version available at www.gsd.harvard.edu/watercolors. Password: lentic-lotic.)

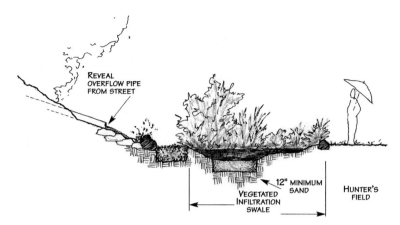

Figure I.4.5 Construction of Hunter Park's swales for stormwater infiltration, storage, and water quality improvement. (From Ferguson, B.K., Pinkham, R., and Collins, T., *Re-Evaluating Stormwater: The Nine Mile Run Model for Restorative Redevelopment*, Rocky Mountain Institute, Snowmass, CO, 1999. With permission.)

adding a cultural role to the recreational park. One type of festival could annually celebrate the watershed with stream "cleanups." The stream is expected to carry 45 cfs during the two-year storm. The stream's meanders are dimensioned for natural "dynamic equilibrium" with the flow. Bioengineering (the use of living plants in combination with nonliving materials to stabilize streams and slopes) is used to protect the banks during 2-year and 10-year storms.

Around the edges of the park, street pavements are narrowed to reduce impervious cover and allow infiltration while adding more parking spaces on permeable edges. The pervious parking stalls are made of concrete pavers with grass, over a gravel bed (Figure I.4.6). The open-celled paver surface and the deep gravel storage basin beneath it adapt the pavement construction to the region's frequent frosts and fine-textured, slowly permeable soil.

In the residential areas all around and above the park, roof leaders, street gutters, and drainage inlets are disconnected from the storm sewer system. Their drainage is diverted into swales and across vegetated slopes in and around the park. Excess runoff remaining in the streets is conveyed to the park's wetlands and swales for treatment.

Stone "traces" through the park mark lines of old mining features. Street trees are added for air and water quality improvement. The combination of strategies preserves and celebrates the natural and cultural history of the area. The improved access to the park promotes its use. Opportunities to learn about the hydrologic strategies are available through interpretative signs and guided tours.

Edgewood Crossroads

Edgewood Crossroads is located near the center of the Nine Mile Run watershed. It is the public center of the Borough of Edgewood, where a historic train station fronts on busy Swissvale Avenue. The old train station is the only one that the famous 19th-century architect Frank Furness designed on this side of the Allegheny Mountains. Across the street are old storefront commercial buildings, a church, and a school. Nearby are Edgewood's town hall, public library, community swimming pool, and numerous well-kept old residences.

Residential streets converge from several directions. Public buses stop at the street intersection; the old railroad bed is slated to become the route of a regional busway.

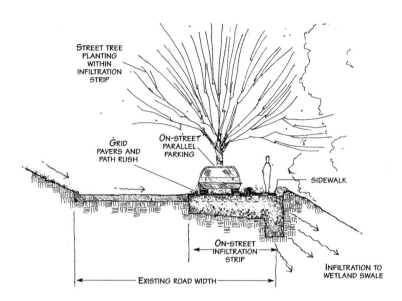

Figure I.4.6 Modification of a Hunter Park street to reduce impervious surface and increase infiltration, while increasing the tree canopy and the availability of parking. Note the gravel-filled storage/recharge bed under the parking stall and sidewalk. (From Ferguson, B.K., Pinkham, R., and Collins, T., *Re-Evaluating Stormwater: The Nine Mile Run Model for Restorative Redevelopment*, Rocky Mountain Institute, Snowmass, CO, 1999. With permission.)

Numerous pedestrians, especially children, move between their homes and community facilities along the sidewalks, across the street intersection, and through the railroad underpass.

Edgewood has dozens of civic groups, and the social closeness almost of a village. In the minds of the local Edgewood people, the cluster of streets, structures, and open spaces around the old train station is the unified center of their community. Protecting and enhancing the sense of community is the central task of any urban design here.

The impervious streets, roofs, and sidewalks of the site and its catchment (Figure I.4.7) generate runoff that ponds up in the street intersection, disrupting pedestrian and vehicular traffic. Eventually, it flushes into storm sewers, carrying oils and other pollutants, while denying recharge of groundwater. As in many parts of the Pittsburgh region, some roof leaders here are connected to sanitary sewers, contributing to sewer overflows downstream.

The proposed design. The design for this site is based on relationships between water systems, urban design, and social values (Figure I.4.8). It integrates the following community issues: reinforcing the social and physical sense of community, preserving public open spaces, reinforcing pedestrian access, eliminating street flooding, and bringing Edgewood into compliance with federal water quality standards by separating storm drainage from the sanitary sewer system.

In the small community park facing the train station, a plaza is developed to be the gateway to a public greenway in the new transit corridor. Here, a prominent stormwater restoration facility integrates stormwater solutions and public education with urban design (Figure I.4.9). The center is a depressed bowl, with a porous block bottom that retains and infiltrates stormwater. During rainfall, about 30 days per year, the depression diverts flood waters off the street and the surrounding plaza; it fills and then slowly drains over a 1- or 2-day period through an infiltration bed beneath the plaza. On dry days, the plaza and the bowl are for communal gathering and play; the wall around the bowl is for

Chapter I.4: Post-industrial watersheds: Retrofits and restorative redevelopment 79

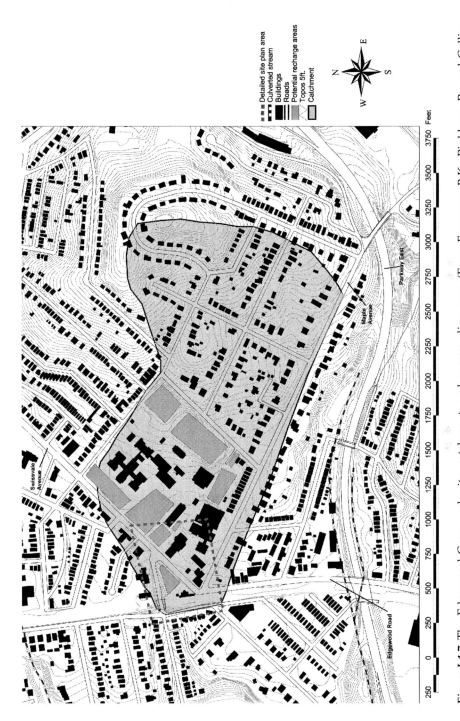

Figure I.4.7 The Edgewood Crossroads site, catchment, and surrounding area. (From Ferguson, B.K., Pinkham, R., and Collins, T., *Re-Evaluating Stormwater: The Nine Mile Run Model for Restorative Redevelopment*, Rocky Mountain Institute, Snowmass, CO, 1999. With permission.) (Color version available at www.gsd.harvard.edu/watercolors. Password: lentic-lotic.)

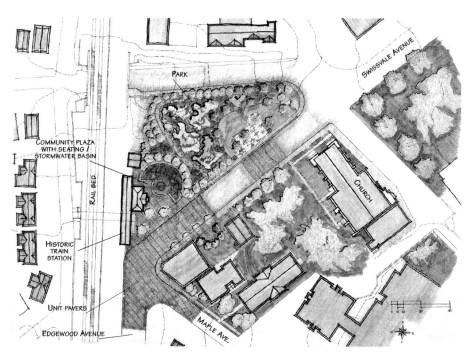

Figure I.4.8 The proposed plan for Edgewood Crossroads' restoration and redevelopment. (From Ferguson, B.K., Pinkham, R., and Collins, T., *Re-Evaluating Stormwater: The Nine Mile Run Model for Restorative Redevelopment*, Rocky Mountain Institute, Snowmass, CO, 1999. With permission.) (Color version available at www.gsd.harvard.edu/watercolors. Password: lentic-lotic.)

Figure I.4.9 The demonstration infiltration basin at Edgewood Crossroads in the proposed public plaza by the train station. The basin provides a community gathering place. During large storms, it captures runoff, which percolates into the subsoil within one to two days. (From Ferguson, B.K., Pinkham, R., and Collins, T., *Re-Evaluating Stormwater: The Nine Mile Run Model for Restorative Redevelopment*, Rocky Mountain Institute, Snowmass, CO, 1999. With permission.)

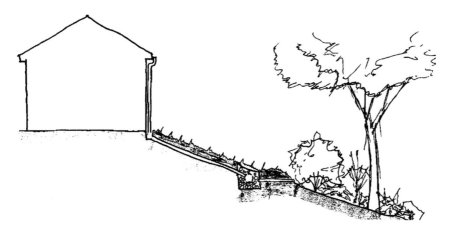

Figure I.4.10 Retrofit of a typical Edgewood residence to disconnect the roof leaders from the sanitary sewers and recharge the roof runoff into the ground using an infiltration trench. (From Ferguson, B.K., Pinkham, R., and Collins, T., *Re-Evaluating Stormwater: The Nine Mile Run Model for Restorative Redevelopment*, Rocky Mountain Institute, Snowmass, CO, 1999. With permission.)

sitting. Permeable unit pavers continue from the plaza across the street intersection, to strengthen pedestrian connections.

The street intersection receives stormwater runoff from a catchment around and uphill from the crossroads. The 73 acres of the catchment are occupied by the community center area, several schools and churches, and numerous, relatively large, Victorian single-family homes. The proposed design manages all stormwater within the catchment by incorporating infiltration measures into lots and landscapes upslope from the site itself.

Open lawns and play fields in the catchment's small parks and institutional grounds are open spaces that can serve the dual purposes of recreation and runoff control. Groundwater recharge beds could be constructed under these areas, while maintaining their surface uses for sports and parks. For example, the retrofit of a playing field to maximize infiltration would include aggregate beneath the turf. If a bed of gravel 18 in. deep with 40% storage volume were provided over the entire 6.2 acres of reasonably available area, it could infiltrate the entire volume of a 2-year storm collected from an area of 18 acres. An alternative construction of preformed "infiltrator" chambers could provide the same capacity.

An additional strategy diverts the runoff of residential roofs into on-lot infiltration basins to significantly reduce stormwater inflows to the sewers (Figure I.4.10). Infiltration and recharge features can be shaped to each individual lot. For example, a large residence has half of its 2,500 ft^2 roof area draining to the front and rear yards, respectively. For each half of the roof, a bed of aggregate or infiltrator chambers with a storage capacity of 208 ft^3 (1,560 gal) would infiltrate all the runoff from all rain events up to and including the 2-year storm. The bed or trench must be properly spaced away from the house to avoid leaking of water into the basement.

Porous pavements at the parking lots of churches and other public places infiltrate additional stormwater (Figure I.4.11). Among the alternatives in porous pavement construction suitable for local soils, frosts, and traffic loads are masonry pavers with open joints, a bituminous mix with open-graded aggregate, or gravel with a layer 4 in. deep of #89 fines over a layer 24 in. deep of uniform-graded aggregate 2 in. in diameter.

Finally, increasing the urban forest throughout the catchment reduces runoff, moderates urban climate, improves air quality, and reduces noise. A dense vegetative structure, such as trees, shrubs, and native ground covers, absorbs more rainwater than a turf slope and is more resistant to erosion during intense storms.

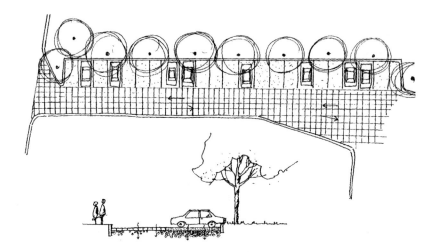

Figure I.4.11 Proposal for porous pavement to be retrofitted into parking areas of Edgewood Crossroads' catchment. (From Ferguson, B.K., Pinkham, R., and Collins, T., *Re-Evaluating Stormwater: The Nine Mile Run Model for Restorative Redevelopment*, Rocky Mountain Institute, Snowmass, CO, 1999. With permission.)

Sterrett School

Sterrett School is located near the headwaters of Fern Hollow, which is Nine Mile Run's largest tributary. It is a middle school in the midst of Pittsburgh's South Point Breeze neighborhood, adjacent to Frick Park. Sidewalks connect homes, shops, and the school grounds safely and conveniently; pedestrians use them actively. The school shares a city block with eight homes (Figure I.4.12).

In the low area between the houses, a combined sewer line follows the course of the original streambed of what was once called Salamander Creek. Drainage inlets between the houses sometimes back up and flood. Local culvert overflows contribute to flooding in the basements of homes. The site's impervious roofs, streets and sidewalks, driveways and parking lots dump runoff into the combined sewer, contributing to polluting overflows downstream in Fern Hollow and Nine Mile Run.

The proposed design. The proposal integrates a variety of stormwater measures with educational programming and neighborhood improvements (Figure I.4.13). For the school building itself, the plan calls for cisterns to collect roof runoff. The 16,000 ft^2 roof generates a lot of runoff — 3,300 ft^3 (25,000 gal) of water during the 2-year storm. Diverting this large volume into the cisterns by disconnecting the school's downspouts from combined sewers has a big effect on downstream overflows and pollution. One possible form of cistern is a transparent "water wall" that would let students monitor the water level in relation to rainfall and water use (Figure I.4.14). Some water can also be permanently stored in the building's attic as a "thermal battery" to moderate indoor temperatures.

From the cisterns, the water from the 2-year storm and all smaller storms during the year is put to productive use on the school grounds, irrigating the school's gardens, greenhouse, and ball field. Water is also put to indoor "graywater" uses such as flushing toilets and urinals. Also, a greenhouse utilizing water collected from the school's roof provides a teaching tool for explaining the water cycle and the role of the school and the neighborhood in the watershed.

Runoff from a storm larger than the 2-year storm, or a rapid succession of smaller storms, will exceed the capacity of the cisterns. Overflow from the cisterns will flow

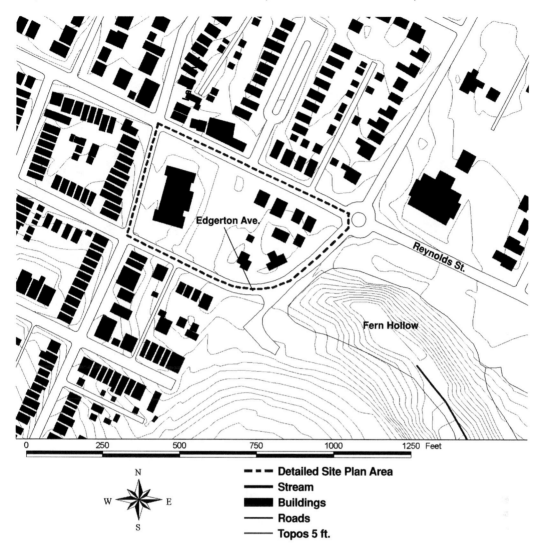

Figure I.4.12 The Sterrett School site and surrounding area. (From Ferguson, B.K., Pinkham, R., and Collins, T., *Re-Evaluating Stormwater: The Nine Mile Run Model for Restorative Redevelopment*, Rocky Mountain Institute, Snowmass, CO, 1999. With permission.) (Color version available at www.gsd.harvard.edu/watercolors. Password: lentic-lotic.)

through the school grounds along an "Art Creek." The creek follows the path of water with a mosaic of tiles, embedding children's poetry and images of animals and leaves.

As water flows farther away from the building, the artificial tiles give way to a meandering water course of earth and plants, a "water garden" that brings water into contact with the ground (Figure I.4.15). The drainage then flows through a vegetated swale with a gravel infiltration bed, ephemeral ponds, and community gardens for residents.

Where Edgerton Avenue crosses the path of the old stream, it has barred the movement of water. Water that would have flowed naturally into the Fern Hollow ravine flows instead in combined sewer lines and is not available to the ravine ecosystem. Closing off this street would cause little traffic disruption, and regrading to remove the street embankment would prevent local basement flooding and restore natural drainage. The water would enter the ravine via a boulder cascade under overhanging willows. A pedestrian bridge would maintain access along the old street alignment.

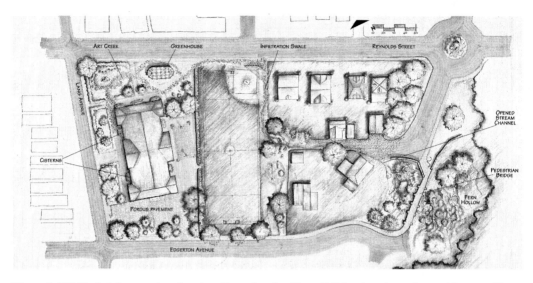

Figure I.4.13 Redevelopment and restoration plan for Sterrett School and nearby residences. (From Ferguson, B.K., Pinkham, R., and Collins, T., *Re-Evaluating Stormwater: The Nine Mile Run Model for Restorative Redevelopment*, Rocky Mountain Institute, Snowmass, CO, 1999. With permission.) (Color version available at www.gsd.harvard.edu/watercolors. Password: lentic-lotic.)

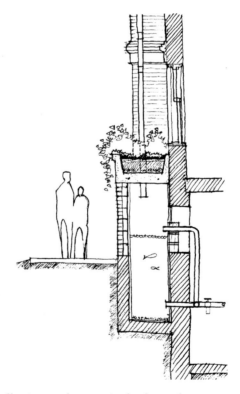

Figure I.4.14 Cisterns for collecting roof waters in the form of a transparent "water wall" alongside the Sterrett school building. (From Ferguson, B.K., Pinkham, R., and Collins, T., *Re-Evaluating Stormwater: The Nine Mile Run Model for Restorative Redevelopment*, Rocky Mountain Institute, Snowmass, CO, 1999. With permission.)

Figure I.4.15 The "water garden" water course at the Sterrett School. (From Ferguson, B.K., Pinkham, R., and Collins, T., *Re-Evaluating Stormwater: The Nine Mile Run Model for Restorative Redevelopment*, Rocky Mountain Institute, Snowmass, CO, 1999. With permission.)

Porous materials replace impervious pavements in the school's parking lot, playground, and sidewalks. The runoff will be further reduced by tree plantings, where the canopy intercepts rainwater during small, frequent storms at the same time it moderates air quality and temperature.

The design reduces the runoff from the site to the level of that from a naturally wooded site. The principal effect comes from disconnecting stormwater drainage from the combined sewer system: diverting the roof runoff into cisterns and vegetated swales, and allowing water to infiltrate as it flows over grass slopes and in broad open swales, instead of buried culverts. These measures, together, disconnect the drainage from 90% of the site area, reducing the 2-year runoff into the culvert to only 40% of its previous volume.

Further runoff reduction comes from specially constructed storage facilities. The design for the infiltration swale illustrates how a quantity of storage is created. A gravel-filled infiltration trench running under the 500-ft length of the swale provides 2,800 ft^3 (21,000 gallons) of storage, based on a depth of 4 ft, a width of 4 ft, and a storage ratio (volume of stored water per total volume of gravel-filled trench) of 0.35. On the swale surface, during large storms, ponding up to 6 in. deep and 12 ft across in a gently banked channel provides an additional 2,000 ft^3 (15,000 gal) of storage. The total storage in the swale is thus 4,800 ft^3 (36,000 gal). Because of the gravel and sandstone substrate, infiltration is probably feasible, but if necessary an underdrain system could be added to release the stored water slowly after the peak of the storm.

The combination of "disconnects" and stormwater storage reduces the runoff to only 26% of its existing amount, an amount equal to that from a naturally wooded site. These calculations do not take into account additional reductions due to mulching, pervious pavements, tree plantings, or bioretention cells (vegetated infiltration basins that capture sheet flow from parking areas, sidewalks, and paved play areas). These are all "extra" restoration capacities.

Regent Square Gateway

The Gateway site is located low in the Nine Mile Run watershed, where the main Nine Mile Run culvert discharges for the first time into an open channel. Here, the borders of

Edgewood, Swissvale, and the City of Pittsburgh converge, by a neighborhood named Regent Square. At the site, Braddock Avenue's on-ramp enters I-376 (the Parkway East), and the abandoned alignment of Old Braddock Avenue abuts an underutilized commercial building.

This seemingly neglected place is in fact an extraordinary focus for the Nine Mile Run watershed and its people, for here the culverted stream first comes into full view in an open channel, and here the historic plan for Frick Park has always foreseen a major eastern public gateway. This is the junction between the upper, developed, urban portion of the watershed and the lower, open, natural part in Frick Park. This highly visible place is the physical confluence of the watershed's stream flows, municipal jurisdictions, and — potentially — community and watershed consciousness.

Local runoff comes through the site from a 64-acre catchment, densely built up with residences (Figure I.4.16). Impervious rooftops and streets comprise 41% of the total area of the catchment. During intense storms, a large part of the local runoff currently bypasses inlets due to the steep slope of Braddock Avenue. Where this runoff reaches the bottom of the site, it has eroded the edges of the Nine Mile Run channel.

The proposed design. The restoration plan locates small ephemeral ponds (500 to 3,000 ft^2 in area) in the neighborhood above the site wherever there is adequate open space, sufficient drainage area, and appropriate soil for infiltration (Figure I.4.17). They filter the runoff that occurs during small storms and the first flush of large storms. Stormwater detention during large storms is not their purpose; high peak flows are allowed to pass through without additional ponding, so they will not combine with relatively long, slow peak flows on the main stream. These ponds will treat runoff during every storm, replenish groundwater to support stream base flows, and allow only the excess water from occasional large storms to enter the stream directly.

The site plan accommodates multiple uses (Figure I.4.18). The large upper story of the building is suitable for profitable reuse as a single-occupant retail store. Runoff from the upper, retail parking lot flows onto grass filter strips to enhance water quality (Figure I.4.19). Excess runoff passes through inlets and perforated pipes into banks of aggregate filter material. The embankments are capped with topsoil for rooting of trees and other vegetation. An underdrain collects the filtered excess and discharges it like "spring" flow to trickle into the Nine Mile Run channel. In this parking lot and throughout the site, plantings of trees and shrubs intercept a portion of rainfall.

The lower levels of the old building can be reused to serve visitors to Frick Park and the greenway. The city's Parks Department could use indoor space for watershed education and research. Private retailers could offer bicycle rentals and food services. At the trail head, the building's facilities could provide trail guide information, trash disposal, and public restroom facilities.

Some runoff from upslope areas will continue to wash over the site during occasional large storms even after construction of small infiltration basins in the neighborhood above. Surface runoff entering the site from Braddock Avenue is diverted into the formerly neglected area between Old Braddock Avenue and the parkway ramp. Here, a series of terraced basins reclaim the area to filter the runoff (Figure I.4.20). The sculptural earth forms symbolize the fluvial processes with which they unite to mitigate the runoff from the urban watershed. Low-flow and first-flush runoff infiltrates into the basins and is filtered. Underdrains collect the filtered water and add it to the "spring" flow for discharge to the stream channel. Larger flows spill gradually over the surface from one terrace to the next, discharging through a spillway before high-peak flows arrive on the main stream.

At the entry to Frick Park, Old Braddock Avenue's remnant trolley tracks disappear under a highway embankment, marking the end of a former era, while Nine Mile Run

Chapter I.4: Post-industrial watersheds: Retrofits and restorative redevelopment 87

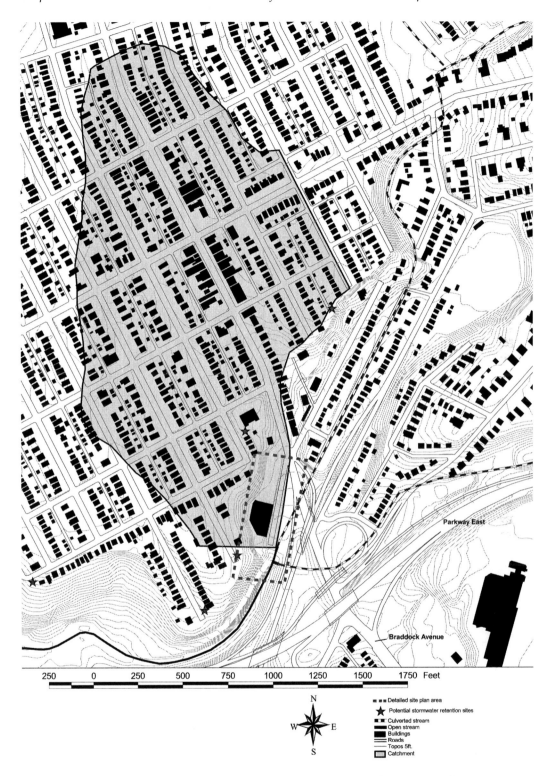

Figure I.4.16 The Regent Square Gateway site, catchment, and surrounding area. (From Ferguson, B.K., Pinkham, R., and Collins, T., *Re-Evaluating Stormwater: The Nine Mile Run Model for Restorative Redevelopment*, Rocky Mountain Institute, Snowmass, CO, 1999. With permission.) (Color version available at www.gsd.harvard.edu/watercolors. Password: lentic-lotic.)

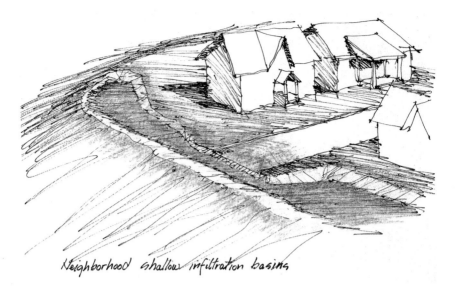

Figure I.4.17 Typical shallow infiltration basin proposed for the neighborhood above the Regent Square Gateway site. (From Ferguson, B.K., Pinkham, R., and Collins, T., *Re-Evaluating Stormwater: The Nine Mile Run Model for Restorative Redevelopment*, Rocky Mountain Institute, Snowmass, CO, 1999. With permission.)

emerges from its culvert, marking the beginning of a new. Filtered "spring" flow trickles across the area from the site's underdrained filter embankments, symbolizing the restoration of the watershed's soil and streams through revitalizing retrofit and redevelopment. Here also is the outfall of the main Nine Mile Run culvert. The headwall of the culvert is retrofitted with scuppers for the trickling spring flow and a spillway for the terraced basin overflow, the surrounding area is redesigned to allow safe low-water access and high-water viewing of the stream, and the channel bed is fitted with sculptural elements that will dissipate the energy of high, wet weather flows emerging from the culvert.

Policies for restorative redevelopment

The charrette identified four overarching policy objectives to support restorative redevelopment. They are as follows:

> *Establish a permanent coordinating body* with the authority and financial security to plan, maintain, and manage the watershed's interrelated infrastructure, natural processes, and urban land uses. The organization must transcend municipal boundaries. It should unite the responsibilities of infrastructure management and ecosystem protection.
>
> *Manage the watershed's sewer and stormwater infrastructure* for efficiency, reduced costs, and reinforcement of natural processes and community vitality. Site-level measures such as those identified by the charrette design teams should be considered part of the infrastructure of the watershed and integrated with conventional infrastructure remediation in order to focus limited community resources on effective systems that produce multiple benefits.
>
> *Restore the watershed's hydrologic and ecological processes* in a manner that utilizes and supports infrastructure rehabilitation and community redevelopment. This includes rehabilitating urban runoff by reconnecting storm drainage with the natural capacities of the watershed. It also includes restoring natural stream, wetland, and forest habitats in critical areas.

Chapter I.4: Post-industrial watersheds: Retrofits and restorative redevelopment

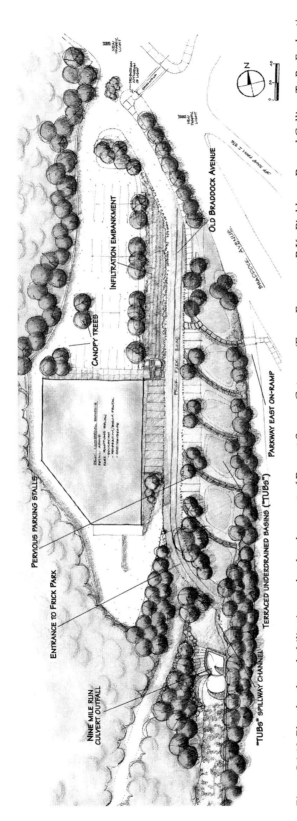

Figure I.4.18 Plan for the rehabilitation and redevelopment of Regent Square Gateway. (From Ferguson, B.K., Pinkham, R., and Collins, T., *Re-Evaluating Stormwater: The Nine Mile Run Model for Restorative Redevelopment*, Rocky Mountain Institute, Snowmass, CO, 1999. With permission.) (Color version available at www.gsd.harvard.edu/watercolors. Password: lentic-lotic.)

Figure I.4.19 Embankment to filter parking lot runoff at Regent Square Gateway: 1) Grass filter strip and inlet; 2) Perforated pipes; 3) Underdrain. (From Ferguson, B.K., Pinkham, R., and Collins, T., *Re-Evaluating Stormwater: The Nine Mile Run Model for Restorative Redevelopment*, Rocky Mountain Institute, Snowmass, CO, 1999. With permission.)

Figure I.4.20 The terraced, underdrained basins, and rehabilitation of parking lots at the Regent Square Gateway site. (From Ferguson, B.K., Pinkham, R., and Collins, T., *Re-Evaluating Stormwater: The Nine Mile Run Model for Restorative Redevelopment*, Rocky Mountain Institute, Snowmass, CO, 1999. With permission.) (Color version available at www.gsd.harvard.edu/watercolors. Password: lentic-lotic.)

> *Enable, support, and require economic revitalization* that reinforces infrastructure management and watershed restoration. Communities, agencies, and developers can structure redevelopment programs and projects in ways that support sewer rehabilitation and restoration of beneficial natural processes.

For each of these objectives, the charrette's policy team prepared detailed rationales and action plans (Ferguson, Pinkham, and Collins, 1999; *Re-Evaluating Stormwater ... Technical Appendix*, 1999).

Status

At this writing, further planning for redevelopment of the charrette sites is proceeding, albeit slowly. The Sterrett School site is not in need of immediate retrofit or redevelopment. Edgewood Crossroads continues to be bound up in debates over the proposed regional mass transit system through the site. Wilkinsburg has officially designated Hunter Park and its surrounding neighborhood as a redevelopment zone and is seeking funding for projects there. Features of the Regent Square Gateway plan have been incorporated into a Pittsburgh Parks master plan. The commercial property there remains for sale.

Meanwhile, the restoration and policy concepts put forward through the charrette and the overall Nine Mile Run Greenway Project are gaining currency in the watershed. Edgewood and the Pennsylvania Environmental Council have secured funding for a stormwater infiltration retrofit demonstration project at a municipal parking lot not far from the Edgewood Crossroads site. Water sensitive designs for several sites within Frick Park and the steel slag redevelopment site were endorsed by watershed citizens and the STUDIO for Creative Inquiry's advisory board in 1999. In 2000, the U.S. Army Corps studied and agreed to these designs, with minor modifications, in preparation for a $7.7 million Corps aquatic ecosystem restoration program in the park and greenway, funded by the City of Pittsburgh and the federal government under Section 206 of the Water Resources Development Act of 1996. Also in 2000, as the STUDIO stepped back from its driving role in Nine Mile Run restoration, interested citizens incorporated a nonprofit Nine Mile Run Watershed Association. Initial funding for the association's director has been secured.

In 1999, the four watershed municipalities entered into a joint agreement to inspect and monitor sewers for illegal connections, develop sewer overflow remediation plans, and implement those plans. This work is proceeding well, and new institutional forms for improved watershed management are under study, including a scheme to change the joint drinking water authority serving the watershed and several other nearby communities into an integrated water, sewer, and stormwater authority.

The projects also have helped catalyze water management initiatives and understandings beyond the watershed boundaries. Ecological restoration is a major component of the Pittsburgh Parks master plan for three major parks in addition to Frick Park. A nationally funded wet weather management demonstration program, 3 Rivers Wet Weather, Inc., has notified Allegheny County communities that it will entertain funding requests for projects incorporating the sorts of techniques illustrated by the charrette designs. ALCOSAN, the regional sanitation authority, is investigating stream daylighting opportunities in conjunction with its efforts to identify and remove stream flows from combined sewers. With funding from the Heinz Endowments, the "Three Rivers 2nd Nature" initiative by 3 Rivers Wet Weather, Inc., ALCOSAN, and the STUDIO for Creative Inquiry is now underway to inventory and assess the ecosystem and community values of Allegheny County's rivers by conducting water quality monitoring, riverbank biotic assessments, a study of public access issues, and a review of tributary streams for restoration and daylighting potential. This effort will likely add to the impetus for restorative redevelopment in the uplands surrounding the rivers.

Patterns of restorative redevelopment

The results of the charrette exemplify certain principles, or patterns, of restorative redevelopment. These patterns of design and process are key concepts for restoring and revitalizing old urban watersheds everywhere (Ferguson et al., 1999). Projects that follow these patterns can have an important effect on sewer overflows; add incrementally to a watershed's long-term, broad-based reduction in impervious surfaces and generation of stormwater; and contribute significantly to the economic and social health of a community.

Patterns of design

Make components multifunctional

Everything that is done in a retrofit or redevelopment project should produce multiple, mutually reinforcing benefits. When a component is multifunctional, it attracts advocates promoting each of its several functions; it attracts broad community and political support. For instance, stormwater has traditionally been moved off city roofs and streets through a single-purpose system of underground pipes. Instead, it can be kept on the surface, recreating a creek that was lost, or infiltrated into the soil to recharge the groundwater and nourish vegetation — in either case providing ecosystem benefits such as habitat for wildlife and stream base flow support, human benefits in experiencing the beauty and wonder of natural systems, and financial benefits in reduced municipal costs of maintaining hidden infrastructure.

Whenever an important component of a project appears to be an undesirable "cost," seek ways to shape it so that it acquires additional desirable benefits. The project and maintenance budget is thereby enlarged as the cost becomes absorbed into the provision of other necessary functions. Multiple functions as various as water quality improvement, employment, housing, separation of storm drainage from sanitary sewers, parking improvements, noise reduction, pedestrian safety, temperature moderation, and social equity can and should be found in the design of every building, street, sidewalk, park, water course, drainage system, residential yard, and institutional landscape.

Use every square inch

Cities are crowded places. The solution to a watershed-wide problem has to be on-site, on every site, because there is nowhere else to go. There is no room to add conventional detention basins or treatment plants. Successful restoration and revitalization depends on utilizing every square inch of a retrofit or redevelopment project for positive, multiple functions. Every component is in the midst of community life and must have positive community benefit in addition to technical function. The cumulative public benefits are enormous. There must be a constant search for restoration and revitalization opportunities on additional sites.

Use freely available natural processes

Freely available natural processes are capable of working for the great benefit of watershed restoration. Vegetated soil absorbs rainwater, and the chemical and microbial processes of the soil capture and degrade most pollutants that may be present. The infiltrated water recharges groundwater tables and restores flows to streams. These processes reduce peak flows and erosion, eliminate sewer overflows, prevent and mitigate pollution, and sustain watershed ecosystems. Taking advantage of them enacts the idea of "green infrastructure," which broadens the conception of a stormwater infrastructure to include the capacities of soil and vegetation to absorb water and filter pollutants. This is a "smarter, cheaper" approach to infrastructure because it puts nature to work and reduces the work humans must do, in contrast to the more active systems of pipes and facilities for conveyance and mechanically dependent treatment.

Use disconnections and reconnections

Sewer overflows are usually the biggest pollutant sources in the watersheds where they exist, such as Nine Mile Run. In such places, the drainage from impervious surfaces should be disconnected from sanitary sewers at every opportunity, no matter how small. To disconnect rooftop drainage, each downspout can be detached from sewers and routed to dry wells, water gardens, and cisterns. To disconnect pavement runoff, the drainage from driveways and walkways can be pitched away from street gutters and onto vegetated soil; large parking areas can be broken up with "infiltration islands" or served by underground storage/recharge beds; street drainage inlets can be detached from combined sewers, and their stormwater diverted into vegetated swales. These techniques can also remove water from storm sewers to improve water quality and restore watershed function.

Drainage that is "disconnected" from sewers in these ways is "reconnected" with its natural path in contact with soil and vegetation. The reconnection with natural processes reduces the volume of surface runoff, filters the pollutants, replenishes the groundwater, and maintains stream base flows. The volume of stormwater, which once seemed a hazard and a nuisance, is turned into a resource and a productive public benefit.

Patterns of process

Cooperate among disciplines

In the process of conceiving and implementing retrofit and redevelopment projects, members of different professions have insight into different problems and opportunities of watersheds and communities, and different types of skills for analyzing and developing them. All professionals need to be members of the project team. History, society, economy, quality of life, art, engineering, and ecology do in fact interact in retrofit and redevelopment projects, because all these processes share the same urban environment. Taking them all into account, as an interdisciplinary team, produces a sound multifunctional result. The choice of individual participants may be important to a project's success. Individuals must be open to the unanticipated insights of members of other disciplines and willing to work with them in design.

Find out what is possible

Diverse, flexible, economical techniques for treating and storing stormwater within urban retrofit and redevelopment projects are being proven in applications throughout the United States. Many techniques useful in the new development context can be adapted and retrofitted into already-built sites or applied upon redevelopment. Developers, public officials, and citizens need to be aware of the alternatives that are available.

Engage the community

Each city and its respective communities have a unique social and political history, style of governance, method of public discourse, and capacity for action. We must carefully define the local application of potential solutions and seek locally integrated forms of innovation. The citizens who must live with the consequences of retrofit and redevelopment projects should have a substantive role in forming them. Collaborative, community-based efforts are key to developing sustainable approaches to issues as broad as sewer overflows, ecosystem restoration, and community development. If functions and benefits in these areas are to be coordinated and maximized, everyone must be involved in the search for solutions. In that process we build cohesive cultural forces invested in long-term success.

Conclusions

The Nine Mile Run stormwater charrette brought together interdisciplinary experts in restorative design and policy from various parts of the country with Pittsburgh natives profoundly experienced in unique local conditions. Their work served as modeling experiments that tested the question: Are these kinds of ideas feasible in the specific conditions of the Pittsburgh region? The results demonstrate that numerous techniques, old and new, can be applied in the Pittsburgh region, specifically in the old urban neighborhoods in ways that are economical, effective, and supportive of economic vitality and quality of life.

As our older cities were built, the cumulative impacts of transforming the landscape mounted, and municipalities replaced natural systems with cost-intensive, conveyance-based infrastructure. Now, when much of the older infrastructure fails to perform to today's or even yesterday's standards, we have an opportunity to reconsider the form and function of the urban landscape — and ultimately integrate gray infrastructure and green infrastructure into a seamlessly operating whole. By adopting the mindset and techniques of restorative redevelopment — that is, by linking stormwater management with urban economic and social development agendas through physical measures that provide multiple benefits — designers and other advocates of water sensitive development can turn old urban watersheds and their waterways from problems into environmental and community assets.

Acknowledgments

The authors thank Bruce Ferguson, who provided considerable assistance with the Nine Mile Run ecosystems and infrastructure charrette and was the lead author of the *Re-Evaluating Stormwater* report (Ferguson, Pinkham, and Collins, 1999). Portions of this paper are based on that report. The 60-plus participants in the charrette and many staff of the authors' organizations — the STUDIO for Creative Inquiry and Rocky Mountain Institute — are also "co-authors" of the restorative redevelopment approach. We thank TreePeople, of Los Angeles, CA, for the inspiration provided by its "Second Nature" charrette and T.R.E.E.S. program. We gratefully acknowledge the generous funding provided to our work in Nine Mile Run by The Heinz Endowments and the Pennsylvania Department of Conservation and Natural Resources. We also thank the following key partners in these efforts: Allegheny County Sanitary Authority (ALCOSAN), Allegheny County Health Department, Borough of Wilkinsburg, Borough of Edgewood, Borough of Swissvale, City of Pittsburgh, Pittsburgh Sewer and Water, and 3 Rivers Wet Weather, Inc.

Literature cited

Ample Opportunity: A Community Dialogue, Carnegie Mellon University, Pittsburgh, PA, 1998.

Condon, P. and Moriarty, S., Eds., *Second Nature: Adapting LA's Landscape for Sustainable Living*, TreePeople, Los Angeles, 1999.

Ferguson, B.K., Lipkis, A., Pinkham, R., Condon, P.M., and Collins, T., The future of old urban watersheds, p. 260-262 in *1999 Annual Meeting Proc. Amer. Soc. Landscape Architects*, American Society of Landscape Architects, Washington, DC, 1999.

Ferguson, B.K., Pinkham, R., and Collins, T., *Re-Evaluating Stormwater: The Nine Mile Run Model for Restorative Redevelopment*, Rocky Mountain Institute, Snowmass, CO, 1999. Available at www.rmi.org as a downloadable file in the "library" section or for purchase in the "bookstore" section.

Lipkis, A., President, TreePeople, Los Angeles, Personal communications, 2001.

Low Impact Development Center homepage, http://lowimpactdevelopment.org/mainhome.html, downloaded April 25, 2001.

Nine Mile Run Rivers Conservation Plan, Carnegie Mellon University, prepared for the City of Pittsburgh; Carnegie Museum of Natural History; Pennsylvania State University; and Commonwealth of Pennsylvania, Dept. Conservation and Natural Resources, Rivers Conservation Plan Program, 1998.

Parsons Engineering Science, Metcalf and Eddy, and Limno-Tech, *Sanitary Sewer Overflow (SSO) Needs Report*, prepared for the U.S. EPA, Washington, DC, 2000.

Pinkham, R., *Daylighting: New Life for Buried Streams*, Rocky Mountain Institute, Snowmass, CO, 2000.

Recycle Your Rain, Flyer from Toronto, Ontario, City Works Services, 1998.

Re-Evaluating Stormwater: The Nine Mile Run Model for Restorative Redevelopment — Technical Appendix, Rocky Mountain Institute, Snowmass, CO, 1999. Available at www.rmi.org as a downloadable file in the "library" section.

T.R.E.E.S. Project Overview, Web page maintained by TreePeople, http://www.treepeople.org/trees/, 2001.

U.S. EPA, *1996 Clean Water Needs Survey Report to Congress,* Washington, DC, Publ. 832/R-97-003, 1997.

U.S. EPA, *Economic Analysis of the Final Phase II Stormwater Rule*, Washington, DC, Publ. 833-R-99-002, 1999.

Response

Raising consciousness through interdisciplinary design workshops

It is important to direct attentions of water sensitive designers not just to issues concerning new developments but also to the restorative redevelopment of older urban regions affected by crumbling or inadequate water infrastructure. As Richard Pinkham and Timothy Collins correctly point out, restoration involves more than technological solutions if it is to succeed; it also must address social and economic concerns in the watershed. When accomplished effectively, redevelopment and retrofit projects can revitalize communities by improving the overall livability of such areas.

The design approach taken in this case study is one of placing small, stormwater treatment and management measures in a dense urban framework. These measures range from disconnecting storm drains, establishing cisterns and water gardens, reconfiguring driveways, adding porous pavements, and daylighting streams. Together, these approaches are those represented in the new paradigm of "low-impact development" as discussed by Larry Coffman in Chapter I.5 and applied by Michael Clar in Chapter I.15, Thomas Liptan and Robert Murase in Chapter I.6, and Robert France and Philip Craul in Chapter I.7. The conceptual design project described in this chapter by Pinkham and Collins is of particular interest in its linking of water management with brownfield redevelopment, in its recommendations for a review of existing institutional arrangements and policy initiatives to increase the chances for success, and in its straightforward message of turning stormwater from being regarded as a nuisance or hazard to being looked upon as a resource (see also the case study described by Glenn Allen in Chapter I.13).

What is perhaps most interesting about this study is the use by Pinkham and Collins of an intense 3-day charrette design workshop with 60 local and regional landscape architects, engineers, architects, artists, planners, policy analysts, and neighborhood citizens in which to raise public consciousness and education about the opportunities for restorative redevelopment. In the end, it is precisely such collaborative, community-based efforts that are the key to developing sustainable water sensitive designs.

chapter I.5

Low-impact development: An alternative stormwater management technology

Larry S. Coffman

Abstract

Low-impact development, or LID, is the general term used to describe an alternative innovative comprehensive suite of lot-level land development principles and practices designed to create a more hydrologically functional urban landscape to better maintain or restore an ecosystem's hydrologic regime in a watershed. This new approach combines a variety of conservation strategies, minimization measures, strategic timing techniques, integrated smallscale site-level management practices, and pollution prevention measures to achieve desired stormwater management or ecosystem protection goals. When these various strategies are integrated into the site design, they create a distributed decentralized management approach. Through the combined cumulative beneficial impacts of all the possible integrated LID site design and management techniques, it is now technically feasible to develop a site with little impact on hydrology or water quality. The basic goal of LID is to engineer a site with as many small-scale retention, detention, prevention, and treatment techniques as needed to achieve the hydrologic functional equivalent to predevelopment conditions.

Background

The need for more effective economically sustainable stormwater management technology has never been greater. With the wide array of very complex and challenging ecosystem and human health protection goals and regulatory requirements to be addressed by stormwater programs, many practitioners are beginning to question the efficacy of conventional stormwater management technology to meet these challenges. Communities are struggling with the economic reality of funding stormwater infrastructure maintenance; inspection, enforcement, and public outreach necessary to support an ever-expanding and aging infrastructure and continued growth. Even more challenging are the exceptionally high costs of retrofitting existing urban development using conventional stormwater management "end-of-pipe" practices to protect the integrity of receiving waters and living resources.

To assist local governments in their efforts to develop more effective economically and environmentally sustainable stormwater management programs, Prince George's County, MD, Department of Environmental Resources (PGDER), with the support of the

U.S. Environmental Protection Agency (EPA), developed a two-volume set of national guidance manuals on the LID approach (PGCDER, 1999). The EPA provided grant funding to assist PGDER in their efforts to develop national guidance manuals to make this technology available to other local governments. This new approach is a significant step toward advancing the state-of-the-art of stormwater management and will provide valuable and useful tools for local governments in their efforts to control urban runoff for both new development and redevelopment.

Prince George's County received the EPA's 1998 first-place National Excellence Award for Municipal Stormwater Management Programs for its pioneering work on LID technology and local LID manual (PGCDER, 1997). Many other efforts are currently underway across the nation to further advance LID technologies such as improving the sensitivity of current hydrology and hydraulic analytical models and developing of new microscale control approaches and practices for highway design, urban retrofit applications, and numerous monitoring efforts.

Some practitioners have found LID's site-oriented microscale control approach to be controversial, because it often conflicts with building codes, challenges conventional stormwater management paradigms, and is perceived by some to accommodate urban sprawl. However, many have found LID's distributed source control technology to be an economical common sense management approach that can be used to achieve superior environmental protection for new development and provide extremely useful new tools to retrofit existing development.

This chapter only briefly outlines the LID approach and its basic theories, philosophy, control principles, and practices. A more detailed explanation on the planning, design, and application of LID technologies is provided in the two-volume national LID manuals (PGCDER, 1999). These two volumes are a good introduction to LID; however, it should be noted that this technology is rapidly evolving, adding new techniques to the lot control principles and practices. Across the country and around the world, an amazing number of new lot-level control techniques are being developed such as multifunctional landscapes features; restoring soil functions; the use of bioretention plant/soil filtration; providing lot-level storage, runoff capture, and use; and modifying timing and pollution prevention. All these techniques go beyond the scope of the original manuals.

For more information on how to obtain copies of the national LID guidance manuals, call Prince George's County's Department of Environmental Resources at (301) 883-5834. It is hoped that the LID national manuals will help to stimulate debate on the state of current stormwater management, watershed protection and restoration technology, and its future direction.

LID in general

LID is a powerful technology that allows development to take place in a manner that can preserve water-related ecological functions/relationships and maintain development potential. LID achieves stormwater management and ecosystem protection goals through the cumulative effects of a wide array of techniques. LID uses new site design planning principles, microscale management practices, and pollution prevention to create environmentally sensitive landscapes that allow the developed area to remain a functioning part of the ecosystems instead of being dysfunctional and apart from the ecosystem.

LID maintains or restores the hydrologic regime and manages stormwater by fundamentally changing conventional site design to create a hydrologically functional landscape that mimics natural ecological hydrologic functions. LID provides numerous tools to maintain the predevelopment volume relationship between rainfall/runoff, recharge, interflow, and evaporation. This is accomplished in five basic steps for new development:

1. Apply conventional conservation planning techniques to define the building envelope. These would include master zoning and environmental features such as streams, wetlands, forests, agricultural/historical preservation, trails, and open spaces.
2. Apply impact minimization strategies to the extent practicable (or allowable) by reducing imperviousness, reducing use of pipes, saving recharge areas, and minimizing clearing and grading.
3. Maintain predevelopment time of concentration by strategically routing flows to maintain travel time throughout the site.
4. Apply distributed integrated management practices to treat, detain, retain, and infiltrate runoff to restore predevelopment conditions. These practices would include use of multifunctional open swales, bioswales, infiltration practices, bioretention (rain gardens) water capture and use (rain barrels), and depression storage in conservation areas.
5. Provide effective public education and socioeconomic incentives to ensure property owners use effective pollution prevention measures and maintain on-site management practices.

LID is a new and creative way of thinking about site design to make every site landscape, roadway, and building feature (green space, landscaping, grading, streetscapes, roads, parking lots, roofs, etc.) multifunctional, multibeneficial, and optimized to manage, treat, or use runoff to maintain/restore hydrologic functions.

The effective use of LID site design techniques can significantly reduce the cost of providing stormwater management. Savings are achieved by eliminating the use of stormwater management ponds; reducing pipes, inlet structures, curbs, and gutters; and resulting in less roadway paving, less grading, and less clearing. Where LID techniques are applicable and depending on the type of development and site constraints, stormwater and site development design, construction, and maintenance costs can be reduced by 25 to 30% compared to conventional approaches.

The creation of LID's lot-level management principles and practices have led to the development of new tools to retrofit existing urban development. Microscale decentralized management practices to recharge, filter, retain, and detain runoff can be easily integrated into the existing green space, parking lots, building design, landscaping, and streetscapes. These integrated management practices, or IMPs, can be constructed as part of the routine maintenance and repair of urban infrastructure, requiring less capital outlay for retrofit compared to conventional large-scale, highly capitalized centralized approaches. LID microscale techniques have been shown to reduce the cost of retrofitting existing urban development. Reducing urban retrofit costs will increase the ability of cities to implement effective retrofit programs to reduce the frequency and improve the quality of combined sewer overflows (CSOs) and improve the quality of urban runoff to protect receiving waters.

Why develop an alternative stormwater technology?

Those jurisdictions that have used conventional "pipe and pond" technology (centralized BMP treatment) over the past 20 years have gained a tremendous amount of experience and insight into the economic and environmental sustainability of a massive stormwater infrastructure. Essentially, practitioners have learned that there are serious economic, environmental, public safety, political, and practical limitations associated with many of the conventional BMPs. LID was designed to address and reduce many of these limitations and burdens, some of which are discussed in the following sections.

Maintenance burdens of a growing, aging infrastructure

As experience is gained with the current management technology, many highly urbanized jurisdictions are beginning to question the efficacy of traditional structural approaches to meet complex environmental objectives. They are also finding it harder to fund the inspection, enforcement, and maintenance of programs necessary for the massive stormwater management infrastructure created by conventional approaches.

Some larger, highly urbanized jurisdictions now have the responsibility for the maintenance, inspection, and enforcement of thousands of BMPs (ponds, infiltration practices, and filters), thousands of miles of pipes and gutters, and tens of thousands of structures (inlets, manholes, and catch basins). This infrastructure, like all urban infrastructures, is growing and aging at the same time. Many of the oldest BMPs have reached their expected service life and are failing. Most jurisdictions have reached the point where they can no longer afford to adequately pay for the upkeep of their stormwater BMP infrastructure. For example, in Prince George's County (population 800,000) the annual stormwater maintenance budget is approximately $6.5 million and rising every year by about $250,000.

Generally, most jurisdictions cannot afford to have proactive maintenance programs for the current suite of conventional BMPs. Maintenance occurs when there are complaints or a total system failure that results in "detectable" damage. Many infiltration or underground treatment systems are never inspected or maintained. Survey and studies of these devices show a failure rate of about 50% after 5 years of operation. In most cases, neither the property owner nor the local jurisdiction can afford to maintain these BMPs and therefore do not. For those jurisdictions that do not have a dedicated funding source for their stormwater programs, the problem of affordability is only compounded, because they cannot successfully compete for resources against police, education, and fire services.

One might ask why we would continue to build treatment systems that we simply cannot afford to maintain? Also, just how effective can current BMPs be in meeting our protection goals if they are rarely or never maintained?

Environmental concerns

Many studies currently demonstrate or strongly suggest that, for example, stormwater management ponds can create their own unique set of environmental impacts. These include problems associated with fish blockages, thermal pollution, groundwater contamination, bioaccumulation of toxics, export of nutrients, sediments and toxics during high flows, and increases in stream erosion. Perhaps the biggest failure of the centralized pond approach is that it does not, and cannot, replicate predevelopment hydrology. Ponds can only be designed to reproduce peak discharges. They do not reestablish the predevelopment rainfall/runoff/recharge volume relationships or maintain the natural frequency of surface discharge. The changes in hydrology resulting from the "pipe and pond" approach accelerates stream channel erosion, changing stream morphology and adversely affecting aquatic habitat structure such as pools, riffles, and shading needed to protect the biological integrity of aquatic biota. Because ponds have their own set of impacts, it is questionable if this control technology can be helpful in maintaining or restoring the ecological integrity of a receiving stream and its biota.

Furthermore, because current management practices only mitigate or lessen the effects of urban development, there is concern about the cumulative impacts of the widespread use of conventional mitigation practices. With the use of conventional management, continued growth will allow increases in pollutant loads and fundamental alterations in a watershed's hydrologic regime. At best, conventional approaches only slow down the rate of change but allow an overall net increase in adverse environmental impacts including pollutant loads and hydrodynamic modifications.

Political problems

There are many public complaints and concerns generated by our current predominant use of management ponds. These complaints deal with issues such as public safety (drowning and mosquito-borne diseases), lack of maintenance (aesthetic), maintenance costs and property owner legal liabilities, i.e., insurance costs. Justifying the continued use of technology that is viewed (real or perceived) by the public as a liability is becoming problematic as political pressure rises to find more acceptable, sustainable, and safer solutions.

Practical problems

A dilemma for local governments is that they are confronted with many protection and restoration goals. They must respond to a wide variety of state and federal regulations and address unique local needs associated with the adverse impacts of urban runoff. Local governments have the difficult task of developing complex multiobjective stormwater management programs. They require multidisciplinary and integrated approaches with the need for as many tools as necessary to meet the desired objective.

New regulatory programs, such as NPDES Phase II and/or TMDLs, now focus on specific targeted issues of concern or compliance requirements. Can conventional technologies meet new goals in a cost-effective and sustainable manner? LID provides many new cost-effective principles and practices that can be added to existing technologies to help "tailor" a program that meets the economic and environmental needs of each community.

Philosophical platforms and organizing principles of LID

Better technology or more restrictive land-use policies

Generally, land-use decisions and development are not typically organized around a set of strong environmental or conservation principles. For the most part, development occurs and is organized around typical land-use principles based on economic needs, individual property rights, public policy, and politics. There are exceptions where environmental laws (wetland and endangered species) can delay development until an acceptable mitigation option is worked out by the regulators or courts. In the end, there is a high degree of certainty that development will occur (perhaps conditioned) but will not be stopped based solely on environmental constraints.

Furthermore, most local governments need continued development or redevelopment to maintain an adequate tax base and are almost always supportive of economic development projects. The problem with relying too heavily on land-use controls (conservation measures alone) to protect natural resources and receiving waters is that they depend on firm and continued political support. The practical reality is that political support is ephemeral, especially in the face of economic development pressures.

Given that development cannot be stopped and that current technology does not reduce new development impacts to predevelopment levels, the impacts of urbanization will continue to increase (perhaps at a slower rate). Development of an appropriate technology that will ensure no net increase in pollutant loads or change in the ecosystem hydrology is needed. The development and use of better technology is key to protecting our receiving waters and ecosystems from continued and rapidly increasing urbanization. Better technology does not mean just better, more efficient BMPs. Technology is defined in the broadest sense as a comprehensive spectrum of planning and design techniques that balance the appropriate level of conservation, minimization, and control techniques to ensure land uses will not impact receiving waters and economic development can continue.

The advantage of a better technology approach is that, once established, it is slow to change and it is application is less susceptible to changes in political points of views or economic development pressures. Generally, technology is apolitical and can be supported by growth, no growth, and conservation proponents.

Are we protecting a watershed or ecosystem?

How you answer this question will have a dramatic effect on your protection priorities, technology, and management strategies. The ultimate watershed protection goal as stated in the Clean Water Act is "fishable and swimmable" waters; however, in our attempts to achieve these goals it has become necessary to understand what an ecosystem is and how it works. The main focus of the Clean Water Act is to protect the aquatic living resources and protect human health. The quality of the receiving waters and the integrity of the aquatic biota are a reflection of and response to what occurs on the land. To improve the integrity of the water and aquatic biota, we must understand the vital environmental processes and ecological functional interrelationship between the uplands and receiving waters.

Watershed-based approach

When we define a watershed, we first place it in the context of topographic boundaries lines defined by geographical and hydrological features. We further define a watershed by other easily measurable tangible features such as acres of land cover types, stream buffer, forests, wetlands, slopes, soil types, etc. Watersheds are compared, evaluated, and ranked by the percentages of these various features such as stream buffer, forest cover, open space, and impervious cover. We begin to develop concepts of what is good or bad based on what is believed to be an appropriate mix of features within each watershed. We may even place limits on impervious surfaces, believing that less is better. We may encourage more open space, believing that more is better. We tend to think of the watershed features in much the same way as we do adequate public facilities necessary to achieve acceptable service or use levels.

When we develop a watershed protection plan, one goal is to protect enough of the watershed features to meet our perceived service levels and uses. Ultimately, the mix and amount of watershed features are determined not by environmental needs but by social, economic, political, cultural, and legal values or constraints. A watershed plan is very much tied to the general land-use or zoning plan. The ecological effects of the plan are considered as part of the development scenarios, with the main objective to minimize environmental impacts by preserving environmental features to the extent practical or politically allowable.

Implementation of watershed plans may be difficult when watershed boundaries cross political boundaries. This is because watershed plans rely heavily on a subjective human construct of what is an acceptable mix of appropriate watershed features and relative to what is politically acceptable. Watershed plans do not start out with the goal of achieving fishable or swimmable waters; they merely strive to reduce impacts. What a watershed plan generally achieves is to define where development is most likely to take place and where environmental resources might best be protected. The degree to which a plan is ultimately followed depends on economic and political factors.

In the end, once a watershed plan is implemented, we may or may not achieve the "fishable and swimmable" goal or be left with enough space to maintain vital ecological functions or ensure the long-term sustainability of the receiving waters' biological integrity. We are unable to achieve our goals through watershed planning alone, therefore, we almost always to have supplement these plans with technological solutions, i.e., BMPs. LID was developed to increase the technological tools necessary to complement watershed plans and better meet environmental objectives.

Ecosystem-based approach

An ecosystem is hard to define in the context of boundaries or measurable features. An ecosystem's boundaries are fuzzy, fluid, and overlapping and differ based on species. Much of ecosystem protection research and debate has centered on trying to define and understand how much space must be conserved to maintain a viable species within an ecosystem. For most species, the space requirement has yet to be clearly defined. Therefore, it is difficult, if not impossible, to know or define an ecosystem in the same physical context as a watershed.

Ecosystem management has been defined by the U.S. Fish and Wildlife Service as "protecting and restoring the natural functions, structure and species composition of an ecosystem that are interrelated" (National Park Service, 1994). From an ecosystem management perspective, to achieve the goal of "fishable and swimmable," it is necessary to protect species, their habitats, and the natural functions that support them. Much attention has been given to saving species and their habitats. State and federal laws provide legal mechanisms for habitat protection. What has not received as much attention in ecosystem management is maintaining the natural functions. Given that it may not be possible to protect enough habitat to maintain natural functions (due to political and economic constraints), one option would be to develop technology that mimics these natural functions within the built environment.

Ecosystem functions include the flow, cycling, and processing of nutrients, energy, and materials (water, sediments, chemicals, and organic materials) through the ecosystem. From the perspective of achieving the fishable and swimmable goal, the hydrologic regime is of primary concern. The flow regime (water cycle) of water from uplands to receiving waters is the primary driving force and mechanism for the transport of materials, energy, and flow of nutrients through the ecosystem. Hydrology is perhaps the most important factor that shapes and defines an ecosystem and the nature of and relationship of the upland to the receiving waters. After all, it is the hydrology that ultimately shapes and defines a desert or a wetland. Significant changes to the nature of the upland will change the ecosystem hydrology and affect receiving waters.

We know much more about ecological functions and processes than we do about the habitat requirements of species. We have monitored, modeled, and studied in detail and know the most about the hydrologic regime, functions, fate, and transport of materials. Numerous models can accurately predict the hydrologic regime for surface and groundwater flow and water quality impacts given the types of soils, slopes, and land cover. From an ecosystem management perspective, maintaining or restoring the ecosystem's natural hydrology regime would be one of the most important goals. The theory of LID is that reproducing the natural hydrologic regime is the one best thing that can be done to protect the integrity of the receiving waters.

The goal of LID from an ecological perspective is to achieve a state of hydrologic homeostasis. This is the state of dynamic self-regulation where the ecosystem is able to maintain essential hydrologic processes and functions to sustain its viability and to avoid changes that would destroy it. LID allows for the recreation of a more natural rainfall/runoff/recharge volume relationships and processes that avoid the extremes in, or flashy nature of, the hydrologic regime typical of conventional site designs. If it is technologically possible to reproduce the hydrologic regime, then it is not necessary to study the ecological consequences of altering the hydrology associated with conventional land development practice and mitigation technology. If we cannot get the hydrology right, what hope is there in achieving the fishable and swimmable goals?

How important is imperviousness?

The amount of total impervious area, or TIA, in a watershed is being used increasingly as a benchmark for determining the "health" of watersheds. Since Klein published one of the earliest papers on this topic (Klein, 1979), numerous authors have presented their views. A modified summary of these findings is presented in Table I.5.1 (Schueler, 1994). The Center for Watershed Protection (CWP) has greatly popularized the concept of a direct relationship between watershed imperviousness and stream health and has reported that stream health impacts tend to begin in watersheds with only 10 to 20% imperviousness (the so-called *10% rule*) threshold. Their theory indicates that sensitive streams can exist relatively unaffected by urban stormwater with good levels of stream quality where impervious cover is less than 10%, although some sensitive streams have been observed to experience water quality impacts at as low as 5% imperviousness, as shown in the study of watershed determinates (Horner et al., 1996). Impacted streams are reported to be threatened and exhibit physical habitat changes (erosion and channel widening) and decreasing water quality where impervious cover is in the range of 10 to 25%. The threshold theory categorizes streams in watersheds where the impervious cover exceeds 25% as typically degraded, having a low level of stream quality and an inability to support a rich aquatic community.

Some researchers and practitioners in urban water resources have questioned this imperviousness threshold theory. A recent study of the relationship between subbasin TIA and the Qualitative Habitat Index (Horner and May, 1999) raises a number of issues. The studies show that, for 31 basins with less than 5% TIA, 14 ranked as excellent, 13 ranked as good, while 4 ranked as fair. No explanation is provided for this variance. Also, for a group of 45 basins with TIA values greater than 25%, 9 were ranked as good habitat quality, 13 were ranked as fair, and 16 were ranked as poor. Obviously, factors in addition to TIA are affecting the ranking of these watersheds because the impervious threshold theory would have categorized all of the watersheds as typically degraded and having a low level of stream quality.

From a statistical standpoint, it can be pointed out that other things being equal, with a large data set, there should be some overall relationship between impervious cover measures and stream quality measures. This type of relationship would also hold for a number of parameters, including population density, but one might expect that these should be noisy, messy relationships as indicated in the Horner/May study, and that a "significant threshold" found in one data set from one setting is unlikely to hold for other settings. If one starts to consider the complexity of controls on stream quality (measured various ways), and the ways in which urbanization impacts these, it is clear that impervious cover on its own is but one of a large number of controls. Figure I.5.1 shows the wide number of factors that can impact the ecological integrity of a receiving stream. Many of these factors are not dependent on the degree of impervious cover.

The pros of using impervious cover as an indicator are that it is simple to measure, planners can work with it, and it makes intuitive sense to a wide range of people. The con is that it is a rather "blunt instrument" but one that does not do justice to the complex physical, chemical, and biological processes we must be concerned with if we are to restore the ecological integrity of watersheds.

Table I.5.1 Summary of Key Findings of Urban Stream Studies Examining the Relationship of Urbanization on Stream Quality

Ref. (Year)	Location	Biological Parameter	Key Finding
1. Klein (1979)	Maryland	Aquatic insects/fish	Macroinvertebrate and fish diversity declines rapidly after 10% TIA
2. Benke et al. (1981)	Atlanta	Aquatic insects	Negative relationship between number of insect species and urbanization in 21 streams
3. Steward (1983)	Seattle	Salmon	Marked reduction in coho salmon populations noted at 10–15% TIA at 9 sites
4. Garie and McIntosh (1986)	New Jersey	Aquatic insects	Drop in insect taxa from 13 to 4 noted in urban streams
5. Pedersen et al. (1986)	Seattle	Aquatic insects	Macroinvertebrate community shifted to chironomid, oligochaetes, and amphipod species tolerant of unstable conditions
6. Jones and Clark (1987)	Northern Virginia	Aquatic insects	Urban streams had sharply lower diversity of aquatic insects when human population density exceeded 4 persons/acre (estimated 15–25% TIA)
7. Steedman (1988)	Ontario	Aquatic insects	Strong negative relationship between biotic integrity and increasing urban land use/riparian condition at 209 stream sites. Degradation begins at about 10% TIA.
8. Limburg and Schmidt (1990)	New York	Fish spawning	Resident and anadromous fish eggs and larvae declined sharply in 16 tributary streams greater than 10% impervious
9. Booth (1991)	Seattle	Fish habitat/channel stability	Channel stability and fish habitat quality declined rapidly after 10% TIA
10. Yoder (1991)	Ohio	Aquatic insects/fish	100% of 40 urban sites sampled had fair to very poor index of biotic integrity scores
11. Schueler and Galli (1992)	Maryland	Fish/aquatic insects	Fish diversity declined sharply with increasing TIA. Loss in diversity began at 10–12% TIA; insect diversity metrics in 24 subwatersheds shifted from good to poor over 15% TIA
12. Luchetti and Fuerstenberg (1993)	Seattle	Fish	Marked shift from less tolerant Coho salmon to more tolerant cutthroat trout populations noted at 10–15% TIA
13. Taylor (1993)	Seattle	Wetland plants, amphibians	Mean annual water fluctuation was inversely correlated to plant and amphibian density in urban wetlands; sharp declines noted over 10% TIA
14. Galli (1994)	Maryland	Brown trout	Abundance and recruitment of brown trout declines sharply at 10–15% TIA
15. Shaver et al. (1994)	Delaware	Aquatic insects/habitat quality	Insect diversity at 19 stream sites dropped sharply at 8–15% TIA; strong relationship between insect diversity and habitat quality; majority of 53 urban streams had poor habitat
16. Black and Veatch (1994)	Maryland	Fish/insects	Fish, insect, and habitat scores were all ranked as poor in 5 subwatersheds that were greater than 30% TIA

Source: Modified from Schueler, T.R., The importance of imperviousness, Center for Watershed Protection, Watershed Prot. Tech., 1, 100, 1994.

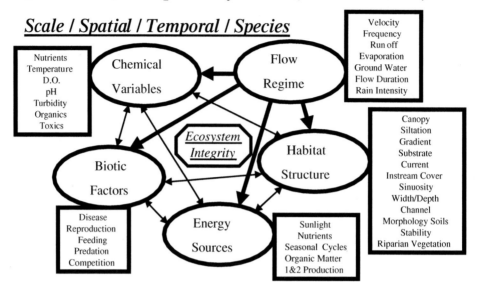

Figure I.5.1 Factors that impact the ecological integrity of a receiving stream.

Furthermore, such a simplistic view of cause and effect of urban impacts does little to help our understanding of the relationship between ecological processes and technology necessary to maintain these processes. Promoters of the impervious threshold theory assume that mitigation of impacts will maintain an acceptable hydrologic regime and maintain the ecological integrity of the receiving waters. Many are not comfortable with this assumption because there is little data to prove that an altered hydrologic regime will maintain a natural homeostasis state to ensure the long-term sustainable vitality of the ecosystem. With LID the goal is to maintain the natural hydrologic regime and its homeostatic state. Intuitively, if the natural hydrologic regime can be maintained, the long-term sustainability of the ecosystem can be assured.

The TIA threshold theory has some appropriate uses as a very rough watershed planning tool, but it would be hard to defend in detail in either a scientific or legal setting (e.g., if one were to write an ordinance setting a limit on imperviousness based solely on TIA, a good set of experts could easily get it overturned).

Alternative approaches to the TIA threshold theory can be obtained by identifying the individual parameters, or metrics, that define stream "health" and better examine the relationship between TIA and these metrics. Some of these watershed impairment metrics include the following:

- Runoff volume
- Bankfull and subbankfull discharge
- Change in bankfull hydraulic geometry
- Channel modifications
- Upstream channel erosion and sediment loads
- Decrease in base flow
- Decrease in wetted perimeter (width/depth ratio)
- Instream habitat structure

- Large woody debris (LWD)
- Stream crossings and fish barriers
- Riparian forest fragmentation, width, and diversity
- Chemical water quality
- Stream temperature
- Aquatic diversity

Many practitioners have viewed these parameters or metrics on an individual basis and have thus developed a series of fragmented techniques, which for the most part have had limited success or failed. For examples and more details of the various metrics that others have examined to relate urbanization to stream degradation, refer to references (e.g., Alley and Veenhuis, 1983; Barton et al., 1985; Beyerlein, 1996; Booth, 1990, 1991, 1993, 1996; Galli, 1991; Graf, 1977; Gregory et al., 1991; Horner et al., 1996; Leopold, 1968; Macrae and Marsalek, 1992; Wigmosta et al., 1994). The impacts associated with urban land-use change, however, can and should be viewed from a hydrologic cycle perspective (i.e., hydrologic functions or regime). A cause-and-effect relationship can be established for the full range of impact metrics, and logical, process-related solutions can be developed.

With LID, reducing the amount of impervious surfaces is only one of many impact reduction techniques. LID's goal is the reestablishment of the natural hydrologic regime and ecosystem functional relationships, so the amount of impervious surface is not as important as how the runoff is managed and functionally restored to the site. LID provides numerous techniques and strategies to manage and reestablish a functional equivalent of the natural hydrology regime independent of the amount of impervious surfaces.

Another troubling implication of the impervious cover threshold theory involves the large expanses of urban areas that have already been densely developed. The impervious threshold theory suggests that it is impossible to restore stream quality in watersheds with high impervious cover; however, a growing number of stream restoration projects throughout Maryland use innovative concepts and techniques based on the application of fluvial geomorphology. These projects suggest that restoration is possible, particularly if adequate stream buffers or riparian zones exist. Several additional projects in Maryland offer promise that, even under adverse conditions, applying sound fundamental analysis of hydrologic functions and processes can restore urban watersheds. These include urban retrofit demonstration projects such as the Bladensburg Port Towns project in Prince George's County and targeted watershed projects such as Sawmill Creek in Anne Arundel County (Arnold et al., 1982).

The principles and practices of LID also allow for the retrofit of existing developed areas by integrating microscale management into the urban landscape. Reducing TIA in existing developed areas alone is neither practical nor effective in retrofitting existing urban areas to reestablish ecological functions or to achieve the fishable and swimmable objectives.

Impacts of urbanization

In general, the degradation process to receiving waters caused by urbanization has the following pattern. Land-use and cover changes, compacted soils, and creation of efficient drainage/conveyance systems using connected impervious surfaces lead to very significant changes in the watershed's hydrologic cycle or regime. These changes consist primarily of a reduction or loss in the initial rainfall abstractions, I_a. These processes include water intercepted by vegetation, water retained in surface depressions, evaporation, and infiltration. The reduction or loss of these initial abstractions, together with accompanying decreases in the time of concentration, T_c (both sheet flow and overland flow), leads to significant increases in both the runoff volume and peak discharge values. These basic hydrologic principles are explained in the Department of Agriculture, Soil Conservation Services technical release series (SCS, 1986).

The receiving streams respond to these hydrologic regime changes by increasing their cross-sectional areas, usually through a combination of channel down-cutting and bank erosion that produces staggering volumes of sediment and results in the destruction and loss of stream habitat and the related reduction and loss of biologic species. In addition, the surface runoff flowing over the impervious surfaces typically displays elevated pollutant levels and temperatures that can be very damaging to most fish species. The reduction or loss of the initial abstractions can reduce groundwater recharge, lower water tables, and reduce or cease baseflow to small streams, particularly during dry periods. A watershed hydrologic cycle is changed by the way we design and construct sites and choose to design our drainage systems, that creates an efficient drainage system that completely alters the natural hydrologic regime.

The key to addressing and controlling these impacts must focus on controlling or at least minimizing the changes to the hydrologic cycle or regime. Simply relying on TIA reduction and conventional detention is not feasible, practical, or sustainable. The challenge of LID technology is to develop a comprehensive toolbox of techniques that allows the recreation of the initial abstraction volumes and frequency of discharges to mimic natural hydrologic processes and thus preclude the traditional impairment associated with urbanization.

The LID approach

Currently, we design and construct every site with one basic overriding goal — to achieve good drainage. In other words, it is important to get runoff off the site as quickly as possible to the conveyance system and centralized BMP treatment device. As a site is developed, its hydrologic functions are first altered on a microscale to create a highly efficient drainage system. The cumulative impacts of these microscale changes results in drastically altered hydrologic regimes that we typically try to mitigate using end-of-pipe management practices.

If we can design sites to achieve good drainage, why not design sites with the opposite objective — to maintain predevelopment hydrologic functions? Can we intelligently engineer sites to replace the microscale hydrologic functions, and would the cumulative beneficial effects result in the preservation of natural watershed hydrologic functions? Can a site be designed in a way to remain a functional part of an ecosystem's hydrological regime or at least more closely mimic natural hydrologic functions? To create a hydrologically functional site, there must be a radical change in our thinking. We need to restore hydrologic functions, not just mitigate development impacts.

Five basic steps for LID designs

The five basic steps of the LID approach are discussed in the following paragraphs. These steps follow a systematic approach to site planning and design. The LID techniques are not necessarily new, but the principle of combining all these techniques in a manner that produces a comprehensive approach of distributed management is new. What is also new is that the LID guidance manuals provide an analytical methodology based on TR-55 (SCS, 1986), which allows one to quantify the hydrologic impacts of the combined affects of all the LID practices.

The objective of LID site design is to manage, recharge, detain, and retain runoff volumes uniformly throughout the site to mimic predevelopment hydrologic functions.

Uniform distribution of small on-lot retention and detention to control both runoff discharge volume and rate is the key to better replication of predevelopment hydrology. The relative change in the frequency and duration of runoff is also much closer to predevelopment conditions than can be achieved by typical application of conventional centralized BMPs such as ponds. Management of both runoff volume and peak runoff rate are both included in the design of controls. This is in contrast to conventional end-of-pipe treatment that completely alters the watershed hydrology to create a new modified hydrologic regime.

The LID site analytical analysis and design approach focuses on four major hydrologic elements. These fundamental factors affect site hydrology and are introduced in the following list. For a more detailed explanation, refer to the LID manuals and the unpublished paper explaining LID hydrology (Cheng et al., 2001).

- Curve number (CN) — A factor that accounts for the effects of soils and land cover on the amount of runoff generated. Minimizing the magnitude of change from the predevelopment to the post-development CN by reducing impervious areas and preserving more natural vegetation will reduce runoff storage requirements and help to maintain predevelopment runoff volumes.
- Time of concentration (T_c) — The time it takes runoff to travel through the watershed. Maintaining the predevelopment T_c reduces peak runoff rates and can be achieved by lengthening flow paths, reducing the use of pipes and paved channels, and conserving natural drainage and depression storage.
- Permanent storage areas (retention) — Retention storage is needed for volume and peak control, as well as water quality control and to maintain the same CN as the predevelopment condition.
- Temporary storage areas (detention) — Detention storage may be needed to maintain the peak runoff rate or to prevent flooding.

Step 1. Conservation. The first step of LID is to minimize or prevent runoff to reduce the change in the CN. This step is similar to traditional techniques of maximizing natural resource conservation, limiting disturbance, and restoring impacted natural resources. This includes considering conservation requirements and watershed plan components such as parks, open space, streams, step slopes, and permeable soils. These are the typical conservation techniques that help to define the buildable area of the site. Conservation techniques also include maintaining natural drainage patterns, topography, and depressions; preserving as much existing vegetation as possible in pervious soils, specifically hydrologic soil groups A and B; and revegetating cleared and graded areas. These measures help to minimize the change in the pre- and post-development CN; see Figure I.5.2 for basic conservation strategies for site planning and design. Chapter I.3 also discusses conservation techniques and benefits.

Step 2. Minimization. Calculation of the LID CN is based on a detailed evaluation of the existing and proposed land cover so that an accurate representation of the potential for runoff can be obtained. This calculation requires the engineer/planner to investigate the following key parameters associated with LID: (1) land cover type; (2) percentage of and connectivity of impervious cover; (3) hydrologic soils group (HSG); (4) hydrologic conditions (average moisture or runoff conditions; and (5) maintaining existing drainage patterns and natural retention features. Reducing the change in CN alone will reduce both the post-development peak discharge rate and volume.

The following are examples of LID site planing practices that can be utilized to achieve a substantial reduction in the change of the calculated CN: narrower driveways and roads

Figure I.5.2 Conservation strategies for site planning.

Minimize Impacts

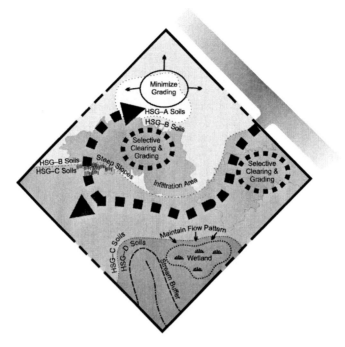

- **Minimize clearing**
- **Minimize grading**
- **Save A and B soils**
- **Limit lot disturbance**
- **Soil amendments**
- **Alternative surfaces**
- **Reforestation**
- **Disconnect**
- **Reduce pipes, curb and gutters**
- **Reduce impervious surfaces**

Figure I.5.3 Example of limited lot disturbance (site finger printing).

(minimizing impervious areas), site finger printing (minimal disturbance), open drainage swales, preservation of soils with high infiltration rates, location of BMPs on high-infiltration soils, disconnecting impervious surfaces to direct and disburse runoff to soil groups A and B, flattening slopes within cleared areas to facilitate on-lot storage and infiltration, and construction of impervious features on soils with low-infiltration rates. See Figure I.5.3 for an example of limited lot disturbance or site finger printing. Chapter I.3 also discusses minimization techniques.

Step 3. Maintaining the predevelopment time of concentration (T_c). The LID hydrologic evaluation requires that the post-development T_c be close to the predevelopment T_c. This is important because LID is based on a homogenous land cover and distributed retention

Maintain Time of Concentration

- Open drainage
- Use green space
- Flatten slopes
- Disperse drainage
- Lengthen flow paths
- Save headwater areas
- Vegetative swales
- Maintain natural flow paths
- Increase distance from streams
- Maximize sheet flow

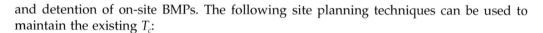

Figure I.5.4 Example of a vegetative swale.

and detention of on-site BMPs. The following site planning techniques can be used to maintain the existing T_c:

1. Maintain predevelopment flow path length by dispersing and redirecting flows using open swales and natural or vegetated drainage patterns.
2. Increase surface roughness (e.g., preserving woodlands, vegetated swales).
3. Detain flows (e.g., open swales, bioretention).
4. Minimize disturbances (minimizing compaction and changes to existing vegetation).
5. Flatten grades in impacted areas.
6. Disconnect impervious areas (e.g., eliminating curb/gutter and redirecting down spouts).
7. Connect pervious areas to vegetated areas.

See Figure I.5.4 for an example of a vegetated swale to slow down runoff to allow more infiltration and evaporation.

Combined use of these techniques, and those to reduce the change in the CN, can modify runoff characteristics to effectively shift the post-development peak runoff time toward that of the predevelopment condition.

Step 4. Maintaining the redevelopment curve number and runoff volume. Once the post-development T_c is maintained at the predevelopment conditions and the change of predevelopment to post-development CN is minimized, any additional reductions in runoff volume must be accomplished through distributed on-site stormwater integrated management practices, or IMPs. The goal is to select the appropriate combination of IMPs that simulate the hydrologic functions of the predevelopment condition to maintain existing CN and corresponding runoff volume. LID design strives to maximize the runoff use and retention practices distributed throughout the site to provide the required volume controls at the source.

Retention storage allows for a reduction in the post-development volume and the peak runoff rate. The increased storage and infiltration capacity of retention BMPs allow the predevelopment volume to be maintained. The most appropriate on-lot retention BMPs include (1) bioretention cells (rain gardens), (2) infiltration trenches, and (3) rain barrels. Other possible retention BMPs include retention ponds, rooftop storage, cisterns, and irrigation ponds. It may be more difficult to distribute these types of controls throughout a development site, but often they are part of the site drainage patterns and have only to be left untouched. Chapters I.6 and I.7 discuss other techniques in greater detail.

Storage Detention & Filtration
"LID's IMP's"

- Uniform Distribution at the Source
  - Open drainage swales
  - Rain gardens /bioretention
  - Smaller pipes and culverts
  - Small inlets
  - Depression storage
  - Infiltration
  - Rooftop storage
  - Pipe storage
  - Street storage
  - Rain water use
  - Soil management

Figure I.5.5 Examples of integrated lot-level management practices.

As retention storage volume is increased, a corresponding decrease occurs in the peak runoff rate in addition to runoff volume reduction. If a sufficient amount of runoff is stored, the peak runoff rate may be reduced to a level at or below the predevelopment runoff rate. This storage may be all that is necessary to control the peak runoff rate when there is a small change in CN. When there is a large change in CN, however, it may be less practical to achieve flow control using volume control only. Figure I.5.5 lists examples of integrated practices that can be used to treat, filter, retain, and detain runoff at the lot level.

Step 5. Pollution prevention. Pollution prevention and maintenance of on-lot BMPs are two key elements in the overall LID comprehensive approach. Effective pollution prevention measures can reduce the introduction of pollutants to LID BMPs, thereby enhancing their ability to reduce pollutant levels and extend the life of the facilities. Public education is essential to successful pollution prevention and BMP maintenance. Not only will effective public education complement and enhance BMP effectiveness, it can also be used as a marketing tool to attract environmentally conscious buyers, promote citizen stewardship, awareness, and participation in environmental protection programs, as well as help to build a greater sense of community based on common environmental objectives and the unique environmental character of LID designs.

Education is the key to effective public participation. With LID techniques all stakeholders (public officials, engineers, builders, realtors, and buyers) must be educated about the positive environmental impacts of LID and its maintenance savings and burdens. With LID, most controls are integrated into the lot landscape features; therefore, property owners will need to know how to maintain these features. This can be achieved through easements, covenants, brochures, and environmental committees. Although additional landscape means more landscape maintenance, there is no major stormwater infrastructure (ponds and pipes and structures) to maintain, and the scale of the maintenance is reduced to what an individual property owner can afford for routine landscape maintenance costs.

Property owners have responded to LID's landscape level of control in two very important ways. First, they feel good about their property's helping to protect the environment. Second, they believe that the additional landscape material adds greater value to their property. Thus, LID controls provide a strong economic incentive to maintain LID practices because property values are also maintained.

Pollution Prevention

> 30 - 40% Reduction in N&P
> Kettering Demonstration Project

- Maintenance
- Proper use, handling and disposal
  - Individuals
    - Lawn / car / hazardous wastes / reporting / recycling
  - Industry
    - Good house keeping / proper disposal / reuse / spills
  - Business
    - Alternative products / Product liability

Figure I.5.6 Pollution prevention strategies.

From 1992 through 1997 PGCDER conducted a public education demonstration project in an 1,150-unit residential area. Monitoring and modeling results showed that it is possible to achieve a 30 to 40% reduction in the use of fertilizer through a very aggressive outreach program. The study further indicated that even more gains could be made if manufacturers provide instructions on their products, suggesting how to limit the use of the fertilizers. Figure I.5.6 summarizes some of the findings of the Kettering study (Coffman et al., 1998).

Other important LID considerations

Potential requirement for additional detention storage

In cases where very large changes in CN cannot be avoided, retention storage practices alone may either be insufficient to maintain the predevelopment runoff volume or peak discharge rates or require too much space to represent a viable solution. In these cases, additional detention storage will be needed to maintain the predevelopment peak runoff rates. A number of traditional detention storage techniques are available that can be integrated into the site planning and design process for a LID site. These techniques include: (1) swales with check dams, restricted drainage pipes, and inlet/entrance controls; (2) wider lower gradient swales; (3) rain barrels; (4) rooftop storage; (5) shallow parking lot storage; and (6) constructed wetlands and ponds. These detention practices can easily be integrated into the site design features.

Where downstream flooding is a problem, additional flow control may be necessary to protect property and ensure safe conveyance. Also, when there is a need to retrofit existing development, additional detention may be needed to control off-site flows using regional ponds.

Determination of design storm event

The hydrologic approach of LID is to retain the same amount of rainfall within the development site as that retained prior to any development (e.g., woods or meadow in

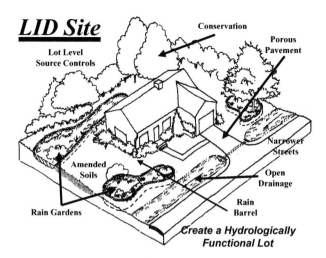

Figure I.5.7 Example of an LID residential lot design.

good condition) and then release excess runoff as the woods or meadow would have. By doing so, it is possible to mimic, to the greatest extent practical, the predevelopment hydrologic regime to maximize protection to aquatic ecosystems and groundwater recharge. This approach allows the determination of a design storm volume that is tailored to the unique soils, vegetation, and topographic characteristics of the developing watershed. This approach is particularly important in watersheds that are critical for groundwater recharge to protect stream/wetland base flow and ground- or surface water supplies. For each watershed, there is a unique amount of runoff that must be retained to mimic the natural conditions. With LID, the volume of runoff to be controlled changes with each site in order to replicate the natural ecological conditions. Figure I.5.7 shows residential LID site design with a number of integrated management practices. Figure I.5.8 shows complete LID residential site design using a variety of techniques.

Integrated management practices (IMPs)

Site design techniques and IMPs can be organized into three major categories, as follows: (1) runoff prevention measures designed to minimize impacts and changes in predevelopment CN and T_c; (2) retention facilities that store runoff for infiltration, exfiltration, or evaporation; and (3) detention facilities that temporarily store runoff and release through a measured outlet. Table I.5.2 lists examples of only some of a wide array of LID IMPs and their primary functions. Placing these IMPs in series (treatment train) and uniformly dispersing them throughout the site provides the maximum benefits for hydrologic controls.

Urban Retrofit

Many highly urbanized areas are almost entirely covered by impervious surfaces such as roadways, parking, sidewalks, and buildings. Control of runoff in urban areas using conventional stormwater management practices is difficult and severely limited due to the lack of open space, cost of land, high construction costs, and high operational and maintenance costs. Typical urban stormwater controls involve construction of expensive end-of-pipe detention facilities, infiltration systems, underground storage, systems to

Figure I.5.8 Example of an LID residential site design.

Table I.5.2 Examples of LID IMPs and Primary Functions

BMP	Prevent Runoff	Detention	Retention	Conveyance	Water Quality	Stream Channel Protection
Bioretention		X	X		X	X
Infiltration trench		X	X		X	X
Dry wells			X		X	X
Roof top storage		X	X		X	X
Vegetative filter strips				X	X	X
Rain barrels		X	X			X
Vegetated swales and small culverts		X		X	X	X
Swales		X		X	X	X
Infiltration swale		X	X	X	X	X
Reduce imperviousness	X					X
Strategic clearing/grading	X				X	X
Engineered landscape	X	X	X		X	X
Eliminate curb and gutter	X				X	X
Vegetative buffers	X				X	X

optimize in pipe storage, and the use of water quality BMPs (filters and hydraulic separators). Conventional stormwater management retrofit approaches (ultra-urban BMPs and end-of-pipe storage) also have limitations due to costs and physical requirements

To adequately address urban runoff problems, cities must have effective low-cost and politically acceptable tools. If dramatic improvements in urban runoff management are to be achieved, it will be necessary to fundamentally rethink current approaches and radically redesign and reengineer urban stormwater management technology.

LID management strategies and practices must be modified to address the unique characteristics of each individual watershed. This makes the protection strategy even more effective as it requires the designer, regulator, residents, and stakeholders to have a better understanding of the complexities and water resource protection needs of each watershed. To apply LID to any land use is simply a matter of developing numerous ways to creatively manage, prevent, retain, detain, use, and treat runoff within multifunctional landscape features unique to that land use. Factors that influence LID strategies and practices include specific land use, soils, climate, rainfall distribution, water/natural resource protection objectives, and the regulatory framework.

Initially LID practices and principles were developed to accommodate typical suburban land uses. During the initial development of the technology it was realized that LID principles such as creating a hydrologically functional landscape, uniform distribution of controls, strategic timing of flows, micromanagement and use of plant soil filter technology such as bioretention have universal applicable regardless of the land use. LID can be applied to address stormwater management goals and water resources objectives for urban, suburban, and rural development.

In order to apply LID to highly urbanized areas, a new set of specific LID urban management principles and management practices must be developed to address the unique landscape features and water resources protection objectives of cities. Each city may have different urban runoff management objectives that are based on compliance with required local, state, and federal regulations and the economic, environmental, and human health needs to protect receiving waters, sensitive environmental features, and living resources. The ability of cities to meet urban runoff objectives will depend greatly on the cost of control measures and the available economic resources. The cost of current state-of-the-art conventional CSO and BMP technologies can be staggering. The potential for cost savings using LID for volume and water quality control to reduce or eliminate these systems in urban areas is tremendous. This is because many of the techniques used in suburban applications are extremely cost effective and can be modified for use in the existing urban infrastructure. Also see Chapters I.1 and I.4 for more information on urban retrofit strategies and techniques.

International microscale experiences and case studies

Conventional stormwater institutions in this country have not encouraged the use of incorporating multiple objectives at small scales. In order to determine how effective sustainable microscale techniques and small-scale multiple objective programs can be, we can look toward the experiences of other countries. The following section will highlight and explore some of the experiences that other countries have had using microscale techniques that are similar to LID approaches for wet weather control. The information obtained in this section is from personal communications with Neil Weinstein, Executive Director of the Low Impact Development Center, and the findings from an unpublished EPA report on the use of LID technology for combined sewer overflow control (Weinstein, 2000). Although many communities in the United States are rapidly accepting and planning for the use of LID and other integrated microscale techniques, a limited number of sites have been constructed using these approaches. Examples of these developments in the United States include Village Homes in California; Somerset and Greenbelt Plaza in Prince George's County, MD; Wonderland Creek in Colorado; and the NEMO project in Jordan Cove, CT. The collection of long-term monitoring and modeling data at these sites is also limited. Because of our limited experience with LID and other microscale approaches, we must look to other regions in order to evaluate the effectiveness of these programs to meet receiving water objectives. We can build on this existing experience to develop a knowledge base of planning, design, and program management tools to help communities evaluate and develop successful LID programs.

European and Asian countries have a significant amount of experience with microscale approaches. Some of the critical environmental and economic factors that have been responsible for the development and acceptance of microscale techniques in other countries include the following:

- In Europe, much of the infrastructure is based on combined sewer systems. For example, England has over 70% combined sewers. The original design capacity cannot account for the amount of infill and increased imperviousness and the systems are surcharging. It is difficult to construct new systems because of the expensive conflicts with other infrastructures and disruptions to the city due to construction. New ways to reduce runoff volumes to these systems must be developed.
- The cost of land in many urbanized and core city areas in Europe and Asia makes the sacrifice of land for developable area for conventional BMPs economically and politically prohibitive. Techniques that provide environmental protection must be incorporated into the infrastructure, and site designs that do have minimal impact on the development are essential.
- Land subsidence in many Asian countries and saltwater intrusion in European countries such as The Netherlands are due to reduction of groundwater levels that result from decreased infiltration capacity of soils and increased impervious cover that has resulted from urbanization. Techniques for groundwater recharge in order to maintain a sustainable water balance must be developed.

International examples (Weinstein, 2000)

National, local government, and watershed compacts have adopted comprehensive strategies that incorporate microscale planning programs, demonstration projects, financing districts, and subsidies in order to improve water quantity and water quality control using microscale and sustainable techniques. Some key studies that have been published in the literature that demonstrate the planning, design, and environmental effectiveness of these programs as well as some of the institutional issues associated with microscale approaches are described next.

German watershed planning using microscale approaches

The Emscher catchment in Germany is approximately 330 square miles with 2.5 million inhabitants residing in over 17 cities. Although traditionally a coal mining and a farming area, over the last 100 years it has evolved into a more industrialized region. The beginning of the transition caused a decision between economic growth and good environmental conditions. Water resources were exploited and watercourses and soils were contaminated. The long-term records have indicated a frequency of flooding corresponding with increased urbanization during the recent years. In addition, over 98% of the infrastructure utilized combined sewers and there were many open systems.

In the early 1990s, a strategy was developed in order to reduce runoff volume and improve environmental quality. The first step was to find strategies to maintain a river ecology that met the aesthetic and recreational needs of the citizens, although the planners recognized the challenge of restoring the system to its predevelopment state due to the intense urbanization. Certain precepts such as not increasing peak flows and volumes and reconstructing straightened channels to a more natural morphology were employed. This approach helped improve the opportunity for the reestablishment of ecological systems and had a minor, 3 to 5%, reduction in flood frequency. The planning for the drainage area was restructured so that subbasins were analyzed in more discrete units that enabled the development of targeted strategies for volume and pollution reduction. The next step was the introduction of source control. Because of the longstanding tradition of efficient conveyance systems, this new concept was at first difficult for many communities within the drainage district to embrace. A competition was held and grants were awarded for communities to incorporate source controls as pilot projects. The emphasis of the project areas was disconnection of imperviousness and incorporation of small-scale simple source controls. In one project area of approximately 400 acres, of which 40% was impervious, 70% of the facilities were constructed by private land owners, 20% were constructed with neighborhood development grants, and 10% were constructed by contractors.

Polls showed that environmental protection was the main reason, and reducing fees and receiving financial incentives for construction secondary. The level of disconnection in this area was 5%. In the seven areas studied, which included this area, the range of disconnectivity achieved varied widely, from 5% to almost 100%. It is estimated that the result of this program will achieve a 10% reduction in the peak of the 2-year interval storm event.

Japanese experimental sewer system

In Japan, the Experimental Sewer System (ESS) has been a highly effective method to control runoff volume. This approach is based on the development of an integrated, highly efficient infiltration approach in order to reduce runoff volumes and increase groundwater recharge in urban areas. From 1983 to 1995, the Tokyo Metropolitan Government built within a 5.5-square mile area 33,300 infiltration pits, 122 acres of permeable pavement, and over 175 miles of infiltration trenches. This approach has also been adopted in many other Japanese cities including Sapporo, Shiogama, Chiba, Yokohama, Naboya, and Amagaski. Representative areas within the ESS were analyzed in order to determine the cost-effectiveness and efficiency of the program. It was determined that the cost of using microscale infiltration techniques such as permeable pavement and infiltration inlets cost approximately 33% less than conventional open pond detention systems and 10% of the cost of storage vaults. This alternative was so much more cost-effective due to the high cost of land and complexity of the existing infrastructure. Perhaps equally impressive was the reduction in monitored storm drain flow volumes of up to 50%. Most important was the reduction in CSO events from 36 to 7. Some of the keys to the success of the acceptance of these programs have been:

- Evaluation of positive effects
- Development of maintenance programs
- Obtaining cooperation of the public
- Evaluating disadvantages
- Evaluating effect on groundwater
- Subsidizing private construction and maintenance
- Providing administrative guidance
- Improvement of Administrative Model District
- Promotion of technology
- Inclusion of infiltration in planning
- Development of a political base

Demonstration projects have been one of the most effective ways to educate people and gain acceptance about this approach and to gain acceptance of this. In Yokohama, model areas were constructed to demonstrate the technology. Within a 15-acre section of the city, 1.8 miles of infiltration pipes (typically smaller than 8 in. with a gravel base), 2.5 acres of permeable pavement, and 10 acres of conventional pavement were constructed as part of the infrastructure. The results, including volume reduction, maintenance, and public acceptance, are continuously being monitored.

Lyons residential infiltration strategies

The area of Lyons, France, has a history of over 50 years of developing infiltration strategies for stormwater runoff control. Lyons is approximately 230 square miles in area with 1.2 million inhabitants. The storm sewer and sewage system infrastructure includes 1,600 miles of sewers, 56 pumping stations, and 9 treatment plants in a combined system. Because of the cost and disruption required to build large-scale relief sewers to reduce CSO events, alternative strategies had to be developed. Although one large-scale interceptor project was planned and built in the 1970s, it was recognized that these systems could not continuously be built and that a reduction in flows to the interceptor was also required. In the 1980s, a strategy of large-scale centralized infiltration ponds and pits with pretreatment devices was incorporated as part of the control strategy. Concerns over large-scale loadings on groundwater, maintenance of the systems, construction of long networks of storm drain pipes to the end-of-pipe facilities showed the limitations of this approach. In the early 1990s, a new decentralized approach to infiltration was proposed as part of the water and sewer master plan. Specific approaches were recommended:

- In housing areas, stormwater from private surfaces must be infiltrated.
- In commercial and industrial areas, stormwater from roofs must be infiltrated.
- Infiltration areas must take place as close as possible to the source.
- Stormwater management must preserve the natural water cycle and not affect the water quality.
- Stormwater facilities must fit into their urban surroundings as well as possible.

Figure I.5.9 **Roof garden.**

As an example of this approach for residential development, houses are required to have private infiltration pits, cisterns, or rain barrels to divert rainwater from roofs away from the public drainage system. Residents are responsible for constructing and maintaining these systems. Overflow from the private systems, as well as water from public streets and driveways, is collected in a similar system along the street before entering the collector system. Education and maintenance programs were included in this approach. Over 2,000 private devices were inspected. The maintenance program also included the development of a standard system for the retrofit and rehabilitation of older systems.

Aesthetic, educational, and social lessons

One of the most powerful lessons of microscale applications is the adaptability, sustainability, and multiplicity of other functions that microscale approaches have. For example, green or eco-roofs, which have been in place for many years in Europe, and are now being used and accepted as stormwater controls for residential areas such as Portland, OR, and have added energy conservation benefits. The roofs have been modified to provide the most energy efficient and water retention properties feasible. Figure I.5.9 shows an example of a roof garden.

Microscale stormwater facilities have been designed as small areas of relief and berms in open space and courtyards, pedestrian streets with rainwater channels, fountains, and pavement relief. The experience of other countries has shown that the opportunities, size, function, and appearance of microscale controls are limitless. These techniques not only constitute stormwater facilities but also have become an expansion of the aesthetic components of the urban fabric. Figure I.5.10 shows examples of rain gardens used in parking lots, which are good examples of multifunctional, multibeneficial, and cost-effective use of space with LID practices.

Once educated about the experience, importance, and potential of these microscale techniques in protecting property values, meeting environmental objectives, and improving the quality of life, communities, stakeholders, engineers, and property owners will initiate programs to develop and implement these strategies as part of their receiving water's protection programs.

Figure I.5.10 Examples of parking lot bioretention (rain gardens).

Costs

LID case studies and pilot programs show at least a 25 to 30% reduction in site development, stormwater, and maintenance costs for residential development. These data were obtained for the Prince George's County LID case study on Patuxent Riding subdivision (PGCDER, 1998). This is achieved by reducing clearing, grading, pipes, ponds, inlets, curbs, and paving. This, in turn, lowers construction costs, allowing builders to add greater value (features) to the property or to be more flexible and competitive in pricing their products.

One of interesting results of LID's on-lot microscale approach is that the stormwater management controls become a part of each property owner's landscape (natural areas, rain gardens, open space, open swales, etc.). This reduces the public burden to maintain large centralized management facilities and reduces the cost and scale of maintenance to a level the homeowner can easily afford — the cost of routine landscape/yard care and pollution prevention.

Prince George's County does not rely on enforcement to ensure maintenance of LID landscape practices. Instead, they believe the economic incentive of maintaining property values will ensure that most property owners will adequately maintain their LID landscape.

Roadblocks to LID

A number of roadblocks must be overcome for the successful implementation of LID. Regulatory agencies, the development community, and the public may all have concerns about the use of new technology. In the development of the PGCDER LID design manual, a multiagency task force spent over two years to address all the concerns and issues. Some of the major concerns include:

1. Develop a hydrologic analytical methodology to demonstrate the equivalence of LID to conventional approaches.
2. Develop new road standards that allow for narrow roads, open drainage, and use of bioretention.

3. Streamline the review process for innovative new LID designs, which allows for easy modification of site, subdivision, road, and stormwater requirements.
4. Develop a public education process that informs property owners about how to prevent pollution and maintain on-lot LID BMPs.
5. Develop legal and educational mechanisms to ensure BMP maintenance.
6. Demonstrate the marketability of green development.
7. Demonstrate the cost benefits of the LID approach.
8. Provide training for regulators, consultants, public, and political leaders.
9. Conduct research to demonstrate the effectiveness of bioretention BMPs.
10. Limited field monitoring data is available to demonstrate the effectiveness of LID in controlling runoff quantity and quality.

Summary

LID is a viable, cost-effective alternative approach to stormwater management and the protection of natural resources. LID is designed to provide tangible economic incentives to a developer to save more natural areas and reduce stormwater and roadway infrastructure costs. LID can achieve greater natural conservation by using conservation as a stormwater BMP. As more natural areas are saved, less runoff is generated and stormwater management costs are reduced. This allows multiple uses of landscape features to achieve environmental, economic, aesthetic, and natural resources benefits.

Additionally, developers have economic incentives to provide better environmental protection by reducing short- and long-term infrastructure costs by reducing impervious areas and eliminating curbs/gutters and stormwater ponds to achieve LID stormwater controls. Reduction of the infrastructure also reduces infrastructure maintenance burdens, making LID development more economically sustainable. LID allows for the same, or in some cases, higher lot yields compared to conventional approaches. Because stormwater management is controlled on each lot using multifunctional landscapes, that portion of the building area that would have been used for stormwater ponds can, in some cases, be used for additional flood control or recovered and used for building, parking lots, open space, or habitat enhancements.

LID promotes public awareness, education, and participation in environmental protection. As every property owner's landscape functions as part of the watershed's ecosystem, they must be educated on the benefits and the need for maintenance of the landscape and pollution prevention measures. LID developments can be designed in a very environmentally sensitive manner to protect streams, wetlands, forests, and habitat and to save energy. The unique environmental protection objectives of an LID development can create a greater sense of community pride based on environmental stewardship.

In the development of the LID hydrologic analysis, NRCS/SCS TR-55 (SCS, 1986) was used because this model is most widely used by site engineers in Prince George's County and throughout most of the country. During the development of LID, it was learned that current analytical models such as TR-55 are not well suited for use with very small watersheds. A significant amount of work is needed to upgrade current models to better quantify the effects of microscale site design control techniques.

One extremely fascinating aspect of LID is that when controlling runoff on a microscale, there exists a whole new world of possible control practices and strategies. So get out of the box of conventional pipe and pond technology, take up the LID challenge, and try thinking small.

Literature cited

Alley, W.A. and Veenhuis, J.E., Effective impervious area in urban runoff modeling, *J. Hydrol. Eng.*, ASCE 109, 313, 1983.

Arnold, C.L., Boison, P.J., and Patton, P.C., Sawmill Brook: An example of rapid geomorphic change related to urbanization, *J. Geol.*, 90, 155, 1982.

Arnold, C.L. and Gibbons, C.J., Impervious surface coverage: The emergence of a key environmental indicator, *J. Amer. Plann. Assoc.*, 62, 243, 1996.

Barton, D.R., Taylor, W.D., and Biette, R.M., Dimensions of riparian buffer strips required to maintain trout habitat in southern Ontario streams, *North Amer. J. Fish. Manage.*, 5, 364, 1985.

Beyerlein, D., Effective impervious area: The real enemy, *Proc. Impervious Surface Reduction Conf.*, City of Olympia, WA, 1996.

Booth, D.B., Stream-channel incision following drainage-basin urbanization, *Water Resour. Bull.*, 26, 407, 1990.

Booth D.B., Urbanization and the natural drainage system — Impacts, solutions, and prognosis, *Northwest Environ. J.*, 7, 93, 1991.

Booth, D.B. and Reinelt, L., Consequences of urbanization on aquatic systems measured effects, degradation thresholds, and corrective strategies, *Proc. Watershed '93 Conf.*, 1993.

Booth D.B., Stream channel geometry used to assess land-use impacts in the PNW, *Watershed Prot. Tech.*, 2, 345, 1996.

Cheng, M.S, Coffman, L.S., and Clar, M.L., Low *Impact Development Hydrologic Analysis*, Prince George's County, MD, 2001.

Coffman et al., Kettering Demonstration Project — Final Report Public Education and Public Participation, Prince George's County, MD, 1998.

Graf, W.L., Network characteristics in suburbanizing streams, *Water Resour. Res.*, 13, 459, 1977.

Galli, J. 1991. Thermal impacts associated with urbanization and stormwater management best management practices, Metropolitan Washington Council of Governments, *MD Dept. Environ.*, Washington, DC, 1991.

Gregory, S.V., Swanson, F.J., McKee, W.A., and Cummins, K.W., An ecosystem perspective of riparian zones: Focus on links between land and water, *Biosci.*, 41, 540, 1991.

Horner, R.R., Booth, D.B., Azous, A.A., and May, C.W., Watershed determinants of ecosystem functioning, *Proc. ASCE Conf.*, Snowbird, UT, 1996.

Horner, R.R. and May, C.W., *Regional Study Supports Natural Land Cover Protection as Leading Best Management Practice for Maintaining Stream Ecological Integrity*, Univ. Washington, Seattle, WA, 1999a.

Horner, R.R. and May, C.W., *Watershed Urbanization and the Decline of Salmon in Puget Sound Streams*, Univ. Washington, Seattle, WA, 1999b.

Klein, R.D., Urbanization and stream quality impairment, *Water Resour. Bull.*, 15, 948, 1979.

Leopold, L.B., *The Hydrologic Effects of Urban Land Use: Hydrology for Urban Land Planning — A Guidebook of the Hydrologic Effects of Urban Land Use*, USGS Circular 554, 1968.

Macrae, C. and Marsalek, J., The role of stormwater in sustainable urban development, *Proc. Canad. Hydrol. Symp.*, Winnipeg, Canada, 1992.

National Park Service, *Ecosystem Management in the National Park Service*, U.S. Dept. Interior, Sept. 1994.

Prince George's County, MD, Dept. Environ. Resour., *Low-Impact Development Design Manual*, PGC, 1997.

Prince George's County, MD, *Patuxent Riding Low Impact Development Case Study*, 1998.

Prince George's County, MD, Dept. Environ. Resour., *Low-Impact Development Design Strategies: An Integrated Approach*, Jan. 1999.

Schueler, T.R., The importance of imperviousness, Center for Watershed Protection, *Watershed Prot. Tech.*, 1, 100, 1994.

SCS, *Urban Hydrology for Small Watersheds*, Technical Release 55, U.S. Dept. Agric., Soil Conservation Service, Eng. Div., Washington, DC, 1986.

Weinstein, N.K.P., *Low Impact Development for Combined Sewer Overflow Control*, U.S. EPA, 2000.

Wigmosta, M.S., Burgess, S.J., and Meena, J.M., 1994. Modeling and monitoring to predict spatial and temporal hydrologic characteristics in small catchments, USGS Water Resour. Tech. Rep. no. 137, 1994.

Response

Thinking big, acting small:
Multi-tasking and the benefits of dispersed micromanagement

Prince George's County, MD, is the originator and leading implementer of low-impact development (LID). As Larry Coffman outlines in this chapter, LID is based on the simple strategy of "nature knows best." In other words, the best we can do is to try to imitate nature as closely as possible when undertaking water sensitive designs. At its core, LID is a system of land development principles and practices based on creating a more hydrologically functional urban landscape to protect water bodies. In terms of scale, LID represents an important shift away from end-of-pipe, centralized, and costly solutions to stormwater management toward decentralized micromanagement at the scale of individual house lots or parking lots.

Coffman identifies a whole suite of conservation strategies, impervious minimization measures, phased timing techniques, and integrated small-scale BMPs that will accomplish these objectives. In terms of design, the goal of LID is to create as many small-scale retention, detention, pollution prevention, and treatment technologies as possible for any site. Also discussed in this chapter is the systematic stepped approach from planning to final design in terms of implementing LID techniques. Another benefit from undertaking such water sensitive management is a reduction in costs compared to conventional, highly engineered technologies. Similar findings are predicted by Jennifer Zielinski in Chapter I.3 and shown by Michael Clar in Chapter I.15. Because LID operates at such small spatial scales, it is very amenable for application to water sensitive retrofits made to aging infrastructure in dense urban settings, as proposed by Richard Pinkham and Timothy Collins in Chapter I.4.

Coffman stresses that, though site-specific in design, LID approaches must be placed in the broader spectrum of comprehensive watershed planning. A very important point that Coffman raises is the controversial nature of LID, representing as it does, a way of turning traditional water management on its head. At least for the near future, until a more general acceptance of the procedures becomes established, water sensitive designers proposing such techniques must recognize that they are challenging the existing paradigm (for example, see the case study described by Robert France and Philip Craul in Chapter I.7). On a broader scale, it is possible that building codes may have to be reexamined before LID can be applied. This underscores an important point raised in this chapter — the need for client education.

In the end, as Coffman informs, LID offers microscale approaches to water sensitive design that are adaptable and sustainable and have multiple functions in addition to stormwater management. He even goes as far as to suggest that with the widespread adoption of LID methods, the very term "BMPs," or best management practices, might be better replaced by "IMPs," or integrated management practices.

chapter I.6

Water gardens as stormwater infrastructure (Portland, Oregon)

Thomas Liptan and Robert K. Murase

Abstract

Water gardens are aspects of the built environment that emulate nature's processes. Water is not a feature unto itself, but an integral element of the site and architecture. Water gardens are a synergetic result of landscape, biology, architecture, and engineering. The principle of integrating water in urban design introduces water as a friendly companion, but always with attention to its potential power and negative aspects. This symbiotic design principle includes soil and vegetation within the urban hardscape. Trees, plants, and soil are employed to function with water in urban spaces previously not used for stormwater management. We might call this a new paradigm, urban-nature, not at the expense of our human habitat . . . but as enhancement of this habitat . . . earth, water, plants . . . all have an artful place with people in the urban context. Documentation of tests, monitoring data, costs, and observations show the viability of the techniques discussed. Application of this new paradigm will result in lower development costs, more cost-effective retrofits, improved livability, and a more natural urban environment.

Introduction

Urban development provides the essentials of human community life. Usually within this human community exists some degree of nature. These human essentials are, to a large extent, forms of impervious surfaces and pipes. Stormwater runoff, the physical phase of precipitation after it falls in the urban community, is almost foreign to the natural environment. The cause of this runoff comes from the impervious surfaces needed within the urban community. To reverse, mitigate, or eliminate the negative effects of stormwater runoff, new ways of designing or retrodesigning the urban community are being explored and tested. Many of these new ideas are actually modern applications of age-old approaches, which for some reason had faded away. Maybe the resurgent interest in bringing aspects of nature back into the community is caused by some of the federal laws governing water, air, and threatened species. Certainly within the Pacific Northwest, water and nature are synonymous with salmon and forests. It is also becoming apparent that the building blocks of nature are also the building blocks of a healthy urban community. These new ideas begin to take shape in the form of urban design techniques, which are methods of integrating water with land and vegetation. Water gardens, or perhaps better stated as eco-gardens, are an aesthetic and cost-effective approach to design with water.

In Portland ideas that regreen or mitigate the effects of impervious surfaces are being developed. These approaches include development of a healthy *urban* forest, revegetation and preservation of riparian corridors and habitat, identification and removal of unnecessary impervious surfaces, improved street designs to reduce environmental impacts, improved zoning codes to reduce hard surfaces, and identification and implementation of green/sustainable building and site design practices. The physical characteristics of impervious surfaces are essentially rooftops and pavement. These surfaces are at best environmental dead zones but are certainly not benign, because they have other than direct water impacts, such as contributing to urban heat island conditions, smog, loss of wildlife and habitat, increased carbon dioxide, and reduced oxygen and photosynthesis.

So what paradigm best describes these new ideas or approaches? Simply stated, it is the careful integration of water with site and architectural design. The applications of design elements allow the urban hydrology to better mimic nature. The design elements are those usually within the landscape architects purview, including soil, plants/trees, rock, and wood with stormwater added.

This chapter introduces scientific information obtained either from direct testing or by recently published technical literature. Next, design techniques are presented with reference to their functions and benefits. Projects that have employed one or more of these techniques are presented. Each demonstration project offers built examples. It is expected that these techniques will improve with each new application. Projects are discussed with a description of the site, techniques employed, and commentary on what works and what does not.

Studies: Pacific Northwest precipitation and hydrology

Portland rainfall analysis

Most precipitation in the Pacific Northwest occurs in the form of small storms. Using U.S. Weather Service data, an analysis of 24-hr precipitation events was conducted. Storms were divided into sizes from 0 to 0.2 in., 0.2 to 0.4 in., 0.4 to 0.6 in., and so on, up to and including the largest storms during the past 16 years. To simplify the analysis at the risk of accuracy, a 24-h event was based on calendar-day rain totals. Figure I.6.1 indicates that, on average, 145 days per year had storm events, or calendar days of rain, of 1.0 in. or less. These storm events accounted for 81% of the average annual rainfall. This characteristic of precipitation is an important part of understanding how to better manage urban runoff and pollutants associated with these numerous events. Limitations of this study include the lack of distinction between calendar days, i.e., 0.2 in. could fall at midnight and another 0.8 in. could fall in the hours immediately following on the next day. Albeit simple, it is not encumbered by the debate of what antecedent dry period should be used to define the beginning and end of a "storm" event. Even so, the information is relatively consistent with other studies, such as Chapter I.1.

Comparison of predevelopment forests versus post-development peak flows and volumes

An analysis of rainfall distribution was conducted to determine the pre- and post-runoff from storm events using 0.1 in. increments, starting with 0.1 in./24 h events and continuing every additional 0.1 in. up to 2.4 in. (the Portland 2-year event). The Santa Barbara Urban Hydrograph (SBUH) method was employed for these calculations. Here again, this

Chapter I.6: Water gardens as stormwater infrastructure (Portland, Oregon)

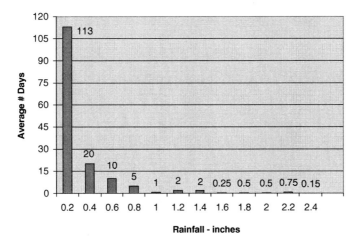

Figure I.6.1 Annual rainfall 1983–1998. Average event frequencies. (Source: BES Hydradata.)

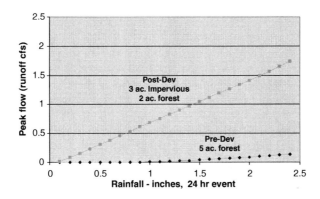

Figure I.6.2 Pre-/post-development comparison. (Source: SBUH method, perv. Cn 70/imp. Cn 98.)

may not be the most accurate method to use, but the city has currently approved it for sizing storm facilities. A hypothetical site of 5 acres was used, with predevelopment assumed to be a forest condition with a curve number CN70. Post-development was 3 acres of impervious surface with a curve number CN98, and the other two acres assumed to be left undisturbed but hydrologically connected to the new storm system. Results for the predevelopment condition indicate that no runoff occurs for storms of 0.0 to 0.8-in. 24-h events. Figure I.6.2 shows that predevelopment peak flows do not begin until at least a 0.9-in. event occurs and then have peak flow rates of up to 0.14 cfs (cubic feet per second) for the 2-year, 24-h event. For post-development conditions, runoff occurs with the first 0.1-in. event. For 0.3-in. events, the post-peak flow is 0.16 cfs and then increases to more than 10 times the predevelopment peak for the 2-year event. Figure I.6.3 shows a comparison of runoff volumes, with predevelopment at *zero* cf. (cubic feet) for all storms up to 0.8 in. and 7,396 cf. for the 2-year event. Post-development discharge volumes were 383 cf. for the 0.1-in. event and 27,810 cf. for the 2-year event. In the predevelopment condition, little, if any, runoff occurs for storms of less than 1.0 in. In the post-development condition, runoff occurs with each event of 0.1 in. or more.

Bolton and Watts (1998) state, "Very little precipitation ends up as overland flow in a mature, undisturbed forest." At the same conference, Beyerlein and Brascher (1998) state that in the Puget Sound area with annual rainfall at 40.7 in., 18.8 in. evapotranspirates and

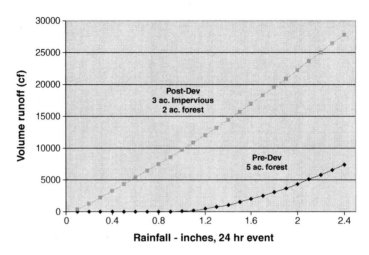

Figure I.6.3 Pre-/post-development comparison. (Source: SBUH method, perv. Cn 70/imp. Cn 98.)

only 0.1 in. precipitation becomes surface runoff. Based on the above SBUH analysis, prior to development, no surface runoff occurs in a conifer forest for over 100 storm events per year, on average. This would also indicate that the energy of water within streams is only affected by surface runoff occasionally. In addition, the pollutants carried into streams from surface flows would also occur occasionally. Surface runoff for post-development conditions occurs more than 100 times a year and carries not only naturally occurring pollutants, but also human-generated pollutants. Energy from the storms is also conveyed to the stream with peak flows equal to the 2-year event occurring with >0.3 in. events. Basically, there's a "whole-lotta" runoff and pollutants that did not previously occur, and it is happening almost every time it rains. Note: conventional detention methods (vaults and ponds) would reduce only the energy of the higher peaks, but not the 100 times per year they occur. Another factor of detention would be the extended duration of time needed to get the higher volumes out of the facility.

In a post-development condition, significant flow, volume, and energy are entering urban streams. Because predevelopment pollutant loadings are zero for these predominant events, then any post-development discharge, regardless of the constituent concentration, would be greater than the predevelopment reference point. In the post-development condition, urbanized sites discharge pollutant loads of many orders of magnitude above predevelopment conditions and occur more often. If there is no runoff, then where does the rainfall go? According to Bolton and Watts, and others, much of it is intercepted in the tree foliage, bark, and branches, which then evaporates. Some rain falls to the forest floor, where it is absorbed and then makes its way into the ground and gradually seeps into streams.

Portland Bureau of Environmental Services (BES) test swales

These swales were constructed to test various design characteristics that, at the time, had not been documented by anyone else (at least not to the knowledge of BES). Issues of concern include pollutant removal or capture efficiencies of different plant species, soils, and flow attenuation and velocities. Additional issues may be tested in the future. Both swales are identical in geometric shape and soil type. Tests to date have been based on a difference of vegetation in each swale. One swale is planted with native grasses and forbes, and the other swale is planted with turf grasses. The turf grass was mowed regularly, and the native vegetation was left to grow naturally. A portion of stormwater runoff from a

Chapter I.6: Water gardens as stormwater infrastructure (Portland, Oregon)

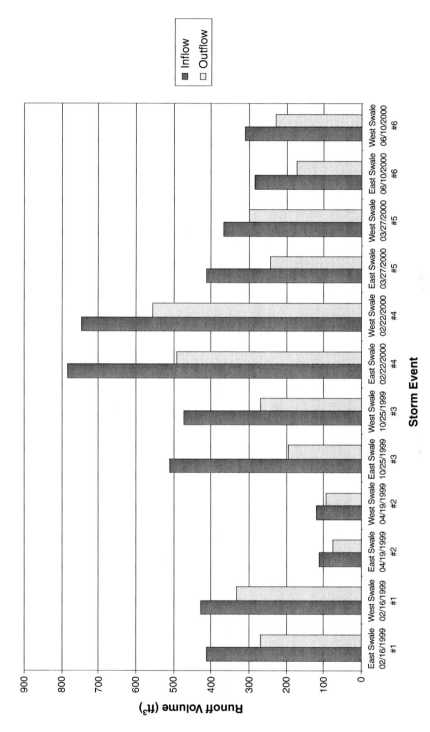

Figure I.6.4 Comparison of test swale inflow and outflow volumes.

	East Swale	West Swale
Grab Parameters		
pH (field)	3	4
Dissolved oxygen (field)	20	25
Temperature	4	3
Conductivity (field)	−13	−3
Total oil and grease	32	31
Nonpolar oil and grease	50	34
Composite Parameters		
Total suspended solids	81	69
Total dissolved solids	28	18
Total solids	60	51
COD	65	52
Total Kjeldahl nitrogen (TKN)	54	40
Total phosphorus	50	38
O-phosphate-phosphorus, DISS	−75	−45
Nitrate-nitrogen	16	8
Hardness	46	33
Cadmium, total	73	61
Copper, total	65	53
Lead, total	72	62
Zinc, total	76	63
Cadmium, dissolved	47	50
Copper, dissolved	52	38
Lead, dissolved	53	36
Zinc, dissolved	64	48

Figure I.6.5 Test swale percent pollutant load change, average six storm events.

50-acre urban area is pumped into each swale with almost identical volumes. Flow meters measure the flow at the end of each swale. Three pollutant samples are taken at each swale every 30 min during a storm event and then combined for analysis. A total of 6 events over the last 2 years have been sampled. Results to date indicate the following.

Runoff attenuation occurs in both swales. The swale with native vegetation retains up to 41% of the flow and the turf swale retains about 27%. No identifiable conditions exist to explain the difference; it is assumed that the native vegetation and lack of mowing allow the swale to facilitate infiltration. This may be due to the robustness of the root systems and presence of more organic material in the native vegetation swale versus the turf grass swale. Figure I.6.4 shows the flow comparison for each storm event.

Pollutant removal efficiency is good during all seasons but is better for warm-season load reductions. Generally, the swale planted with native grasses captures more pollutants than the turf swale, except for O-phosphate-phosphorus. This may be due to accumulation of organic matter in the swale, whereas grass clippings were removed from the turf swale. Recent storms monitored don't show a measurable difference in swale performance even though the turf has not been mowed for over a year and the inflow has been increased from 0.04 cfs to 0.08 cfs. Figure I.6.5 shows that both swales perform relatively well.

Summary

Pollutant removal relative to concentrations is good; for example, average total suspended solids (TSS) removal is 59% for turf and 68% for native vegetation. When loads are calculated based on runoff volume captured in each swale and concentration removal, the TSS removal percentages are 69% for turf and 81% for native vegetation. Vegetation maintenance is not necessarily required. A messy or somewhat natural-looking planting does not indicate the stormwater management functions have been impaired.

Design techniques or "green solutions"

Techniques are being developed nationwide by many who are helping to shape this new paradigm (Chapters I.1 and I.5). Water, soil, and vegetation are purposely introduced into site and building elements previously isolated from each other. Four basic functions result: water soaking into soil/vegetation; water flowing over soil/vegetation and inanimate objects; water transpired by vegetation; and precipitation intercepted by vegetation and evaporated (water even evaporates during dry periods of storm events; see BES tree study, 2001). The combination of these functions is often advantageous to achieve better water distribution. Pollutant capture is achieved by water having to filter through the soil and/or vegetation. Atmospheric pollutants are captured in plant foliage and then trapped in the soil, where many of them have an opportunity to break down. All these approaches allow runoff to be diffused, which also allows pollutants to be distributed in the landscape instead of concentrated. Increases in urban air and water temperature are minimized or eliminated by shade on impervious surfaces and, in the case of eco-roofs, nearly all precipitation is retained during warm-season months (May to October). The physical forms of these techniques are almost infinite, but are described here as:

Eco-roofs — Figure I.6.18
Stormwater planters — Figures I.6.13 and I.6.14
Infiltration garden — Figure I.6.6
Landscape swales — Figure I.6.11
Vegetative filters — Figure I.6.28

Portland demonstration projects

Buckman Heights

A redevelopment project in the combined sewer area was designed in 1996 and opened in 1998. The site's previous use was for a new-car dealership parking lot. The site is 2 acres, with a 150-unit, 4-story apartment building and some surface parking and underbuilding parking. The project has many environmental attributes including car sharing for the tenants. The owner, Prendergast Associates, wanted to do its part to remove the site's runoff from the combined sewer system. The buildings are organized around a main courtyard; the traditional layout is articulated with low seating walls off the sidewalk and two large planting beds designed as landscape infiltration areas to filter and absorb the stormwater from the building's downspouts. The parking areas are designed with care and detail to reduce the presence of the automobiles and absorb the water runoff from the paved surface.

Landscape infiltration

Courtyard. Figure I.6.7 shows two 18 ft × 45 ft infiltration gardens integrated with the site. The gardens were designed to accept runoff from the rooftops and the surrounding courtyard paved areas to flow into the vegetation. The planter area tapers from 6 in. along the perimeter of the surrounding walkways to 18 in. at the center. Moisture-tolerant plants of spirea, iris, Oregon grape, and astibe were planted within a Japanese holly border.

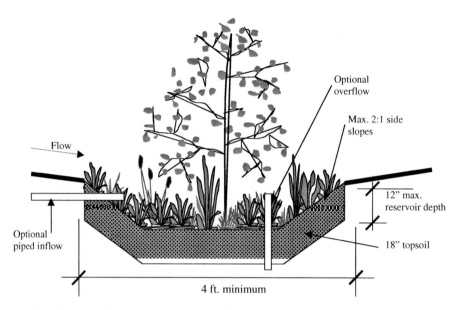

Figure I.6.6 Landscape infiltration.

Section Not to Scale

Description: Landscape infiltration areas can be integrated into the site landscaping. The design can be formal or informal in character. They may be used in courtyards, parking lots, or where other planting areas are available. Although the area is saturated during storm events, infiltration occurs quickly.

Stormwater management function and sizing: The system works by holding runoff and allowing pollutants to settle as the runoff infiltrates. Flow and volume are also managed with these facilities. Allows evapotranspiration and groundwater recharge and retains warm weather runoff. Depending on soil type and infiltration rate, this facility may provide 10- to 25-year event disposal. Using above proportions size at 0.045 × impervious area for the 0.9 in., 24-h storm event.

General specifications: Acceptable soil types A & B. Minimum soil infiltration rate of 2 in./h. Facility storage depth may vary from 2 to 12 in. Filters can be planted with a variety of trees, shrubs, and ground covers, including grasses appropriate for periodic inundation. Freeboard not required.

Figure I.6.8 displays an overflow pipe set 9 in. above the bottom of the basin. Runoff enters the landscape infiltration area and soaks into the soil, except for large storms that flow out the overflow. This keeps the area at a maximum depth of water during storms of 9 in. with percolation rates estimated at 2 in./h; the area is expected to drain within 5 h.

What have we learned? Aesthetically it is a very pleasing design. Based on visual observations, infiltration rates appear to be increasing. Infiltration tests will be conducted this spring to confirm these observations. If true, then the overflow inlet can be raised several inches to allow more runoff to be infiltrated. Vegetation appears healthy and is growing well. Pruning is done annually in the fall to maintain the overall desired aesthetic. From a functional perspective, the pruning could be reduced. No fertilizers or pesticides will be needed, although the entire site has an irrigation system. Piping from the downspouts to the planter has a tendency to clog with sediments from the roofing materials and tree leaves. This requires at least annual cleaning to maintain good drainage; however, during a very intense 1.4 in./3-h storm on October 1, 2000, the pipes were full and runoff

Chapter I.6:   Water gardens as stormwater infrastructure (Portland, Oregon)   133

Figure I.6.7 Buckman Heights courtyard infiltration garden. (Color version available at www.gsd. harvard.edu/watercolors. Password: lentic-lotic.)

Figure I.6.8 Buckman Heights courtyard infiltration garden, with outlet riser. (Color version available at www.gsd. harvard.edu/watercolors. Password: lentic-lotic.)

simply overflowed across the lawn and walkways into the planter. Two alternatives, and probably better designs, would have been shallow surface channels integrated with the walkway design or larger pipes.

Parking lot.   Figure I.6.9 shows the uniquely designed landscape infiltration perimeter. Perimeter landscaping is required by city code; for a project like this, it requires a 5-ft wide strip the entire length of the parking lot, 200 ft. The lot was one of the first to use substandard dimensions with a 20-ft-wide aisle and 17-ft-long × 8.5-ft-wide stalls.

Figure I.6.9 Buckman Heights parking lot landscape infiltration. (Color version available at www.gsd.harvard.edu/watercolors. Password: lentic-lotic.)

Parking is at 90°. Unique about this integrated landscape is the lack of a freeboard (precautionary measure to contain flows). When considering the configuration and grading of the site, a freeboard was not needed to protect property or people. This allowed a smooth transition from the pavement edge into the planting area. Runoff flows into the landscape via curb cuts and then infiltrates into the relatively porous soil. A 2-in. storage capacity is available for runoff as it soaks in or for large storm flows to move by displacement toward two inlets at each end of the strip that drain into underground dry wells. Plantings include only Red Sunset maple trees (*Acer rubrum*, "red sunset") and Oregon grape as a hedge. Approximately 8,000 ft^2 of concrete surface drains into the landscape, which is about 1,000 ft^2 in area. It is estimated that this area will infiltrate the 10-year storm event (3.2 in./24h).

What have we learned? The plantings look good and the runoff is captured in the landscape area. Visual observations to date have indicated the area will hold at least a 2-year storm event (2.4 in./24 h), as occurred in November 1998. In one area the pavement was not sloped at the specified 2% gradient and causes some ponding in the corner of one car stall. Grinding the existing curb cut about ¾ in. will allow the water to drain off. The lack of a freeboard has not posed any concerns. Figure I.6.10 shows one of the curb cuts that must be kept clear to allow flow passage. These curb cuts have not been cleaned since installation and are still relatively unobstructed.

Buckman Terrace

Buckman Terrace is another redevelopment project by Prendergast Associates and is across the street from Buckman Heights. The project was designed in 1997 and opened in 1999. This is a 0.8-acre site with 150 apartment units, all with under-building parking, and a 1,500 ft^2 commercial section in a 4-story structure. The building also has car sharing and numerous other environmental attributes. An eco-roof has been installed on the commercial portion and another at the main entrance of the building. Landscape planters are

Figure I.6.10 Buckman Heights parking lot curb cut. (Color version available at www.gsd.harvard. edu/watercolors. Password: lentic-lotic.)

being used on the east side, and a landscape swale has been installed on the west side of the building. The landscape techniques integrate lush, moisture-tolerant planting with the function of stormwater quality and environmental enhancement.

Landscape swales

Westside swale. Figure I.6.12 shows the westside swale adjacent to the building where all roof downspouts discharge. Approximately 13,000 ft^2 of rooftop drains into a 430-ft swale. The swale is 6 ft wide and 3 in. deep; it has rock check dams every 15 ft. The swale is sized to convey the 25-year storm event flows but also to provide detention of all storms up to the 10-year event. These storms are detained behind the check dams. All flows that exceed the infiltration capacity of the soil discharge to a catch basin at the downstream end of the swale. A pea gravel mulch was used to slow the flow and provide opportunities for sediment deposition and some infiltration. The swale gradient is approximately 2% until it reaches the last 100 ft, where it increases to 4%. Purposefully excluded from the design was the standard 12-in. freeboard. This was done for two reasons: first, the added safety of a freeboard was not needed since there is no possibility of damage to the building or adjacent property; second was aesthetics (i.e., because of the narrow area, a 12 in. deeper swale would have been unsightly and dangerous). Plantings include sedges, miscanthus, spirea, Oregon grape, and Japanese iris.

What have we learned? The plantings are attractive and add considerable interest to this side of the building. Visual observations immediately after a large storm event (1.4 in./3 h) were conducted on October 1, 2000. Design and construction quality is important for best water management. The check dams were incorrectly constructed parallel to the flow (they were supposed to be perpendicular), which caused peak flows to bypass and erode part of the pea gravel on the 4% slope section. The 4% slope section only had grasses planted in a single row and should have had three rows with triangular spacing. The pea gravel has worked successfully as mulch to protect the swale and help filter flows.

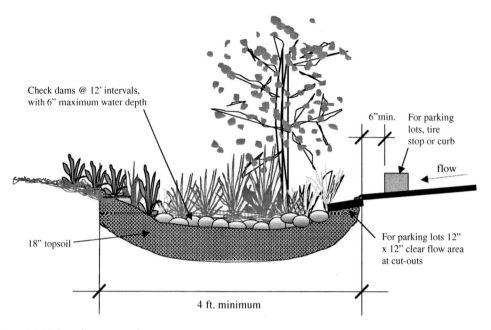

Figure I.6.11 Landscape swale.

Section Not to Scale

Description: Landscape swales are planting areas with a slight depression of up to 6 in. that allow runoff to enter, infiltrate, and flow through. They are usually long and narrow in width, which makes them well suited for parking lots and other narrow landscape spaces. Swales are constructed with a variety of trees, shrubs, grasses, and ground cover, depending on soil.

Stormwater management function and sizing: Swales capture pollutants as runoff is detained and absorbed in the soil, vegetation, and organic matter. Using above proportions, size at 0.05 × impervious area. Detention is provided for storms up to the 10-year event. Swales help mitigate runoff temperatures by retaining most of the runoff in warm seasons. Groundwater recharge occurs as check dams facilitate infiltration.

General specifications: Acceptable for all soil types. Minimum swale length is 20 ft. Maximum slope is 6%. Clay soils shall be amended with 50% sandy loam in the top 18 in. of the swale. Check dams to be of durable, nontoxic materials — i.e., rock, brick, and old concrete. Check dams shall be width of swale × 3–5 in. height. Swales using these design criteria need not bypass larger storms. Liners are not needed unless required for groundwater protection or to protect building foundations. Freeboard not required.

Stormwater planters

Eastside planters. Two designs were used here; Figure I.6.15 shows the north section planter is at grade and somewhat inconspicuous next to the building. Figure I.6.16 shows the raised planters in the south section at 18 to 36 in. above grade. Figure I.6.17 shows that runoff is directed into the planters via scuppers from the roof. Gravel soakage trenches accept water as it filters through the upper soil/vegetation portion of the planter. Plantings include Japanese spurge, iris, vine maple, and Oregon grape. The planters have more surface area than required for Portland conditions, and thus the reservoir space is only 2 in. deep.

Figure I.6.12 Buckman Terrace west side landscape swale. (Color version available at www.gsd.harvard.edu/watercolors. Password: lentic-lotic.)

What have we learned? Aesthetically, the design is equal to the original concepts considered without the use of water. Water is very visible to the tenants as they walk by. Because the soakage trench was within 8 ft of the building foundation, the owner and his consultants expended quite an effort to get building bureau approval. If the facility had been only 2 ft farther away, no special approval would have been required. In tight urban settings like this, the building bureau prefers the "Portland CD" planter. The CD unit is not designed to allow infiltration, other than some incidental amounts. Although very beautiful, the non-native plants might not perform as well as some native moisture-loving species.

Eco-roofs

The entire building has a roof area of approximately 25,000 ft^2, and the building is constructed with sufficient weight capacity to hold an eco-roof. As a test, eco-roofs were placed on two sections. Figure I.6.19 shows a 200-ft^2 eco-roof above the front entrance. A small, 25-ft^2 rooftop above drains into the eco-roof. Figure I.6.20 shows the main eco-roof over 1,500 ft^2 of commercial space, which has full solar exposure. An additional 750 ft^2 of impervious roof drains into the main eco-roof. Figure I.6.21 shows a closeup of Oregon sedum on the entrance eco-roof, which is also planted with sword fern, licorice fern, and white stonecrop. It is on the east side and in the shade of a north-facing wall. Both were planted in March 2000. The commercial eco-roof was planted with two species of Oregon sedum, various wildflowers, native grasses, and a few licorice ferns. Grasses and wildflowers were planted from seed, and mulch was hand broadcast to protect against wind erosion. Figure I.6.22 shows a globe flower (*Gilia capitata*) blooming on the main eco-roof. An irrigation system has not been installed for either eco-roof. The soil profile is 20 lb/ft^2 when saturated and 4 in. deep. An American Hydrotech waterproof membrane and reservoir drain system was used. BES staff specified the soil mix and vegetation.

What have we learned? Grasses and wildflowers achieve a graceful, flowing appearance. It is reminiscent of eastern Oregon or the Midwestern American prairie. During the

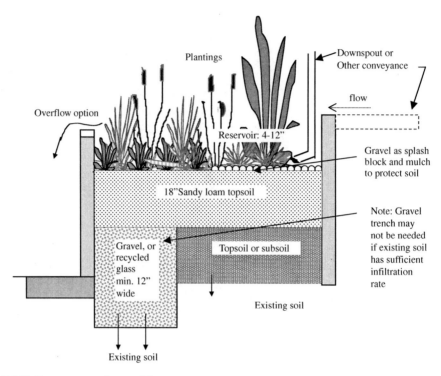

Figure I.6.13 Stormwater planter AB.

Section Not to Scale

Description: Planter AB is designed with a pervious bottom. The planter is used where infiltration is desirable. Planters are excellent for dense urban development.

Stormwater management functions and sizing: Planter AB is designed to allow runoff to filter through the planter soils and vegetation (thus capturing pollutants) and then infiltrate into the native soils (flow control). The planter is sized to accept runoff and temporarily store the water in a reservoir on top of the soil. Reservoir sizing, above the 18-in. topsoil, is for a 0.9-in., 24-h storm event. To calculate, use impervious area square feet (sf) × .045 = reservoir cubic feet (cf) of storage required. The infiltration gravel area can be designed to accommodate any storm event.

General specifications: Acceptable soil types A & B. There are numerous design variations. The planters shall be designed to allow captured runoff to drain out in 2 to 6 hours after a storm event. Plantings shall be appropriate for moist and seasonally dry conditions and can include rushes, reeds, sedges, iris, dogwood, currants, and numerous other shrubs, trees, and herbs/grasses. Topsoil shall have a minimum infiltration rate of 2 in./h. Sand/gravel area may not be required if existing soil has at least 5 in./h infiltration rate. The sand/gravel trench width, depth, and length are to be determined by a qualified professional. Minimum planter width is 30 in.; there is no minimum length or required shape. The structural elements of the planters shall be stone, concrete, brick, wood, or other durable material. If treated wood is used, it shall not leach out any toxic chemicals. Planters within 10 ft of structure will probably require special approval from the local building code agency.

warm season, storm event runoff was visually observed to be very low or nonexistent. The eco-roof had the capacity to hold much of the additional flow from the other roofs. During winter storms, runoff occurs often but is detained. Many of the plants survived

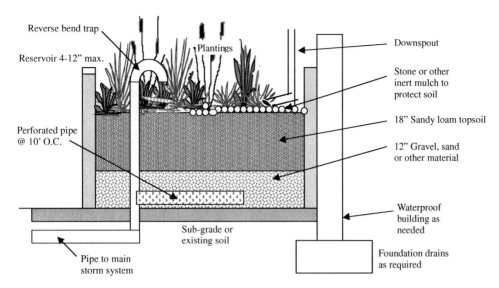

Figure I.6.14 Stormwater planter CD.

Section Not to Scale

Description: Planter CD is designed with an impervious bottom or is placed on an impervious surface. This planter is used where infiltration is not possible or desirable, such as unstable slopes or brownfields. Planters are excellent for dense urban development.

Stormwater management function and sizing: Pollutant reduction is achieved as the water filters through the soil; flow control is obtained by storing the water in a reservoir above the soil. Nominal infiltration can be allowed if soils and other geotechnical issues are addressed. The planter is sized to accept runoff and temporarily store the water in a reservoir on top of the soil. Reservoir sizing, above the 18-in. topsoil, is for a 0.9-in., 24-h storm event. To calculate, use impervious area square feet (sf) × .05 = reservoir cubic feet (cf) of storage required.

General specifications: Acceptable for all soil types. There are numerous planter design variations. The planters shall be designed to hold water for no more than 4 to 6 hours after a storm event. Plantings shall be appropriate for moist and seasonally dry conditions and can include rushes, reeds, sedges, iris, dogwood, currants, and numerous other shrubs, trees, and herbs/grasses. Minimum planter width is 18 in.; there is no minimum length or required shape. Topsoil shall have a minimum infiltration rate of 2 in./h. Sand/gravel shall have a minimum infiltration rate of 5 in./h. The structural elements of the planters shall be stone, concrete, brick, wood, or other durable material. If treated wood is used, it shall not leach out any toxic chemicals. Irrigation is optional, although plant viability shall be maintained.

or reseeded with only one hand watering. Although no maintenance was conducted this year, it appears the grasses will need to be mowed at least once a year. Should the owner prefer a different plant association, sedum might be added at some future time. It is very important to assure good vegetation coverage, especially over the lightweight soils to prevent wind erosion.

On another eco-roof project, the Hamilton Apartments in downtown Portland, almost an inch of soil was lost to wind erosion. Depending on the initial planting scheme, cover crops, such as common clover, may provide excellent soil coverage, which happened on a section of the eco-roof. Water from air conditioning condensate is a possible source of

Figure I.6.15 Buckman Terrace stormwater planter at grade. (Color version available at www.gsd.harvard.edu/watercolors. Password: lentic-lotic.)

Figure I.6.16 Buckman Terrace raised stormwater planters. (Color version available at www.gsd.harvard.edu/watercolors. Password: lentic-lotic.)

free, nonpotable water for irrigation. Condensate flows were significant during the hottest part of the summer, with flows measured at 12 oz/min in the afternoon and 6 oz/min in the late evening. This might prove to be a free source of irrigation water, if considered during the design phase. On this project, BES is testing to determine characteristics of planting methods, measurement of runoff flows and precipitation, and viability of soil and vegetation. Other issues may be addressed in the future. Figure I.6.23 shows the

Figure I.6.17 Buckman Terrace raised planter and scupper. (Color version available at www.gsd.harvard.edu/watercolors. Password: lentic-lotic.)

southeast quadrant of the eco-roof. Figure I.6.24 shows one of several stonecrop species in bloom last August. Figure I.6.25 shows some of the moss that colonized certain areas of exposed soil and helped reduce wind and soil erosion. Lightweight soils must be fully covered to prevent erosion. A Garland Co. waterproof membrane and planting design was used on this project.

Bureau of Environmental Services (BES) Water Pollution Control Laboratory

The project site was previously a 6-acre industrial site along the east bank of the Willamette River and adjacent to a city park on its north side. It was used for industrial activities for almost 100 years. The upland neighborhoods are a mixture of old residential and industrial land uses, with new residential conversion and infill occurring rapidly. The area was served by a combined sewer system that overflowed into the Willamette River about 75 times annually. BES installed a $5.4 million separated stormwater pipe system for 50 acres that drains to the water garden. At the time, treatment of separated stormwater was not required, but it was decided that a water quality pond would be constructed to reduce pollutants entering the river. Neighborhood citizens were very concerned that BES was going to create a problem pond; however, after considerable effort by the citizens and BES, a mutually agreeable plan was prepared, and a water garden, as it was eventually named, was constructed.

Simultaneously, BES was preparing to build a new Water Pollution Control Laboratory on another portion of the site. The new building would have laboratory, office, and field operations space, as well as a public meeting room for the neighborhood and others. The building footprint is approximately 12,000 ft^2, with parking for 60 cars, and 1.5 acres was allocated for the water garden. About one acre still remains vacant for potential expansion. The parking lots were designed with landscape swales (see subsection on BES test swales). The project was designed in 1995 and opened for use in 1997. Combining the best of the artistic and the utilitarian, the design transformed this once-industrial site into a meaningfully sculpted landscape, integrated with ecological processes. Each component of the site speaks to the inherent poetry of water and its role in our environment.

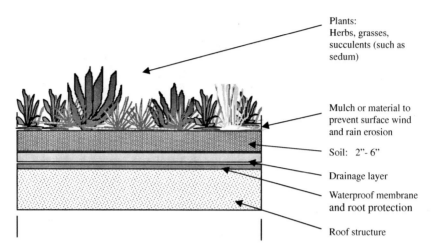

Figure I.6.18 Eco-roof (green roof).

Section Not to Scale

Description: An eco-roof is a lightweight roof system of waterproofing material with a thin soil layer and protective cover of vegetation. The eco-roof can be used in place of a traditional roof.

Stormwater management function and sizing: The eco-roof captures and then evapotranspirates 10 to 100% of precipitation, depending on the season. Roof runoff rates are significantly retarded because the rain must first soak through the soil before running off the roof. An eco-roof provides peak flow detention for storms up to the 10-year event. Eco-roofs mitigate runoff temperatures by retaining most of the runoff in warm seasons. Groundwater recharge can occur where roof drains flow to landscape areas. Sizing is equal to the square footage of eco-roof.

General specifications: Quality waterproof material appropriate for eco-roof application. Soil of adequate fertility and drainage capacity at depths of 2 to 6 in. Building structure adequate to hold an additional 10 to 25 psf weight. Self-sustaining vegetation, without the need for fertilizers or pesticides. Soil coverage to prevent erosion must be established immediately upon installation, by using mulch or protective blanket or vegetation mats (sod). Ninety percent plant coverage to be achieved within 2 years. Temporary irrigation to establish plants is recommended. Permanent irrigation systems using potable water may be used, but the water application shall not exceed 0.25 in. every 10 days for June–September season. Irrigation is not needed from October to May. Projects are encouraged to have alternative means of irrigation, such as cooling tower condensate or other nonpotable sources. For roof slopes greater than 10%, measures (such as geotextile webbing, and sleepers) shall be used to prevent soil slippage.

The water garden (pond)

From the very beginning, this facility had to be special. Attention to aesthetics and integration with the neighborhood was essential. The upland catchment contains approximately 40 acres of impervious surfaces and 10 acres of mixed pervious surfaces. Some of these include gravel yards for storage of heavy equipment. The pond was designed to accommodate the peak flow from a 0.83-in./24-h storm event, with a diversion structure to bypass large storm flows directly to the river (bypass flows are assumed to carry lower pollutant concentrations). Figures I.6.26 and Figure I.6.27 show the focal point of the design, a 1-acre pond formed from two converging circles. Elements include a circular stone wall to house the pond outlet structure, a 100-ft-long rock-filled concrete shute that conveys flow, yet provides an artful sculpture during dry weather, and a lushly planted pondscape that is integrated with the building landscape design and the adjacent park. The curvilinear flume in the upper cell is reminiscent of a glacial moraine. It slows the

Figure I.6.19 Buckman Terrace entrance eco-roof. (Color version available at www.gsd.harvard. edu/watercolors. Password: lentic-lotic.)

Figure I.6.20 Buckman Terrace main eco-roof. (Color version available at www.gsd.harvard. edu/watercolors. Password: lentic-lotic.)

stormwater while directing it into the detention cells. The cells are planted with a variety of aquatic and emergent plant material that naturally facilitates sedimentation and biofiltration of pollutants. Circular weep holes on both sides of the flume uniquely display the flow of water. The plantings include a mix of native and nonnative species, and include Oregon ash, red alder, red maple, several grasses, redtwig dogwood, Douglas spirea, Oregon grape, and numerous wetland species. An observation platform was designed as

Figure I.6.21 Buckman Terrace entrance eco-roof Oregon sedum (*Sedum oreganum*). (Color version available at www.gsd.harvard.edu/watercolors. Password: lentic-lotic.)

Figure I.6.22 Buckman Terrace main eco-roof globe flower (*Gilia capitata*). (Color version available at www.gsd.harvard.edu/watercolors. Password: lentic-lotic.)

an extension of the main spine of the building and extends over the water. Three monitoring stations were set up to measure flows and pollutants to determine pond efficiency.

What have we learned? Pollutants and flow have been monitored over 2 years for a total of 11 storm events. Generally, pollutant removal is good all year for most concentrations except cadmium, copper, and zinc. Some of these pollutants are seeping in from subsurface flows and surface flows during high-intensity storm events. Phosphorous and

Figure I.6.23 Hamilton Apartments eco-roof southeast quadrant. (Color version available at www.gsd.harvard.edu/watercolors. Password: lentic-lotic.)

Figure I.6.24 Hamilton Apartments eco-roof stonecrop. (Color version available at www.gsd.harvard.edu/watercolors. Password: lentic-lotic.)

nitrogen are also high on occasion. The pond remains wet all year, but summer discharges are small to nonexistent. Evaporation and soakage into surrounding soils are significant. Moderate amounts of excess irrigation water and unknown nonstormwater flows from the pipe system flow into the pond. Because the pond does not have a constant summer flow, however, the pond water occasionally becomes anaerobic and objectionable odors have been recorded. This problem remains to be addressed; one solution may be to drain

Figure I.6.25 Hamilton Apartments eco-roof moss. (Color version available at www.gsd.harvard.edu/watercolors. Password: lentic-lotic.)

Figure I.6.26 Water garden stone wall outlet structure. (Color version available at www.gsd.harvard.edu/watercolors. Password: lentic-lotic.)

the pond every summer. BES tried to accurately measure inflow, outflow, and bypass, but the subsurface and surface flows cannot be measured by the flow meter in the intake pipe. These uncontrolled sources also compromise the pollutant removal information. Education is needed in the neighborhood to reduce pollutants at their sources.

After construction an inspection of the diversion structure showed that the concrete dam had been installed in the wrong place and low flows were not entering the pond.

Figure I.6.27 Water garden curvilinear inlet flume. (Color version available at www.gsd.harvard.edu/watercolors. Password: lentic-lotic.)

The dam was immediately removed. It is very important to assure good design and construction quality control. Check the engineering and construction details. Concerns about warm-season water temperatures impacting the river are still being investigated, however, the pond does not discharge very much in the warm season. Sediments are accumulating in the rock flume and vegetation is starting to grow. It has been decided to let this continue and try to determine if an unmaintained pond loses its efficiency, although aesthetic pruning will continue. Many undesirable and potentially hazardous materials have been removed from the pond, including hypodermic needles, rat poison, trash, and debris. Use of a forebay to trap these things would have been desirable.

Another issue — the pond vegetation and presence of water — has been attractive to wildlife. Ducks are always around, and at least two families have nested there. The ducks sometimes feed in the bottom sediments where the pollutants are being trapped. It is yet to be understood whether this is a problem to the wildlife. Domestic dogs also like to play in the pond. Raccoons and other wildlife have been observed. Another note concerning fish and flooding: during the 1996 flood, the pond became a backwater for the river flows. Carp and potentially other fish found the pond and, as the flood receded, they were isolated from the river. When the hot summer of 1998 occurred, all the fish died. It is unknown whether any currently threatened species might have used the pond for refuge during the flood, but they surely could not survive the summer conditions. Designers should keep this is mind when proposing facilities near natural water bodies.

Vegetative filters

Figure I.6.29 demonstrates the use of scuppers as roof runoff cascades to the vegetative filter located on the south side and adjacent to the building. Most of the building has a metal roof of about 12,000 ft^2 that directs flow into a gutter with several steel scuppers. These scuppers allow the runoff to freefall into the garden area below. Large stone and rock are at the impact point of each freefall to diffuse energy and spread the water into the plantings. Planters have a mixture of lush ornamental, native, and wetland plants and are lined with crushed stone to provide a visually unique image year-round. Runoff is

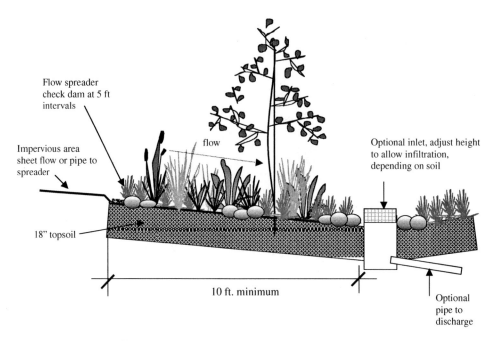

Figure I.6.28 Vegetative filter.

Section Not to Scale

Description: Vegetative filters are gently sloped areas. Stormwater enters the filter as sheet flow from an impervious surface or is piped and then converted to sheet flow using a flow spreader.

Stormwater management function and sizing: Flow control is achieved using the relatively large surface area and a generous proportion of check dams. Pollutants are removed through filtration and sedimentation. Using above proportions, size at 0.6 × impervious area.

General specifications: Acceptable for all soil types. Filters shall be a minimum of 20 ft. × 10 ft. Maximum slope is 10%. Check dams shall be of durable, nontoxic materials — i.e., rock, brick, and old concrete. Check dams shall be the width of the filter × 3–5 in. height. Filters designed using these criteria will not need to bypass larger storms. Runoff shall enter the buffer as predominately sheet flow. Check dams and flow spreaders are required. Filters can be planted with a variety of trees, shrubs, and ground covers, including grasses. Freeboard not required.

detained in the planters where some infiltration occurs. Large storm events have runoff that overflows to a catch basin.

What have we learned? It works great and looks good, too. Unfortunately, the planting areas are unnecessarily deep to allow for a nonessential freeboard. If a failure occurred at the catch basin inlet, the excess runoff would flow into the parking lot drains — an important reason to discontinue use of the freeboard, where they are not needed, is to allow more surface area for runoff to spread within the planting areas. The addition of a few check dams and raising the inlet grate a few inches will help achieve this goal.

The Oregon Museum of Science and Industry (OMSI)

This is an 18-acre redevelopment site, with a 100,000 ft² museum and exhibit space and parking lots for 700 cars. The project was Portland's first major demonstration of on-site parking lot stormwater management using 10 landscape swales to convey, infiltrate, and

Figure I.6.29 Roof scuppers and vegetative filter. (Color version available at www.gsd.harvard.edu/watercolors. Password: lentic-lotic.)

filter runoff. These swales were used in place of the originally proposed, conventionally raised landscape medians. Based on visual observations, as the soil and vegetation "mature," significant infiltration now occurs and runoff is only associated with large storm events. (See subsection on BES test swales.)

This project first came to the attention of the Portland Bureau of Environmental Services (BES) in 1990 when plans were submitted for review. At that time, neither the city of Portland nor the state of Oregon had specified site design requirements for stormwater quality discharges to the Willamette River; however, BES was gearing up to address forthcoming NPDES stormwater regulations and a combined sewer overflow problem of significant proportions, so clean rivers were, and still are, of concern. Following a review of the preliminary OMSI site plans, BES staff approached OMSI with the unprecedented request that it voluntarily redesign the parking lots and landscape to capture pollutants in stormwater runoff. This proposal did not affect the layout of the buildings and parking. The BES suggestion was to adjust site grading and change the already proposed landscape medians to accept rather than shed runoff.

OMSI was very interested in the environmental approach, but the nonprofit organization was under a tight budget and timeline. It agreed to change the design on the condition that the existing schedule be met with no overall increase in costs. The OMSI consultant team determined that even with the redesign fees, the related construction cost savings would result in a net reduction of project costs. Taking the environmental approach would actually be less expensive to construct.

The parking lots were initially redesigned to include grass/turf swales to filter out pollutants as the runoff traveled through them. OMSI took this a step further and required the landscape architects to improve the design to detain water longer and to incorporate native and wetland vegetation. OMSI considered this "mini linear wetland" concept a more attractive and educational approach for its new facility. Interpretative signs were installed to educate the public about the benefits of the swales and wetlands in improving water quality. The city, for its part, had established a special team from various bureaus to assist in moving the project smoothly through the city approval process. This took an

Figure I.6.30 OMSI Willamette riverbank. (Color version available at www.gsd.harvard.edu/watercolors. Password: lentic-lotic.)

enormous effort, because no existing policies or codes existed which allowed developers to use water quality site design techniques.

Figure I.6.30 shows some of the restoration and pedestrian improvements along the Willamette River bank. The site's western boundary is along the riverfront, which was stabilized with rock riprap and included intensive plantings of native riparian vegetation.

Landscape swales

Figure I.6.31 shows one of the ten swales. The swales are 6 ft wide and vary in length from 100 to 250 ft, for a total length of 2,330 ft. They were originally designed as biofilters but have continued to exhibit good infiltration characteristics. Check dams were installed every 50 ft to slow the flows and encourage infiltration. The parking dimensions were modified to allow more space for plantings and minimize impervious surfaces. (Special approval was required; minimum stalls were 9 ft by 18.5 ft.) Stalls are 8½ ft wide by 16 ft long with bumpers designed to overhang the curbs. Parking is at 90° and the aisle is 24 ft wide. Recently, the Portland city council approved a new parking lot code that allows 20-ft aisles and 9-ft by 16-ft stalls. Landscape space in front of the stalls is required at 6- to 8-ft widths to allow for stormwater management. No parking spaces are lost, because the code just provides more room for landscape and less for cars.

What have we learned? In 1996, the BES prepared a water quality audit and estimated that the bioswale system captures 50% of the average annual total suspended solids (TSS) loadings from the site. It was also estimated that with minor design improvements, such as additional check dams and more curb cuts, TSS capture would increase to 90% of the average annual site loading. Curb cuts were installed at 30 ft on center. Performance could be improved by installing curb cuts at 10 ft on center. These improvements have not been implemented because a major portion of the site was modified when a public street was constructed. Many of the improvements may still be made in the future. Visual observations continue to show that the swales allow much of the runoff to infiltrate.

Figure I.6.31 OMSI landscape swale. (Color version available at www.gsd.harvard.edu/watercolors. Password: lentic-lotic.)

Swale maintenance is incorporated into the normal operations of the site. The landscape medians were always intended to be a landscape feature of the site, which would require maintenance. The bioswales only require a little more attention. Curb cuts must be checked and cleared at least once a year. City staff have observed that some curb cuts were poorly constructed and/or located and need modification to reduce build-up of sediments, which blocks runoff from entering the swales. In hindsight now all swales were unnecessarily designed with 12-in. freeboards. Because the bioswales are oversized, it will take several decades before any accumulation of pollutants will need removal, if ever. The parking lot has significant use and lots of trash. Wind blows the trash into the swale vegetation, where it is trapped. This has been considered a desirable aspect of the swales, because the trash is somewhat camouflaged until OMSI staff removes it.

The success of OMSI demonstrates that water quality and stormwater runoff measures can fit into constrained spaces, save construction costs, and have an attractive appearance. The owner documented a savings of about $78,000 in construction costs. These savings were achieved through the reduction of pipe, manholes, and catch basins.

Custer Park

Custer Park is a small neighborhood park with a playground, open lawn, and softball and soccer fields. This was a 4-acre park renovation project, with high demand for a full-sized soccer field. Before the park was developed in the 1950s, a small seasonal creek ran through the site. Several seasonal springs also contributed to the small creek flows. The original 1950s park design called for the piping of the creek to drain the springs and stormwater runoff from an adjacent residential neighborhood upland of the park. The Parks Bureau's desire was to have the water out of site and out of the way, but Parks has been struggling with wet turf ever since. In 1996 and 1997, BES, Parks, and Murase Associates worked with very little budget and redesigned the site to daylight a 400-ft section of piping. Appearing as a daylighted swale, it runs adjacent to a new pathway and is planted with

Figure I.6.32 Custer Park landscape swale. (Color version available at www.gsd.harvard.edu/watercolors. Password: lentic-lotic.)

native hydrophytes and riparian shrubs and trees. A series of stone weirs slows the velocity of the flow, allowing sediment to drop out, while the plantings provide filtration. The swale terminates in a detention pond that removes much of the remaining pollutants from the flow. Although the swale provides improved water quality and makes visible those parts of the hydrologic cycle that are usually hidden, it also introduces habitat that is beneficial for wildlife and adds an additional layer of interest to park users. The completed project responds to the neighborhood desire to return nature to the park and provides a larger playing field as well.

Landscape swale

The site terrain has a grade change of 20 ft over a distance of 500 lineal ft. It was determined that only the city "water quality" storm event would be used to determine the bioswale design. These relatively small storm events would not pose as much risk to the park, and the larger storms are allowed to continue to flow within the original pipe system. A diversion structure was installed to direct small storm flows into a swale cut into the slope along one side of the park. At the same time the soccer field was extended up to the edge of a pedestrian path that runs adjacent to the bioswale on the other side. Figure I.6.32 shows the swale from the downstream end looking up toward the park.

What have we learned? It works great and looks good, too. Plantings were not sufficient to allow adequate coverage and establishment in the first two years but now are doing fine. The swale included an unnecessary 12-in. freeboard, which diminishes the aesthetics. Because the project is within an open area and a high flow diversion occurred upstream, the freeboard was overdesigned. A positive aspect of the excessive depth of the swale is its capacity to accumulate sediment for many decades without the need for maintenance.

Wrap-up

Water is the main theme of this chapter, but as many philosophies espouse, all things are connected. The essence of this chapter is to present techniques that help the urban environment function in a more natural way. The reason for doing such is to reduce negative impacts caused by human development. Although the case studies are within the city of Portland, the principles are universal and can be applied to any region. The success of these projects is proof that ecological design not only benefits the environment of humans and wildlife but also often costs less to implement and sustain. Perhaps, we can "have our cake and eat it, too." It is no easy task, however; many institutional barriers and professional mindsets must be overcome. Design, research, demonstration projects, and education are all key elements in helping to bring these and other new approaches to the professional community.

Acknowledgments

The authors thank Ryan Retzlaff, Emily Hauth, Nicci Lambert, Emily Brown, Elizabeth Liptan, Scott Murase, and Sue Brantley for their assistance in the preparation of this chapter.

Literature cited

Beyerlein, D. and Brascher, J., Traditional alternatives: Will more detention work? in *Salmon in the City,* Tom Holz, Amer. Public Works Assoc., Washington Chapter, 45, 1998.

Bolton, S. and Watts, A., Results from forest hydrology studies: Is there a lesson for urban planners? in *Salmon and the City,* Tom Holz, Amer. Public Works Assoc., Washington Chapter, 49, 1998.

Portland, Fifth Stormwater Monitoring Report (July 1, 1999–June 30, 2000), National Pollutant Discharge Elimination System (NPDES) Municipal Separate Storm Sewer System Discharge Permit Number 101314, Annual Compliance Report No. 5, prepared for Oregon Dept. Environ. Quality, submitted by City of Portland, Multnomah County, Port of Portland, Nov. 30, 2000.

Portland, Bureau of Environmental Services, Tree Monitoring Report, Portland, OR, Mar. 12, 2001.

Disclaimer

The discussion in this chapter is not intended to substitute for professional advice applicable to specific project circumstances. Design approaches are offered to facilitate understanding of the concepts and must be considered in terms of the project, local building codes, and regional climate. Readers are urged to seek professional assistance before applying any of the techniques of this chapter. The techniques and other information presented may not represent the latest, approved approaches of the City of Portland, OR.

Response

Letting it soak in

"It," here, refers to both the rainwater and the message about how to effectively manage it. This chapter by Thomas Liptan and Robert Murase identifies significant elements in the paradigm shift now taking place in stormwater management of moving away from centralized and expensive infrastructure toward localized and dispersed micromanagement. Whereas larger-scale reconfigurations of impervious surfaces are important, as, for example, discussed by Jennifer Zielinski in Chapter I.3, one very significant element of low-impact development championed by Larry Coffman and Michael Clar in Chapters I.5 and I.15, respectively, is attention directed to the small, but numerous, impervious surfaces of roofs and parking lots.

Liptan and Murase make the point that rain gardens are both an aesthetic and cost-effective approach to managing stormwater in urban settings. In particular, they invoke a clarion call for water sensitive designs to integrate earth, water, and plants with the site and the surrounding architecture. In this concept of "urban nature," stormwater becomes a "friendly companion" to be celebrated as a design feature rather than tucked away as a nuisance. The results of engaging in such a process, as Liptan and Murase state, will be lower development costs, more cost-effective retrofits, improved livability, and a better natural urban environment for all to benefit from.

This chapter goes on to offer a set of cogent messages for water sensitive designers:

1. Water gardens need not be restricted to only high-visibility locations (such as Portland's Water Pollution Control Laboratory), because they can also offer ecological and aesthetic benefits when placed in association with much more mundane and utilitarian sites such as apartment blocks and service parking lots (in Chapter I.7, Robert France and Philip Craul show designs for one such a "green" parking lot).
2. As Bruce Ferguson stresses in Chapter I.1, designing in the absence of good, site-specific hydrological knowledge (as, for example, the fact that most precipitation falls in small storms) is unwise at best, and possibly foolhardy at worst.
3. Landscape architects need to do a much better job in undertaking experiments to determine what attributes of which water sensitive designs work, not merely look, best.
4. A great need exists for hubris to be replaced by humility in terms of honestly sharing information and experiences about the lessons learned from various design projects — both successes and mistakes.

chapter I.7

Retaining water: Technical support for capturing parking lot runoff (Ithaca, New York)

Robert France and Philip Craul

Abstract

Water sensitive design, as part of a larger framework of integrated watershed management, is, due to its inherent complexity, a daunting and onerous task for anyone to engage in independently. Just as water is an integrator across physical landscapes, it can also be an integrator across professional landscapes. Because few individuals have the breadth of ability to both fully comprehend *and* effectively design innovative stormwater management systems, complicated problems necessitate the formation of diverse, interdisciplinary teams. This chapter briefly outlines some of the attributes suggested by a team of technical experts (from the disciplines of hydrology and soil science) that should be considered in the landscape architecture of a combined bioretention-wetland system designed to treat parking lot runoff at the same time as providing a scenic amenity at a high-visibility location.

Introduction

Remedying environmental development pressures

Sprawling patterns of community growth and transportation systems reduce vegetative cover and produce large areas of impervious surfaces (roads, parking lots, driveways) that distort watershed hydrology, and produce rapid, high-volume stormwater runoff patterns (Horner et al., 1994; Chapter I.3). Impervious surfaces typical of urban and suburban development areas prevent the infiltration of rainwater into soils (Chapter I.1). As a result, groundwater supplies cannot be recharged, streams degrade, and flooding is increased, with associated habitat reduction (Schueler, 1995). As the runoff moves over impervious surfaces, it collects contaminants, increasing pollution in streams and wetlands. In addition, impervious surfaces retain solar energy, raising air and water temperatures, which in turn negatively impact aquatic communities.

An increasing body of scientific research conducted in many geographic areas and using varied techniques supports the theory that impervious land coverage can be a reliable indicator of stream degradation (Richman et al., 1997; Chapter I.3). In particular,

Bannerman and Dodds (1992) document the contribution of parking lots to runoff pollution problems, demonstrating that for commercial and industrial land uses, "parking lots are a critical source of stormwater pollution . . . account[ing] for approximately one-fourth to two-thirds of the suspended solids, total phosphorous, total copper, and total zinc loads in the commercial and industrial areas studied."

The first portion of a storm cycle has the greatest impact on water quality (Chapter I.6). Rapid, small storms disperse the most highly concentrated contaminants (oils, metals, and other toxic substance), thereby producing the most negative impacts. Therefore, treating the first flush of runoff is key to controlling nonpoint source pollution. This implies that the primary focus in stormwater management practices should be at the source of the runoff. Supporting this is the fact that the more distant runoff treatment efforts are placed from the source, the more effort in terms of cost and maintenance is required to operate them (Richman et al., 1997).

Design strategies important to stormwater management include the following:

- Minimization of directly connected impervious areas
- Maximization of permeability
- Employment of access streets
- Plans for alternative modes of transportation
- Integration of the drainage system(s) with natural landforms and topography

Directly connected impervious areas are those paved or roofed surfaces that drain into a catch basin or other conveyance structure. If runoff is collected and concentrated in a series of drainage structures as it is transported, no filtration by soil or organic matter occurs. Additionally, the speed and volume of the water flow are increased, thereby increasing erosion and flood potential, contributing a cumulative impact on stream systems. Subsurface flow and filtration of runoff are accomplished by the use of pervious areas, depressions, and swales in conjunction with drains. Foregoing single-use drainage ways in favor of multifunctional waterways is also recommended.

Maximization of permeable areas not only improves water quality, it also can significantly reduce development costs by eliminating or reducing the need for underground conveyance stormwater systems. This can be accomplished with the use of permeable pavement surfaces, reduction in the footprint size of new buildings, cluster development, and shortened or shared driveways (Center for Watershed Protection [CWP], 1998; Chapters I.1, I.3, I.5, I.15).

A number of so-called best management practices (BMPs) can be employed by landscape architects, engineers, and designers to minimize human impacts on the coastal environment (Schueler et al., 1992; Chapter I.4). Frequently in the past, corrections for the adverse effects of stormwater runoff on aquatic ecosystems were not accounted for in site planning. It is now recognized that an often underused environmental tactic is the integration of drainage and filtration into the initial stages of the site design process (Chapters I.5, I.15). Conventional storm drain systems divert water beneath the ground surface and fail to integrate with surface topography. If instead, drainage and filtration considerations are addressed in the initial planning process, then proper environmental considerations can be integrated into the landscape in an aesthetically pleasing and economically sound fashion through minimizing expensive earthwork.

Techniques for managing stormwater drainage include

- Extended detention ponds: These ponds detain stormwater runoff for a short period of time, allowing pollutants to settle out.
- Stormwater wetlands: These constructed, shallow pools support growth of wetland plants and maximize pollutant removal through plant uptake.

- Infiltration swales and basins: Reservoirs are created out of shallow trenches that are lined with porous material that enables the filtration of cleansed stormwater into the water table.
- Multiple-pond systems: These pond systems combine a number of pond designs such as extended detention, permanent pool, shallow wetland, and infiltration, providing reinforcement of pollution removal abilities.
- Sand filters: This technique diverts the first flush of runoff into a sand bed, at which time the filtered water is then collected in an underground drainage system and conveyed back to a stream or water body. This method is particularly well suited to treat parking lot runoff.
- Grassed swales: Swales are frequently used for their pollution removal potential; runoff is collected and filtered in concave earthen depressions.
- Filter strips: These level vegetated strips of land area (or enhanced natural buffers) intercept overland sheet flow running off from development areas. Although the vegetative cover enhances pollutant removal, filter strips are not especially effective in treating high-velocity runoff and should be used either in conjunction with other management techniques or in low-density development areas.

The best solutions for stormwater management in a given design project may combine several of these categories, depending on the individual objectives and factors (EPA, 1990; Schueler et al., 1992). For example, existing water quality, size of site, type of development, economics, amount of time necessary for implementation, and limitations such as space requirements, soil type, etc. often vary. While the long-term benefits of these strategies are not known in detail due to their relatively recent invention, these practices are widely regarded by the scientific community to be effective in reducing the environmental damage caused by development (e.g., Horner et al., 1994; Schueler et al., 1992). One particular type of stormwater management BMP beginning to receive attention, particularly in relation to parking lot designs, is the use of bioretention swales.

Bioretention swales and green parking lot design

Our urban centers are increasingly planned and built for the benefit of cars, rather than, and sometimes even at the expense of, people. For example, accommodating transportation-related activities can contribute up to three-quarters of the total impervious surface coverage within urban watersheds (Richman et al., 1997; Prince George's County [PGC], 1999), the remainder being associated with buildings. Perhaps no more substantial contributor to impervious surfaces exists than parking lots. Consequently, a literature is rapidly developing that investigates procedures to reduce the amounts and mitigate or alleviate the effects of stormwater runoff from parking lots (Schueler, 1995; Claytor and Schueler, 1996; PGC, 1999; Chapters I.5, I.15).

"Bioretention" is a method to manage and treat stormwater runoff by using plant materials and a conditioned planting soil bed to filter runoff temporarily stored in shallow swales (PGC, 1993, 1999; Claytor and Schueler, 1996; Chapter I.5). Water purification occurs through both physical filtering and biological (plant and microbe) uptake processes (PGC, 1993, 1999; Claytor and Schueler, 1996). Additional benefits include wildlife attraction. Several projects in Maryland, Oregon, and Colorado have successfully employed bioretention swales in commercial parking lots (Thompson, 1996; Chapters I.4, I.5, I.15).

Although bioretention swales and retention basins are based on functions of natural riparian and forest plant communities, they still require regular inspections and maintenance. They should be inspected on a semi-annual basis for the first year following all major storm events, and subsequently by annual inspections (PGC, 1993; Claytor and

Schueler, 1996). In such cases, soil should be tested, erosion problems corrected, regular mulching implemented, woody vegetation pruned, inflow areas checked for clogging, and built-up sediments removed.

In addition to the goal of creating an effective stormwater management plan for integrated parking lot design, other goals that are guiding principles in bioretention include:

- Decrease the amount of impervious surface on the site by increasing the amount of greenspace.
- Enhance the pedestrian character of the location.
- Provide opportunities for visitors and residents to experience and learn of the functioning of the natural environment.
- Minimize the environmental impacts of the parking lot while providing the same amount of parking.

Project site description

A site design proposed by Childs Associates for Cornell University's new Ornithology Laboratory at Sapsucker Woods in Ithaca, NY, is based on an ecological approach in the creation of stormwater wetlands and bioretention swales. Prior to establishment of the Sapsucker Woods Sanctuary in the 1950s, the 12-acre site, including its created lake, barn, and fire pond, had been a sheep farm. Today, the relatively flat site, with a perched water table, has naturally evolved to a wooded wetland predominately made up of invasive, nonnative species. A new building and accompanying parking lot for 190 cars is planned (Ulrich, 2000), which will make a substantial intrusion on the site, both physically and visually (Figure I.7.1). These built elements have the potential to significantly impact prime open space along the lakeshore, increase site runoff dramatically, cut off views and vistas of the lake, and reduce both the visual quality of the site and its use as a wildlife habitat for birds.

The design concept for the site has been drawn from the natural topography and spatial features (Childs Assoc., 2000). The new building and the parking areas are envisioned as "islands" floating within the wet woodland and meadow context of the site, similar to the existing island in the middle of the abutting lake. The parking areas will occupy discrete spaces or "islands" that are enfolded and absorbed into the larger natural context of the site. Changes in vegetation and topography will be used to define the parking "islands." The perimeter of each "island" will have a dense vertical edge of wetland trees and shrubs that will provide wildlife habitat and contrast to the vegetation at the interior of the "island" (Childs Assoc., 2000).

The site plan includes the design of the vehicular and pedestrian circulation systems, the creative management of stormwater runoff, and the design of soils and plantings for wetland mitigation and bioretention areas. All the stormwater runoff from the building and the parking areas will be filtered through bioretention swales that will run between the designed "islands" (Figure I.7.2). Vegetation and soil construction in the bioretention swales will remove contaminants from the parking lot runoff. The cleansed water will then flow into a newly created wetland approximately 1.5 acres in size and will eventually drain into the existing lake.

The concept of parking islands surrounded by wetlands represents an ecological approach to design that utilizes strategies to direct and filter runoff into newly created treatment areas that will double as wildlife habitats. The overall intention is to improve both the visual and ecological values of the site. This design concept is further intended to integrate the new, large laboratory building into its larger context and to express the

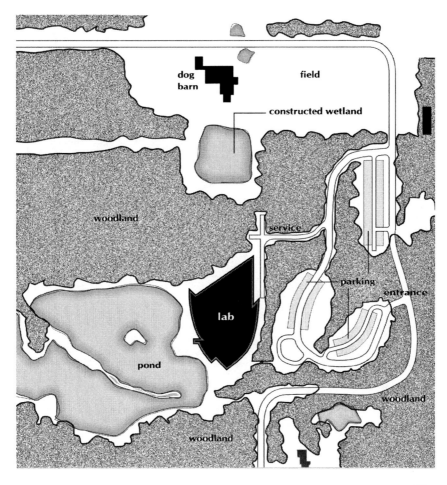

Figure I.7.1 Schematic design plan showing intended site development. (Adapted from preliminary design plans by Susan Childs Associates.)

Ornithology Laboratory's "commitment to the conservation of natural ecosystems" and desire to add to the "the complement of trails that traverse the 220-acre wooded setting" (Ulrich, 2000).

Wetlands

Wetland enhancement of the northern pond

Motivated by need for adherence to wetland mitigation laws to compensate for wetlands unavoidably lost during site construction, the plan calls for enhancement of the farm pond situated several hundred meters to the north of the development site in order to increase its wildlife potential. For construction, the following principles from France (2002) are suggested to achieve effective wetland design:

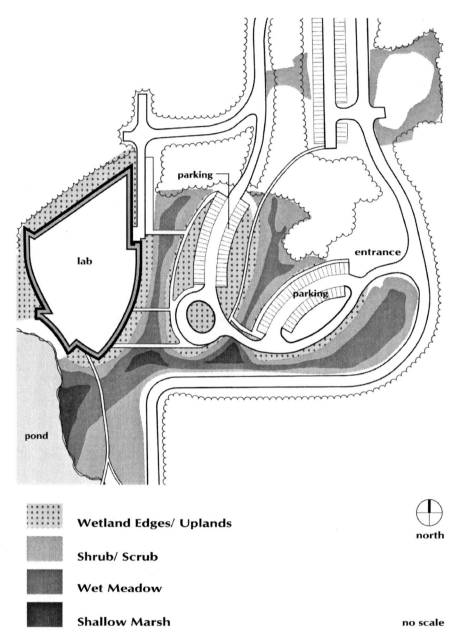

Figure I.7.2 Detailed schematic site design plan showing new ornithology laboratory building and parking lot "islands" surrounded by bioretention swales and treatment wetland. (Adapted from preliminary design plans by Susan Childs Associates.) Dark gray = shallow marsh; medium gray = wet meadow; light gray = shrub/scrub wetland; and stippled area = wetland edges and "island" uplands (see Table I.7.1 for plant list).

Grading

Shape and blend in with natural contours of the landscape; i.e., ridges should be kept as promontories or islands and depressions removed to build embayments.

Soil should be saved for later shoreline sculpting such as island construction.

Once the new wetland edge has been shaped, the compacted soil should be broken up before breaching the final edge to the existing pond.

Soil
:   At least 4 to 6 in. of soil is required for shallow rooting aquatic plants.

Islands
:   They should be at least 4 m^2 or 0.02 ha in size, located 15 m apart and separated from the shoreline by at least 15 m of permanently flooded water.

    Low, cross-shaped islands are ideal as waterfowl habitats because their irregular shape and increased edge enable establishment of distinct territories, thereby maximizing space utilization.

Orientation
:   In cold climates like Ithaca, NY, maximizing southern exposure through an east–west alignment can be beneficial for overwintering waterfowl using emergent vegetation as shelter.

Shoreline
:   Creating irregular shorelines can increase their length by up to 10 to 20% per unit wetland area, which will provide benefits to waterfowl by increasing locations for nesting and resting.

Slope
:   Shorelines should be shallowly graded (3:1 to 5:1) to provide easy access for wildlife and to limit erosion caused by wave contact with the toe of the bank.

Planting
:   For areas greater than 1 ha (as in this case), it will be more economical to use a hydroseeded mixture.

    The seasonal timing for seeding is not particularly critical, though spring is to be preferred.

Wetland creation at the southern construction site

The goals for creating this wetland are twofold. First, the area created will go toward compensating for wetlands unavoidably lost during site construction. Second, this wetland will provide water quality treatment by receiving stormwater from the bioretention swales draining the parking lots and building roof, and in so doing, perform final "polishing" before the stormwater is discharged into the lake. For construction, the following principles from France (2002) are suggested to achieve effective wetland design:

Marking, access, and excavation
:   Careful surveying is required to delineate location of the selected site in relation to access and erosion control measures during construction.

    Excavation should proceed from the center toward the intended land edge, with soil saved for later shoreline and riparian landscaping.

Sidewalls
:   They should be sloped no steeper than 2 horizontal:1 vertical.

Site preparation
:   Once the wetland has been shaped and graded, the compacted soil should be broken up before shallow flooding to enable the material to settle and a level bed to become established.

Soil
:   At least 4 to 6 in. of soil is required for shallow-rooting aquatic plants.

    Increasing the organic matter content of bottom sediments will help to bind contaminants.

Planting
:   Given the estimated size of this wetland (about 2 acres), it will be more economical to hydroseed a large portion of it, using direct plantings as the core (in such a case, it will be necessary to have a dry establishment period).

Residence time
: This is in the order of about 1 week to ensure effective removal of most contaminants.

Length
: About 100 ft is effective for removal of most contaminants.

Depth
: For stormwater wetlands, effective allocation of total area should be in the order of 50% shallow (0- to 1-ft depth) marsh, 30% deep (1- to 3-ft depth) marsh, and 20% deep water (3- to 6-ft depth) including open-water pools.

: Deep zones should be arranged perpendicular to flow direction to limit the chance that water can short-circuit the intended pathway.

: Because a mixture of flooding regimes expands wetland functions, interspersion of shallow and deep water areas will increase the efficiency of contaminant removal and attractiveness for wildlife.

Slope
: The slope should not exceed 0.5 to 1% in order to maintain depth differences of less than 3 in. from inlet to outlet.

: Steep vertical banks should be avoided for safety purposes.

Length:width ratio
: Between 3:1 and 4:1 is most efficient for contaminant removal.

Water distribution
: Use of multiple input points along the length (termed "step-feeding") is a strategy to distribute the contaminant load evenly.

The following design particulars are suggested for the southern stormwater wetland:

- Sidewalls graded down to a V-notched central gully, which is 2 ft lower than the parking lot elevation.
- The bulk of the wetland (most significantly, the area "upstream" of the central bioretention swale) will impound water to a depth of no more than 1 ft during flood events and may therefore not need to be graded so deeply.
- Three regions of greater depth (at least 1 ft below the summer watertable) that will be year-round openwater areas; these three regions (two situated at the mouths of the inflowing bioretention swales, the other sculpting the lakeshore inland) will be shaped to receive, concentrate and then redistribute stormwater.
- For marsh areas use a 4-in.-thick layer of organic bottom soil overlaying impervious clay.
- A 1% grade along complete length will ensure that the approximate 700-ft travel distance is more than adequate for a water residence period of about 5 days maximum (based on calculations of estimated flow velocity derived from the hydraulic radius and Manning's surface roughness coefficient).
- Hand plantings (emergents) will fringe all three openwater pools, whereas the rest of the wetland will be seeded with a regional mixture; riparian land–water interfaces will have a variety of water-tolerant hardwoods.

Bioretention swales at the parking lot site

The goals of the bioretention swales are to provide the first line of defense for mitigating contamination that will arise from parking lot runoff. Other attributes of such swales are to convey filtered stormwater to the wetland where it may be detained and further treated, to provide an aesthetic amenity, and to provide wildlife attraction. For construction, the following criteria from Prince George's County (1993) are suggested to ensure effective treatment of the parking lot runoff:

Grass filter strip
 Provides vegetative pretreatment of sediment laden runoff sheetflow
 A 1% minimum grade to 4:1 maximum grade
 Between 20 and 100 ft wide
Bioretention swale size
 Minimum area of 5 to 7% total drainage site
 From 10 to 25 ft wide and at least 40 ft long
Ponding area
 A maximum of 6 in., which drains within 4 days
Mulch layer
 Functions as a contaminant filter, protects soil from desiccation, and serves as a habitat for microorganisms
 Organic material such as shredded wood chips or commercial compost
 Depth of 3 in.
Planting soil
 Major purpose is to sustain plant growth
 Secondary purpose is to trap contaminants
 Should be of a depth of 2 to 4 ft
 Loamy sand mixture, etc., with no more than 25% clay content
 Underlying 1-ft sand bed increases the infiltration capacity and provide aeration for the plant roots
Plant material
 Entraps contaminants through growth and ET
 Designed to mimic a terrestrial forest community's functions
 Provides a host of other site amenities, including wildlife attraction
Hydrology
 Installation of an overflow storm drain inlet (2 in. above max ponding depth) with a simple piping system to ensure that excess water will be carried offsite and thereby prevent any flooding of parking areas.
 Bedding underdrains in pea gravel will assist site draining.

Planning for incorporation of bioretention swales for stormwater treatment necessitates estimation of the size needed to hold the quantity of runoff expected. We used the following hydrological assumptions:

1. Common storm event
    Total site area = 91,000 ft^2
    Existing RCN = 70
    Proposed post-development RCN = 84
    Design storm = 3 in. (2-yr, 24-h event)
    Design water depth = 3 in.
    Calculated runoff based on land characteristics is about 2 in. (assuming that no pervious pavement is used)

About 24,000 ft^2 of bioretention area would be required for water quantity storage. As our planned site is to have 77,000 ft^2 of combined bioretention and wetland area, runoff from the design storm can spread out to a depth of about 1 in. This suggests that flooding is unlikely to pose a problem. With this knowledge, it becomes possible to plan the elevation of the new parking lot accordingly.

Calculations for water quality improvement found that a smaller bioretention area would be required for runoff treatment. Therefore, the 0.2 in. of first-flush runoff (which will contain the majority of the contaminants) can easily be treated in the combined bioretention-wetland system area. For drainage estimations, the water residence time (based on estimated flow velocity derived from hydraulic radius and Manning's surface roughness coefficient, ET, and infiltration rates) was calculated as being less than 1 day for the 2-yr, 24-h storm event.

2. Extreme storm event
    Design storm 7 in. (100-yr, 24-h event)
    Other details as described previously

About 64,000 ft^2 of bioretention area would be required for water quantity storage. Working backward using the available treatment area for the proposed site design found that an absolute minimum of 2.5 in. of freeboard would be required to contain this runoff (this may have implications for constructed soil depths).

The following design particulars are suggested for the bioretention swales:

- The 50- to 55-ft sidewalls are graded down to a V-notched central gully, which is 3 ft lower than the parking lot elevation.
- The central gully is widened by about 4 ft at the depth of 3 ft below parking lot grade.
- An underdrain is installed ending in the wetland, which is surrounded by pea gravel.
- Above this is 6 in. of sand with a high infiltration capacity.
- Above this, and extending out to the margins of the overall swale, is 1 to 1½ ft of planting soil.
- A vertical pipe and overflow drain are installed at an elevation of 2 in. above max ponding depth.
- A 3-in.-deep mulch layer of woodchips is placed over the trough, blending into a 2-in. layer of compost spreading out the sloping flanks.

Plants

Given the high visibility of this site, the landscape architects paid careful attention (Childs Assoc., 2000) to plant selection. Two criteria were integral in the selection of candidate vegetation: First, for those species to be associated with the wetland and bioretention swales, their ability to withstand inundation as well as provide phytoremediation of contaminants were paramount. Lists for potential bioretention species from Maryland (PGC, 1993) were compared with lists of native wetland plants for New England to select those mesic species known to thrive in this north-temperate bioregion. Second, given that the site is a center for professional ornithological study and recreational bird watching, priority was given to species with known abilities to attract avifauna. A screening matrix was developed to identify those species able to fulfill both selection criteria; a selection of these species is listed in Table I.7.1. A comprehensive landscaping quality assurance document was produced, covering such items as sources of trees and shrubs, their acceptable sizes, details about fertilizers, mulches, and erosion control materials, site preparation and planting, and maintenance (Childs Assoc., 2000).

Table I.7.1 Proposed Plant List for Wetlands and Bioretention Swales around the Parking Lot "Islands" at the Cornell Ornithology Laboratory

1. Shallow marsh	3. Shrub/scrub wetland
Sweet flag	Trees
Water plantain	Red maple
Bearded sedge	Green ash
Tussock sedge	American larch
Turtlehead	Swamp white oak
Spike rush	Black willow
Yellow water lily	Small trees
Royal fern	Box elder maple
Arrow arym	River birch
Pickerel weed	Shrubs
Northern arrowhead	American cranberry
Common three-square	American cranberry bush
Burreed	
Skunk cabbage	
2. Wet meadow	**4. Wetland edges/upland ("island" edges)**
Shrubs	Trees
Speckled alder	Red maple
Red chokeberry	Black gum
Buttonbush	White pine
Silky dogwood	Eastern hemlock
Red oster dogwood	Small trees
Winterberry	Shadblow
Spikebush	Gray birch
Swamp rose	Cockspur hawthorn
Pussy willow	Black cherry
Common elderberry	Shrubs
Ground covers/perennials	Sweet pepperbush
Marsh marigold	Gray dogwood
Fox sedge	Witch hazel
Joe pye weed	Bayberry
Blue flag	Highbush blueberry
Cardinal flower	
Sensitive fern	
Cinnamon fern	
Fowl bluegrass	
Narrow-leaved cattail	
Common cattail	
Blue vervain	

Note: Additional plants were proposed for the upland parking lot "islands" (contact Childs Associates).
Source: Adapted from Susan Childs Associates, Boston, MA.

Soils

The general situation

Unless careful attention is paid to soil creation, landscape projects will be only marginally successful in terms of sustainability (Craul, 1992, 1999). The major portion of the construction for the new Cornell Ornithology Laboratory and its associated structures occurs in a recognized wetland environment. It is therefore necessary to construct a series of bioremediation areas among the structures and parking areas. Their purpose is to remove the various pollutants contained in the runoff from rooftops and parking lots before it enters the natural drainage system. To ensure success in the construction of bioremediation wetlands, greater attention must be given to the details of soil and plant specifications and installation than in routine landscape projects. That process is described next.

The soils of the bioremediation site

The soils of the site are identified as the Erie-Ellery channery silt loam complex and the Ellery-Chippewa-Alden channery silt loam complex in the Tompkins County Soil Survey Report (Neeley, 1961). The Ellery series has since been merged into the Chippewa series and is not discussed further.

- The Erie series consists of very deep, somewhat poorly drained soils formed in loamy till. They have a fragipan layer starting at depths of 10 to 21 in. below the soil surface. These soils are of uniform slope and are on footslopes and broad divides in glaciated uplands. Permeability is moderate above the fragipan and slow in the fragipan and substratum.
- The Alden series consists of very deep, very poorly drained soils on upland till plains in depressions and low areas in the landscape. They are formed in a thin, silty colluvial mantle overlying glacial till. They are on upland till plains in depressions and low areas in the landscape. Slopes range from 0 to 3%.
- The Chippewa series consists of very deep, poorly drained and very poorly drained soils formed in compact till deposits. These soils are in upland depressions. A dense fragipan is present in the subsoil. Permeability is moderate above the fragipan and slow or very slow in the fragipan and substratum. Slope ranges from 0 to 8%.

Two soil pits were excavated on January 15, 2000, to confirm the soil conditions. Supplemental piezometers were installed to refine the determination of the fluctuating perched water table present in these soils during the dormant season (it may be present in the growing season during wetter-than-normal years). The brief soil profile descriptions are given in the list that follows.

- Soil pit 1:
    - Ap-0-7 in.; very dark brown mucky silt loam; weak, medium granular structure; friable; common fine roots
    - E-7-16 in.; brown silty clay loam; weak, fine subangular blocky structure; few medium, faint-yellowish brown mottles; very friable; many fine roots
    - Bg21-16-25 in.; yellowish-brown silty clay loam; weak, coarse subangular blocky structure; with common medium, distinct gray mottles; friable; water drainage at roots and their channels
    - Bg22-25-28 in.; yellowish-brown silty clay loam; weak, medium platy structure; common medium, prominent gray mottles; firm; no roots
    - Bg3-28-37 in.; yellowish brown silty clay loam; moderate to strong, medium platy structure; very common medium, prominent gray mottles; very firm; no roots
    - Cg-37-52+ in.; gray loam; massive and firm; no roots

- Soil pit 2:
  - Ap-0-8 in.; dark grayish-brown silt loam; weak, fine granular structure; very friable; common fine roots
  - Bg21-8-16 in.; yellowish-brown silt loam; weak, medium subangular blocky structure; many coarse prominent gray mottles; firm; few fine to no roots
  - Bg22-16-23 in.; yellowish-brown silty clay loam; weak, medium platy structure; many coarse, prominent gray mottles; firm; no roots
  - Bg3-23-30 in.; yellowish-brown silty clay loam; moderate to strong platy structure; many coarse, prominent gray mottles; firm; no roots
  - Cg1-30-42 in.; yellowish-brown loam; very firm breaking into strong medium platy peds; many medium distinct mottles; drier than above horizons; no roots
  - Cg2-42-47+ in.; yellowish-brown loam; weak to moderate medium platy structure; common medium distinct mottles; firm to friable; no roots

The implications of these soil conditions for a bioremediation site are:

- The native soils are already classified as wetland soils being poorly to very poorly drained.
- The subsoil horizons of these soils contain either a fragipan (a dense layer impermeable to water which perches water at a shallow depth) or firm horizons that have slow permeability. This is also a disadvantage, however, in that it may cause ponded water at high elevation during wet seasons, requiring structural works to be elevated or filled to higher than desired final grade.
- The vegetation present on the undisturbed portions of the project site is of wetland species composition and does not require significant modification to meet wetland criteria.

The designed soil profiles

Grass and vegetation filter strip soil profile
   This profile (Figure I.7.3) occurs over the sculptured native subsoil, but generally above the maximum high perched water level or the shallow ponding area level. It extends from the gravel parking lot to the bioretention areas.
   The profile consists of a topsoil horizon and a subsoil, both of which are derived from other soil materials. The specifications are given in the next section.
   The native subsoil (that below the topsoil, the former plow layer, of 7 to 8 in.) is sculptured so that a 1% grade is created extending from the longitudinal centerline of the parking area to the edge of the bioretention area or the approximate elevation of the shallow ponding area within the bioretention area.
   The subsoil depth is given as "residual" because the total depth of the soil profile will decrease approximately 2½% from the appropriate subsoil depth at the edge of the parking area to 0 at the edge of the bioretention area.
The bioretention area and adjacent filter strip soil profile
   This profile (Figure I.7.4) is constructed as the rooting soil for the bioretention plants in the loamy sand planting bed and the adjacent filter strip on either side. The bioretention area will be saturated through most of the dormant season and will have shallow ponding during much of that season.
   It is expected to dry out in the growing season, especially during the latter portion of July and August, and early September in dry years. Normally, it will become saturated by November.

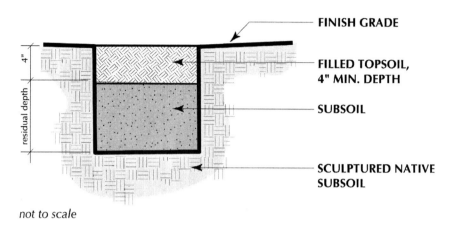

Figure I.7.3 Schematic of the grass and vegetation filter strip soil profile.

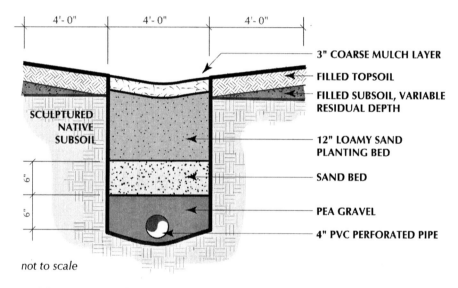

Figure I.7.4 Schematic of the bioretention area and adjacent filter strip soil profile.

The profile consists of a coarse mulch layer over a specified loamy sand planting bed.

The profile vertical dimension is provided by the excavation of the natural subsoil to the specified depth. The soil and mulch layer are installed in the excavation.

The bottom of the planting bed trench is graded longitudinally at 1% to facilitate saturated flow through the bioretention trench and eventually to the outlet.

The trench bottom is also tapered or V-notched to the centerline in cross-section, at a right angle to the longitudinal grade, to facilitate saturated water flow to the center of the trench. The depth of the trench is 24 in. throughout.

The soil horizon specifications

The grass and vegetation filter strip topsoil
The soil of the filter strip should aid in reducing the runoff velocity and filter particulates from the parking area runoff. The topsoil must have the characteristics of

Table I.7.2 Range in Percent Passing Sieve Sizes for the Bioretention Planting Bed Loamy Sand Soil

Sieve Size	Percent Passing Range
10	100
18	88–100
35	70–80
60	25–35
140	13–23
300	10–20
Silt	8–18
Clay	2–12

high infiltration rate, high permeability, relatively high organic matter, and sufficient depth, coupled with the subsoil, to support vigorous growth of plants.

The existing topsoil (former plow layer or Ap horizon) on the construction site is to be stripped and stored for use as a soil mix material.

The existing topsoil has a texture of silt loam, which renders it unsatisfactory for the grass and vegetation filter strip topsoil. It must be modified to at least loam (40% sand) or coarser texture by the addition of a coarse sand. A sand content of 60% or more may be necessary to offset the silt plus clay content of the topsoil. The existing silt loam soil must be tested for particle size distribution (ASTM D-422-63) to determine the appropriate sand–soil mix ratio. A probable ratio may be 2:1. Percolation and texture tests will be required for the final mix and must be approved by the landscape architect.

The sand specified for the bioretention sand bed may be used as the coarse sand amendment to the existing topsoil.

The final soil mix should be free of any stones, gravel, glass, plastic, masonry, drywall, asphalt, rubber, metal, or any other type of foreign debris.

The grass and vegetation filter strip subsoil

The soil material employed as the filled subsoil in the grass filter strip must have moderate permeability and fertility with adequate organic matter content and sufficient strength to support both plants and the topsoil.

No acceptable material exists on site for use as the subsoil for the grass and vegetation filter strip areas. Therefore, a soil material must be imported.

The unamended loamy sand specified for the bioretention area planting bed may be used as the grass and vegetation filter strip subsoil.

The final soil should be free of any stones, gravel, glass, plastic, masonry, drywall, asphalt, rubber, metal, or any other type of foreign debris.

The bioretention area mulch layer

Chipped or shredded and screened clean hardwood mulch, with 100% passing a ½-in. screen. It should not contain any debris such as stones, gravel, masonry, drywall, glass, plastic, asphalt, rubber, metal, or any other type of foreign debris.

The bioretention area planting bed soil

The planting bed soil must be a sandy soil that exhibits a high infiltration rate (at least 2 in./h).

Texture: the specified particle size distribution is given in Table I.7.2.

Organic matter content: the planting bed soil should have an organic matter content of about 2% by weight and may range from 1.5 to 3.0% (approximately 5% by

Table I.7.3 The Particle Size Distribution of the Bioretention Area Sand Bed[a]

Sieve size (no.)	Percent passing	Size content, %
10	100	—
18	90	10
35	70	20
60	21	49
140	13	8
300	10	3
Silt	4	6
Clay	—	4

[a] Actual values may be ± 2.5%.

Table I.7.4 Particle Size Distribution of the Gravel Drainage Layer (AASHTO #8)

Sieve Size (in. or no.)	Percent passing
½ in.	100
⅜ in.	85–100
#4	10–30
#8	0–10
#16	0–5

volume). Many natural sand deposits will have an organic matter content approximating this percentage. Fertilization may be necessary to meet the nutrient requirements given below.

Others: the pH range for the planting soil should be 5.5–6.5; magnesium content: 35 lb/acre; phosphorus content: 75–100 lb/acre; potassium content: 85 lb/acre; soluble salts: not to exceed 500 ppm.

The soil bed should be free of any stones, gravel, glass, plastic, masonry, drywall, asphalt, rubber, metal, or any other type of foreign debris.

The bioretention area sand bed

This horizon is primarily a sand and acts as an additional facilitator to water drainage and treatment.

The texture of the sand bed should be a sand with the particle size distribution given in Table I.7.3.

It should be free of any stones, gravel, glass, plastic, masonry, drywall, asphalt, rubber, metal, or any other type of foreign debris.

The bioretention area gravel drainage layer

The drainage layer is primarily to aid drainage of excess overflow. The 4-in. perforated PVC pipe with an overflow standpipe prevents high-water levels from reaching the parking lots and associated areas during severe, intense storms. The material is commercial "pea gravel," which may be #7 or #8 by AASHTO designation for coarse aggregates. The particle size distribution for #8 is given in Table I.7.4.

Installation and sequence of the soil

Installation and the sequencing of the soil work involves some details that require attention to ensure that the soil is placed properly:

Soil must be placed in lifts not exceeding 12 in. in thickness.

Work should begin with excavation of the bioretention trench, laying of pipe, and backfilling. Work should proceed outward from the trench to the parking surface.

Minimum compaction is achieved by routine travel over the soil by lightweight soil-handling equipment. Excessive compaction is to be avoided by repeated traffic over the same area or by frequent turning. Multiple, offset linear travel patterns provide minimum desirable, but not excessive, compaction.

Soil must not be handled, placed, tilled, or traveled over when it is wet or the moisture content exceeds field capacity (the water in the soil voids glistens in light on close examination and/or the soil exhibits thixotropy when shaken).

The bioretention mulch is best placed after the planting of trees and shrubs has been accomplished. Any ground cover specified as plugs may be installed once the area has been mulched. Ground cover established by seeding or consisting of grass should not be covered with mulch.

It is necessary to sequence the installation of sediment and erosion controls to minimize the contamination of the planting soil with silts and other fines (clay and organic matter). Silts and fines will have a tendency to clog the planting soil and impair the functions of the bioretention area.

Sediment controls are installed around the area to be disturbed before grading and around the bioretention area before the excavation of the trench for the planting soil.

Post-installation maintenance

Post-installation maintenance of the bioremediation areas is necessary for successful continued functioning:

Biannual mulching is recommended. The previous mulch may be removed and discarded to an appropriate disposal area or retained if it is decomposed. The mulch is replaced to a depth not to exceed 3 in.

Dead or poor, low-vigor plants and trees within the bioremediation area should be promptly replaced as part of a comprehensive landscape maintenance program. The operation is necessary to maintain the function of the bioremediation area. Lack of growing plants can greatly reduce the effectiveness of the bioremediation area for which it is designed. Continuous inspection for insect and disease outbreaks or physical damage or injury to the vegetation should be a routine part of the maintenance program.

Any obstructions to drainage of the bioremediation area should be promptly eliminated.

Damage or disturbance to the bioremediation area soils by burrowing animals should be immediately rectified and the animals eliminated or removed from the area.

Education

Wetlands, once referred to as "wastelands," are now regarded as important centers of biodiversity, as purifiers of contaminated water, and as regulators of watershed hydrology. Recently, wetlands have received increased recognition because, given all these accredited roles in addition to their admired beauty and attraction for recreational purposes, they can also serve as important vehicles for environmental education (France, 2002). The recent book published by the Association of State Wetland Managers and the EPA, titled *Guidebook for Creating Wetland Interpretation Sites Including Ecotourism*, is witness to the fact that "wetland interpretation sites are invaluable to secondary and university science and ecology educational efforts" (Kusler et al., 1998).

In addition to the obvious opportunities for passive education in the form of interpretive signage, etc., at least two ideas are worthy of serious consideration for formal research projects concerning wetland function. Both of these research projects could be easily initiated and implemented by wetland scientists at Cornell University and run by volunteers and summer students. The purpose of this section is to introduce these two research topics as potential elements to be incorporated into the education program at the Ornithology Laboratory, not to develop the actual scientific sampling programs necessary for their successful implementation.

Wetland enhancement of northern pond — effects on avian biodiversity

As part of adherence to wetland mitigation laws to compensate for wetlands unavoidably lost during site (mostly building, not parking lot) construction, the plan is to enhance the existing farm pond in order to increase its wildlife potential. An oft-acknowledged need exists for more information to be made available about how the physical conversion of farm ponds to wetlands will increase the attraction of these improved habitats for wildlife. With this in mind, comparative bird counts of species abundance, visitations, and nesting sites could be conducted before and after enhancement.

Parking lot development and bioretention facilities — effects on vernal pool hydrology and biointegrity

Whereas most wetlands are important from a landscape-biodiversity perspective, vernal (temporary) pools have been recognized as being of extreme significance in this regard due to the absence of predaceous fish and consequent inhabitation by amphibians. Although the design plan calls for avoidance of one such critical vernal pool habitat in the careful placement of the parking lots and bioretention swales, no scientific information is available as to how wide the protected vegetated buffer area should be in such circumstances. With this in mind, comparative amphibian counts could be made before and after site development for several years. In addition, because it is the hydrology of vernal pools that is so critical to their continued existence, several piezometers could be installed in order to monitor water level fluctuations before, during, and after construction of the nearby parking lots.

Acknowledgments

This landscape architecture project is currently being designed by Antonia Bellata Osborne and Susan Child of Childs Associates, whom we thank for allowing us to participate in the conceptual design phase as well as providing support documents for this technical paper. Lisa Cloutier aided in writing a portion of the introduction, and Matthew Tucker prepared the figures.

Literature cited

Bannerman, R. and Dodds, R., *Sources of Pollutants in Wisconsin Stormwater,* Wiscon. Dept. Natur. Resour., Madison, 1992.

Center for Watershed Protection, *Better Site Design: A Handbook for Changing Development Rules in Your Community,* Center Watershed Protect., MD, 1998.

Childs Associates, Site work quality document for the Cornell Laboratory of Ornithology, 2000.

Claytor, R.A. and Schueler, T.R., *Design of Stormwater Filtering Systems*, Center Watershed Protect., MD, 1996.
Craul, P., *Urban Soil in Landscape Design*, John Wiley & Sons, New York, 1992.
Craul, P., *Urban Soils: Applications and Practices*, John Wiley & Sons, New York, 1999.
EPA, *Urban Targeting and BMP Selection. An Information and Guidance Manual for State Nonpoint Source Program Staff Engineers and Managers*, U.S. EPA, Chicago, 1990.
France, R., *Designing Wetlands. Principles and Practices for Landscape Architects and Land-Use Planners*, W.W. Norton, in press, 2002.
Horner, R.R. et al., *Fundamentals of Urban Runoff Management: Technical and Institutional Issues*, Terrene Inst., Washington, DC, 1994.
Klute, A., Ed., *Methods of Soil Analysis, Parts 1 and 2*, Monograph no. 9, 2nd ed., Amer. Soc. Agronomy, Madison, WI, 1986.
Kusler, J.A. et al., *Guidebook for Creating Wetland Interpretation Sites Including Wetlands and Ecotourism*, Assoc. State Wetland Manag., New York, 1998.
Neeley, J.A., Soil survey of Tompkins County, New York. USDA-Soil Conservation Service and Cornell Agricultural Experiment Station, Sup. Docum., U.S. GPO, Washington, DC, 1961.
Prince George's County Maryland, Dept. Environ. Resour., *Design Manual for Use of Bioretention in Stormwater Management*, Prince George County Publ., MD, 1993.
Prince George's County Maryland, Dept. Environ. Resour., *Low-Impact Development Design Strategies. An Integrated Approach*, Prince George County Publ., MD, 1999.
Richman, T. et al., *Start at the Source. Residential Site Planning & Design Guidance Manual for Stormwater Quality Protection*, Tom Richman Assoc. Publ., CA, 1997.
Schueler, T., *Site Planning for Urban Stream Protection*, Center Watershed Protect., MD, 1995.
Schueler, T., Kumble, P., and Heraty, M., A current assessment of urban best management practices. Techniques for reducing non-point source pollution in the coastal zone, U.S. EPA, 1992.
Thompson, J.W., Let that soak in, *Land. Arch.*, 11, 60, 1996.
Ulrich, C., Building on the past. A lab for the 21st century, *Living Bird*, 19, 24, 2000.

Response

To build an oxymoron:
A green parking lot

For a landscape architecture firm engaged in a stormwater management project, recognizing the point when outside consultation by scientists is required, is an important realization that the best solutions to environmental problems result from the marriage of art and science together. As quoted by C. Caudwell in Garrett Ekbo's seminal *Landscape for Living*, "Art is the science of feeling, science is the art of knowing. We must know to be able to do, but we must feel to know what to do." This chapter by Robert France and Philip Craul, similar to those by Richard Pinkham and Timothy Collins (Chapter I.4), Thomas Liptan and Robert Murase (Chapter I.6), Robert France and Kaki Martin (Chapter I.12), and Nicholas Pouder and Robert France (Chapter I.16), demonstrates the strength shown by successful pairing between disciplines. This echoes the call for greater interdisciplinary design in Chapter I.17 by Wendi Goldsmith.

Three important messages are put forward in this chapter. The first is the detailed presentation of the mostly descriptive, but also rudimentally quantitative, methodology needed to assess the potential for any site of applying bioretention techniques, an important tool of the emerging paradigm of low-impact development as discussed elsewhere in this book (e.g., Chapter I.1 by Bruce Ferguson, Chapter I.5 by Larry Coffman, Chapter I.6 by Thomas Liptan and Robert Murase, and Chapter I.15 by Michael Clar). The complexity of these issues is what may, depending on the availability of in-house expertise, necessitate scientific consultation, as was the situation for this project led by S. Childs Associates.

The second take-away message from France and Craul's work is recognition that a "green" parking lot is not really as oxymoronic as it sounds. We have to abandon the image of parking lots as mere functional appendages to development projects and instead begin to look upon them as worthy extensions of the aesthetically designed landscape. This, more than perhaps anywhere else, is the nexus where science and art can mesh for effective stormwater management. Given the high visibility of some building sites, such as the ornithology laboratory described in this chapter, it is unrealistic to expect them to be surrounded by ugly and environmentally dysfunctional fields of concrete when so much more can be accomplished.

The last, and perhaps most important, message developed by this chapter is the attention France and Craul give to assessing existing soils along with developing recommendations for constructing new soils. More than any other chapter in this book, with the possible exception of that by Wendi Goldsmith (Chapter I.17), it educates us about the need to remember that, in order for water sensitive design to work effectively, careful consideration must be given to the interactions with water upon the physical landscape of the soil.

chapter I.8

A productive stormwater park (Farmington, Minnesota)

Diana Balmori

Abstract

The Farmington, MN, waterway provides drainage for a suburban development of nearly 500 homes approximately 25 miles from downtown Minneapolis. It is an open water drainage system, replacing the usual underground pipe line that frequently floods streams with its sudden discharges. In addition, the waterway provides 91 acres of planted green space, as well as a system of connected streams that widen into larger ponds for recreational use. A park, by becoming a drainage system, acquires a function that justifies its creation. The idea of establishing such a productive park could be implemented in many communities. This chapter describes the design decisions and goals involved in winning city planners' approval for the project. An appendix suggests vegetation compatible with climate and local conditions in addition to soil and water regimes.

Introduction

Suburban development produces problematic amounts of water runoff from the large, impermeable surfaces of its new roads, roofs, and driveways (Chapters I.1 and I.3). This water, dumped all at once into existing rivers and streams, produces sudden floods that erode riverbanks and can potentially harm plant and aquatic life.

The Farmington, MN, waterway was designed to run alongside a suburban housing development of 486 midpriced homes built by the Sienna Development Corporation of Minneapolis. The project grew out of our desire to try a different approach to standard drainage systems. The Farmington waterway was created as a drainage system alternative to the usual underground straight-line water pipe. It functions successfully as a drainage project, as shown in the recent peak flooding years of 2001 and 1998, when the Mississippi River crested and remained above its floodplain for at least three months in each of those years. In addition, and perhaps more important, the project creates public space — a park — as part of the drainage solution. Because the waterway project makes the green space useful, thus paying for itself, we call it a "productive park."

Figure I.8.1 The Farmington, MN, site in winter. (Photo by Tom Hammerberg. © Copyright Balmori Associates.)

The site: Midwest prairie

The area near Farmington is completely flat (Figure I.8.1). In winter months the temperature often dips to well below zero. A slight covering of snow seems glossy, reflecting the low sun of winter. Around the site, a few farmhouses surrounded by acres of farmland and fields of dry cornstalks stand in the snow. On the edge of the planned development, a north–south road bisects the rural terrain. On the opposite side is the town of Farmington, which has a population of 12,500. According to local residents, other nearby developments flood every spring.

Farmington, a small farming town, is about 25 miles northwest of Minneapolis, close enough to the nearby urban area to create a high demand for housing. Rod Hardy, director of the Sienna Development Corporation, proposed building a development of some 500 houses over the course of 10 to 15 years, with the initial phase totalling approximately 170 houses. In 1993, several months after the summer of the worst flooding of the Mississippi River in 28 years, William Moorish, director of the Center of American Urban Landscape at the University of Minnesota, proposed to the town that we would design a different kind of drainage system for the planned Farmington development, which the town had not approved at that point. The developer's initial proposal was standard, directing the increased runoff into a single pipe, which would then empty into the nearby Vermillion River and continue on into the Mississippi River, about 20 miles from Farmington.

Additional water from an increased number of impermeable surfaces in the new development would no doubt exacerbate the already dangerous risk of flooding. Hydrologists, landscape architects, and ecologists have suggested alternate drainage schemes, such as trying to slow the runoff among vegetated edges to clean it gradually and, in large storms, store it until the peak of the storm has passed. When the moment comes for town boards or city planners to approve a drainage system, however, the water goes into pipes more often than not due to the common and understandable reluctance to try something new and unproven. In addition, such a slow-down system of managing runoff would probably require using more land, which a development corporation is not likely to purchase. Funding the project thus becomes a concern.

On the other hand, what if the space around a body of water or wetland system became public space, an amenity for the whole town? William Moorish of the University of Minnesota suggested just such a proposition to the town board in Farmington. Happily, the small farming town had a capable town planner who supported the idea. Because new solutions require broad training and a willingness to rethink old problems within new constraints, Yale University hydrologist Paul Barten was added to the project team and charged with exploring alternatives to standard drainage systems. Although little money was available to carry out our drainage solution ideas, the Farmington prairie waterway project was one way of trying out new ideas in a real-world setting. The town of Farmington and the Sienna Development Corporation eventually provided funding.

The municipal approval process

The town of Farmington made the development of an innovative drainage system a condition for approval of the development plan. The town planner supported the idea of a prairie waterway/park throughout the lengthy approval process, and the town board, mostly comprised of local farmers, did not object, but expressed some degree of apprehension over the need to buy land exclusively for the water system. The land was purchased with city money recovered through taxes paid on the property once the houses had been sold (tax increment financing). The taxes were established by the town as contributions by the developer to provide needed facilities, but the city also had to buy additional land beyond the development to allow access for the water to flow to the Vermillion River.

The design process and landscape aesthetic

We presented two schemes for a surface water drainage system to the Farmington town board and the developer. The first scheme, which the town board and the Sienna Development Corporation much preferred, made the waterway a naturalistic stream (Figure I.8.2). The second scheme, which we preferred, created a series of geometric interconnected ponds (Figure I.8.3) that cleaned and retained the water based on the system used in cranberry bogs. The town board and developer raised objections to the gridded bog system, claiming that it seemed unnatural. The ideal of the "natural" in an artificial drainage system comes from the 18th-century Romantic landscape aesthetic. Because it is important in any new work to find an aesthetic appropriate to it, the use of a well-known, human-created system, such as the time-tested system of cranberry bogs, seemed appropriate to the site and the project requirements. Creating an appropriate aesthetic was not one of the primary concerns of either the town or the developer, however, and truthfully belongs instead to the aesthetic concerns of the designer. Still, being able to implement our stream-like scheme was also acceptable to us, especially because convincing the town and the developer to accept an open drainage system was a big step forward.

Our goals, as presented to the Farmington Town Board and the Sienna Development Corporation, included the following:

- Design the waterway as a civic amenity that would contribute to the long-term prosperity of the city of Farmington, MN.
- Recognize that the waterway is a created landscape, not a "natural" one, and that its form can announce it as a cultural endeavor shaped by natural systems.
- Create better drainage for the watershed.
- Give the residents of the development, which they now call Park Place, a viable outlet for basement water.

Figure I.8.2 The winning scheme made the waterway a naturalistic stream. (Drawing by Bill Coyne and Joseph Marek. © Copyright Balmori Associates.) (Color version available at www.gsd.harvard.edu/watercolors. Password: lentic-lotic.)

- Help manage the housing growth of the city of Farmington in ways that would benefit the developer, the current community, and the future residents of Farmington.
- Create a stormwater pond and wetland system for the proposed Planned Unit of Development (PUD) and keep floodwater from impacting nearby housing.
- Create a design that has high ecological integrity and becomes a place of high biological diversity.
- Minimize construction and maintenance costs.
- Design for the recreational needs of the community of Farmington.

Chapter I.8: *A productive stormwater park (Farmington, Minnesota)* 179

Figure I.8.3 The scheme that was not chosen created a series of geometric interconnected ponds based on the system used in cranberry bogs. (Drawing by Bill Coyne and Joseph Marek. © Copyright Balmori Associates.) (Color version available at www.gsd.harvard.edu/watercolors. Password: lentic-lotic.)

Design changes to the development's layout

The development's layout was modified in order to reduce the stormwater runoff (Chapters I.5 and I.15). Street widths were narrowed, driveways were shortened, and at

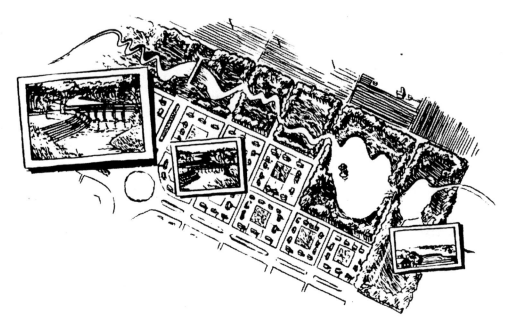

Figure I.8.4 The park drainage system with an overview from Spruce Street, the main avenue. A new development is on one side and agricultural fields border the other side. (Drawing by Fran Leadon. © Copyright Balmori Associates.)

the center of each block a depression was made to gather the water from spring rains running off the roofs and lots. These latter areas at the center of each block were thickly planted with trees (e.g., red maples) that can survive with roots in standing water for part of the year. These trees, through their foliage (evapotranspiration), would get rid of the water in these depressions.

Planting turned out to be the most troublesome part of the project. Basically, neither the developer nor the city had sufficient funds to allow planting the center of the blocks or the public green space created by the new water system. We proposed several inexpensive solutions, including gathering seeds from local vegetation and subsequent planting of those seeds by local schoolchildren as part of their science curriculum. We also proposed that the Town Board contact the Minneapolis Department of Transportation (MIN-DOT) to request permission to reuse plants that were being removed to widen roads after discovering that there was just such a project not too far from the site. A cover of winter rye was put in to control erosion. Eventually the town implemented a 10-year plan for investing $10,000/year in small trees. As of this writing, four years of planting (largely willow, hackberry, plum, and ash trees) have taken place.

The park

The park-like landscape for the growing city of Farmington consists of a riparian waterway, a civic lawn on axis with the downtown area, and playing fields. The wetlands associated with this urban waterway provide a recreational component for those interested in observing wildlife or using green public spaces. In addition, of course, they store stormwater. The form of this landscape is revealed through the rise and fall of water levels during and after storms (Figure I.8.4).

At the end of Spruce Street, the main thoroughfare in this development, a civic connection is made by a semicircle of upland lawn, which is terraced down to a pond

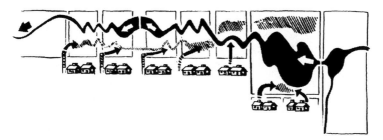

Figure I.8.5 Water movement in the drainage system. Hatched areas are locations for storing and delaying entry of water. (Drawing by Fran Leadon. © Copyright Balmori Associates.)

allowing views of the waterway, as well as access to it in both directions. Viewers can watch the moon rise over the farm fields from this curved terrace. Tree rows line the streets of the housing development and continue into the park, establishing a series of "rooms" roughly the same size as the city blocks (i.e., about 400 ft^2). Reminiscent of windbreaks, these frames give way on the inside of the "rooms" to native and naturalized plants (see list in Appendix A) that are determined in their distribution and type by the specific conditions in which they grow. Between each pair of tree rows, separating the planted "rooms" in much the same way that the roads separate the blocks of the housing area, is a roadway of tall grasses that provides shelter to the small animals that move through this landscape. Certain areas of the park were kept isolated to encourage a wildlife habitat, but pathways and boardwalks allow access to many parts of the site, including one of the wetlands. An 8-ft trail surrounds the green system, runs by, and crosses the waterway.

The waterway

Three separate water systems are used in the Farmington prairie waterway (Figure I.8.5):

1. A swale system to treat the stormwater runoff from the new housing development that lies immediately to the west of Park Place
2. A pond and channel system designed to move and clean the water over a large area before it enters the Vermillion River
3. Type I (periodically inundated) and Type II (sedge meadows) wetlands intended to make up for wetlands lost to the proposed development and to provide temporary storage for elevated groundwater levels

Inputs: Sources of water

Two housing developments to the south of the Park Place development in Farmington, which are known as Castle Rock and Henderson, will also be served by the drainage system because their stormwater will enter the pond at the southern end of the Park Place site. Both new developments are reported to have drainage problems; basements in the homes flood each spring. This pond reduces the velocity and turbulence of the stormwater and settles out the coarser particles. From here, the water continues to a larger pond that is surrounded by an aquatic bench of emergent wetland plants 10 to 15 ft wide with a depth from 6 to 12 in.

The channel leaving the pond has the sinusoidal curved edges produced by running water. The channel also slows the water, giving it greater opportunity to become cleansed as it moves through the vegetation. The great conductivity of the local soils means that groundwater rises and falls quickly during and after storm events. Under these circumstances, the great width of the channel, which varies between 80 and 110 ft, helps to reduce the velocity

Figure I.8.6 First stage of the housing development at Park Place, made up of 167 units. (Photo by Charles Tooker. © Copyright Balmori Associates.) (Color version available at www.gsd.harvard.edu/watercolors. Password: lentic-lotic.)

and scouring of the waterway. The vegetated, expanded streambed traps particulates and takes up nutrients.

Finally, runoff from the Sienna Development housing itself is sent into stormwater swales just east of the neighborhood. These wide stormwater swales are set above the groundwater level and are temporarily inundated after a storm. The water moves very slowly southward over sedges and grasses that help to clean the water before it subsides or moves north through subsurface flow into the channel and on to the Vermillion River. This doubling-back of the water flow gives increased residence time, and therefore increased treatment, before the water enters the channel. During periods of base flow, the channel has a small amount of water in it. During medium-sized storms, the channel water rises to fill a shelf on either side of it. The stormwater swales become inundated from both road runoff and the temporarily elevated groundwater. The water remains below the level of the recreational pathways and the upland edges of the park. In addition, the water has not spread overland into the mitigating wetlands.

During major storm events, the wetlands at the south end of the site accommodate overflow from the ponds. The mitigating wetlands to the north temporarily store the higher groundwater, but only in 100-year events is the berm between the wetlands and the channel breached. The berm separating the stormwater swales from the channel is set at an elevation that will keep them separate even during the 100-year event. On September 15, 1998, just as the construction of the drainage system was completed, a 100-year flood event took place. The drainage system worked exactly as it was supposed to, the berm was breached, and the mitigating wetlands stored the higher waters.

The Farmington housing development, started in 1990, was intended to be built in three 5-year phases. But the project was so successful, the built houses sold out quickly during the first phase of 167 houses. By 2000, seven phases totalling 486 new houses had been completed. An interesting footnote: the third phase was partially purchased by the builders of the project. Aerial photographs show the three stages of building (Figures I.8.6 to I.8.11). The waterway can be clearly seen in each photo because it preceded the building of houses in the development.

Chapter I.8: *A productive stormwater park (Farmington, Minnesota)* 183

Figure I.8.7 The second stage of the housing development. (Photo by Bordner Aerials. © Copyright Balmori Associates.) (Color version available at www.gsd.harvard.edu/watercolors. Password: lentic-lotic.)

Figure I.8.8 Third stage of the housing development of homes priced at $137,000 to $153,000. (Photo by Bordner Aerials. © Copyright Balmori Associates.)

Conclusion

The Farmington, MN prairie waterway project and 486-unit housing development was finished in the fall of 2000 and was completely sold out by that time. Residents who purchased the $150,000 homes are pleased not only with the fact that they do not have to worry about

Figure I.8.9 View of the third stage of the Park Place housing development with water system. (Photo by Design Center for American Urban Landscape. © Copyright Balmori Associates.) (Color version available at www.gsd.harvard.edu/watercolors. Password: lentic-lotic.)

Figure I.8.10 The 486-unit, completed Park Place development in 2000. (Photo by Design Center for American Urban Landscape. © Copyright Balmori Associates.)

possible flood damage to their homes, but also that they can enjoy the beauty and proximity of the park, pond system, and waterway at the edge of the development. Children play in the ponds and connecting streams. Fishermen catch bass. Families spread blankets for picnics near the water's edge. Joggers and cyclists frequently spot deer, woodpeckers, chipmunks, and other wildlife as they use the paved paths and green spaces (Figures I.8.12 to 8.15).

In addition, neighbors from two nearby developments also use the green space and waterways created. The drainage system in the Farmington Park Place project, which was completed in 1998, has worked well since its timely inauguration with a 100-year flood in September of that year. Then in April 2001, as the Mississippi River again swelled over its banks and flooded large areas, the drainage system prevented substantial flood damage to the immediate area. Some local residents are completely unaware that the waterway/park

Chapter I.8: A productive stormwater park (Farmington, Minnesota)

Figure I.8.11 The finished Prairie Waterway/Park Place project in 2000. (Courtesy of City Planning Department.) (Color version available at www.gsd.harvard.edu/watercolors. Password: lentic-lotic.)

Figure I.8.12 A view of the series of interconnected ponds and streams at Park Place with houses in the background. (Courtesy of William Morrish.)

system they enjoy for recreation is also responsible for solving the stormwater and drainage problems that would likely have existed without it. The productive park is now 4 years into its planting and is on its way to becoming established. In this case, our goal to combine a successful drainage system to control stormwater with usable parkland alongside a stream/pond area indicates that function need not take precedence over a public amenity like green space. The Farmington waterway project shows that a stormwater

Figure I.8.13 Planted wetlands help to clean and slow runoff at Park Place. (Courtesy of City Planning Department.

Figure I.8.14 The gridded street pattern and drainage improvements as suggested to the Sienna Development Corporation. (Courtesy of William Morrish.)

solution, such as a productive park, can indeed work successfully and improve the quality of life for local residents.

Acknowledgments

Design, Diana Balmori (Principal), Patricia Crow, and William Coyne, Balmori Associates, Inc., Landscape and Urban Design, New Haven, CT

Hydrologist Paul Barten, Ph.D., Yale Univ. School of Forestry and Environmental Studies, New Haven, CT

William Moorish and Tom Hammerberg of the Design Center for American Urban Landscape, Univ. of Minnesota, Minneapolis, MN

Figure I.8.15 Residents who purchased the homes are pleased by the beauty and proximity of the 91-acre park, pond system, and waterway at the edge of their development. (Courtesy of William Morrish.)

Developer Rodney D. Hardy, V.P., Sienna Development Corporation, Minneapolis, MN
Richard Krier and James R. Hill, Inc., Planners/Engineers/Surveyors, Burnsville, MN
Ronald P. Peterson, Peterson Environmental Consulting, Inc., Edina, MN
John Corrigan, Soil Conservation Service, Farmington, MN
Glenn R. Cook, Mustafa Emir, and John Smyth of Bonestroo, Rosene, Anderlik and Associates, Engineers and Architects, St. Paul, MN
John Never, MN Soil and Water Conservation, Minneapolis, MN
Dave Ballman, Corps of Engineers, St. Paul, MN
Ken Brooks, MN Dept. of Natural Resources, St. Paul, MN
Susan Galswich, Society for Wetland Scientists, Des Plains, MN
Jim Perry, and Erville Gorham, Univ. of Minnesota, Minneapolis, MN
Neil Diboll, Prairie Nursery, Westville, WI
Pat Lynch, MN Dept. of Natural Resources, St. Paul, MN
Bill Penning, MN Dept. of Natural Resources, St. Paul, MN
Organizations including The Nature Conservancy, Ducks Unlimited, and the Freshwater Institute could be useful in planting or monitoring.

Literature cited

Boon, B. and Groe, H., *Nature Is Heartland: Native Plant Communities of the Great Plains*, Iowa State Univ. Press, Ames, IA, 1990.
Kusler, J.A. and Kentula, M.A., *Wetland Creation & Restoration: The Status of the Science*, Washington, DC, 1990.
Mitsch, W.J. and Gosselink, J.G., *Wetlands*, Van Nostrand Reinhold, New York, 1993.
Minnesota Dept. Natural Resources, *Minnesota's Native Vegetation: A Key to Natural Communities*, State of Minnesota, Dept. Natural Resour., St. Paul, 1993.
Schueler, T.R., *Design of Stormwater Wetland Systems: Guidelines for Creating Diverse and Effective Stormwater Wetlands in the Mid-Atlantic Region*, Metropolitan Washington Council of Gov., Washington, DC, 1992.

Appendix A: Guidelines

Up to this point, I have given a general description of all the parts that went into designing a water system to serve as both a drainage system and a public space, i.e., a productive park. What follows is a synthesis of the main guidelines given to the Farmington Town Board and the Sienna Development Corporation to enable them to carry out the project.

Vegetation

Planting recommendations

Because of the difficulty in predicting post-construction water levels, a final Planting Plan for the Prairie Waterway can be done as soon as the waterway has been dug and the conditions for planting are known. It is possible, however, to set forth some guidelines for planting at this time. Guidelines for plant choice:

- Use material compatible with soils, climate, and local conditions (see Plant Lists). This criterion is of the utmost importance.
- Use material initially that will stabilize banks and prevent erosion. Thereafter, plant with as much local material as possible, because that material has proven to adapt to local circumstances well and is available free of charge, minus labor, which can be voluntary.
- Plant according to hydroperiods — use the right plant for the specific water regime of a given location. (See Wetland Vegetation section that follows.)
- Plant many trees. They provide shade for the waterway, which is important for many reasons including bank stabilization, spatial interest, habitat for birds and other animals, variation from farm fields, windbreak, evapotranspiration, and help in creating a sense of scale.
- Plant shrubs, which will provide shade for small animals, habitat and food for birds, protection of wetlands from windblown particles, enclosure of space, and edge definition.
- Plant grasses and sedges, which provide protection from soil erosion and from drying out, a valuable source of food for wildlife, the basis for rich prairie soil, a medium that slows the progress of the water and catches fine particles, and shelter that allows animals to get around safely.

The range of plant material types will provide a healthy environment for water and a diversity of wildlife. The next step after initial planting to stabilize the waterway is planting trees, shrubs, and permanent grasses, as well as wetland plants.

A number of things will ultimately determine the planting patterns in the Prairie Waterway park, including soils and water regimes, waterway function (such as stability of the form of the waterway, its function as a habitat, and treatment of stormwater), recreational and aesthetic concerns, and maintenance issues.

Wetland vegetation

This list of wetland vegetation excludes *Typha spp.* and *Phragmites communis*, because they are of little value to wildlife. William Mitsch holds that local soil underlain with clay is best: limestone-based gravel rather than silica-based gravel should be used for retention of phosphorus and anaerobic conditions are better for denitrification.

Primary species

Phalaris arundinacea, Sagittaria latifolia, Scirpus fluviatilis, Scirpus americanus, Scirpus acutus, Scirpus validus, Sparganium eurycarpum, Caresx lacustri, and *Carex spp.*

Secondary species

Acorus calamus, Cephalanthus occidentalis, Hibiscus moscheotos, Hibiscus laevis, Nuphar luteum, Peltandra virginicaí, Pontederia cordata, Saururus cernus, and *Zizania aquatica palustris.*

Soils and water regimes

Soils

As shown in the Dakota County Soil Survey, soils in the area of the Prairie Waterway Park Place are made up of Cylinder (129) and Marshan (252) soils.

Cylinder 129

The pH of the Cylinder soils becomes more alkaline with greater depth. The range is from pH 5/6 to 8.4. Windbreaks and environmental plantings for this soil include (8–15î) Siberian crabapple, tatarian honeysuckle, Siberian peashrub, lilac, (16–25î) eastern red cedar, hackberry, northern white cedar, bur oak, white spruce, Scotch pine, and (26–35î) green ash.

Habitats most compatible with Cylinder soils are open-land wildlife and woodland wildlife, because the soil is good for grain and seed crops, grasses and legumes, wild herbaceous plants, hardwood trees, and coniferous plants. Cylinder oils allow only fair conditions for wetland wildlife.

Marshan 252

Marshan soils range with depth from pH 5.6 to 7.3. Windbreaks and environmental plantings for this group include (8–15î) redosier dogwood, northern white cedar, American plum, purpleosier willow, tatarian honeysuckle, (16–25î) amur maple, white spruce, (26–35î) hackberry, golden willow, silver maple, green ash, and (35î+) eastern cottonwood.

Habitats most compatible with Marshan soils are open-land wildlife and wetland wildlife because the soil is good for grain and seed crops, grasses and legumes, wild herbaceous plants, wetland plants, and shallow water areas. Marshan soils will allow only fair conditions for woodland wildlife.

Once the waterway has been built, it is intended that Wetland types will be:

Type I: Periodically flooded wetlands
Type II: Sedge meadows
Type V: Ponds with an aquatic bench of emergent wetland plants at the edge

Waterway function

Stability

A cover crop such as winter ryegrass or millet (in summer) should be seeded immediately after excavation to stabilize the soil. It is recommended that the topsoil removed from the excavation of ponds and wetlands be stockpiled and redistributed in the wetlands because this soil will contain seedbanks that should flourish in the new waterway. This has been found to be a successful way of planting constructed wetlands. Whole plants and plugs of 8 to 10-cm diameter from existing local wetlands have the greatest chance of success. Buying plants from a nursery is also successful, but expensive.

Habitat

Varied habitats are included in the waterway because of the different types of wetlands planned. Type I wetlands contain water for a short time after heavy rains or snowmelt and are vegetated with bottomland hardwoods and grasses. Nineteen species of reptiles and amphibians, 11 species of mammals, and 83 species of birds are attracted to this habitat.

Type II wetlands are inland fresh meadow (sedge meadows > 0). These are shallow depressions without standing water but waterlogged within a few inches of the surface during the growing season. They are vegetated with grasses, sedges, and rushes. Seven species of reptiles and amphibians, 10 species of mammals, and 47 species of birds frequent this habitat.

Type V wetland is inland open fresh water with emergent vegetation restricted to an aquatic bench at the edge. Fifteen species of reptiles and amphibians, 12 species of mammals, and 74 species of birds inhabit this area.

Stormwater treatment

Four types of areas related to stormwater treatment will require planting: upland, stormwater swales, the channel basin, and the wetlands. Upland area vegetation, although above the high-water level, protects the wetlands from windblown particles. It is a vital buffer zone that separates the wetlands from windblown particles. In addition, its roots will help stabilize the banks of the channel and break the rainfall, allowing it to filter into the ground, where it hits, rather than washing into the waterway. It also performs a vital role in evapotranspiration. The trees that would be recommended for this area should be long-lived, because they provide the most visible structure of the waterway planting. Shrubs should be dense to block windblown particles during all seasons.

Stormwater swales and the dry NURP will contain sedges and grasses to increase residence time of the water and to provide plenty of surface for catching fines. The main requirement of these sedges and grasses is that they survive varied water conditions, from fairly dry to inundated. In addition, they need to perform well in taking up nutrients. The channel basin planting will also need to survive frequent inundation, as the channel water will frequently rise to fill this area. A dense root system, quickly established, to prevent erosion, will be important.

See the sections that follow for recommendations on planting the wetlands.

Maintenance issues

The path used for bicycling will also allow small maintenance vehicles to get around the site for general grooming of the area. If most trees and shrubs are well adapted to their site, and there is plenty of diversity in the planting, little maintenance will need to be done other than pruning to open up view or remove branches that enter the pathway. The areas that will require somewhat more maintenance are the ball fields and the lawn areas overlooking the ponds. The grass will need to be kept short for families who plan to play and picnic here. The pathway system will also allow access to the water for periodic monitoring.

Aesthetic and recreational concerns

Whether walking, jogging, bicycling, picnicking, playing in the streams, or running in the grass, residents benefit from a diverse, well-planted landscape. Planting can assist with the perception of scale. A user's enjoyment increases if one can locate oneself in the landscape.

Planting for all color and early spring greening will draw people to the park for longer periods of time and increase their enjoyment.

Civic involvement and education

The following are some suggestions of projects related to the waterway and groups that might be interested in becoming involved.

Horticultural organizations in and around Farmington may know of a regional expert in species identification who would be able to make an appraisal of the wetlands to be replaced. Some of the plants from these wetlands may be incorporated into new wetlands.

Local genotype planting material will succeed better than material from another location. A local Heirloom Seed Group and local nurseries would be possible sources of such material.

Most service organizations, such as Rotary Clubs, Veterans' groups, or Girl Scout and Boy Scout troops, plan projects for their communities. Investment in the park/waterway project could take many forms, including donated time or the donation of special elements such as picnic tables, benches, or bridges.

The Waterway will be a destination for school groups, 4-H clubs, the Photography Club, or garden clubs. Some projects that these groups might find interesting include:

- Making and installing swallow and bat boxes to control mosquito populations near the waterway
- Collecting seeds from local plants to use in the waterway or removing plugs from existing wetlands to plant in the new wetland
- Planting trees as a town-wide effort — the trees that outline the "rooms" can be planted with proper supervision as soon as the waterway seeding has taken hold. Various organizations could choose one or two specific rooms to plant.
- Monitoring — this would be a possible enrichment to the curriculum of a science, geography, or social studies teacher as a short- or long-term project for students. The Freshwater Institute offers seminars to educate teachers about monitoring. A short-term project could entail monitoring the runoff from the housing areas and determining the effects of lawn fertilizer on the waterway, for example.

Other types of projects include wildlife monitoring, including birds, or testing the quality of water where it enters the system, compared with the exit point. James Perry at the University of Minnesota might also be interested in using the Park Place waterway as a study area for graduate students.

Response

Successfully marrying form and function in stormwater management

This chapter by Diana Balmori demonstrates the successful incorporation of many elements of stormwater management described in other chapters in this book: low-impact development approaches (Larry Coffman and Michael Clar in Chapters I.5 and I.15, respectively), detention ponds (Desheng Wang in Chapter I.2), swales and streams (Glenn Allen in Chapter I.13 and Robert France and Philip Craul in Chapter I.7), and wetlands (James Bays in Chapter I.14).

Balmori's strategy in the Minnesota project outlined in this chapter most closely matches, in both spirit and execution, that of Catherine Berris in Chapter I.9 for a site in British Columbia. The water sensitive design that Balmori presents perfectly encapsulates the paradigm shift occurring in wetland creation in terms of movement from functional stormwater wetlands to multipurpose stormwater wetland *parks*. In this new paradigm, wetlands are no longer seen as just means to improve environmental conditions; they are also regarded as an excuse to create green spaces to improve sociological conditions. Recreation, aesthetics, and other human amenities, once marginalized as afterthoughts, are instead important elements incorporated into the overall water sensitive design from the very start. As this chapter states, stormwater function need not take precedence over creation of a park with corresponding benefits to the quality of life for nearby residents.

Three other features of this wetland park project are worthy of attention:

- The requirement for a large, multidisciplinary team was integral to project fruition.
- Innovative financing in terms of taxes to the developers helped to raise funds.
- The interesting approach of fielding real, viable alternatives, any of which would have been appropriate for selection, instead of stacking the cards with inappropriate, substandard alternatives other than the obvious one the public is "forced" to choose, is noteworthy.

chapter I.9

A stormwater wetland becomes a nature park (British Columbia, Canada)

Catherine Berris

Abstract

The study area is a 57-acre park surrounded by suburban development in Abbotsford, a small city in the Fraser Valley of British Columbia, Canada. The original purpose of the project as defined by the client was to provide a stormwater detention facility to minimize downstream flooding of agricultural lands, with a path surrounding it. The landscape architects expanded the purpose to include a variety of ecological and recreational objectives.

The site consisted of overgrown farmland with a small, channelized creek inhabited by two endangered species of fish. The design and construction resulted in a lush, diverse nature park. Some of the design features and methods included:

- Relocation of the existing creek to include meandering and a larger cross-section
- A large constructed wetland for storage of 10 acre-ft of water to a 5-ft depth
- Retention of the best existing vegetation
- Shoreline designed for variety and edge effect with peninsulas and islands
- Wetland, riparian, forest, and meadow plantings, with extensive use of bioengineering
- Landforms and planting designed to shade the creek to the highest degree possible
- Major and minor path system, with some islands inaccessible to humans
- Six structures to provide use and interpretive opportunities, with design reflecting regional context, and built in an environmentally sensitive manner
- Phasing of construction over 3 years to satisfy environmental authorities

Introduction

Stormwater runoff has become a pervasive problem in suburban settings (Chapters I.1 and I.4). The removal of original forest vegetation, stripping of top soil, paving of roads and parking areas, and construction of housing have changed the hydrological cycle.

During storm events, the rate of runoff is extremely high, causing erosion along streams unsuited for the strong flows. In the winter when the ground is saturated, storm events often cause flooding because there is no capacity to hold the excessive amount of runoff. Conversely, in the summer, seepage and gradual runoff from forested areas are no longer

Figure I.9.1 Channelized creek through previous farm field prior to construction. Fishtrap Creek, Abbotsford, BC. (Photo by Catherine Berris Associates.) (Color version available at www.gsd.harvard.edu/watercolors. Password: lentic-lotic.)

available. Stream flows are very low, or nonexistent, water temperatures rise, and water quality decreases.

It has been shown that the total watershed impervious area and riparian forest integrity are measurable indicators of the ecological health of urban watercourses. In British Columbia, ecological health is rated based on the following criteria, among others: excellent (streams that support wild salmon, particularly coho); good (hatchery salmonids and some species of wild salmonids); fair (trout and hatchery fish only); and poor (no fish or coarse fish only, e.g., stickleback and sculpins) (Page et al., 1999; Kerr, Wood Leidal Assoc., 2001).

New approaches to stormwater management are attempting to reduce the extent of impervious surfaces and to promote new forms of development that direct water back into the ground as close to its source as possible (Gibb et al., 1999; Chapters I.3 and I.5). This desire to "cure the disease of urban runoff" (Chapter I.1), however, is not always possible in settings that have already been fully developed.

As in several other chapters in this book (e.g., Chapters I.2 and I.8), this chapter describes a stormwater management system that has been incorporated into a developed area. It presents a story about palliative care. It describes how a stormwater management wetland was introduced into a community to address the flooding and habitat problems caused by suburban development.

The context

The project was initiated in 1990. The site was a 57-acre area of overgrown fields containing a small channelized creek (Figure I.9.1), some scrubland, and a small area of mowed grass with a path installed by the Parks Department (Figure I.9.2). Fishtrap Creek drains 7,500 acres of urban upland and agricultural lowland south to the Nooksack River in Lynden, WA.

The problem was frequent flooding of rich agricultural lands. The cause was the suburban development in the upper watershed, development undertaken in the only way known at the time. The land was cleared, roads installed, storm sewers constructed, houses built, and then street trees (sometimes) added.

Figure I.9.2 Small park area near creek prior to construction. Note mowed grass park, walkway, and flooding of creek. (Photo by Catherine Berris Associates.) (Color version available at www.gsd.harvard.edu/watercolors. Password: lentic-lotic.)

The purpose of the project, as the client defined it, was to provide a stormwater detention facility. The project function was to attenuate post-development flows from 1- to 10-year storm events lasting 1 to 24 h. A predevelopment runoff of 92 ft^3 per day (which became the outlet control flow) and a post-development runoff of 512 ft^3 per day were predicted for a 10-year, 1-h duration storm (Hicks et al., 1993). The governing detention of 10 acre-ft was determined for 10-year, 6- and 12-h duration storms.

Engineers were hired to design the stormwater management system. Fortunately, early in the process, the engineer called a landscape architect and said, "The City has purchased some land for a stormwater detention facility. They would like you to design a path to go around it. By the way, two endangered species of fish live in the creek system, so the environmental agencies will have some input into the project." That started a planning, design, and construction process that spanned five years.

The project had significant requirements. The need to store 10 acre-ft of stormwater meant that most of the existing vegetation had to be cleared. The creek supports cutthroat trout and two endangered fish species: the Salish sucker and Nooksack dace. Environmental authorities required the project to be sensitive to the habitat needs of those fish. The main requirements were to allow instream construction only during the 2-month period least critical to the fish life cycle, to minimize siltation, and to limit water temperature increases in the future. Because of the concerns about potential impacts on fish habitat, a 2-year delay was required between completion of the first phase — the south basin — and construction of the north basin.

The primary landscape architecture design concept was to give people that little bit of Thoreau that we all crave. Many people moved to Abbotsford for a small-town, "rural" experience. It is a town surrounded by agricultural fields, streams, and hedgerows. Within the community, however, those bird sounds and the ability to wander through a natural space had become scarce. The purpose of the park design therefore included:

- The development of a park with a variety of natural habitats, to bring back a semblance of the nature that was there before the land was cleared for agriculture
- Promotion of use by wildlife, for the benefit of all species, including the human one

Figure I.9.3 Site during construction. Extensive removal of material was required for stormwater storage. (Photo by Catherine Berris Associates.) (Color version available at www.gsd.harvard.edu/watercolors. Password: lentic-lotic.)

- Provision of nature-oriented recreational opportunities, like bird watching, walking through clean air, and appreciation of nature's sights and sounds
- Provision of ecological and cultural interpretive facilities, to enable learning about the natural systems on the site and its past history of early pioneer homesteads

Wetland design

The basic approach to the site design included relocation of the existing creek to include meandering and a larger cross-section. The channel was designed to provide a 6-ft depth at low water, in order to remain cool for fish and to maintain hydraulic efficiency. The meandering was introduced in order to provide a more natural appearance than the previously constructed drainage ditch.

Adjacent to the creek is a large constructed wetland capable of storing the excess stormwater that overflows the creek banks. During storms the water level can rise 5 ft, and then it is gradually released through a control structure downstream of the wetlands. Riparian slopes around the wetland range in slope from 3:1 to 5:1. This grade provides stability from erosion and the opportunity to support riparian vegetation easily, and it allows for a return to the existing grade within a reasonable distance, thereby enabling the protection of the best existing vegetation. Even with these slopes, most of the site had to be cleared and excavated (Figure I.9.3).

The process of designing the shoreline configuration around the creek and the wetland involved a significant collaboration among the engineers and landscape architects. The required storage volume for stormwater was a major limiting factor. From a design that began as a large oval, every "pushing" of the shoreline toward the wetland to provide habitat, lookout points, and shade for the water required a respective "pulling" of the shoreline back in another location (Figure I.9.4). Space was extremely limited, and consideration was given to provision of appropriate buffers for residents and protection of important existing habitat, such as a raptor nest.

The resulting configuration includes variety and a significant amount of "edge" with peninsulas and islands. The landforms and planting were designed to shade the creek to the highest degree possible; however, effective cooling was difficult due to the size of the wetland and the extent of the area to be planted.

Figure I.9.4 Plan of north basin. Note meandering creek surrounded by wetland, small patches of retained forest, and proposed walkways and structures. Ball fields were preexisting. (Photo by Catherine Berris Associates.) (Color version available at www.gsd.harvard.edu/watercolors. Password: lentic-lotic.)

Enhancement of water quality through trapping, transformation, and storage is an important function of the wetlands. Sediment traps were built on all urban storm flow inlets to the wetlands to capture heavier sediments.

Planting

Planting in the park included wetland, riparian, forest, and meadow plantings, with extensive use of bioengineering. The park involved revegetation of a very large area. About 35,000 trees, shrubs, and ground cover plants and 66,000 wetland and riparian plants were installed. The plants are almost all native, with imported species only in the more manicured portions of the park. The growth rate has been outstanding.

Wetland test plots of different species were established immediately after the excavation was complete. Because of the depth of water inundation, cattails were by far the most successful wetland plant. Although extensive planting of cattails can limit diversity, in this case the environmental agencies were concerned that vegetation in the wetland become established quickly, and cattails offer high value for biofiltration. One challenge was that a severe winter resulted in freezing of the wetland after a fall planting. The ice pulled numerous wetland plants from the ground.

The riparian slopes were planted almost exclusively with live stakes and brush layering of willow and dogwood. Their success was primarily dependent on the slope material. Where there was native peaty soil or an imported growing medium, the live cuttings established extremely well. In locations with a high proportion of gravel on the slopes, the survival rate was lower. In some areas, the growth of willows was so significant that views of the creek and wetland along major stretches of the paths, as well as views from benches, were blocked. In these areas, the city was advised to clear view windows after several years.

All areas of the site above high water were hydroseeded. Trees and shrubs were planted within the grass areas. This method resulted in some loss of shrub material,

Figure I.9.5 Entry pier extending over creek as seen across new wetland. (Photo by Simon Scott.) (Color version available at www.gsd.harvard.edu/watercolors. Password: lentic-lotic.)

because shrubs were outcompeted by the grass, but the eventual appearance achieved the desired effect of naturalized scrubland. In other projects, a ring of mulch around each tree and shrub helped to increase the success rate of plants. The alternative of mulching beds around shrubs results in a more formal appearance, especially in the short term, and it also requires a high level of maintenance in the first few years; however, more shrubs survive.

The tree planting program was very successful, and there was a high survival rate. Some trees were planted in monoculture groves, with others as mixed forest species. The attempt to establish a grove of Garry oak, which is native to a site not far from the study area, was not successful. The site is probably too low-lying and peaty for that tree, even though the Garry oaks were planted on the upper portion of a hillside.

Wildflowers were planted with the grass at several key locations. Native wildflower seeds were acquired from a specialized supply company. The wildflowers were very successful in the first year, but as is often the case, they became less profuse over time due to competition from the grasses. Lupines have remained a highlight year after year.

Park amenities

A major paved path meanders along one side of the wetland, and a secondary gravel path provides access on the other side. The paths form three major loops, one around the south basin, and another around the north basin, with a pedestrian bridge dividing the latter into two smaller loops. Some of the islands have small gravel paths onto them. Others have wildlife habitat as the primary goal, and these islands are inaccessible to humans.

Six structures within the park support various types of viewing and access and incorporate interpretive information. The design of the structures reflects the regional context, with barn-like roof forms and timber construction. To address the environmental sensitivity of the site, all timbers were precut and pressure-treated off-site. The structures over the water were built on piles to minimize ground disturbance. The structures were designed to blend into the setting, to appear "light" from a distance yet solid from nearby.

The entry pier (Figure I.9.5) is located near the parking lot. It extends over the wetland and is open to the sky and the water in the center. An orientation map and interpretive information are on a sign where the walkway extends onto the pier.

Chapter I.9: A stormwater wetland becomes a nature park (British Columbia, Canada) 199

Figure I.9.6 Picnic shelter on peninsula with pedestrian bridge crossing creek on right. (Photo by Catherine Berris Associates.) (Color version available at www.gsd.harvard.edu/watercolors. Password: lentic-lotic.)

Figure I.9.7 Deck on old railway embankment ends in lookout over park. (Photo by Catherine Berris Associates.) (Color version available at www.gsd.harvard.edu/watercolors. Password: lentic-lotic.)

A picnic shelter (Figure I.9.6) is located on a peninsula central to the north basin, near the pedestrian bridge. The picnic shelter has space for gathering along its side and has an open grass area for associated play. A patch of domestic blueberries, which are a remnant from the farm use, is located near the picnic shelter.

A deck located on an old railway bed (Figure I.9.7) provides the greatest opportunity for historic interpretation. The deck was designed to mimic the railway pattern and provides an excellent overlook because it is on a raised embankment that previously led to a trestle. Old piles from the trestle found during construction were placed vertically within the wetland as wildlife habitat. Laborers from the local correctional center participated in building a stairway onto the railway grade.

Figure I.9.8 Reading shelter. Note interpretive signage on structure and seating steps overlooking wetland. (Photo by Simon Scott.) (Color version available at www.gsd.harvard.edu/watercolors. Password: lentic-lotic.)

Figure I.9.9 Boardwalk to island with interpretive sign panel and bench. (Photo by Catherine Berris Associates.) (Color version available at www.gsd.harvard. edu/watercolors. Password: lentic-lotic.)

Other structures include (1) a small "reading shelter" (Figure I.9.8) that provides a stopping place in sun or rain and gives a view the length of the wetland and (2) a boardwalk over a wetland area to an island (Figure I.9.9). Small tables and benches are distributed in special locations throughout the park.

Interpretive signs throughout the park (Figure I.9.9) explain (1) the overall concept and design of the park, (2) the vegetation, fish, and wildlife and their ecological importance, and (3) the rural history of the area. The signs are almost all integrated into the structures.

Figure I.9.10 Fallen tree left in park provides habitat, play environment, and display of root system. (Photo by Catherine Berris Associates.) (Color version available at www.gsd.harvard.edu/watercolors. Password: lentic-lotic.)

Park management

An important aspect of the park development has been the ongoing management and maintenance. Several trees on the edge of existing forested areas fell during construction due to windthrow. The trees were left in place, where they provide wildlife habitat, climbing opportunities, and a magnificent display of a root system (Figure I.9.10).

Close contact is maintained with the city regarding the extent of mowing in the park. Parks maintenance crews are accustomed to mowing all grass areas. At this park, because the trees and shrubs were planted within hydroseeded areas, it is critical that there be no mowing throughout most of the park. This policy was implemented, and a swath about 3 ft wide immediately adjacent to the main pathway is now mowed, mainly for security (visibility) reasons and to keep the area clear for service and emergency access. This has allowed naturalization to occur throughout most of the park.

Guidelines on weeding are also followed within the park. A naturalized appearance is desired, so volunteer native plants are welcome. For example, during establishment, numerous red alders seeded in one area. Only nonnative, aggressive plants are to be removed, e.g., Scotch broom and Canada thistle. Fortunately, no major invasions of such plants have occurred, and purple loosestrife has not found its way into the park.

Park use

The project supports a high level of use by surrounding residents. As soon as paths were built, they were being used. Only 2 years after completion of the landscape work, the park had a very natural appearance. As the years pass, previously scrubby areas gradually become more forest-like as growth takes place. Many people who visit the site report that it appears completely "natural." The paths through the park are now part of a larger greenway network.

Use by wildlife is also high with inhabitants including turtles, small mammals, and numerous birds such as waterfowl, herons, and owls. The local naturalist club was concerned about the project before construction. Now the naturalists are delighted with the outcome and they use it as an example of a nature-friendly park design. The overwhelming sensations in the park, especially in spring, are the natural fragrances and the constant chorus of birds, providing a sharp contrast with the surrounding developments and road traffic.

In the initial design stages of the project, some concerns were raised about safety, given the 3:1 slopes into the wetland. Children are occasionally seen playing at the edge of the wetland, but there have been no safety problems. The thick, tall growth on most of the riparian slopes makes it difficult to reach the wetland in most locations. Where the vegetation is not thick, the grass is deep and the slopes are relatively easy to climb (i.e., not slippery), so children can readily scamper up the bank.

At several points during construction, neighboring residents expressed concerns about the park development. Site meetings were held with residents to resolve all concerns. Any neighboring resident wanting a fence was offered one. The city now receives many positive compliments on the park from neighbors and visitors, and there is a high level of community pride in this "engineered" facility that has such natural, peaceful qualities.

Summary

This project does not reduce stormwater runoff. It has served to ease the pain that the "disease" of runoff has caused in this local area. The wetlands manage stormwater as intended. The park has also improved the quality of life of visiting humans and wildlife, providing a "poetic response to the prosaic issue of stormwater retention" (Weder, 1997).

The intent of the project was to develop a different type of park, focused on providing people with the opportunity to experience and learn about nature in the city. It has succeeded in providing a sanctuary for fish and wildlife and shows how an environmentally sensitive design can be incorporated into a community.

Acknowledgments

The author thanks the client, the City of Abbotsford Engineering Department, particularly Ed Regts, for the enthusiastic support for all aspects of this project. The primary consultants were Dayton and Knight Ltd., engineers; Catherine Berris Associates Inc., landscape architects; Brad Cameron, architect; and Envirowest Consultants Ltd., environmental consultant.

Literature cited

Gibb, A. et al., Best management practices guide for stormwater, prepared for Greater Vancouver Sewerage and Drainage District, Burnaby, BC, 1999.

Hicks, R.W.B., Regts, E., and Kelly, H.G., Urban wetlands for multidisciplinary stormwater management, *Eng. Hydrol. Symp.*, Hydraulics Division/ASCE, July 1993, San Francisco, 1090, 1993.

Kerr, Wood Leidal Associates Limited, Description of Ecological Health Rating System, North Vancouver, BC, 2001.

Page, N. et al., Proposed watershed classification system for stormwater management in the GVS&DD area, prepared for Greater Vancouver Sewerage and Drainage District, Burnaby, BC, 1999.

Weder, A., Creating a natural diversion, *Azure*, March/April, 32, 1997.

Response

Naturalized design:
Triumph of imagination and innovation

The stormwater wetland project that Catherine Berris describes in this chapter is a signature success story highlighting the triumph of the new paradigm of wetland park creation (as described elsewhere in this book — James Bays in Chapter I.14, Diana Balmori in Chapter I.8, and Robert France and Kaki Martin in Chapter I.12) over the old paradigm of highly engineered detention basin construction. The choice of the word "creation" over "construction" is informative here, for although a thing constructed can be appreciated after it is built, a thing created can be appreciated before it is even designed. Berris' perseverance in shifting the original plans for a simple, single-purpose detention basin to the development of a multipurpose wetland park is a model lesson in the valued role that landscape architects can play in water sensitive designs.

Elements of this project that are particularly meritorious include:

- Phased development
- Integration into the existing residential neighborhood
- Restoration of the "semblance of nature" such that recent visitors unfamiliar with the wetland's creation imagine it to be completely natural (as is the case for most modern visitors to Olmsted's famous Back Bay Fens in Boston)
- Creation of wildlife habitat such as islands for waterfowl or shaded trees to regulate temperatures for salmon
- Acknowledgment of human history
- Great attention paid to recreation opportunities and to allowing people to come into close contact with the water (easier to accomplish in less-litigious Canada than in the safety-paranoia legal environment of the United States)
- Use of architectural structures of a style endemic to the region
- Need for education of city parks maintenance crews with respect for curbing their desire to mow all grass right to the water's edge (also an issue experienced in the project that Nicholas Pouder and Robert France describe in Chapter I.16)

chapter I.10

Wetlands-based indirect potable reuse project (West Palm Beach, Florida)

Larry N. Schwartz, Lee P. Wiseman, and W. Erik Olson

Abstract

As the population in the city of West Palm Beach grows, and demands on water resources for urban use increase, the development of alternative sources of water supplies becomes increasingly important. The city has developed a program to use highly treated wastewater from its East Central Regional Wastewater Treatment Plant (ECRWWTP) for beneficial reuse including augmentation of its drinking water supply. To protect and preserve its surface water supply system and to develop this reuse system to augment the water supply, the city purchased a 1,500-acre parcel, referred to as the Wetland Reuse Site. The Wetland Reuse Site is one component of the Wetlands-Based Indirect Potable Reuse Project. The other component of the project is the Standby Wellfield. The Wetland Reuse Site and the Standby Wellfield sites consist of wetlands and uplands dominated by *Melaleuca*. An important goal of the project was to develop an Advanced Wastewater Treatment facility. The facility produces reclaimed water that, when discharged, will be compatible with the hydrology and water quality at the Wetland Reuse Site.

Introduction

A great deal of research has been performed documenting the ability of wetlands, both natural and constructed, to provide consistent and reliable water quality improvement. Wetlands are effective in the treatment of BOD, TSS, nitrogen, phosphorus, pathogens, metals, sulfates, organics, and other toxic substances (Kadlec and Knight, 1995). Several factors are important in determining the appropriate design of a wetland treatment system. For natural wetlands these include the type of wetlands as defined by the dominant vegetation and soils, the direction and extent of surface water flow to and from the wetland, the location and type of downstream water bodies, the presence of protected species, and the regulatory requirements. Although wetland treatment systems are designed to maximize water quality improvement, the appropriate design will establish a hydroperiod to encourage and maintain valuable wetland communities and wildlife habitat. Therefore, the ancillary benefits that wetland treatment systems provide become a valuable educational and public resource, as indicated in Chapter I.14.

The city of West Palm Beach's water supply system includes a surface water allocation from Lake Okeechobee to the M Canal, via the L-8 Canal, through the Water Catchment

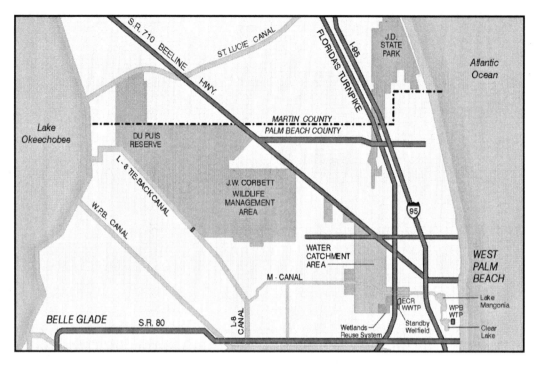

Figure I.10.1 Hydrologic features in northern Palm Beach County.

Area (WCA) to the city's water treatment plant at Clear Lake (Figure I.10.1). As part of the Everglades restoration program being implemented by the South Florida Water Management District (SFWMD), the timing, location, and quantity of water releases to canals from Lake Okeechobee will be modified. More water will be directed toward the Everglades for restoration and less water will be sent to the canals (SFWMD, 1999). This translates into less water available for the city's water supply.

Concurrent with increases in potable water demand, the average rate of wastewater flow from the ECRWWTP is expected to increase to approximately 60 mgd by the year 2010. Presently, all of the wastewater effluent from the ECRWWTP is injected over 3,000 ft into the ground (boulder zone) at the facility using six deep wells. Rather than continuing to dispose of the wastewater effluent from this facility, the city of West Palm Beach, with the assistance of Camp Dresser and KcKee, Inc. (CDM), developed the Wetlands-Based Indirect Potable Reuse Project.

In order to protect and preserve its surface water supply system, and to develop an appropriate reuse system to augment the water supply, the city of West Palm Beach, in conjunction with Florida Communities Trust, purchased land immediately east of the WCA and west of Florida's Turnpike (Figure I.10.1). This parcel, referred to as the Wetland Reuse Site, is one component of the Wetlands-Based Indirect Potable Reuse Project. The other component of the project is the city's Standby Wellfield. The Wetland Reuse Site covers an area of 1,415 acres and consists of a combination of wetlands (wet prairies and forested hammocks dominated by *Melaleuca*) and uplands. The Standby Wellfield site covers an area of 323 acres and consists of similar wetlands and uplands dominated by *Melaleuca*. An important goal of the project was to develop an Advanced Wastewater Treatment (AWT) facility at the ECRWWTP. The facility will produce reclaimed water that, when discharged, is compatible with the hydrology and water quality at the Wetland Reuse Site.

Development of the wetlands-based indirect potable reuse project

Advanced wastewater treatment constructed wetlands demonstration project

In 1992 the city of West Palm Beach implemented an AWT/constructed wetlands demonstration project. The Florida Department of Environmental Protection (FDEP) defines AWT as producing an effluent quality of total suspended solids (TSS), 5-day carbonaceous biochemical oxygen demand ($CBOD_5$), total nitrogen (TN), and total phosphorus (TP) goals of 5, 5, 3, and 1 mg/L, respectively. This 150,000-gal/day demonstration facility was constructed in 1995 and was in continuous operation for over one year. Effluent from the ECRWWTP was treated to AWT levels at the demonstration facility using a solids contact unit for phosphorus reduction and deep-bed denitrification filters for nitrogen reduction. High-level disinfection was provided before discharge to the constructed wetlands. Planting, application of reclaimed water, volunteer colonization of vegetation, and maintenance for removal of undesirable species resulted in the establishment of a viable wetland ecosystem in the two constructed wetland cells.

The pilot AWT facility met the treatment goals as well as all of the state and federal drinking water standards (except for iron), Florida groundwater guidance concentrations, and surface water quality standards (except for total residual chlorine and coliforms). During the year of operation of the demonstration project, TP concentrations in the AWT effluent were in excess of the TP concentrations at the Wetland Reuse Site. A coagulation process optimization study was conducted to evaluate the feasibility of achieving appropriate reclaimed water TP concentrations for discharge to the Wetland Reuse Site.

The first phase of the optimization study focused on the iron salt coagulation process to maximize phosphorus removal and minimize iron floc and soluble iron carryover on a bench-scale (jar-test) basis. The bench-scale study included testing of three coagulants (ferric sulfate, ferric chloride, and polyferric sulfate) at varying dosages, optimization of pH conditions, and examination of several flocculation and sedimentation polymers for floc conditioning. Alum was also originally considered as a coagulant but was disregarded due to concerns with possible toxicity effects and problems associated with dewatering and disposal of alum sludge. Based on results of the bench-scale study, operational criteria for demonstration scale operation were recommended to optimize and evaluate the iron coagulation process. Results of the demonstration-scale study indicated that, during a significant majority of the time, the existing treatment achieved effluent parameter concentrations at or below the anticipated permit compliance limits for iron, TSS, $CBOD_5$, and TN, and the operational goal of 0.05 mg/L for TP. Based on the performance of the demonstration project, the new reclaimed water production facility was designed as an AWT process with coagulation/clarification (for additional phosphorus removal) and deep-bed denitrification filters followed by ultraviolet (UV) light disinfection.

Baseline monitoring of wetlands

As required by the FDEP, baseline monitoring was performed at the Wetland Reuse Site from July 1996 to August 1997. The purpose of this monitoring was to establish baseline conditions in the wetlands prior to reclaimed water application and to determine the appropriate quality of the reclaimed water that will be applied to the Wetland Reuse Site. In addition to the monitoring of background hydrology, groundwater quality, and surface water quality, the baseline-monitoring program investigated sediment quality, vegetation, fish, and the presence of listed threatened and endangered plant and animal species.

Groundwater samples from the Wetland Reuse Site and the Standby Wellfield met the requirements for drinking water except for iron. Iron was detected in excess of the Secondary

Drinking Water Standards of 0.3 mg/l at all of the wells, but not in excess of the Class III surface water quality criteria of 1.0 mg/l. TN concentrations in the wetlands ranged from 0.67 mg/l to 3.85 mg/l, with an average value of 1.36 mg/l. The concentrations of TP were low throughout the wetlands, ranging from <0.01 to 0.13 mg/l, with an average value of 0.027 mg/l.

Hydrologic modeling

A hydrologic model capable of simulating both groundwater flow and overland flow was constructed and calibrated to assess the hydrology, hydrogeology, and potential hydraulic conveyance characteristics within the project area. The groundwater flow computer code used in this study is MODFLOW (McDonald and Harbaugh, 1984). The SFWMD Wetland Package (Restrepo and Montoya, 1997) was used to represent overland (sheet) flow in MODFLOW, the associated hydroperiods, and the interaction between groundwater and surface water within selected wetlands in the modeled area.

Model simulations were performed using water levels and flows at the end of an average wet season followed by a 7-month dry season based on a 1-in-10-year drought condition. Additionally, the water levels and flows were evaluated with current and projected pumping at the Standby Wellfield. The model indicates that maintenance of viable wetlands (i.e., no extended wet or dry periods) on the Wetland Reuse Site and the Standby Wellfield and aquifer recharge to augment the water supply can be achieved. Reclaimed water will initially be applied to the Wetland Reuse Site at a rate of 2 in./week, which corresponds to a reclaimed water flow of approximately 6 mgd over 770 acres of the 1,415-acre site. The results of the modeling indicate that up to 6 mgd of reclaimed water can be applied to the Wetland Reuse Site without producing more than an 8-in. average rise in surface water levels in the wetlands over the 1996–1997 baseline hydroperiod.

A one-mile reclaimed water distribution header will be constructed along the WCA berm (Figure I.10. 2). The reclaimed water will sheet flow across the site to a collection ditch where it will be pumped to the Standby Wellfield for aquifer recharge. The collection ditch and containment berm will parallel the Turnpike Canal and will sever the existing surface water connection. Routing of flow from the Wetland Reuse Site to the Standby Wellfield and more frequent operation of the Standby Wellfield are necessary for project implementation. It is anticipated that the Standby Wellfield will be pumped at an average rate of 7 mgd to increase recharge and provide raw water to the M Canal. The existing raw water main for the Standby Wellfield extends to the M Canal.

Preliminary design

AWT design goals

Consideration of demonstrated treatment technologies and the results of activities to optimize the removal of phosphorus from the AWT discharge indicate that the high-quality reclaimed water facility can be designed to consistently meet permit compliance limits for a TN and TP of less than 3.0 mg/L and 0.1 mg/L (on an annual average basis), respectively. The high-quality reclaimed water facility should achieve operational goals for a TN and TP of less than 2.0 mg/L and 0.05 mg/L (on an annual average basis), respectively, to minimize change in the wetland vegetation communities. Permit compliance levels, project AWT reclaimed water goals, and backgroundwater quality at the Wetland Reuse Site are presented in Figure I.10.3. A commitment to construction and operation of a high-quality reclaimed water facility has been provided to meet these stringent discharge requirements.

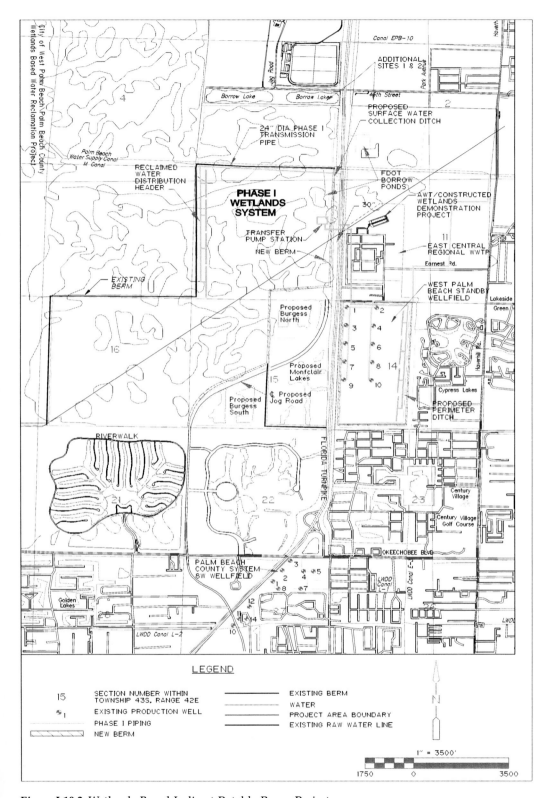

Figure I.10.2 Wetlands-Based Indirect Potable Reuse Project.

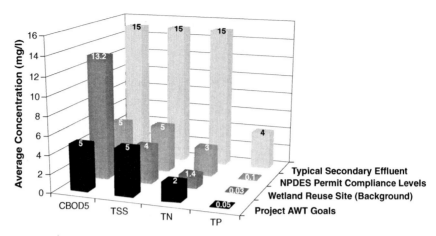

	CBOD5	TSS	TN	TP
■ Project AWT Goals	5	5	2	0.05
Wetland Reuse Site (Background)	13.2	4	1.4	0.03
NPDES Permit Compliance Levels	5	5	3	0.1
Typical Secondary Effluent	15	15	15	4

Figure I.10.3 Comparison of secondary and AWT effluent qualities and permit compliance levels to backgroundwater quality at the Wetlands Reuse site.

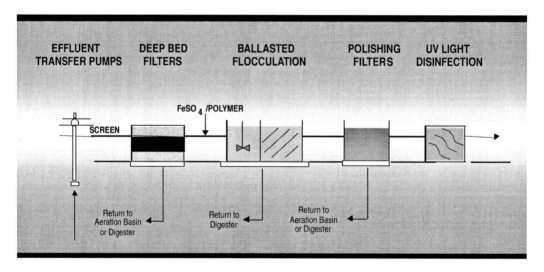

Figure I.10.4 Process schematic for the 10-mgd AWT facility.

The design of this facility was planned to produce a constant flow of 10 mgd of reclaimed water. A schematic process flow diagram for the AWT design is provided in Figure I.10.4.

Four transfer pumps were planned to convey 10 mgd of ECRWWTP effluent to the influent of the splitter box for the deep-bed denitrification filters. Methanol would be added as required to ensure optimal denitrification efficiency in the filters. By moving the denitrification reactor ahead of the phosphorus removal step, more effective and consistent denitrification should be achieved at reduced methanol feed requirements. Effluent from the denitrification filters would flow into a ballasted coagulation, flocculation, and sedimentation process for phosphorus removal. The ballasted process consists of two 5-mgd ACTIFLO™ process trains with the clarification stage sized for 30-gpm/ft^2. A ferric sulfate dosage of 15 mg/L and an anionic polymer dosage of 1 mg/L will be added to improve the flocculation process. The clarified effluent from the ACTIFLO process will flow into a polishing filter. The recommended polishing filter design is an automatic backwash (ABW) traveling-bridge filter (2 units at 5 mgd, each at a maximum filtration rate of 4 gpm/ft^2). Traveling-bridge filters are economical and are in service at numerous wastewater treatment plants for TSS removal. The polishing filter will provide additional floc removal, protect against ACTIFLO upset conditions, and provide additional solids removal. The polishing filter should normally be lightly loaded with solids and should not appreciably increase side stream flows back to the ECRWWTP. The AWT facility will use ultraviolet (UV) light irradiation for disinfection. Since the FDEP does not have regulations addressing disinfection methods other than using chlorination, the California Administrative Code, Title 22 (1978), has been used as guidance.

Wetland reuse system

For the Wetland Reuse Site, a 30-in.-diameter reclaimed water transmission pipe will be extended from the ECRWWTP west to an 8-in. HDPE discharge header located on the west side of the Wetland Reuse Site. Existing containment berms are located along the north and west perimeters of the Wetland Reuse Site. Additional containment berms will be constructed along the east and south boundaries of the Wetland Reuse Site. A low-head transfer pump station will be installed in the southeast corner of the Wetland Reuse Site to convey wetlands surface water to the Standby Wellfield. This transfer station will draw from a collection ditch constructed on the eastern boundary of the Wetland Reuse Site.

Summary

Development of the Wetlands-Based Indirect Potable Reuse Project required substantial vision and effort. The city of West Palm Beach purchased the Wetland Reuse Site in order to protect and preserve its surface water supply system and develop an appropriate reuse system to augment the water supply. The AWT-constructed wetlands demonstration project was implemented to evaluate the feasibility of full-scale implementation of the Wetlands-based Indirect Potable Reuse Project, and a baseline-monitoring program was completed for the Wetland Reuse Site. A study was performed to optimize the coagulation process to achieve appropriate TP concentrations in the AWT effluent. Consideration of demonstrated treatment technologies and the results of activities to optimize the removal of phosphorus from the AWT discharge at the demonstration project indicate that the high-quality reclaimed water facility can be designed to consistently meet permit compliance limits and operational goals for TN and TP.

A commitment to construction and operation of a high-quality reclaimed water facility has been provided to meet strict discharge requirements. The results of modeling indicate that maintenance of viable wetlands on the Wetland Reuse Site and augmentation of the water supply through aquifer recharge can be achieved. The key is proper operational management of the entire system, which includes flow to the Wetland Reuse Site, conveyance to the Standby Wellfield, and pumping from the Standby Wellfield to the M Canal.

Literature cited

California Administrative Code, Title 22, Environmental health — Wastewater reclamation criteria, State of CA Dept. Health Serv., Sanitary Eng. Sect., Berkeley, CA, 1978.
Kadlec, R.H. and Knight, R.L., *Treatment Wetlands*, CRC Lewis Publishers, Boca Raton, FL, 1995.

McDonald, M.G. and Harbaugh, A.W., A modular three-dimensional finite-difference groundwater flow model, U.S. Geol. Survey Tech. Water Resour. Investig. Book A3, 1984.

Restrepo, J.I. and Montoya, A., MODFLOW wetland module final report, prepared for the South FL Water Manage. Dist., 1997.

SFWMD (South FL Water Manage. Dist.), *1998 Everglades Annual Report*, 1999.

Response

Treating wastewater with innovative technology

South Florida can, in many ways, be regarded as a microcosm for both ailments and innovative solutions with respect to water management. Daniel Williams outlines various aspects of regional watershed management in Chapter II.11, and James Bays discusses a wetland creation project in Chapter I.14. In this chapter, Larry Schwartz presents a case study from the same area that is informative in its creative use of science and engineering as integral components toward achieving water sensitive design.

Faced with growing water shortages due to redirection of flows to the Everglades as part of the restorative efforts there, West Palm Beach decided to halt its practice of deep-well injection of wastewater and instead to treat it on the surface through use of wetlands. Schwartz describes various advanced technologies that effectively "reclaim" the treated water for future use. One of the ancillary motivations of this shift in treatment approaches was the understanding that the surface wetland, upon completion, would provide benefits to wildlife and be used as an educational and public recreational resource.

The most important lesson gleamed from this study for water sensitive designers is the recognition that without solid and defensible scientific data, the effective design of treatment wetlands may be problematic, if not impossible. It is absolutely critical that a comprehensive predesign monitoring program be established for projects, especially those implementing innovative technologies. Otherwise, it is impossible to determine the relative quality of the discharge "reclaimed" water following wetland treatment. Further, as Schwartz mentions, biological assessments must also be addressed and the use of hydrological modeling applied as a design tool.

chapter I.11

Restoring urban wetland — Pond systems (Boston, Massachusetts)

Clarissa Rowe

Abstract

The Hall's Pond Sanctuary, located in the Town of Brookline, Massachusetts, is an exceptional natural area within a dense urban environment. Its relationship with the adjacent residential areas, the caring and commitment of the Friends of Hall's Pond, its history as a spark for the development of the Massachusetts Audubon Society, as well as the beginnings of the national conservation movement all contribute to its importance. Visitor pressure, unchecked invasive vegetation, and increased development and runoff have acted in concert to degrade the pristine nature of the sanctuary. The challenge of the Hall's Pond Sanctuary Restoration Project was to improve water quality and to expand, reintroduce, upgrade, and integrate carefully a number of habitat types into a cohesive whole, while providing sensible opportunities for visitor enjoyment and interpretation of a rare and valuable urban wild.

Hill's Pond, located in Menotomy Rocks Park in Arlington, Massachusetts, is another example of a wetland and pond restoration in a dense suburb of Boston. Hill's Pond, three times larger than Hall's Pond, is also the collection point for storm runoff from the surrounding watershed. In 1991, the Town of Arlington let a dredging contract that proposed cleaning up the pond, creating a wetland swale, and placing the pond dredge throughout the park. The challenge of the Hill's Pond project was to get a stalled dredging project completed and to redesign elements of the engineering project to better suit the public park setting.

Introduction

Over the past 100 years, the suburbs of Boston have changed from farmlands and large tracts of family land to densely populated communities that support Boston's economy and position as one of the country's major cities. The farmlands and family lands have been replaced by dense residential development. Luckily, in these two case studies there were individuals and town committees that recognized the land had natural value that could enhance the lives of future generations. In Brookline, which abuts Boston to the west, Minna Hall, a founder of both the land conservation movement and the Massachusetts Audubon Society, was concerned about protecting the pond and the surrounding land for future generations. She offered to donate her family land, about 2 acres, to the town as a nature sanctuary, but her offer was refused. It was not until 1975 that the town's citizens, along with the Conservation Commission, acquired the land in order to prevent its threatened development and subdivision. At the end of the 19th century in the northwest town of Arlington, a small group of prominent citizens, members of the Arlington Improvement Association, saw the value of securing public parkland for the town. As a result, Menotomy Rocks Park was founded in 1896. Hill's Pond, which abutted the original Park, belonged to the George Hill family, which had received title to the land from a 17th-century royal land grant. In 1884 farmer Hill transformed the former Little Island Swamp into an irrigation pond for his farmlands. The town subsequently acquired Hill's Pond in 1924.

Unfortunately, by the 1990s it became apparent that the municipal owners of both Hall's and Hill's ponds had not been effective stewards of their ponds and natural areas. Fortunately, in both cases, dedicated groups of nature lovers formed organizations to oversee the care of the land. In the early part of the 20th century, municipal budgets allowed a large municipal work force, which often included people with skill in arbor care and an understanding of natural systems. In the 1980s, however, Massachusetts passed a tax-cutting measure that allowed municipal budgets to grow by only 2.5% per year. Consequently, money could be spent only on the essentials, such as schools, health, and safety. By the early 1990s, many Massachusetts parks were in dire need of capital improvements and maintenance, and both these projects had been relegated to the bottom of their towns' capital improvement plans. Park commissions were focused primarily on providing active recreation for their towns. It fell to the conservation commissions to be the stewards of the natural and passively used areas. In both case studies, the conservation commissions realized the problems with their ponds and began the process to clean them up. Because the conservation commissions' budgets were minimal, cleanup of the ponds was really accomplished by the vocal park advocates, who never stopped putting pressure on town officials.

The author was involved in both projects. Brown and Rowe, Landscape Architects and Planners, was the prime consultant for the design phase of the Hall's Pond Restoration Project. The author, in her capacity as president of the Friends of Menotomy Rocks Park, served as a pro bono landscape architect for the Hill's Pond project.

The pond science and engineering services for both Hall's and Hill's ponds were provided by ENSR's Dr. Ken Wagner. Dr. Dennis Lowry, also of ENSR, was the wetland scientist for the Hall's Pond project.

Environmental issues at Hall's Pond, Brookline, Massachusetts

Situated in a low-lying and busy urban area, Hall's Pond Sanctuary, Amory Woods Conservation Sanctuary, and Amory Playground represent valuable open space for the people and wildlife of Brookline, MA. Few other significant public open spaces are located in this section of the town, and none with a body of water, which further adds to the sanctuaries' beauty and ecological importance (Figure I.11.1). "A Plan for Hall's Pond Sanctuary," prepared under the auspices of the Brookline Conservation Commission in 1996, outlined several goals for the pond (Giezentanner and Eunson, 1996). The two sanctuaries needed to be reconnected and the wetland between them reconstructed (Figures I.11.2 and I.11.3). Invasive vegetation and water pollution caused by groundwater and storm runoff needed to be brought under control to improve the water quality of the pond. Other goals called for developing a new path system (Figure I.11.4), creating new entrances and gateways, erecting new fences, and reintroducing native plants.

An understanding of the sources of water and contaminants is essential to improving and managing any pond (Figure I.11.5). Generally, three primary water sources are available:

Figure I.11.1 Hall's Pond. (Photo by Alison Richardson.)

Figure I.11.2 Edge of Hall's Pond. (Photo by Charles Mayer.)

Figure I.11.3 Long view of Hall's Pond. (Photo by Charles Mayer.)

direct precipitation, groundwater seepage, and surface runoff. Sources of water and contaminants for Hall's Pond had been evaluated in a past study (Metcalf and Eddy, 1986) and confirmed by later testing (ENSR, 1999). Precipitation, often a minor influence and not especially amenable to control, did not appear to be a critical factor for Hall's Pond. The other two sources were more influential and manageable.

Figure I.11.4 Boardwalk at edge of Hall's Pond. (Photo by Clarissa Rowe.)

Figure I.11.5 ENSR personnel testing water and sediment in Hall's Pond. (Photo by Clarissa Rowe.)

Hall's Pond lies within a highly urbanized watershed. The pond's drainage area covers 107 acres, which can be divided into three subbasins (Figure I.11.6). Subbasin A, which is served by a closed drainage system, covers 89 acres and drains to the pond through a 4 ft × 4 ft box culvert, which is the major drainage inlet to the pond. Subbasin B contributes

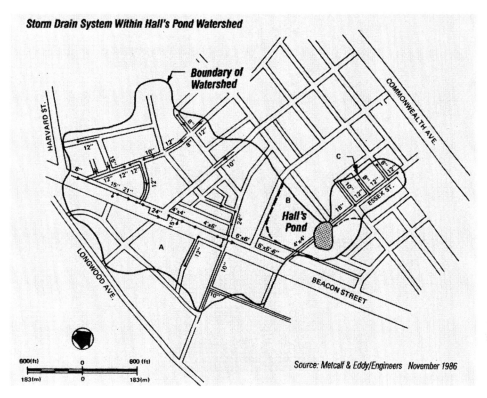

Figure I.11.6 Storm drain system within Hall's Pond watershed, image from *A Plan for Hall's Pond Sanctuary*.

10 acres of overland drainage from the area immediately surrounding the pond, including Amory Playground's ball fields. Subbasin C contributes 8 acres of drainage from several streets north of the pond. The pond drains out through a 60-in. circular metal pipe that runs out to Amory Street and down to the Charles River. The pipe consists of varying diameters on its route to the river. At times the Charles River backs into Hall's Pond.

Groundwater tends to be a fairly stable influence and, in the case of Hall's Pond, appears capable of sustaining the water level at an acceptable level, even during a very dry summer. The pond is shallow, but no reports have been made about the pond going dry during summer droughts. Evidence of substantial groundwater inputs was apparent to the ENSR team. Prior to the routing of stormwater into the pond, groundwater was probably the main influence on water level and quality. Water quality is still affected by the groundwater, as evidenced by the iron and manganese oxides in the pond. These are typically the result of high levels of dissolved iron and manganese in low-oxygen groundwater being exposed to oxygen after seeping into the pond.

The water quality of surface runoff is highly variable in most cases; for Hall's Pond it appears to be the dominant influence on pond hydrology and pondwater quality. Runoff from adjacent ball fields and tennis courts or sanctuary grounds poses some threat because of town maintenance practices, but management is entirely feasible. The discharge of runoff from the stormwater drainage system at Beacon Street, a busy, six-lane commuter road to Boston, has a major impact on water level and quality.

Improvement of water quality is extremely difficult to achieve in urban situations, and very costly. Hall's Pond has been serving as a small detention system for highly contaminated urban runoff for decades. The runoff may account for most of the undesirable features of the pond, including the decline of amphibian populations through either

Figure I.11.7 Invasive vegetation near outflow structure in Hall's Pond. (Photo by Clarissa Rowe.)

toxic reactions or siltation of egg masses during the critical spring breeding season. The alternative to stormwater management would be diversion of the inflow pipe to the pipe around the pond, which is achievable in this case.

Restoration approach to Hall's Pond

Significant enhancement of the habitat potential of Hall's Pond requires a substantial overhaul, involving diversion or purification of stormwater, possible dredging, and edge treatment. There was a desire to reintroduce biological communities, including fish, reptiles, amphibians, aquatic invertebrates, and plants, after the water quality improved and the native vegetation was reintroduced (Figure I.11.7). Although not inexpensive, the restoration program envisioned by the town is feasible and could markedly improve conditions in the pond. Properly implemented, such a program can minimize future maintenance needs, fostering a stable and natural environment for habitat and passive recreational pursuits.

Wetland between Hall's Pond and Amory Woods

The concept of an expanded wetland area within the sanctuary has definite merit, but its development needs considerable thought and discussion within the context of goals and site limitations. The suggested area for the new wetland requires substantial grading if it is to support wetland plants and discourage invasive, non-native nuisance species. Use of wetland as a fringe around the pond has appeal both for filtering the pond water and for habitat enhancement. Location of the wetland and its vegetation features will depend to a large extent on future hydrology, which will be affected by any stormwater management or diversion.

Creation or enhancement of wetlands should be viewed within the context of other, more fixed features such as the pond, garden areas, drainage features, desirable trees, and access points. Wetland-related goals were achievable in multiple areas because of the shallow depth to groundwater and the amount of runoff available. Clear goals are essential, along with some priority for potentially competing objectives (Figure I.11.8).

Vegetation management is also critical in wetland areas. Cattails are present and, although welcome, are expanding; a quantity of jewelweed is present, and the possibility exists for phragmites to take over in the future.

The Friends of Hall's Pond have been the driving forces for the project and the steward of the sanctuary for over a decade. Once the town completes the project, the Friends group expects to take over most of the maintenance of the sanctuary again. With the Brookline Conservation Commission, the Friends of Hall's Pond determined the program for the sanctuary, Hall's Pond, and the expanded wetlands into the Amory Woods Sanctuary.

Wetland creation represents an exciting opportunity for community involvement, and Brown and Rowe designed and organized the effort in a way that would allow volunteers to reduce the impact to the construction budget. Once the work that needed heavy machinery is completed, the Friends group can do some of the planting, replanting, and maintenance of the Sanctuary. The wetland fringe plants can be installed in small sizes and, within 3 years, will reach sizes that give form to the area and will begin to provide habitat value and filtration and aeration to the water. See Appendix A for a list of the wetland plants that will be introduced for the new wetland planting.

Control invasive vegetation and reestablish native plants

Although the preservation of key landscape features and associated vegetation is viewed as an important component of this project, current vegetative features require major alteration to achieve project goals. Change focuses on two key elements — alteration of vegetation to maximize habitat and prevention of nonnative nuisance plant invasions.

Many of the larger, older willow trees need maintenance, but most can be saved even as the wetland is expanded. Much of the understory, however, is dominated by invasive species or plants of limited habitat value and needs substantial clearing and replanting. Although careful design and implementation should establish a fairly permanent vegetation structure, control of existing invasive vegetation and prevention of future successful invasions will require some form of physical or chemical management technique. Physical techniques are more labor-intensive but are amenable to directed volunteer labor and do not carry the stigma associated with many chemical techniques. If physical techniques are employed, great care must be given to removing all roots so that the plants do not grow back. It should be kept in mind, however, that chemical controls have progressed greatly over the last decade to a point at which highly selective and narrowly applied chemicals could be used to open target areas or to control future growth (Figure I.11.9).

Community process

The Town of Brookline was strongly committed to a lengthy and open community process during both the planning and the design phases of the work. During the initiation of the Sanctuary Plan, 20 public meetings about the sanctuary's future were held. During the design phase, 12 public meetings were held. The majority of the public was in favor of the project, but a vocal minority, which included some Brookline Town Meeting members, was opposed to it. Their opposition focused on the wetland being a "swamp," a mosquito-breeding center

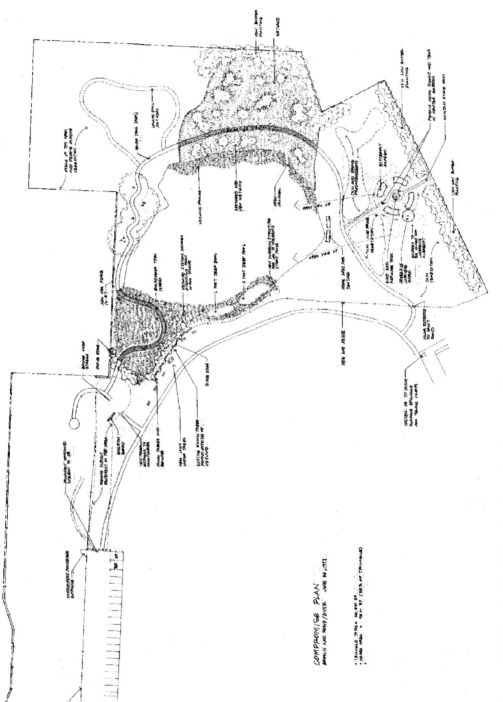

Figure I.11.8 Plan for Hall's Pond produced by Brown and Rowe, Inc.

Figure I.11.9 Boardwalk in wet area at Hall's Pond. (Photo by Clarissa Rowe.)

that would cause disease. Dog walkers also commented that the new sanctuary would exclude dogs in order to protect the other fauna. Some opposition to the project resulted from the proposal to close off a shortcut from the elite, surrounding neighborhood to a nearby elementary school, but this opposition was usually cloaked in "the swamp as a disease-ridden hazard" language instead of the inconvenience of having to walk an additional four blocks to school.

Environmental issues at Hill's Pond, Menotomy Rocks Park, Arlington, Massachusetts

The goals of this project are similar to those for Hall's Pond, except that Menotomy Rocks Park is a public park, not a nature sanctuary. Hill's Pond, part of a wonderful system of water bodies in the Alewife Reservation, is a stopping place for many species of migratory birds. Pedestrians, dog walkers, children, fishermen, skaters, and passive nature lovers use the town park year-round.

In the 1890s, Farmer Hill had turned a swamp on his farmland into a manmade pond of about 3 acres (Mattheisen, 1996). The town acquired the pond in 1924, and it quickly became a centerpiece for winter skating and summer fishing activities. By the 1980s, the pond was in an advanced stage of eutrophication, a condition caused by the quality of the storm runoff from the surrounding residential community and by the presence of invasive, nonnative vegetation in the pond (Metcalf and Eddy, 1986).

Restoration approach to Hill's Pond

Town entities spent considerable time determining how to solve the water problems in the pond. Conservation Commission members pushed to have new stormwater technology solve the problems of runoff. A grease and grit Vortechnics trap (Figure I.11.10) and daylighted stone swale were two elements of the new technology. Both elements cleaned the storm runoff water before it reached the pond itself.

Unfortunately, park users and abutters were not consulted during the engineering design process. When the final construction drawings were reviewed by the park users,

Figure I.11.10 Vortechnics Box installation at Hill's Pond. (Photo by Clarissa Rowe.)

Figure I.11.11 Park users consult with Town Selectmen of Arlington. (Photo by Clarissa Rowe.)

vocal opposition surfaced (Figure I.11.11). The large stone runoff swale was to be located in one of the only flat, sunny, usable areas of the park. The swale that would slow the runoff before it reached the pond was about 35 ft long and 8 ft wide, and lined with large rock boulders. Park users thought that the rocky swale would be an ugly, dangerous addition to the heavily used park area.

Dredge Hill's Pond

In the early 1990s, the town started dredging the 3-acre pond (Figure I.11.12). Dredging the pond was the major task of the project. Significant problems developed during the dredging project. In the winter of 1991, the project was halted and the park was closed to the public. Neighborhood opposition intensified during the dredging construction because

Figure I.11.12 Hill's Pond dredge. (Photo by Clarissa Rowe.)

Figure I.11.13 Hill's Pond dredge on park areas. (Photo by Clarissa Rowe.)

the town closed the park. Rainy weather badly stalled the work. Soon it was apparent that the initial testing underestimated the depth of the dredge, and piles and piles of wet dredge were placed all over the well-used level areas of the park. No plans were included in the construction package for removing any of the dredge from the park (Figure I.11.13). The engineering company had recommended comprehensive testing for the depth of the material in the pond, but the town decided that this testing was not within its budget and thus authorized only limited testing.

During that winter, the Arlington selectmen appointed a 93-citizen committee to study the dredging problem (Figure I.11.14). Despite its size, this committee worked with the town entities to redesign the project: to move the "filtering" wetland area into a corner of the pond; to assist in the construction overview of the project; and to find a loam contractor that would remove the dredge for his loam operations at little cost to the town. The Friends

Figure I.11.14 Hill's Pond fully dredged. (Photo by Clarissa Rowe.)

of Menotomy Rocks Park was born and remains active in advocating for the park in a climate of municipal belt tightening.

Luckily for the Hill's Pond project, the pond dredge was organic and clean because the primary drainage into the pond was from the surrounding residences and residential streets. The only notable contaminant in the water was winter road salt, which could be easily leached out of the soil. Groundwater was not a factor in Hill's Pond. Several small springs do appear to feed into the pond, but they do not really affect the water level. After the dredging project, the pond had to be filled with water from a newly dug well in the park. A pond outfall to control the pond water level was installed (Figure I.11.15). A water line from the well to the pond was installed so that the pond could be fed and aerated during the summer months. The Friends of Menotomy Rocks Park organized, helped pay for, and installed the wetlands planting for the pond (Figures I.11.16, I.11.17, and I.11.18).

In the spring of 2001, only 6 years after the dredging was completed, Hill's Pond was treated again with various herbicides to control invasive weeds. Pond maintenance needs to be an ongoing line item in any municipal maintenance budget.

Install "filtering" wetland

During the redesign process, the wetland was relocated into a corner of the pond itself. The purpose of the wetland was to slow the stormwater from surrounding hills and have the sediment in the runoff settle into a "settling basin," allowing the wetland plants to "filter" or change the nature of the sediment before the water entered the larger pond. An earthen berm, which is usually just above the water level of the pond, separates the stilling basin from the rest of the pond (Figure I.11.19). The Friends group now has years of water data that document the success of the basin in cleaning the water before its goes into the bigger pond. After dredging, the pond was restocked with sunfish and bass so that it remains a popular fishing spot. The Friends group maintains the basin by weeding out any invasive plants, like purple loosestrife, that arrive with the birds. The original planting plan also included cattails, which are pruned back yearly as they attempt to spread into the larger part of the pond (Figures I.11.20 and I.11.21).

Figure I.11.15 Hill's Pond outflow chamber. (Photo by Clarissa Rowe.)

Figure I.11.16 Friends of Menotomy Rocks Park install bioengineering "logs" and plants. (Photo by Clarissa Rowe.)

Figures I.11.17 Planting the wetland at Hill's Pond. (Photo by Clarissa Rowe.)

Figures I.11.18 Planting the wetland at Hill's Pond. (Photo by Clarissa Rowe.)

Conclusion: Understanding the natural systems

The lessons learned from these two urban wetland projects are myriad. Most concern making sure that any municipality engaging on a pond or wetland restoration has an adequate design, construction, and maintenance budget for the work involved.

1. The municipal pond owners should test the pond sediments and surrounding soils at the beginning of any design or engineering effort. Both towns in this case study tried to cut costs during the investigative stage of the project. In both instances, this approach cost them substantial money later in the project. In one project, the pond sediment was actually an asset; it could be sold to a maker of topsoil because it was so rich in organic content and free of contaminants. In the other project, the

Chapter I.11: Restoring urban wetland — Pond systems (Boston, Massachusetts) 229

Figure I.11.19 Earthen berm separates the stilling basin from Hill's Pond. (Photo by Clarissa Rowe.)

Figure I.11.20 Wetland plantings at Hill's Pond thrive 3 years later. (Photo by Clarissa Rowe.)

pond sediment exceeded state standards for contamination and its removal was cost-prohibitive. This finding caused a major rethinking of the project.

2. The pond projects, because they were in urban/suburban areas and were well watched, needed the services of a pond scientist, a wetlands scientist, and a landscape architect throughout the design process. The balance of the disciplines was necessary to cover all aspects of the projects. In Hill's Pond in Arlington, the town again tried to save money by hiring only a pond scientist and civil engineer. The resulting design put an innovative engineering solution right in the middle of one of the most heavily used areas of a popular public park. A landscape architect on the team would have suggested a better location for the swale. The Town of Brookline did not have the funds to continue the services of the landscape architect and scientists during the construction phase. The mix of professions meant that

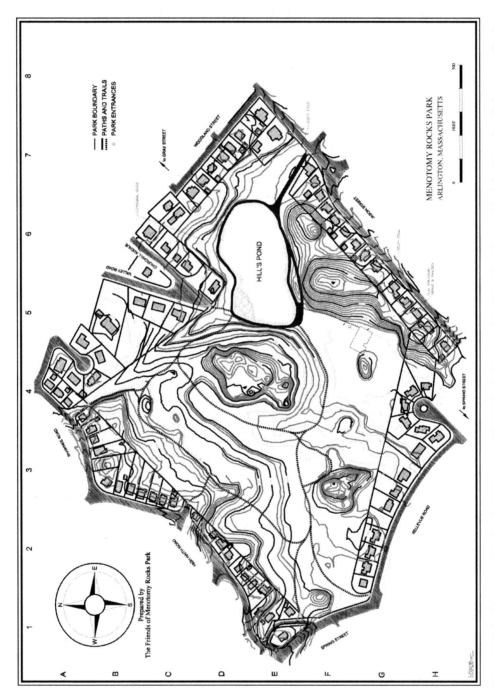

Figure I.11.21 Topographic plan for Menotomy Rocks Park. (Prepared by Tom Gonsiorowski of the Friends of Menotomy Rocks Park.)

the project costs were a lot higher than the standard design and engineering services.
3. In both towns, the clients were neither familiar with the delicate nature of wetland restoration and pond science, nor did they understand how the natural systems of their ponds really worked. Both projects were done under the bailiwick of the Departments of Public Works. Neither department had extensive experience with pond maintenance and construction around delicate wetland areas. In both cases the volunteer boards of the conservation commissions spearheaded the renovation projects but then ran out of or had no funds when the project required almost full-time involvement on the town's part.
4. A comprehensive vegetation management plan should have been developed as part of the design process.
5. The public process is an important step in educating the general public about the long-term health of their water resources. In Arlington, where the public process was not well advertised, the public outcry set the project back by months. In Brookline, the public process required an additional three years of public meetings, and even after that lengthy process, there was still a vocal majority against the "swamp" project. Because of these two public processes, both ponds now have very active and enlarged Friends groups that act as the chief stewards for the continued health of their water bodies. Both still have an uphill battle convincing the financial stewards of their towns to invest in the continued maintenance of these municipal assets. Taking care of these town ponds is a great deal more complicated in the early 21st century than it was "when Farmer Hill and a few mules or horses created Hill's Pond in the mid-19th century and which, then, [the pond] mostly looked after itself," as one of the Friends of Menotomy Rocks Park said recently.

Acknowledgments

Clarissa Rowe, one of the founders of Brown, Richardson and Rowe, Inc., and a resident of Arlington, has been actively involved in providing pro bono services for the Hill's Pond project. Dr. Ken Wagner, the pond scientist from ENSR and the Engineer of Record for the town, continued to work closely with the Friends group after ENSR's contract with the town expired. The Friends also set up an ongoing public education program for the park and the pond. The Friends group won the 1995 North American Lakes Management Society's award for volunteer technical excellence, with special recognition going to Friends' leaders Benjamin Reeve, Town Manager Donald R. Marquis, and Ms. Rowe.

ENSR (155 Otis Street, Northboro, MA 01532) provided the pond science, wetland creation, and engineering services for the Hall's Pond project. Ken Wagner was the pond scientist and Dr. Dennis Lowry was the wetland specialist. Emilie Stuart and Nina Brown of Brown, Richardson and Rowe, Inc., were the editors for this chapter.

Literature cited

ENSR, Notice of intent for Hall's Pond Sanctuary restoration, prepared for the Brookline Dept. Public Works, 1999.

Giezentanner, B. and Eunson, D., Massachusetts Audubon Soc. Environ. Extension Serv., A plan for Hall's Pond Sanctuary, prepared for the Town of Brookline Conservation Commission and Friends of Hall's Pond, Brookline, MA, 1996.

Mattheisen, D., *Menotomy Rocks Park — A centennial history*, prepared for The Friends of Menotomy Rocks Park, Arlington, MA, 1996.

Metcalf and Eddy Engineers, Feasibility study of Hill's Pond, Menotomy Rocks Park, report to the town of Arlington, Arlington, MA, Dec. 1986.

Metcalf and Eddy Engineers, Final report to town of Brookline on diagnostic/feasibility study of Hall's Pond, Brookline, MA, Nov. 1986.

Appendix A

Hall's Pond wetland plant list

These plants will be planted during the Hall's Pond Restoration program:[a]

Arrowhead (*Sagittaria latifolia*)
Atlantic white cedar (*Chamaecyparis thyoides*)
Arrowwood (*Viburnum dentatum*)
Buttonbush (*Cephalanthus occidentalis*)
Burreed (*Sparganium americanum*)
Bluejoint grass (*Calamagrostis canadensis*)
Blueflag iris (*Iris versicolor*)
Bulrush (*Scirpus validus*)
Black willow (*Salix nigra*)
Wood reedgrass (*Cinna arundinacea*)
Chokeberry (*Aronia melanocarpa*)
Cardinal flower (*Lobelia cardinalis*)
Cinnamon fern (*Osmunda cinnamomea*)
Highbush blueberry (*Vaccinium corymbosum*)
Mountain holly (*Nemopanthus mucronatus*)
Marsh fern (*Thelypteris palustris*)
Fowl mannagrass (*Glyceria canadensis*)
Marsh marigold (*Caltha palustris*)
Meadow sweet (*Spirea latifolia*)
Sensitive fern (*Onoclea sensibilis*)
Pickerelweed (*Pontederia cordata*)
Red maple (*Acer rubrum*)
Royal fern (*Osmunda regalis*)
Speckled alder (*Alnus rugosa* or *A. incana*)
Spicebush (*Lindera benzoin*)
Sedges (*Carex crinita* and *C. stricta*)
Sweetflag (*Acorus calamus*)
Sweet gale (*Myrica gale*)
Shadbush (*Amelanchier canadensis*)
Sweet Pepperbush (*Clethra alnifola*)
Soft rush (*Juncus effusus*)
Winterberry holly (*Ilex verticillata*)
Water lily (*Nuphar luteum*)
Water plantain (*Alisma trivale* or *A. plantago-aquatica*)
Wild raisin (*Viburnum cassinoides*)

[a] Wetland plant list by Dennis Lowry of ENSR and Clarissa Rowe of Brown and Rowe, Inc.

Response

Project development through concerned citizenry

In urban settings, ponds become valued for the sense of space and solace they provide in contrast to the landscapes surrounding them. For just as stormwater rapidly courses off impervious pavement, urban lifestyles also seem to rush by atop such surfaces. As Clarissa Rowe discusses, water sensitive design works best from a bottom-up process of engaging concerned citizenry at early, and indeed all, stages of project development. The two ponds described in this chapter suffer from the same problems as those covered by Nicholas Pouder and Robert France in Chapter I.16, and by Thomas Benjamin in Chapter II.4. As in those other studies, Rowe takes pains to state that ultimate solutions lie in integrating site-specific restoration measures within a framework of more comprehensive watershed management. The two case studies outlined in this chapter demonstrate the strength of commitments of local stewardship and the role of conservation commissions in mobilizing restorative actions.

Another interesting element in this story is the manner in which created wetlands were used to help mitigate inflowing problems; i.e., construction of a berm in the inflow embayment to create a small wetland comprising a settling basin and filtering plants for treating the nutrient-laden water before it merges with that of the pond itself. This approach of identifying and then targeting an "area of concern" is similar to that promoted by Gail Krantzberg and Judi Barnes in Chapter II.6 with respect to the entire Great Lakes Basin.

Finally, Rowe's recounting of the — what one would have hoped to be antiquated — perception of wetlands by pond abutters as being disease-ridden swamps harboring all sorts of ills and hazards, indicates that no matter how clever our water sensitive designs may be, public education is necessary. In other words, if the oft-repeated mantra of the real estate profession is, "location, location, location," that of the water sensitive design profession should be, "education, education, education."

chapter I.12

Water connections: Wetlands for science instruction (Wichita, Kansas)

Robert France and Kaki Martin

Abstract

Wetlands provide wonderful opportunities for directly experiencing the workings of nature. By engaging curiosity with exploration, created wetlands can function as outdoor laboratories for experiential science instruction. This chapter describes one such project in Wichita, KS, where a functional wetland is integrated physically into a playground while at the same time fostering environmental learning. Water from a reflecting pool surrounding the learning building is circulated through a treatment wetland for purification through nutrient removal before being discharged back into the pool. Another important element of the project is the role water plays as a physical integrator between the architecture and the landscape components of the site development. Due to the high visibility of the project, careful attention was paid to plant selection, planting, maintenance, and the circumvention of grading problems during construction.

Introduction

An anecdote from a recent wetland conference is very illustrative of the problems underscoring the discipline of treatment wetland design. A well-known speaker who was in the process of updating a major guidebook about wetland construction dodged several questions from the audience concerning the possibility of designing for multiple purposes. Defending himself, this individual stressed that the guidebook he was developing for the EPA was concerned with what he referred to as "blue-collar wetlands," those for which, because of obvious space limitations, it was impossible to create functions (presumably for wildlife benefits, aesthetics, or educational opportunities) other than those intended as engineering solutions to water quality improvement. Such an attitude, especially coming from an individual charged with the important task of writing an official government document, is worrisome given its demonstration of limited design vision.

Unfortunately, such an attitude has been the status quo in the construction and engineering of created wetlands for decades. What is particularly alarming is that it obviously continues to play a blinkering role in restricting vision, as witness to the government official cited previously. It is time to bury such a pessimistic view of the potential for innovative wetland creation. In reality, the difference between a utilitarian blue-collar wetland and a multipurpose "white-collar wetland" has really much less to do about

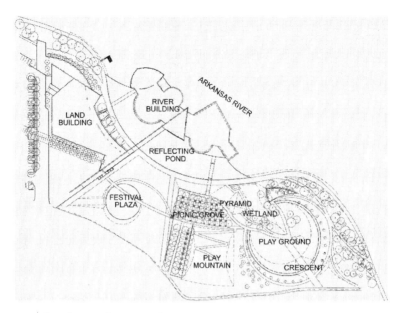

Figure I.12.1 Site plan for Exploration Place, Wichita, KS, showing building surrounded by the Arkansas River and a reflecting pool connected to a treatment wetland within the playground.

absence of space as it has to do about absence of imagination (France, 2002). Today, numerous wetlands combine functional utility with other ancillary benefits (e.g., Chapters I.10, I.13, and I.14). Specifically, the opportunity for creating wetlands to provide natural laboratories for science education is immense (Kusler et al., 1998; France, 2002). For example, the justification for constructing one such school wetland was explained as, "Natural objects inspire a child's curiosity, evoking a passion for learning that stimulates all the senses" (Hammatt, 2001).

Study site description and wetland justification

Located at the confluence of the Little Arkansas and Arkansas rivers, the 98,500 ft^2 Exploration Place building dramatically occupies a 20-acre site at the rivers' edge adjacent to downtown Wichita. Exploration Place has become not only a significant addition to the Wichita skyline but also an innovative learning center that brings together people of all ages to a site where exploring and creativity of all kinds are encouraged.

Due to the spatial adjacency of the reflecting pool and the building and the unobtrusive quality of the dam walls at each end of the pool (Figure I.12.1), from many vantage points one is easily convinced that the Arkansas River flows in its natural course between the two buildings. This feeling is most notable when crossing the interior "bridge" from the Land Building, which houses the ticket booths, administrative offices, café, and museum store, to the River Building, which houses the exhibits. The 41,000 ft^2 pool wraps the landside edge of the River Building and meets the river at either end with concrete retaining walls. During spring flooding or at other times following heavy rains, the water levels in both the river and the pool rise blending the two bodies of water. Most of the time, however, water from the river and the constructed pool do not mix but are separated by the retaining walls on each end such that they only appear to be connected.

The construction of the pool was not complex. The depth of the pool is 3 ft at the deepest points, with the grade sloping to a maximum depth of 6 in. at the perimeter for safety purposes. The pool bed is underlain with an 18-in. clay lens over well-compacted earth with 6 in. of sand over the top to unify the surface. The dam walls at either end do not serve any structural purpose, acting only as retaining walls for the pool. Pools this large and this shallow, with no inherent circulation, have a well-known tendency to accumulate significant levels of nutrients and consequent algal blooms (Tourbier and Westmacott, 1992). Given the importance of visual quality, it was important that the aesthetic of the pool remain intact and consistent. As a result, water quality needed to remain high to prevent unacceptable blooms. At first, the architectural team and the client team considered aerators as their only method to achieve the goal.

Park space

The expansive landscape associated with Exploration Place (www.explore.org) was conceived as a programmatic extension of the inside exhibits and activities. In its physical form, the landscape plays upon and reinforces the strong forms of the architecture. The landscape was designed and developed with the missions of Exploration Place at the forefront: education and scientific exploration for all ages. The designed landscape incorporated, through the spatial relationships between landforms, a compelling landscape that incorporated the goals and desires of the client. Specific features include the "Hill," a dramatic earth mound that utilized the excavated earth from the building construction to make a 18-ft-high point on the site for everything from star gazing to kite flying; a generous picnic area under a grid of trees as well as the earthwork "Crescent," which wraps around the "Kids Explore" playground. The Crescent is an ideal place from which to view the playground as well as the 7,458 ft^2 wetland that extends out from its center.

The wetland was envisioned as another site feature that could be utilized for educational purposes but also provide a strong aesthetic component to the site, primarily through the use of more showy water loving plants such as *Iris versicolor* (blue flag). A water flow connection between the wetland and the reflecting pool (the other water system on the site — discussed earlier) was not considered in the initial design phases though their physical proximities clearly led to the final decision to create a functioning, interconnected water system.

Wetland connections

Water is the means by which the strong architectural form of the building and the structured landscape components of the project meet. In physical terms, the broad end of the misshapen triangular wetland excavation meets the outside of the reflecting pool arc and is separated by only a bituminous concrete bike path (Figure I.12.2).

Reflecting pools such as that at Exploration Place are notoriously prone to nutrient buildup and consequent proliferation of algae. Given the summer temperatures to be expected in this region and the low N:P ratios found in many nearby water bodies, the reflecting pool will have a high propensity to develop nuisance blooms of nitrogen-fixing blue-green algae. Even if new nutrient loads are reduced, the pool would still be susceptible to eutrophication due to resuspension of bottom sediments. Given the shallow depth of the reflecting pool in relation to its length of openwater fetch, empirical relationships (France, 1997) indicate that wind-induced mixing will be a frequent event.

The other problem of reflecting pools occurs as a result of the death, settling out, and decomposition of phytoplanktonic algae. To counter the ensuing anoxia, large fountains are often used to oxygenate the water. In the Midwest and West, these frequently take the form of expensive water jets that notably detract from the site aesthetics. Such water jets will also serve to recirculate bottom sediments, thereby possibly sustaining algal bloom conditions.

Figure I.12.2 Image of hydrological connections between the treatment wetland and reflecting pool. Water from the latter enters the wetland through the right channel, swings around the central berm, and then exits back to the pool via the large channel in the middle of the photograph. Note how the building seems to "float" on the water of the reflecting pool and river. (Color version available at www.gsd.harvard.edu/watercolors. Password: lentic-lotic.)

As a result of some brief discussion about the specifics of the site including the size of each water body and the construction technique used to build the reflecting pool, as well as the overall goals of the Exploration Place, it became clear that by hydrologically connecting the two water bodies it would be possible to make a dramatic contribution in two ways. Not only would reliance on aerators in the reflecting pool be reduced but also a more substantial educational element to the park would be added by making the wetland truly functional rather than being merely "aesthetic." Our plan is therefore based on bridging — both physically and conceptually — the bike path separating the building–reflecting pool side and the park–land works side. There is a comprehensive literature demonstrating the role of constructed wetlands in improving water quality (e.g., Campbell and Ogden, 1999; France, 2002 and references therein). A secondary analysis of nearly 100 case studies (France, 2002) indicates that reductions in the range of 95% for phosphorus and of 90% for nitrogen are possible in constructed wetlands. By circulating the water from the reflecting pool, through the wetland, and back into the reflecting pool, water quality will be improved and the number of aerators reduced.

Wetland design

System elements

Whereas the perimeter form of the intended wetland did not change in the revised process of connecting the two systems, the interior of the wetland did. In order to increase the flow length of the water, it was necessary to locate a berm down the center of the wetland. This resulted in the desired effect of doubling the water flow length and because the berm originates from the bike path, access was created to allow teachers or museum guides to actually bring small groups into the wetland (Figure I.12.3).

Long linear wetlands are ideal for nutrient removal due to maximizing opportunities for contact of water with plants and associated epiphytes. In this case, the berm prevents flow short-circuiting and therefore goes toward ensuring efficient nutrient stripping. A

Figure I.12.3 Construction of the central berm designed to increase the flow path distance for nutrient removal in the treatment wetland. This area will also function as an avenue to bring students into the wetland for study. (Color version available at www.gsd.harvard.edu/watercolors. Password: lentic-lotic.)

secondary analysis of constructed wetlands (France, 2002) revealed that distances between 60 to 120 ft are sufficient for the effective removal of most contaminants. The central berm creates a treatment distance of 318 ft for this otherwise small wetland. This is a more than adequate distance for reducing the amount of nutrients expected in the system.

Water enters the constructed wetland through a PVC pipe that perforates the concrete wall of the reflecting pool. A small submersible pump is located at the far end of the reflecting pool to aide in the movement of water from one end of the 584-ft-long reflecting pool to the wetland outflow pipe. A pretreatment sediment forebay 3.5 ft deep serves to dissipate the energy of the incoming flow. This allows sedimentation to be confined to a single region rather than running the possibility of clogging wetland planting beds. The nearby pedestrian pathway provides easy access for maintenance should dredging become necessary in the future.

Because a mixture of flooding regimes expands wetland functions, interspersion of shallow and deep water areas will increase the efficiency of contaminant removal and wildlife attractiveness. Edges of the pretreatment forebay are planted with narrow-leaved cattail (*Typha angustifolia*) and common 3-square bulrush (*Scirpus pungens*). Water from the pretreatment forebay moves over a small earthen dam that prevents the water from leaving the forebay until the water level rises high enough in the forebay to begin to overflow the dam, sending the water into the shallow wet meadow zone. In this zone water is no more than 4 in deep. Sedges and flags dominate in this zone, including sweet flag (*Acorus clamus*), blue flag (*Iris Versicolor*), fringed sedge (*Carex crinita*), and fox sedge (*Carex vulpinoidea*). The grade drops, allowing the water to gradually become deeper water. Soft rush (*Juncus effusus*) and rice cutgrass (*Leersia oryzoides*) are planted in this transition zone that culminates in a second pool dominated by a stand of giant burreed (*Sparganium eurycarpum*). This area marks the midpoint of the wetland circuit. Continuing from the burreed stand, the grade slowly drops again and the planting is pushed to the edges, allowing for more open areas of water. Broad-leaved arrowhead (*Sagittaria latifolia*), arrow arum (*Pelandra virginica*), spike rush (*Eleocaris paustris*), and lizard's tail (*Saurus cernuus*) cover the edges of the outer wetland edge and the berm edge in this zone, allowing for a continued

Figure I.12.4 Newly planted wetland showing the importance of grading to plant survival. (Color version available at www.gsd.harvard.edu/watercolors. Password: lentic-lotic.)

area of open water. The end of this linear zone culminates in the final micropool. This pool, situated at the discharge outlet, further reduces the release of sediments and floating organic matter back into the reflecting pool. As the case for the pretreatment sediment forebay, the proximity of the pedestrian pathway will allow for servicing of this area should the need arise. Soft-stemmed bulrush (*Scirpus validus*) provides the vegetative cover here. The berm is planted with a dense shrub layer of button bush and two specimen trees, *Quercus bicolor* and *Cornus amomun*, that will eventually provide some shade. The heavy cover on the berm not only provides additional wildlife cover but also discourages access by visitors when specific wetland activities are not occurring. And, in order to stabilize the steep banks of the wetland, all edge areas were planted with prairie cordgrass (*Spartina pectinata*).

Planting list and strategy

Our criteria for screening potential wetland plants included:

- Demonstrated success in the bioregion of the western plains
- Attractiveness — a visually diverse palette for humans
- A structurally complex habitat and food source to wildlife
- Demonstrated performance in improving water quality
- Ease of availability

Our selected list contained 2 submerged plants and 17 emergents in addition to 5 shoreline edge or wet meadow plants, and 5 woody shrubs and tree species.

Implementation

Grading

Often, a major factor affecting wetland success is the creation of precise elevations. These will influence the hydrologic residence time and thereby the performance of the wetland in terms of water quality improvement. Grading is also important for the survival of wetland plants, some of which require a narrow inundation range in order to achieve maximal growth and, therefore, phytoremediation (Figure I.12.4).

Close supervision of site work is essential for accurate adherence to design guidelines. The best plans in the world may never bear fruition if the site work crew does not follow plans properly or if less-than-ideal weather conditions during construction prevent the site work from being as precise as is required. In the present case, torrential rains led to saturated ground conditions in Wichita during the final grading of much of the site, including the wetland area. More than once, the weather not only delayed work but also caused work that had originally been done correctly to either become substandard or be lost all together. For example, one corner of the wetland became silted up during one such storm event. Although the final grading was correct and had been approved, the subsequent rain and fact that the slopes of the wetland had yet to be planted or sufficiently protected against erosion meant that considerable corrections needed to be made before the clay liner was laid down. Further, weather conditions and a well-thought-out work schedule are important not just for the moving and compacting of earth to happen efficiently and correctly but also for the installation of the clay liner. If it is too hot and dry, the liner can dry up and become cracked if water is not almost immediately placed in the basin. Conversely, if it is too wet, the manipulation of the clay liner can be unnecessarily difficult for work crews.

Planting and maintenance

Although landscape construction crews will have considerable experience with terrestrial trees, their knowledge about the particulars of successful planting of wetlands is often limited. To counter this possible lack of experience, we provided a brief outline for the contractor undertaking the plantings.

Timing

For seeding, timing is not particularly critical. With nondormant spring planting, however, timing of planting may be one of the most important elements in project success, such that if project delays negate spring plantings, it is often wise to delay planting for an entire year rather than attempt to do so one month late, which is outside the ideal success window. For fall planting, dormant plants will have greater overwinter survival and be ready to grow in the early spring during freshets.

Plant material

All nondormant plants should be healthy with vigorous leaf, stem, and root systems, free of disease or substantial damage. Potted plants should have well-developed root systems such that if they can be easily removed from their containers they should be rejected.

Storage

During transit and prior to installation, plants must be kept covered, moist, cool, and out of the elements. If not planted on the day of delivery, all plants should be draped in moist matting such as straw. Tarps may need to be erected during the sorting period to block sun and drying wind in order to prevent desiccation. Staggered arrivals will allow for planting to occur within 2 days. For dewatered plantings, it is essential that the entire process be completed within 5 days.

Layout

Demarcated areas must be clearly laid out with identifying stakes (including upslope locations) prior to arrival of plants so as to minimize vegetation storage time. Obviously, a good understanding of the expected water depths will enable identification of the correct zonation of planting areas tailored for each species.

Spacing

A spacing interval of about 3 ft for emergents and 1.5 ft for small flowering plants should be adopted to allow for spread. The planting density and stock size should be increased to compensate for expected losses in high-energy systems (the alternative in the present situation is to reduce, or better yet, stop, the flow of water from the reflecting pool until the plants are healthily established). Some plants such as *Typha* are aggressive and may need to be physically restricted within buried *in situ* contained pots. Plantings should be extended upslope at least 3 ft above the expected water level.

Fertilizer

Some experts favor fertilization for all wetland plantings; the decision should be based on anticipated water nutrient levels, expected rate of vegetative spread, and depth of water. Most deepwater plantings probably do not require fertilization. For others, placement of a tablet at the bottom of the planting hole rather than being applied aqueously to the entire area will limit development of algal blooms. For nondormant plants, the mixture should be at least 19% N, 6% P, and 12% K.

Water

There are pluses and minuses for dewatered plantings over plantings in the expected water levels. If planting in dewatered conditions, there may be a tendency (1) to place the materials at soil depths that are too shallow and/or (2) not to firmly plant into the soil such that once water is introduced, the plants will float free. Peat pot plants should be inserted at least 1 in. (but not more than 2 in.) deeper than what they were grown at in the nursery to ensure root/rhizome coverage. In most cases, 2- to 3-in. (not more than 4-in.) depths are ideal for littoral plantings. Gradual flooding is required to protect the liner, limit suspension of sediments, and allow plants to become acclimatized. Plantings should be allowed to become well established for at least 6 weeks before contaminated water is introduced into the system and access permitted for herbaceous animals. Some plants may require water fluctuations, but limit drawdown periods until they become well established.

Plant establishment period

In most traditional landscaping, the plant establishment period is typically one year, during which the contractor guarantees the survival of the plant material. Such a procedure should also be insisted on with respect to wetland plantings (Dunne et al., 1998). The "one-year" establishment period should extend from the time of planting through August of the second summer (i.e., generally more than a 365-day year). At the end of this period, areas with less than 20 to 50% aerial cover or less than 50 to 80% plant survival shall be deemed deficient and in need of replacement plantings or other maintenance. Sometimes it is possible to assess wetland planting success at a much earlier date (see below). The desired goal expected in the subsequent year after wetland planting should be the establishment of viable populations of a wide variety of species. Once the plantings become established, it is important to recognize that succession will take over and little or no subsequent management will be required.

Herbivory

Once emergents are established, the attractiveness of the area to geese will be reduced. In the meantime, however, it may be necessary to install a protective lattice network of strings tied to wooden stakes in areas where waterfowl and rodents may be present.

Drawdown

In future years, a spring drawdown may be a good idea to "kick start" the system by allowing increased light penetration to reach the plants.

Project tending

As quoted in the wetland design manual of France (2002): "Failed wetland creation/restoration projects abound due to opaque goal aspirations, poor engineering and ecological design, inadequate background data collection, absence of adaptive management, the erroneous perception that project termination occurs with planting, and a cavalier process of critical evaluation"; and "Wetland creation by no means ends with simply filling in a hole with water over which plants seeds are sprinkled or vegetation is hand-planted." Therefore, the following essential duties need to be addressed for successful implementation of any wetland creation project:

- Long-term responsibilities and funding
- Routine schedule for cleaning and checking inlet and outlet water control installations and pumps
- Mowing and embankment inspection
- Depth of sediment accumulation and maintenance schedule for dredging the sedimentation forebay
- Water level and hydrological budget recalculation once emergent plants become mature and begin to restrict water flow (their intended job for chemical cleaning)
- Periodic water quality monitoring of both reflecting pool and treatment wetland
- Periodic survey of plant species composition, density and viability, and overall wetland health

Additionally, plant resources and availability were an issue that the landscape contractor needed considerable guidance with. Plant resources were researched and determined by the landscape architect. In this case, bare root plant material was shipped next-day air from the Midwest. This required coordination and precise scheduling on both ends.

Replanting

For the present case, a diminishing project budget did not allow for weekly or sometimes even bimonthly visits during construction. Trips were planned for efficiency. In some cases this meant not being on-site to oversee significant moments of the construction process from beginning to end. The planting of the wetland was one such moment. Approximately one month after the planting was done, it was obvious that there had been significant plant loss. Although the reason for the failure is not exactly known, it is believed that the wetland was not filled with water quickly enough to support that plant health. Of the approximately 724 plants placed, 500 required replacement. By the time all parties agreed to do a second planting, plant availability had dwindled on some of the original specified species. In this case, three substitutions were made. The second planting proved to be far more successful, and a digital camera sent daily images to the designer to ensure proper progress.

Post-construction

Within days of the flooding of the wetland, even without a very high plant survival rate, the wetland was active. Insects and birds of varying types were almost immediately audible and visible in the wetland. Today, the wetland is a functionally healthy and aesthetically visible component of the overall site, and has become a popular destination for students and their teachers or parents.

Future education program

With time, this wetland, primarily designed for water quality improvement, could play an additional role in outdoor education as part of the Exploration Place program. One important aspect characteristic of educational wetlands is the manner in which they bring students into a close relationship with nature. It is easiest to imagine this occurring at the present site if there were a raised boardwalk partway along the central berm, perhaps augmented with an overhanging observation platform. A series of small step-down platforms along the complete length of the wetland could offer students access to the water for chemical sampling, in order to monitor how effective the system is in reducing nutrient concentrations.

Second, although often underfunded, creation of an education program utilizing illustrations and accompanying text on interpretive signs can be a good way to connect people with wetlands. In the present case, such outdoor education could be supported by establishing a teaching program inside the museum. Information such as brochures, slide shows, videos and models, and nonstructural elements, such as school science projects and guided tours, would go a long way toward promoting the goals of this project. Although most interpretive programs do a reasonable job in conveying information about the plants and animals present in wetlands, there is a need, especially in cases such as this, of raising public awareness about the wider benefits that wetlands contribute to watershed processes, in particular with respect to their function as purifying kidneys in water quality improvement.

Literature cited

Dunne, K.P., Rodrigo, A.M., and Samanns, E., Engineering specification guidelines for wetland plant establishment and subgrade preparation. Wetlands Res. Prog. U.S. Army Corps Engineers, 1998.

France, R.L., Land–water linkages: influences of riparian deforestation on lake thermocline depth and possible consequences for cold stenotherms. *Can. J. Fish. Aquat. Sci.* 54, 1299, 1997.

France, R.L., *Designing Wetlands: Principles and Practices for Landscape Architects and Land-Use Planners.* W.W. Norton. In press, 2002.

Hammatt, H. 2001. Mind games. *Landscape Architecture Magazine*, Jan. 2001.

Kusler, J.A. et al., *Guidebook for Creating Wetland Interpretation Sites Including Wetlands & Ecotourism*, Assoc. State Wetland Manag., 1998.

Tourbier, J.T. and Westmacott, R., *Lakes and Ponds*, Urban Land Inst., Washington, 1992.

Response

Project development of interpretive wetlands

As the project described by Robert France and Kaki Martin in this chapter illustrates, an incredible didactic role can be played by water sensitive design. Whether approached as small-scale demonstration projects such as that described by Nicholas Pouder and Robert France in Chapter I.16; as incredibly beautiful projects at high-visibility sites such as those presented by Glenn Allen in Chapter I.13, and Thomas Liptan and Robert Murase in Chapter I.6; or as large-scale, completely functional projects with detailed interpretive signage, such as that outlined by Catherine Berris in Chapter I.9, water is an element that easily lends itself to instructional use for learning about how the environment functions and how humans interact with nature. The opportunities for developing interpretive wetlands for environmental education are enormous and something sorely needed to raise the image that many in the public still have about such noxious "swamps" of ill repute (see Chapter I.11 by Clarissa Rowe).

Another salient message put forward by France and Martin which is of importance to water sensitive designers, is the sobering realization that no matter how good one imagines a design to be, unless time is spent carefully supervising the contractors' physical implementations of that design upon the landscape, the best intentions can fall far short. In this respect, "project development" from conception to final implementation (i.e., the "cradle-to-grave" mentality endorsed by those dealing with the environmental sustainability of commercial products) becomes critically significant for measuring lasting project success.

chapter I.13

Constructed wetlands and stormwater management at the Northern Water Feature (Sydney Olympic Park)

Glenn Allen

Abstract

The redevelopment of the Homebush Bay area is one of the largest urban/industrial renewal projects in Australia to date and has resulted in a sporting, commercial, and residential area of exceptional quality. The Homebush Bay site for the Sydney Olympics constitutes 770 hectares (1,900 acres) of what were once originally mudflats and mangrove wetlands. Over the last 100 years, the area had been largely taken over by landfill and industrial operations. The master concept design, by Hargreaves Associates (with the Government Architect Design Directorate), focused on environmental rescue — recreating the landscape for public use. One of the major gestures of that reclamation is the use and reuse of water on the site (the Blue Move). Central to this move is the capturing of the site's stormwater runoff, its storage and cleansing in a series of water quality control ponds and created wetlands, and its reuse in the recycled water system of the site. Hargreaves' Northern Water Feature is a key component of this water cycle infrastructure and embodies the water quality goals of the environmentally sustainable development.

Introduction

The redevelopment of Sydney's Homebush Bay for the Sydney 2000 Olympics is one of the largest urban/industrial renewal projects in Australia to date, and it has resulted in a sporting, commercial, and residential area of exceptional quality. Hargreaves Associates' master concept design for the site focused on environmental rescue — to reclaim what was an abused and abandoned industrial landscape for public use. One of the major components of that reclamation is the capturing of the site's stormwater runoff, its storage and cleansing in a series of water quality control ponds and created wetlands, and its reuse in the site's recycled water system (Figures I.13.1 and I.13.2). The Northern Water Feature is a key element of this water cycle infrastructure and embodies the water quality goals of this environmentally sustainable development.

The 770-hectare (1,900-acre) site originally consisted of mudflats and mangrove wetlands. Over the last 100 years, however, the area was largely taken over by landfill and

Figure I.13.1 The Northern Water feature. (Photo by John Gollings. Courtesy of Hargreaves Associates.)

Figure I.13.2 Aerial view of the Northern Water Feature with Olympic Plaza beyond. (Photo by Olympic Coordination Authority. Courtesy of Hargreaves Associates.)

industrial operations, including the State of New South Wales Abattoir, the Newington Landfill, and the New South Wales State Brick Works. The original mangrove and wetland edges had all but disappeared by 1996. In addition, an indicator species of environmental quality native to the area, the green and gold bell frog (*Litoria aurea*) has over the last few decades dwindled to tiny numbers (50 individuals on the Homebush Bay site). Its habitat

Figure I.13.3 Master concept design — The Red Move. (Courtesy of Hargreaves Associates.)

is now concentrated in this and other similarly abandoned industrial sites, and it has been declared an endangered species.

The master concept design (1997) for the site was developed by Hargreaves Associates in association with the Government Architect Design Directorate for the Olympic Coordination Authority (OCA). The design created for the Homebush Bay site included an Urban Core of 300 hectares (741 acres) of major Olympic venues, exhibition, and commercial facilities united by a major act of place making. Central to the master concept design are three key design "moves" that give form and coherence to the project's public spaces.

The Red Move is Olympic Plaza, the central urban space. A 9.5-hectare (23.5-acre) open space addresses the major venues and buildings. This paved open space flows seamlessly like a bold carpet over the heart of the public domain, uniting the site, and accommodating huge crowds. This is now the heart of Homebush Bay, a civic place of arrival, a place of ceremony and procession, and one of Australia's greatest civic spaces (Figure I.13.3).

The Green Move is a landscape framework of "fingers" and parks that stretch through the site, connecting Olympic Plaza directly to the surrounding Millennium Parklands. Five east–west green fingers, each with its own distinctive character that range progressively from the urban to the natural, were developed. In addition, a large, open green park was created as a counterpoint to Olympic Plaza, preserving significant existing stands of eucalyptus and fig trees (Figure I.13.4).

The Blue Move is the use of water on the site, made visible as an ordering element of the urban core. The Fig Grove at the center of the Urban Core and the Northern Water Feature at the north end of Olympic Plaza celebrate the collection, cleansing, and reuse of stormwater runoff while symbolically connecting the site and its systems to the wetlands, creek, and river systems beyond (Figure I.13.5).

Crucial to the Blue Move is the OCA's development of an approach to the management of drainage and stormwater as part of an overall water cycle management strategy for Homebush Bay. The strategy deals specifically with those aspects of the water cycle related to the generation, treatment, and disposal of stormwater. Its objectives included the facilitation and procurement of the water cycle infrastructure; the implementation of ecologically

Figure I.13.4 Master concept design — The Green Move. (Courtesy of Hargreaves Associates.)

Figure I.13.5 Master concept design — The Blue Move. (Courtesy of Hargreaves Associates.)

sustainable guidelines and practices; the use of innovative and technically feasible approaches; and the completion of this infrastructure within the designated time frame and budget.

The OCA mandated that ecologically sustainable development be the underpinning of the environmental philosophy of the Sydney 2000 Olympics as a whole, and the Homebush Bay development in particular. The environment strategy for Homebush Bay stated that its identified goals were as follows:

The conservation of species

Ecosystems

That the remaining natural ecosystems of Homebush Bay be protected throughout the ongoing use of the site

That the work incorporate appropriate stormwater management practices to ensure the protection of the natural ecosystems of Homebush Bay

That the new landscape works enhance the native species biodiversity of Homebush Bay

People

That after redevelopment, Homebush Bay offer a high quality of life to those who live or work at the site, and a highly desirable recreation destination

That Homebush Bay become recognized as a benchmark in the evolution of environmentally sensitive urban development

Conservation of resources

Water

That the new works at Homebush Bay incorporate a reduction in the demand for potable water from Sydney's main supply

That a "reclaimed" water scheme be instituted at Homebush Bay

Energy

That the Homebush on-site works minimize the long-term increase in demand of energy from sources that are nonrenewable or emit greenhouse gases in energy generation or consumption

Construction materials

That the Homebush Bay works minimize the use of materials that deplete natural resources or create toxic pollution in their manufacture, use or disposal

Topsoil

That importation of topsoil to the Homebush Bay site be minimized

Pollution control

Air

That the on-site works minimize negative impacts on Sydney's air quality and avoid ozone-depleting substances

Noise

That measures be taken to minimize the impact of noise at the Homebush Bay site

Light

That the on-site works minimize the impact of night lighting on both environmental conservation areas and residential areas

Water

That the on-site works result in improvement in the quality and quantity of water entering Homebush Bay and the Parramatta River from the site

That the natural groundwater be recharged wherever possible, and that all stormwater be treated for removal of gross pollutants and nutrients

Soil and sediment

That the on-site works remediate the results of polluting activities of the past and ensure the protection of soil and sediments within the developed area

Waste management

That all development maximize the appropriate use of recycled material and reduce waste generation

The Homebush Bay water cycle strategy's aims, based on the environment strategy, were, then, to:

- Achieve the drainage design standards necessitated by the creation of the Public Domain
- Integrate stormwater pollution control structures into the drainage system
- Develop affordable and economical solutions
- Visually integrate drainage structures into the built environment
- Preserve and enhance landscape and heritage values
- Provide a safe and habitable environment for both local species and site users

As part of this effort, Hargreaves Associates brought together a multidisciplinary design team to develop the concept and the design and undertake the construction of the Northern Water Feature. Critical to the success of the project was addressing a number of significant design challenges, including treating runoff from a 100-ha catchment of potentially high-peak population density, particularly during the Olympics; reducing average annual pollutant export from the Public Domain catchment prior to discharge into a receiving waterway by 70 to 90%; creating new habitats for threatened and endangered species (*Litoria aurea*) in an area of contaminated soils and landfill reuse; and minimizing the environmental impact on the adjoining Haslams Creek and other sensitive neighboring sites, such as the Royal Australian Showgrounds (Figure I.13.6).

Since the initial formulation of the drainage and stormwater management strategy, the original concepts for stormwater infrastructure and, in particular, the water quality control ponds to be located at the outlets of the western, eastern, and southern catchments were refined. The team further developed the strategies and systems as the design transformed the former "western catchment water quality control pond" into the dominant landscape element that is now the Northern Water Feature.

Physically, the Northern Water Feature frames the northern end of Olympic Plaza, stepping down in arching granite terraces to meet a newly created wetland at the edge of Haslams Creek (Figure I.13.7). Here, 10-m-high arcs of water fan down the terraces, making visible the cleansing through the marsh of the site's stormwater runoff (Figure I.13.8). The challenges of the stormwater management strategy have been met in a newly created freshwater marsh and transformed into an emblem of the environmental

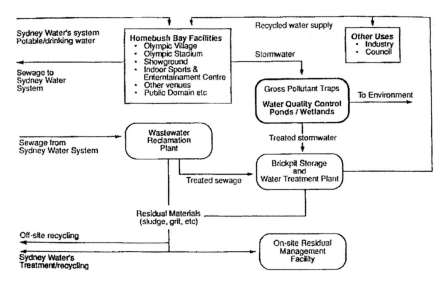

Figure I.13.6 The water cycle management strategy for Homebush Bay. (Courtesy of Hargreaves Associates.)

strategy of Homebush Bay. The water feature provides a series of curvilinear fingers with a number of open water areas, protruding headlands, and long continuous stretches of macrophytes.

The primary stormwater quality treatment features of the Northern Water Feature include the creation of a pool and wetland system (Figure I.13.9) with a low-permeability clay liner to keep the water separated from the contaminated fill and the groundwater beneath. Excavation for the wetland involved the recycling and reuse of these highly variable landfill wastes and construction fill and the installation of a deep leachate system using a custom-developed method of installing polyethylene liners (Figure I.13.10). The creation of the wetland included construction of a deep inlet pond with open water and fringing macrophytes, with submerged inlets to the inlet pond; the splitting of frequent flows up to approximately a three-month average recurrence interval (ARI) runoff and discharge at the head of the two arms of the inlet pool (to promote water circulation in the inlet pool) (Figure I.13.11). It also involved the construction of a deep outlet pool with open water and fringing macrophytes as well as the rehabilitation of the existing spillway. The design provided the provision to pump "first flush" to the brickpit storage pond for reuse, to recirculate water from the outlet pond to the inlet pond, to supply makeup water to the fountain, and to irrigate the surrounding landscape. Efficient water and water plant management necessitated the creation of control structures that enable water levels to be controlled during macrophyte establishment and subsequent maintenance, and the provision of a grassed overflow spillway in the event of severe blockage of the primary outlet spillway during major storm events (Figure I.13.12).

In addition, the feature includes the creation of some of the deepest continuous deflection separation system (CDS) units ever built to intercept gross pollutants from the catchment. These large, nonmechanical, nonblocking, high-flow rate, screening chambers strain out the larger pollutants from the runoff and are located 12 m below the plaza just upstream from the storm drain outfalls into the wetland. The units are configured so as to use the water flow of the runoff event to control the water-borne pollutants in a circular flow within the unit, settling them out into a sump or floating to the top of the water within the unit.

254 Handbook of water sensitive planning and design

Figure I.13.7 Olympic Plaza and the Northern Water Feature. (Courtesy of Hargreaves Associates.)

Figure I.13.8 Twelve-m-high jets of the Northern Water Feature. (Photo by John Gollings. Courtesy of Hargreaves Associates.)

Figure I.13.9 Created Wetland Ponds, showing curvilinear "fingers" of macrophytes and open water areas. (Courtesy of Hargreaves Associates.)

Edge treatment (Figure I.13.13) of the wetland pond is primarily a carefully constructed, vegetated landscape. At the edge, stepped terraces from 3 to 7 meters wide allow progressive establishment of wetland species around the perimeter of the pond.

A shallow, 1:50 grade for the bank adjacent to the pond edge addresses issues of public safety and liability by creating a safe, gentle transition to deeper open water (Figure I.13.14). The slope also ensures adequate drainage to avoid still areas of water and allow effective mosquito management. Strategically placed gabions (at and beneath the permanent water level) direct water flows through the system (Figure I.13.15).

Soft landscape edge treatments reflect the environmentally sustainable design principles that form the basis of the wetland design and limit the reliance on expensive imported materials. Graded zones of aquatic and terrestrial plants (Table I.13.1) stabilize the water's edge physically and visually record the environmental processes operating in the ponds.

The fluctuating water level and changing habitat describe the moisture gradient from water to land as an educational narrative. This narrative of environment and water cleansing is expanded in a public art installation. In "Osmosis," by Ari Purhonen (Figure I.13.16), the color transition of the artwork (from red, through the color spectrum, to violet), perceptible through the steel grate decking of a 112-m-long wetland viewing pier, interprets the water and wetland systems below.

The Northern Water Feature meets its multiple objectives, including stormwater quality treatment, aesthetic, and landscape roles, and is now a key component of the water cycle infrastructure for Homebush Bay (Figure I.13.17). It is a signal landscape and a construction that has become a memorable urban development project that exemplifies the OCA's goals of environmentally sustainable development.

Acknowledgments

The permission of the Olympic Coordination Authority to outline the drainage and stormwater management strategy, the master concept design, and to discuss the Northern Water Feature and its role in the water cycle management strategy for Homebush Bay is gratefully acknowledged. The views expressed in this chapter are those of the author and are not necessarily the views of the Authority.

Figure I.13.10 The Northern Water Feature — wetland plantings and fountain jets beyond. (Photo by John Gollings. Courtesy of Hargreaves Associates.)

The Northern Water Feature design team

The design team included Hargreaves Associates, with Gavin McMillan, local landscape architect; Anton James, local landscape architect; Shaffer Barnsley, local landscape architect; Willing and Partners–Geoff Sainty & Associates, environmental engineers; Woodward–Clyde, environmental engineers; Ove Arup, civil/structural engineer; Barry Webb & Associates, lighting consultants; Sydney Fountains/Waterforms, fountain mechanical consultants; Fluid Flow, irrigation design consultant; and Ari Purhonen, artist.

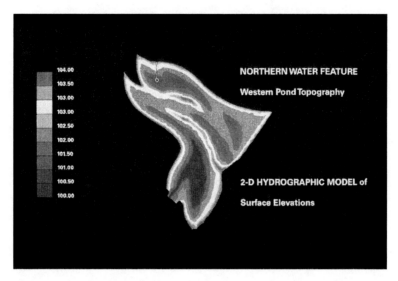

Figure I.13.11 Hydrographic model of surface elevations. (Courtesy of Hargreaves Associates.)

Figure I.13.12 Wetland edge plantings — the Northern Water Feature. (Courtesy of Hargreaves Associates.)

Table I.13.1 Macrophyte Plantings of the Northern Water Feature

Wetland Plant Schedule	Spacing
Baumea articulata Jointed twig rush	4 per m/2
Balboschoenus fluviatilis Marsh club rush	4 per m/2
Eleocharis acuta Common spike rush	4 per m/2
Eleocharis Tall spike rush	2 per m/2
Gahnia sieberiana Red-fruited saw sedge	1 per m/2
Lepironia articulata Lepironis	2 per m/2
Phylidrum lanuginosum Wolly frogmouth	3 per m/2
Schoenoplectus mucronatus Schoenoplectus	5 per m/2
Schoenopolectus validus River clubrush	4 per m/2

Figure I.13.13 The wetland and fountain. (Courtesy of Hargreaves Associates.)

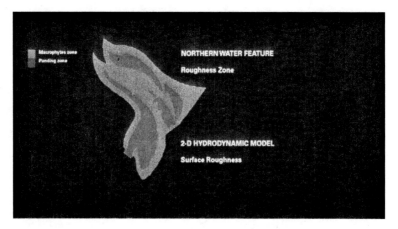

Figure I.13.14 Surface roughness in the Northern Water Feature. (Courtesy of Hargreaves Associates.)

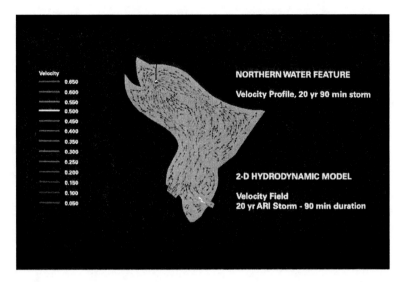

Figure I.13.15 Flow patterns in the Northern Water Feature in a 20-year ARI flood. (Courtesy of Hargreaves Associates.)

Figure I.13.16 "Osmosis" — public art installation by Ari Purhonen on the observation pier. (Photo by John Gollings. Courtesy of Hargreaves Associates.)

Figure I.13.17 The Northern Water Feature at night. (Photo by John Gollings. Courtesy of Hargreaves Associates.)

Literature cited

AGC Woodward-Clyde Pty, Ltd., Statement of environmental effects, the Northern Water Feature, prepared for the Olympic Coordination Authority, 1997.

Allen, G., Public domain design workshop — Process, *Landscape Aust.*, 22, 2000a.

Allen, G., The Northern Water Feature, *Landscape Aust.*, 22, 2000b.

Allen, G. and McMillan, G., *Sydney 2000, the Northern Water Feature; Sustainable Architecture White Papers*, Earthpledge Foundation, Brown, D.E., Fox, M., Pelletier, M.R., Eds., 2000.

Hargreaves Associates (with Government Architect Design Directorate), The Homebush Bay site — Master concept design, prepared for the Olympic Coordination Authority, 1997.

Olympic Coordination Authority, Homebush Bay development guidelines Vol. I — Environ. strat., 1995.

Phillips, B.C., Allen, G., and Listowski, A., Constructed wetlands — Their role in the water cycle management strategy for Homebush Bay, *Proc., Hydrastorm '98 Conf., IEAust*, Adelaide, 1998.

Willing & Partners Pty, Ltd., Stormwater management strategy for Homebush Bay, final report, prepared for Sydney Water and Olympic Coordination Authority, 1995.

The State Government of New South Wales, Sydney regional environ. Plan No. 24 — Homebush Bay.

The Sydney Olympics 2000 Bid Ltd., Environmental guidelines for the Summer Olympic Games, 2000.

Response

Highly visible water:
Recreating a landscape for public use

As Glenn Allen demonstrates, in this chapter, a valuable role that water sensitive design can often provide is one of environmental connection and education, especially if undertaken at sites of high visibility. For despite great advances in public education about the multiple roles wetlands play in watershed functioning, there is still the popular perception that such "swamps" are blights where creation, instead of destruction, is simply ludicrous (see, for example, Chapter I.11 by Clarissa Rowe). As long as wetland creation remains trapped in the construction of "blue-collar" functional wetlands (cf. Robert France and Kaki Martin in Chapter I.12), they will continue to be unrecognized and regarded with derision. The project described in this chapter illustrates that the educational benefits of creating "white-collar" interpretive wetlands at sites of high visibility (Indeed, it is hard to imagine anywhere being more visible than an Olympic Park!) will be important for the conscious raising that should aid in establishment of interpretive wetlands at other, less visible, locations elsewhere, such as those described by France and Martin in Chapter I.12, by James Bays in Chapter I.14, and by Catherine Berris in Chapter I.9.

The new water features created at this site truly connect existing natural waters and provide a conceptual and spatial integration for the overall landscape design (such a strategy was also adopted on a smaller scale by France and Martin in Chapter I.12). The role of landscape architecture is strongly felt in the design of this project, for not only was water regarded from an environmental perspective such as for water quality improvement, habitat creation for endangered species, and conservation reuse, it also was celebrated as a beautiful and worthy element in its own right through the use of sculptures and fountains. This represents a very important direction in recent water sensitive design: the engineered exposure and artistic acknowledgment of hydro-infrastructure.

chapter I.14

Principles and applications of wetland park creation

James S. Bays

Introduction

Natural and constructed wetlands have been used for decades for stormwater and wastewater treatment in all climates across North America and the world. By design, treated effluent can be carefully applied to either natural forested wetlands or shallow vegetated constructed marshes, where biological, chemical, and physical treatment processes in the wetland water, soils, and vegetation provide consistent and sustainable pollutant removal. Increasingly common are treatment wetlands designed to provide multiple benefits, including aquifer recharge, creation and maintenance of undeveloped "green space" within urban areas, wildlife habitat, and educational and passive recreational opportunities for the community.

Wetland "parks" can be designed to blend seamlessly into a neighborhood, public facility, or local park network, frequently serving as a separate destination to bird watchers, nature students, and members of the public interested in passive recreational activities such as hiking and sightseeing. Well-used wetland parks (e.g., Chapters I.8, I.9, I.12, and I.13) are designed to provide a variety of marsh, aquatic, transitional, and terrestrial habitats, with ample opportunity for bird photography, nature study, and environmental education. Valued environmental functions of wetlands then become accessible and visible to a general public that ordinarily might not take notice. Project owners, sponsors, and operators then benefit from an appreciative public.

Successful designs for wetland park facilities result when a project design team exhibits the following attributes:

- Experienced technical staff, including landscape architects, wetland scientists, civil and water resource engineers, and public involvement specialists
- Communicative and solution-oriented project leaders motivated to understand the perspectives of the different technical disciplines employed
- Enthusiastic support of project owners who become directly involved or issue clear direction to enable the project team
- Meaningful public interest and involvement predicated on objective presentations of project siting and design issues that provide opportunities for public responses to be incorporated before and during the final design phase

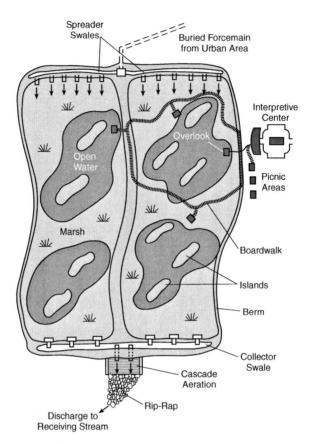

Figure I.14.1 A conceptual wetland park. In addition to water distribution and collection features and vegetated habitats, wetland parks include public parking and access, interpretive facilities, a boardwalk and trail system, and wildlife observation facilities.

- Operation and maintenance training and support that educate the user on ecosystem processes, project design objectives, and troubleshooting as well as describe the physical attributes of the system

This chapter presents an overview of the feasibility assessment and planning, design and construction, and operation, maintenance, and monitoring issues required to design a wetland park using wastewater or stormwater as a water source. Successful examples of wetland park designs are provided.

Planning and feasibility assessment

Wetland parks provide a unique public amenity. Emphasis is placed on developing passive recreational uses, such as hiking, bird watching, and photography, or environmental education, through signage, guided tours, or other direct involvement. Park features that meet these needs include adequate parking, trails, an interpretive center, and an accessible boardwalk into the wetland (Figure I.14.1); however, wetland parks are rarely created solely for this purpose. Instead, a wetland park project is generally propelled by the need to achieve one or more of a wide range of environmental planning objectives: flood control, water quality improvement, water storage or aquifer recharge, fisheries enhancement, wildlife habitat creation or restoration, and maintenance of open space. Beneficial public use of such projects has historically been viewed as important but ancillary to the original design intent (Knight, 1992, 1997). As a result, constructed wetlands designed as parks must balance site aesthetics and public-use features while meeting fundamental hydrologic, ecological, and engineering requirements.

Careful planning, design, installation, and maintenance of a wetland project determine whether or not the wetland will perform its intended functions. Project success reflects the level of consideration given to maintenance of the post-construction water regime and desired vegetative community during the planning, design, and construction. Useful general references for wetland planning and design include Kusler and Kentula (1990), Marble (1991), Hammer (1992), WPCF (1990), USDA (1992), Mitsch and Gosselink (2000), Kadlec and Knight (1996), NCHRP (1996), and U.S. EPA (2000).

The creation of wetland parks varies from project to project. The public relations benefit of the restoration or creation of a diverse, productive wetland habitat should be readily apparent, but experience indicates that an informed, enthusiastic, and balanced perspective on the part of the wetland owner can help overcome stakeholders' potential concerns regarding liability, maintenance, costs, and performance. Initially, a feasibility analysis that objectively examines the potential technical, regulatory, and economic benefits and constraints of the project should be performed. A positive outcome of this critical step can lead the way to regulatory acceptance, final design, and construction.

The feasibility study should begin with the clear definition of the desired function(s) of a planned wetland. These functional goals can be described further as a series of objectives that include measurable criteria in the form of a performance standard. Performance standards should specify the desired future conditions for hydrology, soils, vegetation species composition and cover, and habitat features of the wetland.

Wetland technology is still in a developing phase; therefore, predicting wetland functional performance with precision requires the use of published hydrologic, water quality, and ecological design tools and a familiarity with successful wetland design experiences. Existing and known potential constraints to successful wetland construction and operation should be taken into consideration when assessing whether project objectives will be met.

Conceptual wetland design criteria are then developed that vary in detail depending on the goal of wetland construction. All require careful consideration of wetland type, configuration, size, water source, soils, and vegetation. Table I.14.1 summarizes design criteria typically considered when planning constructed wetland projects.

Preliminary locations for wetland projects are then developed based on an analysis of identified alternative locations and the extent that they satisfy stated siting requirements, or criteria. Stated goals and objectives of the project should be balanced with site-specific constraints. Criteria for locating a wetland will vary depending on whether a wetland is being constructed to replace or restore lost ecological functions or enhance existing wetland functions, as in a mitigation wetland, or whether a wetland is being constructed or enhanced to provide a new ecological function, as in a constructed or natural wetland treatment system.

The importance of the following wetland site selection criteria can vary between sites, depending on the wetland goal:

- Proximity to desired location
- Availability of sufficient contiguous area
- Availability of suitable long-term wetland water source
- Favorable site hydrogeology
- Acceptable site geotechnical constraints
- Presence of existing or potential limiting land use, natural wetlands, protected species, and historical or archaeological resources on or adjacent to site

Table I.14.1 Typical Wetland Design Criteria

Criterion	Purpose	Examples
Type	Identifying the dominant vegetation types desired will direct subsequent water regime analysis and soils characterization.	Emergent marsh — tidal or freshwater Forest — tidal or freshwater Vernal pool — seasonal Riparian woodland
Configuration	The physical layout of the wetland project guides the design of water distribution and public access features.	Location and distribution of habitats associated with different vegetation types and water depths Number and orientation of component basins or cells Important physical features, including depth, volume, islands, side:slopes, and others
Size	The area of a wetland project can be determined by project objectives and subsequently modified by land cost, availability, and proximity to project location.	Water quality — models can be used to predict the minimum area necessary to achieve a specified goal for a given hydraulic loading rate and pollutant concentration Habitat — minimum area requirements can be set by compensatory mitigation goals (i.e., a ratio of created, restored, or enhanced wetland area to wetland area displaced by a project; a minimum depth and size may also be set as a function of species requirements Hydrologic storage — minimum area requirements may be set by the desired storage volume for a specified storm event or flood return frequency
Water source	A suitable long-term wetland water source must be seasonally available and appropriate in quality. Possible sources include stormwater, treated municipal and industrial effluent, and intercepted surface and groundwater flows.	The proposed site must have the appropriate hydrology and topography for the intended wetland type. Excavation, diking, and grading of the site to create suitable topography for wetlands are possible options; however, for wetlands not dependent upon effluent as a water source, a connection to a permanent, natural groundwater or surface water source must be within reach for successful creation.

- Potential ease and cost of acquisition of ownership rights, easement, or other controlling interest
- Ease of access for construction and maintenance
- Availability of sufficient construction materials and labor resources

Relevant features of these criteria that may affect project feasibility are described in Table I.14.2. Categories of data useful for assessing site suitability and project feasibility are listed in Table I.14.3.

Determination of the water regime and water balance is a critical component of the feasibility planning phase of the project. To the extent possible, the water balance analysis should establish the water needed to meet the targeted hydroperiod or to determine the area that can be supported with the available water supply. Figure I.14.2 illustrates the conceptual requirements of a wetland water balance that can be used to structure a

Table I.14.1 Typical Wetland Design Criteria (continued)

Criterion	Purpose	Examples
Soils	Soils should be suited to support wetland vegetation and to support the desired hydrology of the wetland. The hydrologic properties of the site soils should be understood before concluding the project design. The geotechnical properties of the soil will influence side:slope ratios and submersed angle of repose.	Infiltration rates — may greatly influence the seasonal water budget, and consequently the need for an engineered soil or synthetic liner, which adds significantly to cost; this can be determined by site permeability testing Soil type — the ability of roots to penetrate the wetland soils will influence the ability of the wetland to support the intended species composition Source — salvaged wetland or upland topsoils may be used to facilitate the establishment of wetland vegetation; the results of seed bank surveys and the potential for invasion by undesirable species should be given full consideration In constructed treatment wetlands, topsoil use should be considered as an option, but it is not necessary as long as the exposed soils are capable of supporting the planted vegetation. Berms should be constructed from stable materials and protected by erosion control materials and methods.
Vegetation	Wetland vegetation should be selected Tolerance of inundation and oxygen-poor, reduced environments Native origin Rate of growth and biomass accumulation Diversity	Vegetation should have a tolerance of prolonged inundation, low oxygen concentrations in the water and soils, and rapid dense growth to shade surface waters and reduce algal production. Planting stock should originate from the project region. Planting centers can range widely depending on the desired rate of growth and the potential for invasion by undesirable species; this may mean a center spacing of 3 to 10 ft for most wetlands, but as small as 2- to 3-ft centers for wetlands where rapid growth is desired. Vegetative diversity in mitigation wetlands should be encouraged by planting multiple species and using topsoil as mulch where feasible.

Note: Wetland design must take into consideration type, configuration, area, water source, soils, and vegetation.

hydrologic analysis. The level of detail needed for performing a water balance analysis varies widely between projects and depends on the level of detailed information available and the objectives of analysis.

The area of the wetland can be determined as a function of wetland area required to meet ecological goals established through a compensatory mitigation analysis, or in the case of water quality improvement, through the use of empirical treatment wetland sizing models (e.g., Kadlec and Knight, 1996). Wetland area will vary according to the flow to be treated, the quality of the water to be treated, and the functional treatment goals. As described earlier, the availability of sufficient land to meet wetland project objectives is a critically important feasibility factor.

Table I.14.2 Wetland Site Selection Criteria

Criterion	Relevance
Proximity	*Compensatory mitigation* — If designed to mitigate for total or partial loss of function, the wetland may need to be constructed in the vicinity of the original wetland (e.g., "on-site vs. off-site"). Sites for mitigation projects are frequently in the same watershed as the impacted wetland and a similar size or larger. *Stormwater treatment* — Wetlands designed for stormwater treatment may need to be located at an appropriate topographic elevation in order to maximize gravity flow. *Municipal/industrial treatment* — Natural and constructed wetland treatment systems may need to be designed on or adjacent to the location of the pollution source in order to minimize land and pumping costs, and to control or limit public access.
Area	Successful restorations are positively correlated with project size. Sufficient contiguous area should be available to allow the wetland to be constructed at one location to minimize construction, operation, and maintenance costs and minimize nonnative species invasion.
Wetland water source	Sufficient availability of a suitable long-term wetland water source is critically important to the construction and viability of wetlands. The proposed site must have the appropriate hydrology and topography for the intended wetland type. Excavation, diking, and grading of the site to create suitable topography for wetlands are possible options; however, a connection to a permanent, natural groundwater or surface-water source must be within reach for successful creation. The constructed wetland's viability will be determined by the seasonal and annual availability of wastewater effluent.
Soils and hydrogeology	Site hydrogeology should be favorable for wetland construction. Excessively drained soils may not be suitable for wetland construction without the installation of an aquitard of clay or other materials of low-hydraulic conductivity. Shallow depths to the surface of bedrock may also constrain wetland excavation. Suitable soils must also be present or available for transfer to the site. Hydric soils can differ considerably from mesic and xeric soil in organic matter content, fertility, pH, structure, and texture. Saturation of upland soils over time will not guarantee the development of a suitable wetland soil. Use of wetland soils from donor in-kind wetlands is an effective technique for establishing native vegetation in mitigation wetlands, but consideration must be given to the potential introduction of undesirable or poorly controllable plant propagules.
Geotechnical attributes	Wetland berm and substrate materials should be suitable for wetland construction and not lead to excessive erosion, sediment loss, or potential for failure under normal design extremes.
Limiting land uses and other siting constraints	Human land use may constrain the suitability of a wetland construction location. Care should be taken to locate the wetland in areas with compatible zoning and other land uses in full recognition of the wetland design goals. The presence of natural wetlands, protected species habitats, and historical or archaeological resources on or adjacent to site may pose additional significant design constraints.
Land ownership	Sites not currently under the ownership of the project owner will need to be assessed for ease of acquisition of ownership rights, easement, or other controlling interest. Wetlands are land-intensive, so land costs will significantly affect the total project cost.

Table I.14.2 Wetland Site Selection Criteria (continued)

Criterion	Relevance
Access	Each site should be evaluated for existing and potential ease of access for construction and future maintenance. Local land-use regulations should be consulted to identify possible constraints to construction and maintenance traffic.
Materials	Availability of sufficient construction materials and labor resources should be evaluated within a regional context in order to minimize project cost and to maintain standards of quality for materials. The availability of skilled contractors, plant nurseries, and acceptable wetland construction materials should be assessed.
Data collection	Sufficient data should be collected from each proposed construction site(s) to respond to the information needs of site selection criteria and to evaluate the potential for successful wetland permitting, construction, and operation.

Note: Wetland goals need to conform to site constraints for greatest potential for project success.

Table I.14.3 Potential Data Requirements for Wetland Feasibility Analysis

Category	Attributes
Climate	Temperature — annual and monthly averages
	Precipitation — annual and monthly totals
	Evaporation — annual and monthly totals
Soils and geology	Classification
	Water table depth and variation
	Hydraulic conductivity
	Cation exchange capacity
	Bedrock depth and type
	Aquifer depth, type, and confinement
Topography and hydrography	Surface elevation contours
	Drainage basins
	Surface waters
	Seasonal flow and inundation patterns
Vegetation and wildlife	Cover types
	Habitat types
	Occurrence of nonnative species
Cultural	Local and regional land uses
	Proximity to recreational facilities
	Property ownership
Regulatory	Wetland jurisdiction and extent
	Protected species occurrence
	Historical and archaeological sites
Soil and water quality	Specific conductance and pH
	Temperature, salinity, and dissolved oxygen
	Nutrients
	Metals
	Organic compounds

Note: Data needs for wetland planning will vary by site. These categories yield useful information for site selection, wetland design, and construction.

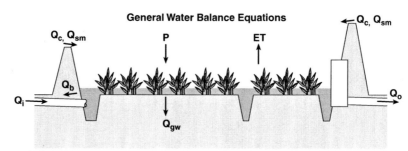

General Water Balance Equations

Dynamic water balance

$$dV/dt = Q_i - Q_o + Q_c - Q_b - Q_{gw} + Q_{sm} + PA - ETA$$

where

V = water storage in wetland, m³
T = time, d
Q_i = input wastewater flow rate, m³/day
Q_o = output wastewater flow rate, m³/day
Q_c = catchment runoff rate, m³/day
Q_b = bank loss rate, m³/day
Q_{gw} = infiltration to groundwater, m³/day
Q_{sm} = snowmelt rate, m³/day
P = precipitation rate, m/day
ET = evapotranspiration rate, m/day
A = wetland surface area, m²

Periodic water balance

$$\Delta V_{stored} = V_i - V_o + V_c - V_b - V_{gw} + V_{sm} + V_r - V_e$$

where

ΔV_{stored} = change in water storage in wetland, m³
V_i = input wastewater volume, m³
V_o = output wastewater volume, m³
V_c = catchment runoff volume, m³
V_b = bank loss volume, m³
V_{gw} = volume of infiltration to groundwater, m³
V_{sm} = snowmelt volume, m³
V_r = rain volume, m³
V_e = evaporated volume, m³

A sample spreadsheet table is presented below:

Assumptions for example

1. All units in m³/day.
2. Example wetland is in Florida and has a synthetic liner (therefore, Q_{gw}, Q_{sm}, and Q_b = 0).
3. January rainfall total = 10 cm.
4. January measured evapotranspiration total = 5.5 cm.
5. Wetted wetland surface area = 10,000 m².
6. Total catchment area within berms = 14,000 m².
7. Runoff coefficient, C = 0.95.

Time period	Q_i	Q_o	Q_c	Q_b	Q_{gw}	Q_{sm}	P	ET	dV/dt
Jan	100	75	1.2	0	0	0	3.2	1.8	27.7
↓	—	—	—	—	—	—	—	—	—
Dec	—	—	—	—	—	—	—	—	—
Total									27.7

Notes:

1. For existing systems, Q_i and Q_o are usually measured parameters. For systems in design phase, Q_o can be estimated by entering values for the other parameters and assuming steady-state ($dV/dt = 0$) conditions.
2. Input precipitation and ET data into appropriate cells. Data can be found on Web sites having statewide or local climate data. Some examples of appropriate Web sites for climate data include http://www.srcc.lsu.edu/ for the southern United States and http://www.wrcc.sage.dri.edu/ for the western United States. ET calculated as 80% of Pan evap. or input actual data if available.
3. ET = ET monthly total × wetted area/# days per month. P = P monthly total × wetted area/# days per month.
4. Q_c = runoff coefficient × (catchment area − wetted area) × P.

Figure I.14.2 Conceptual wetland water balance approach. The detail in a wetland water balance is frequently limited by the availability of appropriate site-specific data, but monthly time steps are the minimum needed in order to determine if there will be adequate water to maintain seasonal stage elevations.

Success of the planting phase of the project will depend on the careful matching of plant species to the soils of the selected site and the proposed topography and hydrology. Species differ in their tolerance to soil waterlogging and flooding. Duration, seasonality, and depth of flooding are also important factors in a species ability to survive wetland condition. Variations in microsite topography provide substrate diversity for establishing seedlings; microsite topography is closely correlated with seedling distribution and regeneration success. Soil fertility can offset some of the more stressful aspects of hydric conditions, extending the range for some marginally tolerant species. Therefore, selection of species adapted to site conditions is critical to the overall performance of the wetland.

Critical components of any wetland feasibility study with the intent of creating a public-use park are to inform and involve the principal stakeholders in the project and to conduct public charettes or informational workshops describing the project's benefits and constraints. Depending on local interest and funding, a productive approach is to charter a public-use planning process that will bring potential users and neighbors to the process to determine optimal and desirable uses for the project, and to provide an opportunity to correctly inform the public on all aspects of planning, construction, and operation. The need to give full consideration to a detailed public information program cannot be over-emphasized; experience indicates that the general public is not familiar with the many benefits associated with a well-designed constructed wetlands park but will become enthusiastic when provided examples of similar successful projects.

Wetland design and construction

Successful wetland construction, enhancement, or restoration requires a carefully prepared set of plans and specifications, a construction management approach, an experienced project team, and a schedule and plan for monitoring construction progress and implementing corrective actions. The design of wetland construction projects typically requires a wide range of technical disciplines, including civil, water resource, geotechnical, and environmental engineers, computer-aided design and drafting technicians, wetland biologists, and landscape architects. Final designs should include a detailed water balance, the basis for wetland sizing, a description of the hydraulic conveyance and profile, a grading plan, a detailed planting scheme, and public-use design features.

Wetland construction requires skilled field labor, such as earthmoving, planting, and structural construction, as well as a general contractor whose main role is to coordinate and oversee construction, and a planting contractor, whose role is to cultivate and/or arrange for delivery of all planting materials, conduct and coordinate planting activities, and arrange or conduct post-construction maintenance and monitoring. The project design engineer or owner's representative should arrange to review the construction progress and be available to resolve situations during construction different than those considered during design.

The contractor should be willing to meet with the project owner or representative on a routine basis during the course of the project to discuss status and review project problems and solutions. The contractor should provide project status reports on schedule during the course of the project.

Contractor selection is critically important to every wetland project. Contractors should be able to demonstrate prior successful wetland construction experience. Contractor staff should include a person with background in wetland creation/restoration design with practical wetland construction experience and familiarity with local hydrologic and edaphic conditions. The contractor or contractor's insurer should be able to secure a performance bond equal to the cost of construction, planting, and a period of maintenance and monitoring.

Nuisance and exotic plants should be controlled during wetland grading and planting. Trash and litter should be prevented from accumulating in the wetland. Wetland vegetation should be irrigated or kept watered as needed during the first year's initial dry season if not inundated to design depths. Water control structures and culverts should be kept free of debris and soil, and repaired if broken.

Wetland construction plans and specifications should be sufficiently detailed for bidding purposes, engineering and biological review, and verification of "as-built" conditions. As a minimum, wetland construction plans should include a table of contents, a detailed location map, a sheet key index, and a table of quantities. Individual sheets should include a compass arrow, scale bar, date of preparation, and a record of reviewers and revision dates. Table I.14.4 summarizes the types of information useful to include within a wetland construction document

Wetland operation, maintenance, and monitoring

"As-built" drawings should be prepared and certified by the earthwork contractor or general contractor prior to installation of planting materials and submitted for approval and acceptance by the project engineer. Final "as-built" drawings, which should be prepared at the conclusion of construction, must verify design elevations, water depths, and elevations and extent of planting zones. These should be submitted with an initial monitoring report (sometimes referred to as a "time zero" report) at the completion of the project, which would include descriptions of the major wetland plant communities, densities, species, and photographs taken at a sufficient number of stations to adequately cover the project. Original Mylar or other media should be annotated and prints certified by a licensed surveyor. Variations from design, and their rationale, should be noted on the plans.

Post-construction monitoring can be performed to measure and evaluate whether a wetland has attained its intended goals. Detailed descriptions of necessary monitoring activities can be included in the construction, design, or permitting documents. Sampling methods, frequency, and monitoring station locations should be described in sufficient detail to permit monitoring to be conducted by qualified individuals unfamiliar with the project. Monitoring plans should include descriptions of methods and goals of collecting data on water levels and plant species' cover and diversity. Photographs of the wetlands should be taken at fixed locations as part of the post-construction monitoring process.

Additional data that may be collected will depend on the goal of wetland construction. Periodic biological surveys of vertebrate and invertebrate communities may be performed to document wildlife habitat and ecological productivity in the wetland. Water quality sampling may be performed to document pollutant assimilation, organic matter production and export, and sediment retention. Flood retention and groundwater recharge functions may be documented by installation of monitoring wells, and water stage and rainfall recorders. Specialized input from biologists, hydrologists, hydrogeologists, and engineers should be sought before designing and implementing any monitoring.

Wetland performance after construction can be determined by comparison of measured wetland conditions at selected time intervals against specific criteria. Criteria to be measured should reflect project goals. For example, specific criteria for a forested mitigation wetland might include mean tree height, basal area, density, and frequency; percent cover by planted and volunteer plant species; minimum, maximum, and average depths or hydroperiod; and occurrence of selected aquatic wildlife. Specific criteria for a treatment wetland might include target effluent concentrations and expected pollutant removal efficiency, as well as other indications of wetland condition, such as percent cover by planted and volunteer plant species.

Table I.14.4 Typical Components of Wetland Construction Documents

Information Type	Description
Aerial photograph	If available, construction plans should include current aerial photographs at a scale sufficient to completely show the outline of the project work area on one or more sheets. Locations of key landmarks, water bodies and drainage pattern, wetlands and other restricted or protected areas (i.e., endangered or threatened species) should be indicated. Larger-scale aerial photographs may be used as a background for the detailed plan set if interpretive clarity is not sacrificed.
Scale	A scale of 1 in. = 90 ft or larger (e.g., 1 in. = 50 ft) is recommended.
Topography	Wetland construction plans should be overlaid on a topographic map of existing site elevation contours. A 1-ft contour interval is recommended as a minimum contour interval. A smaller interval, such as 0.5 ft or less, may be more useful in developing site plans for tidal wetlands with a broad but shallow slope. Benchmark locations and elevations should be clearly indicated.
Cross-sections	Typical cross-sections of all earthwork should be prepared to scale clearly indicating all design elevations, slopes, and dimensions. The number of cross-sections should be sufficient to identify typical and atypical conditions.
Geotechnical data	Locations of test borings and soil pits should be identified within the plan set so that they may be relocated, if desired. Soil profile illustrations should be identified and presented within the plan set and should include information on soil chroma profile elevations, observed water elevations, depth to underlying rock, and presence of potentially significant aquitards, such as naturally formed layers of clay or less permeable materials.
Jurisdictional wetlands delineation	Jurisdictional wetland boundaries should be clearly and accurately identified on the site topographic map as negotiated with the pertinent regulatory agencies.
Hydrologic data	Plans should indicate existing and expected water levels, identify adjacent water bodies, and establish major surface drainage patterns at the construction site. All elevations should be made relative to National Geodetic Vertical Datum (NGVD) or other datum predetermined by convention, or an elevation conversion should be supplied. Sufficient information should be provided to determine seasonal elevations of receiving waters. *Nontidal wetlands* — Site hydrological data should include seasonal high and average water elevations determined from vegetative indicators, soil indicators, or hydrological monitoring data for existing wetlands, if any, and at adjacent upland sites or perennial streams, which may provide an indication of the adjacent groundwater table elevation. A series of 0.75- to 1.0-in.-diameter monitoring wells in representative soil types within the wetland may be installed and read over a period long enough to describe a normal seasonal range. *Tidal wetlands* — Tidal wetlands should include mean high water (MHW), mean low water (MLW), and spring high tide (EHWS), and spring low tide (ELWS) relevant to the locale of the construction site. *Water supply during construction* — Provision should be made on a site-specific basis to divert water temporarily to provide inundation to the wetland during construction by either temporary or permanent structures, such as pumps, irrigation pipes, siphons, diversion channels, or other method appropriate to the site.

Table I.14.4 Typical Components of Wetland Construction Documents (continued)

Information Type	Description
Planting specifications	Plans should indicate zones or areas to be planted. A planting list should be prepared for each wetland zone that includes quantities, elevation ranges, propagule description (e.g., bare root, container grown nursery stock, field harvested specimens) and acceptable planting conditions. Special considerations or requirements should be noted and described in sufficient detail. These should include fertilizer specifications, preplanting conditioning, geographic constraints on plant sources, performance, and irrigation requirements. Plants should be planted at intervals sufficiently dense to assure rapid growth of vegetative cover. Plants should be commercially available within the project area. *Plant materials* — Most planned wetland projects require the introduction of plant species for successful creation and restoration of a wetland. Only healthy stock should be used. Local seed sources are preferable to nonlocal sources due to acclimation to local climate and other factors. Many commercial nurseries specialize in the production of woody and herbaceous species suitable for wetland restoration projects. If field harvested plants are used, the planting contractor should be able to provide proof of permission to sample from the nursery wetland. *Plant propagules* — Planting stock can be found in several forms. Woody seedlings can be either containerized or bare root. Containerized seedlings are grown in a small tube-shaped container that produces a small, compact root system contained in a small plug of potting medium. Bare-root seedlings are harvested from the nursery bed mechanically, are root pruned, and should be free of associated soil. Herbaceous stock is usually produced as containerized seedlings. Planted stock may be priced per unit for large, high-quality containerized seedlings and saplings or per bundle of 500 or 900 for woody bare root seedlings. *Care of plant stock* — Care of the planting stock after delivery at the planting site is as critical as selection of appropriate stock. Stock should be kept in cold storage until planting and protected from sunlight and drying. This is especially important for bare root stock. Stock may be stored under tarpaulins or in coolers.
Vegetation maintenance	Control of exotic or nuisance plants should be required within the wetland during and after construction. Construction documents should list nuisance or nonnative plant species that will require control. Details on control methods should be provided for expected nuisance species. Control of herbivory by animals may be required and should be anticipated in the construction and monitoring phases. Provisions should be made for irrigation during construction with available water sources for mitigation wetlands and effluent for constructed wetlands.
Land use	Locations of restricted areas, structures, utility lines, or other infrastructure within or adjacent to the construction area should be indicated. Special construction restrictions or contractor coordination requirements should be indicated.
Erosion and sediment control	Construction plans should indicate the location, quantities, and maintenance of acceptable and appropriate sediment control methods and slope stabilization techniques. Possible sediment barriers include staked haybales, geotextile silt-screens, sod, and plant seeding. Barriers should be placed at the construction periphery and within the wetland in such a manner as to minimize sedimentation and erosion of wetland berms or edges.

Table I.14.4 Typical Components of Wetland Construction Documents (continued)

Information Type	Description
Grading plan	A grading plan should be included with the plan set that identifies the location, elevations, and dimensions of project earthwork. The plans should include sufficient information on radii, turning points, and baseline offsets for the contractor to accurately locate and build the wetland. Plans should specify soil quality requirements, soil sources and disposal areas, and means of transporting soil. Grading specifications should indicate the allowable tolerance in wetland grade elevation. Constructed wetlands require strict adherence to wetland grade specifications, while mitigation wetlands should not be graded to a completely uniform wetland soil elevation.
Site preparation	Construction plans may include removing the top 1.5 to 2.0 ft of substrate from the project site and stockpiling of that material to use as cover for the site to provide a seed bank or propagule source. This measure requires careful consideration and analysis of the existing seed bank and potential for introduction of nuisance or nonnative species.
Water supply and flow control	Plans should show the location of surface water inflows, whether open channel, pumped, or piped, and all outflow weirs, hydraulic control structures, surface water conveyance, seasonal high-water overland flow, and pipe outflows. Profile sheets should provide invert elevations of all culverts, structures, seasonal overflows, and weirs. Detail sheets should provide information sufficient to construct all inflow and outflow control structures, including typical plan and profile views, and descriptions and possible manufacturers of all items not constructed specifically for the project.

Note: The extent of the information required will depend on the project, but this table provides helpful guidance on any constructed wetland project.

Corrective action should be taken if monitoring indicates that performance criteria are not being met, or if other indications are found that the wetland is not functioning as designed. Mitigation and constructed wetlands performance can be adversely affected by inundation less than or greater than the design requires. Flow, residence time, pollutant removal efficiency, and compliance with wetland discharge standards may be adversely affected. Wetland vegetation may be adversely effected. Possible solutions may include changing the volume, quality, or timing of water deliveries to the wetland, the invert elevations of water control structures, the wetland grade elevation, and the species of vegetation to be planted. Corrective actions for mitigation wetlands should be coordinated with permitting agencies.

Selected case histories

Four wetland projects have been selected to illustrate how wetlands designed for water quality improvement can be designed to provide unique, high-quality user experiences as recreational parks and environmental educational facilities (Figure I.14.3). Conceptual details and operational histories are provided for the following projects:

- Green River Natural Resource Area, Kent, WA
- Sweetwater, Tucson, AZ
- Victoria Wetlands, Victoria, TX
- Wakodahatchee, West Palm Beach, FL

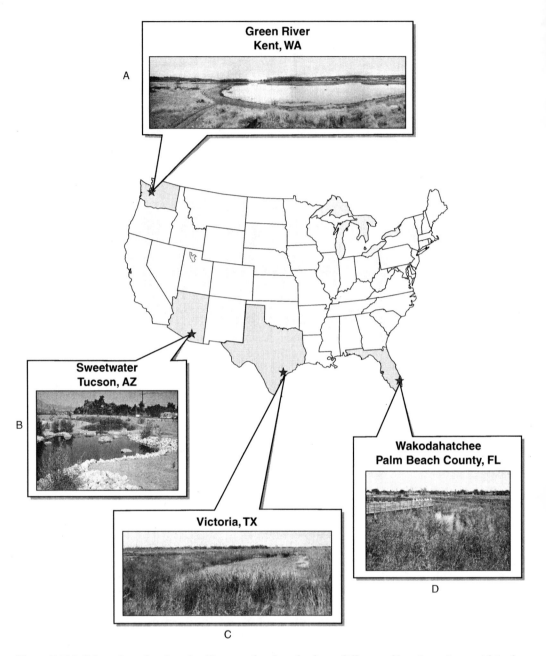

Figure I.14.3 Selected wetland parks. Four wetland parks from different climatic regions within the United States combine water quality improvement, wildlife habitat creation, water reuse, and public use and recreation.

Table I.14.5 summarizes and compares the multiple design objectives of the four selected wetland case histories.

Chapter I.14: Principles and applications of wetland park creation

Table I.14.5 Selected Wetland Planning Goals and Objectives

Functional Objective	Green River Natural Enhancement Area, Kent, WA	Sweetwater, Tucson, AZ	Victoria Wetlands, Victoria, TX	Wakodahatchee, Palm Beach County, FL
Control flooding	Reduce flooding by storing a significant fraction of Mill Creek spring flow	Not a design criterion	Not a design criterion	Not a design criterion
Recharge local groundwater	Not a design criterion	Discharge about one million gal/day of effluent to recharge basins	Not a design criterion	Maintain existing pond infiltration to shallow groundwater
Restore native wildlife habitat	Create emergent and transitional marshes and open ponds with native plant species	Create emergent marshes and wooded riparian habitats with native plant species and exclude exotic plants	Create emergent marshes with native plant species	Create emergent marshes with native plant species and exclude exotic plants
Enhance native fisheries	Create summer habitat for migrating native salmon	Nonnative fish excluded from project	Native fish community has become established	Not a design criterion
Preserve open space	Maintain open, green space in a rapidly urbanizing area south of Seattle	Create green space in marginal lands	Create a native wetland ecosystem as a buffer	Maintain green space in densely populated eastern Palm Beach County
Reuse existing facilities and materials	Polishing marshes constructed in abandoned wastewater lagoons	Replaced existing wastewater ponds	Recycled lumber used in boardwalk and interpretive center; operational windmill used to supply water to interpretive center	Constructed from existing percolation ponds
Create public-use and recreational facilities	All-access hiking and bicycle trails; shaded viewing towers; public-access parking	All-access trails, shaded gazebo, informational kiosk, and marsh overlooks; public-access parking	Accessible trails, boardwalk, informational kiosk, bird-viewing blind, and marsh-viewing tower; public-access parking	All-access boardwalk, shaded gazebos, informational kiosk, and multiple wetland overlooks; public-access parking

Table I.14.5 Selected Wetland Planning Goals and Objectives (continued)

Functional Objective	Green River Natural Enhancement Area, Kent, WA	Sweetwater, Tucson, AZ	Victoria Wetlands, Victoria, TX	Wakodahatchee, Palm Beach County, FL
Improve discharge water quality	Reduce pollutant load from runoff from developed watershed	Reduce suspended solids and nutrients before application to recharge basins	Treat nutrients and pollutants in biologically treated industrial wastewater to meet discharge quality	Treat nutrients and pollutants in secondary municipal effluent to meet discharge quality limits

Note: These well-utilized public-use and recreational facilities balance functional objectives with site aesthetics and public use.

Green River Natural Resource Area, Kent, Washington

Introduction

The Green River Natural Resource Area (GRNRA) is an innovative stormwater management and habitat enhancement project in Kent, WA. The project integrates the needs of people and nature in a rapidly urbanizing landscape. The Green River Natural Resources Area project transformed an abandoned sewage lagoon system into a combined stormwater detention and enhanced wetland facility that provides a rich diversity of wildlife habitat. The 304-acre site is one of the last remaining open tracts of land in the Kent Valley and incorporates state-of-the-art techniques of wetland creation and enhancement, urban wildlife management, and stormwater treatment. With its adjacent public park and trail system, the site is one of the largest man-made, multi-use wildlife refuges in the United States. Useful background information, maps, and site details may be found on the city of Kent Web site (City of Kent, 2000) and in CH2M HILL (1999).

Background

The project area is in the northwestern portion of the Kent Valley and is bounded generally by the Green River to the west, South 212th Street to the north, 64th Avenue to the east, and the Puget Power pedestrian/bike trail to the south (Figure I.14.4). Kent used the five original lagoon cells for sewage treatment from 1969 to 1973, when construction of Metro's Cross-Valley Interceptor diverted all sewage flows to the Renton treatment plant. The lagoons then became a fairly stable pond system, maintained by a relatively impermeable clay liner. The northern cell was overgrown with vegetation; the two central cells were normally filled with water but became partially dry in the fall, creating mudflat habitat; and the two large southern cells (each 16.5 acres in size) contained up to 3 ft of water.

Before construction, the site contained a number of wetland and upland habitat types. Over 200 species of mammals and birds were observed at or near the lagoons, which served as a nesting, feeding, and brooding area for many species that use the Green River corridor and the nearby east–west powerline corridor as travel routes.

Planning for the new facility began in 1979 and focused on the site's potential as a regional stormwater detention facility that could help to alleviate increases in stormwater flows in Mill Creek. The City of Kent Public Works Department has funded the construction and continued operation of the facility since 1996. Grants from the King County 1993 Regional Conservation Futures Acquisition Program and Metro's Regional Shoreline Improvement Fund for $500,000 each helped purchase additional land, bringing the

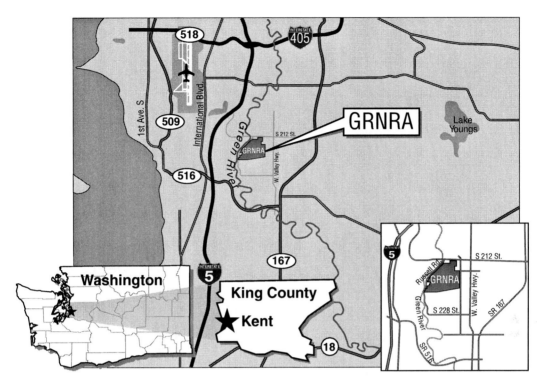

Figure I.14.4 Location of the Green River Natural Resources Area, Kent, WA. Located southeast of Seattle in Kent, WA, the GRNRA has two public gates that allow parking and pedestrian access to the site along Russell Road, and another may be accessed along the south bike path by bike or foot. City of Kent Public Works, 220 Fourth Ave S., Kent, WA 98032. Operations, 5821 S 240th St., 253-856-5600.

project's total size to 304 acres. Total construction cost for the entire 304-acre property, including wetlands, upland restoration areas, public recreational and interpretive facilities, and land purchase was estimated at $10.6 million.

Beneficial features

Flood control. The expected storm flows in the Mill Creek drainage basin in a 100-year flood event are 560 cubic feet per second (cfs). The GRNRA provides sufficient storage to reduce flows in the lower portion of the Mill Creek to 180 cfs — a 68% reduction. The project is therefore a critical element in preventing flooding in downtown Kent and nearby industrial areas, where certain low-lying streets, parking lots, and warehouses used to be inundated nearly every year during high-flow events in the spring and winter months.

Water quality improvement. The water quality of Mill Creek, which lies in a highly urbanized area, was heavily impacted by untreated stormwater. The GRNRA was configured to protect high-quality habitat at the site while substantially improving water quality in the lower reaches of Mill Creek. Stormwater runoff entering the site passes through an extensive treatment system, including two presettling ponds and a 20-acre constructed wetland, reducing sediment loads and urban pollutants from upstream (Figure I.14.5). Water then enters the main lagoon for additional treatment prior to draining back into Mill Creek. In addition, provisions have been made to supplement flows during the critical low-flow periods in summer using three groundwater extraction wells. This flow

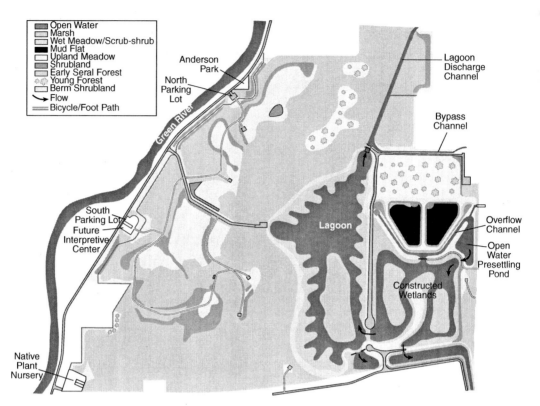

Figure I.14.5 Plan view, Green River Natural Resources Area. The GRNRA comprises over 300 acres of natural and created upland and wetland habitats, including treatment marshes, a storage and polishing pond, final polishing wetlands, and numerous natural habitats and islands.

augmentation has the potential to improve oxygen levels and temperature conditions in lower Mill Creek.

Wildlife habitat restoration. The project's 304 acres provide habitat to an estimated 165 bird and 53 mammal species and serves as a nesting, feeding, and brooding area for many species that use the Green River corridor as a travel route. The creation of a large emergent marsh/open-water wetland, in addition to increases in scrub-shrub, forested wetland, and improved upland habitats, is expected to increase the diversity of wildlife habitat, particularly for waterfowl. Habitat value will be maintained through formal preservation status, active management, and specific limits on public access. The GRNA is already considered a "hotspot" by local birders.

Fisheries enhancement. The design of the GRNRA specifically addressed elements identified as critical limiting factors in a comprehensive fisheries evaluation of the entire watershed. For example, the stormwater detention pond was designed to provide critical habitat where coho fingerlings can overwinter without being swept out of the watershed in a storm event, as could happen in the main channel of Mill Creek. In addition, the detention pond was sized to accommodate the full fisheries potential of the watershed.

Preservation of open space. The Green River Valley is a highly urbanized area, with much of the land surface covered by asphalt or rooftops. Through funding obtained from King County, the city was able to purchase over 200 acres of fallow farmland adjacent to the abandoned sewage lagoons and incorporate this land into the GRNRA project site, thereby saving one of the last open areas in the region from development as an industrial park.

Reuse of existing facilities and excavated materials. A main feature of the project, and an impetus behind its development, was the reuse of a 65-acre sewage lagoon that was abandoned in 1973. Since that time, the five-cell lagoon had become a fairly stable pond system, with the two largest cells containing up to 3 ft of water year-round and two other cells shifting seasonally from water to mud flat habitat. The final design reused the existing embankments and followed existing land shapes to reduce project costs and minimize construction time, thereby limiting disruption to the species already in residence. A total of 600,000 yd^3 of material were removed to create the current wetland configuration and then reused to create on-site uplands.

Public education and recreation. With its nature walks, three wildlife viewing towers, and a bike path along the west and south sides of the site, the facility provides extensive recreational as well as educational opportunities (Figure I.14.6). The facility also provides ongoing opportunities for hands-on involvement in the management and maintenance of the facility — for example, in constructing habitat, maintaining vegetation, and monitoring water quality and wildlife populations. Volunteers have built and installed bird boxes, planted native plant species, and conducted bird counts. A 5-acre native plant nursery, constructed on the southwestern corner of the site, will propagate the thousands of native plants for placement within the GRNRA and other natural areas within the city and will provide a home for plant salvaging projects. The nursery will also serve as an education and training facility for youth and volunteers in Kent. In addition, the site's master plan allows for construction of an Environmental Interpretive Center, which will be a focal point for citizens' involvement and education on environmental issues throughout the Green-Duwamish watershed.

Sweetwater Wetlands, Tucson, Arizona

Introduction

The Sweetwater Wetlands and Recharge Project (Figure I.14.7) is a 24.3-ha (60-acre) operational facility built by Tucson Water to combine functional elements such as effluent treatment, recharge, and research with a natural park setting that offers educational and wildlife viewing opportunities to the community (Karpiscak et al., 1993). The wetlands were developed with significant assistance and input from citizens, students from elementary through high school grade levels, and numerous environmental and community organizations.

Background

The Roger Road Wastewater Treatment Facility produces around 42,000 acre-ft of secondary-treated effluent annually. Most of this is discharged to the Santa Cruz River. In May 1994, the Arizona Department of Environmental Quality (ADEQ) filed a suit against the city of Tucson, alleging violations of state monitoring and reporting requirements. The city and ADEQ negotiated a settlement that, among other things, committed the city to designing and building an experimental wetland/recharge facility to treat backwash filter water from the city's Reclaimed Water Treatment Plant. The wetland would be constructed within an area including an existing sludge pond. The facility would be designed to include wildlife habitat and educational as well as recreational amenities.

Figure I.14.6 Public-use facilities, Green River Natural Resource Enhancement Area. Three wildlife viewing towers are connected by a graveled trail accessible by foot, bicycle, and wheelchair.

The city's reclaimed water treatment plant's filters are periodically cleaned by backwashing. The backwash water then is recycled through the county's treatment plant for reprocessing, at an annual cost of about $100,000. Instead of being reprocessed by the plant, the backwash water now is treated in the Sweetwater Wetlands.

The backwash water is first conveyed to two relatively small (1.2-acre) but densely vegetated treatment wetlands for settling of solids before entering the wetland ponds. Additional solids removal takes place through microbiological transformations in two wetland ponds totaling 17 acres downstream of the settling wetlands (Figure I.14.8). The backwash water is treated to meet or exceed secondary standards. The wetland effluent then flows to two recharge basins totaling 6 acres. About 300 acre-ft of backwash water will be treated annually for recharge.

Chapter I.14: Principles and applications of wetland park creation

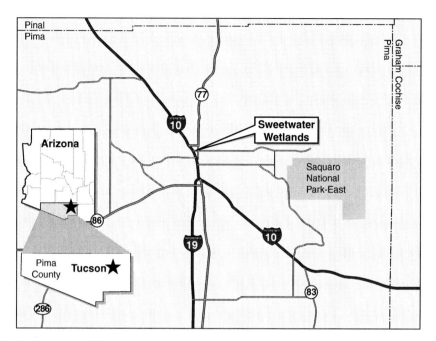

Figure I.14.7 Location of the Sweetwater Wetlands, Tucson, AZ. Located northwest of Tucson on Rogers Road, the Sweetwater Wetlands are accessible by car and bicycle year-round.

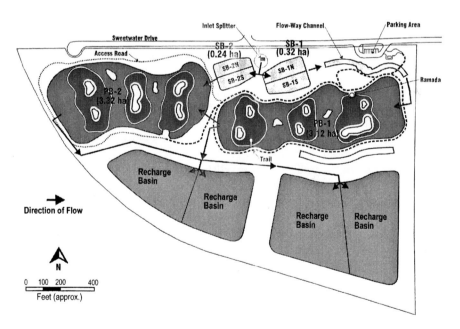

Figure I.14.8 Plan view, Sweetwater Wetlands. The Sweetwater Wetlands include 30 acres of treatment marsh, polishing, and habitat wetlands and 50 acres of recharge basins.

Estimated total construction cost was about $1.7 million, with about $600,000 earmarked for public-use amenities. The facility design was finalized in late 1995. Construction was completed in early autumn 1997.

Beneficial features

Water quality improvement. Backwash water from the city's Reclaimed Water Treatment Plant, as well as secondary effluent from Pima County's Wastewater Treatment Facility, are introduced into four densely vegetated settling ponds covering 0.7 ha (1.8 acre). The water flows from the ponds into two 3.0 ha (7.5 acre) free-water-surface wetland cells where natural biological processes further treat the water. The water then is released into four recharge basins (5.7 ha [14 acre]) where it filters through the ground. The recharged water is later recovered and reused to irrigate parks, golf courses, schoolyards, and street medians in the Tucson area. Figure I.14.8 provides an overview of the wetland configuration.

The wetland system was designed to reduce suspended solids to less than 8 nephelometric treatment units (NTU) and BOD to less than 7 mg/L. A second alternative consisted of a treatment wetland system to treat backwash water from the filtration plant. The average suspended solid load is 275 mg/L, and the target load is less than 20 mg/L. Available data indicate that Sweetwater is meeting these goals.

Wildlife habitat restoration. Since April 1997, the emergent plants have grown rapidly and the density has increased greatly. Additionally, new shoots have grown back after each winter. A wildlife study funded by the Arizona Game and Fish Department (AGFD) and further observations by the Tucson Audubon Society have documented over 120 species of birds at the Sweetwater Wetlands since the spring of 1997. Prior to construction of the Wetlands, background surveys noted less than 20 bird species as being present at the 24.3-ha (60-acre) site. The AGFD study has documented not only seasonal species changes but also a shift in populations with time as the Wetlands develops. Studies also have documented the nesting of species such as black-necked stilts and black-bellied whistling ducks. Mammals observed at the site have ranged from field mice to bobcats.

Public education and recreation. This project included an extensive public information and participation task to ensure public input. The city's consultant, along with a citizens' advisory committee appointed by the city council, developed the types of educational and recreational amenities appropriate to this unusual facility. In addition, three public open house meetings were held to solicit input from the general public. As a result of this process, the wetland became an urban park.

The Sweetwater Wetlands includes extensive passive public-use elements to provide the visitor with a variety of experiences relating to water management and ecological issues (Figure I.14.9). A public parking area is designed to accommodate cars as well as school buses. All the public-use facilities are wheelchair accessible. A six-sided kiosk is the initial focal point for visitors. A ramada provides shade, seating, and a place for groups to meet. Restrooms are available near the kiosk, served by their own subsurface flow treatment wetland. Extensive landscaping of the area includes the marsh of the wetland, a transitional marsh around the edge of the wetland cells, and an upland zone outside and above the transitional community. A looping trail system provides numerous viewpoints and interpretive signage to inform the visitor. Interpretive signage is provided at the kiosk. Native arid landscape plantings are incorporated throughout the facility. Since the public opening in early 1998, Sweetwater has attracted more than 40,000 human visitors, as well as nearly 200 species of birds, mammals, reptiles, and insects.

Mosquito control research has been extensive and detailed at this facility; bat houses are included as a natural control for adult mosquitoes, and native fishes and invertebrates are encouraged. A commonly used predator of mosquito larvae, the mosquitofish (*Gambusia affinis*), has not been implemented at this wetland project, unlike other areas, because

Chapter I.14: Principles and applications of wetland park creation 285

Figure I.14.9 Public-use facilities, Sweetwater Wetlands. Shaded gazebo overlooks, paved trails and bridges, and informational signage provide access to a variety of habitats for visitors to the Sweetwater Wetlands.

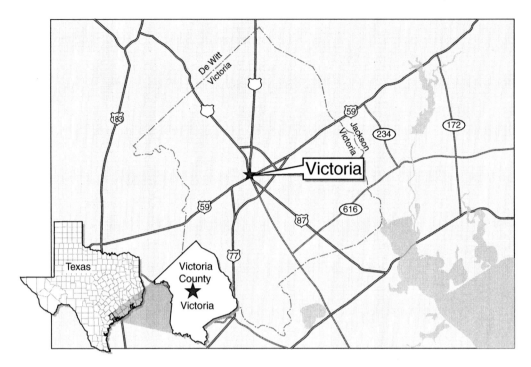

Figure I.14.10 Location of the Victoria Wetlands, Victoria TX. The Victoria Wetlands are located at 2695 Bloomington Rd., N., Victoria, TX 77905, Contact Amy Hodges at (361) 572-2137 (amy.e.hodges@usa.dupont.com) or John Snyder, DuPont Wetland Educator, (361) 572-1153 (wetlands@txcr.net), http://www.lwv.org/where/protecting/webwalk/drill_down7.html.

of the strong desire of the local environmental community to limit the distribution of species non-endemic to the region. Natural aquatic predators of the mosquitoes, including arthropods and dragonfly and damselfly nymphs, although numerous, have not been able to keep up with the mosquito larvae population, and active measures for its control have been implemented. These have included innovative aerial distribution of safe mosquito larvicides using a remote-controlled helicopter.

DuPont Victoria Wetlands, Victoria, Texas

Introduction

At DuPont's nylon intermediaries manufacturing facility in Victoria, TX, a 53-acre treatment wetland park and wildlife habitat area has been designed and constructed (DuPont, 2001). Designed to provide final polishing of biologically treated process wastewater, the wetland also provides a high-quality wildlife habitat and a scientific educational facility for public schools (Figure I.14.10).

The wetland was designed with three stages that focus on water quality improvement at the front end (Stages 1 and 2) and habitat and public use in the final stage. Features include five parallel cells and transverse deep zones for efficient solids removal in Stage 1; two parallel cells with more sinuous embankments and deep zones for effective hydraulic efficiency and habitat/aesthetic values in Stage 2; and a single, sinuous cell with large open-water areas, mixed marsh, and shallow littoral shelf areas for habitat and public use in Stage 3 (Figure I.14.11). Total project cost is estimated to be $3 million.

During the project design, DuPont staff and their consultants worked to address questions from the Education Advisory Group, the Texas Natural Resources Conservation

Chapter I.14: Principles and applications of wetland park creation

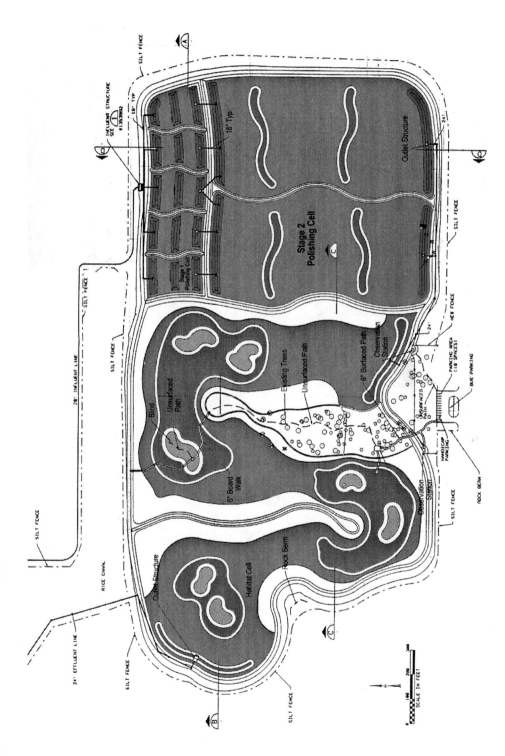

Figure I.14.11 Plan view, Victoria Wetlands. The Victoria Wetlands comprise over 53 acres of created treatment marshes, polishing wetlands, habitat marshes, preserved upland mesquite bosque, and numerous natural habitats and islands.

Figure I.14.12 Public-use facilities, Victoria Wetlands. A large, open-air interpretive facility, bird blinds, shaded overlooks, and numerous foot trails support an intensive wetland training program.

Council, and citizens. The third-stage wetland area includes an outdoor/open laboratory building for science education, a blind for bird watching, a wheelchair-accessible path and overlook for the general public, and a boardwalk through marsh vegetation (Figure I.14.12).

Beneficial features

Water quality improvement. Available monitoring data indicate measurable treatment by the wetlands of nitrate in the polished effluent from the manufacturing facility. Wetland discharge permit criteria, established through the National Pollutant Discharge Elimination System (NPDES) permit, have not been exceeded since the wetland commenced operation (Cole et al., 2001).

Figure I.14.12 (continued)

Wildlife habitat restoration. The wetland wildlife community includes a total of 195 bird species have been documented at the wetland, along with a full complement of frogs, reptiles, amphibians, and mammals. The average bird population density of 10/acre (25/hectare) is reported to be consistent with that found in other treatment wetlands (Cole et al., 2001). The remote location, dense but diverse habitat, and secure facility provide an important sanctuary to wildlife year-round.

Twenty species of common wetland plants were initially installed in the wetlands, including cattail (*Typha* sp.), giant bulrush (*Scirpus californicus*), hardstem bulrush (*S. acutus*), tall Olney's bulrush (*S. americanus*), short Olney's bulrush (*S. pungens*), pickerel-weed (*Pontederia cordata*), and softstem bulrush (*S. validus*) (Cole et al., 2001). Natural introduction of plant species by wildlife and wind and water movement has increased the total species present to 48 (Cole et al., 2001).

The nutria (*Myocastor coypus*), a nonnative aquatic rodent that is increasing in abundance throughout the southeastern United States, has become an important management concern. This aquatic herbivore has consumed considerable portions of the standing vegetation crop within this wetland project; current management options include reducing water levels to reduce the suitable habitat available, trapping the animal, and humane methods of removal. This example of animal herbivory points to the need to develop an adaptive operation and maintenance plans that allows unconventional management issues to be addressed.

Fisheries enhancement. Fish populations representative of normal riverine fish communities are present in the wetlands and have remained diverse and stable since operation began in 1997. Mosquitofish and bluegill (*Lepomis macrochirus*) are dominant within this community.

Preservation of open space. A lesser benefit in this rural setting than the other wetlands described in this paper, the open, prairie-like aspect of the wetlands creates an attractive setting and is consistent with the Victoria Plant's general interest in making beneficial uses of its property.

Reuse of existing facilities. The reuse of an old windmill as a pump for water to the interpretive center, as well as the careful site planning necessary to conserve a mesquite bosque and existing stands as habitat islands, reflect the care taken to ensure that the ecological effect of constructing the wetland remained positive.

Public education and recreation. DuPont has entered into a cooperative arrangement with the local school system and the regional education center to hire a full-time teacher who implements a curriculum program for use by elementary, middle, and high school students. Local teachers within a 150-mile radius of Victoria have integrated this educational facility into their required science curriculum at the various age levels. The Dupont-Victoria site is typically booked for visiting students a year in advance. The site has been estimated to host 2,000 to 2,500 students per year. As of April 2001, more than 9,000 4th-grade through 12th-grade students have participated in the wetland educational program. This extensive overall program reaches well into the community and involves representatives from at least 25 outside groups ranging from the Texas Parks and Wildlife Department and the U.S. Natural Resources Conservation Service to the Audubon Society and Texas A&M University.

Wakodahatchee Wetlands, Palm Beach County, Florida

Introduction

In 1996, the Palm Beach County Water Utilities Department (PBCWUD) in Florida converted nine existing wastewater percolation ponds into constructed treatment wetlands at a former regional wastewater utility located in densely populated, suburban Delray Beach (Figure I.14.13). Appropriately, the wetlands were named the Wakodahatchee, or Seminole Indian language for "created waters." Wakodahatchee covers a total of 56 acres of created wetlands populated with native plant and animal species created from existing wastewater percolation ponds. Opened to the public in 1996, this park-like setting provides not only a significant environmental resource but is highly valued by regional nature lovers and wildlife photographers (Bays et al., 2000).

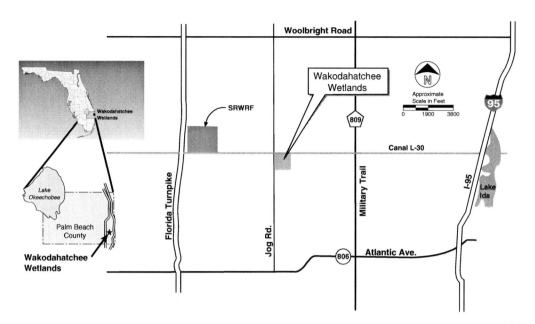

Figure I.14.13 Location of the Wakodahatchee Wetlands, Palm Beach County, FL. The Wakodahatchee Wetlands are located at 13026 Jog Road in suburban Delray Beach, FL, on the east side of Jog Road, between Woolbright Road and Atlantic Avenue. The wetland is open to the general public from sunrise to sunset, seven days a week. Tours of the Wakodahatchee are conducted on the second Tuesday of every month at 9 a.m., and on the third Wednesday of the month at 4 p.m. during winter months. The tour lasts approximately one hour, and the tour group size is limited to 20. Space on the tour must be reserved in advance. For reservations or questions about the wetland, please phone (561) 641-3429.

Background

Originally conceived in 1992 as a demonstration of the beneficial reuse of reclaimed water, the Wakodahatchee Wetlands are an important component of the 30-million-gal per day (mgd) Southern Region Water Reclamation Facility (SRWRF). The SRWRF treats the effluent to a secondary quality after which it is disposed of by deep injection well (DIW), a common practice in Florida. Prior to the construction of the SRWRF facility, however, wastewaters were disposed of through percolation ponds and a deep injection well at the System No. 3 Regional Utility since the 1970s. As part of the recent Phase II expansion of the SRWRF, the construction permit required the conversion of the System 3 percolation ponds to treatment wetlands as an effluent reuse demonstration.

The percolation ponds at the former System 3 Regional Utility were modified between November 1995 and October 1996 at a cost of $2.85 million into eight emergent marsh wetlands, interspersed with broad areas of open water and islands of wildlife habitat. Wetland area totals 39 acres, with individual wetland cells ranging from 2.3 acres to 10.9 acres. Most of the marsh area is designed to operate at an average depth of 0.5 ft, but may be operated normally at depths to 1.5 ft. Total volume of water within the wetland is ~20 million gal at the normal pool elevation. The average length to width ratio is ~3:1. The total site hydraulic loading rate varies from 0.9 cm/day at flows of 0.5 mgd, to 4.6 cm/day for flows of 2 mgd. Nominal detention time varies from a maximum of 40 days at 0.5 mgd of flow to 10 days at 2 mgd. A total of 28 deep zones and 8 habitat islands is included. Snags and perching posts were added for wildlife use.

Secondary effluent is pumped from the SRWRF to a splitter box, which distributes the flow to six parallel ponds (Figure I. 14.14). Effluent from ponds B, C, D, E, and F flows

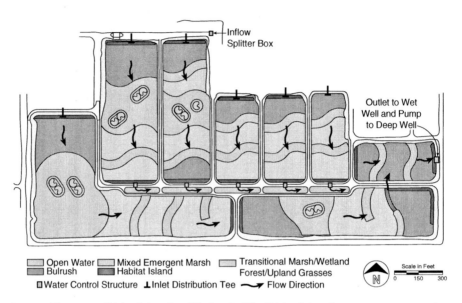

Figure I.14.14 Plan view, Wakodahatchee Wetlands. The Wakodahatchee comprises over 50 acres of emergent marshes, swamp forests, deep ponds, islands, and transitional and upland habitats.

to a collector channel and then to terminal collector pond I. The sixth parallel pond (AG) discharges in series to pond H, which in turn discharges to pond I. All wetland effluent is then pumped to an on-site DIW for disposal during the first year of operational monitoring. Pending state water quality regulatory approval, the treated wetland effluent will be discharged to the adjacent L-30 Canal for surficial aquifer recharge.

Beneficial features

Water quality improvement. The Wakodahatchee were designed to reduce PBCWUD reliance on DIW disposal of effluent by using treatment wetlands to improve water quality of the effluent to where it could be discharged to the regional surface water canal system. Available performance data for the period from November 1996 through September 1998 indicate that, at an average flow rate of 1.4 mgd, the wetland reduces total nitrogen in the SRWRF secondary effluent from 24.5 mg/L to 8 mg/L, and total phosphorus from 2.0 mg/L to 1.0 mg/L. These data match reasonably well the expected wetland performance estimated using the first-order, area-based treatment model (Kadlec and Knight, 1996).

Hydrologic storage and recharge. A stated design objective of the Wakodahatchee Wetlands was to recharge local waster supplies through infiltration to the surficial aquifer. Available hydrologic records for the wetland indicate that, of an average inflow rate of 1.4 mgd, about 0.5 mgd of effluent was lost to infiltration and evaporation over the period from November 1996 through September 1998.

Wildlife habitat restoration. The Wakodahatchee were designed to create significant wildlife habitat accessible to the public in a park-like setting. About 70% of the wetland area is vegetated by native emergent, forested, and transitional wetland species designed to emulate native south Florida wetland plant communities. Emergent marsh zones are composed of softstem and giant bulrush, duck-potato (*Sagittaria lancifolia*), arrowhead (*Sagittaria latifolia*), spikerush (*Eleocharis cellulosa*), fireflag (*Thalia geniculata*), and pickerelweed. Herbaceous species planted at the upper edge of the marsh zone include sawgrass (*Cladium jamaicense*), Fakhahatchee grass (*Tripsacum dactyloides*), and Gulf muhlygrass (*Muhlenbergia* sp.). Forested species also planted at marsh edge include cypress (*Taxodium distichum*), pond apple (*Annona glabra*), Carolina willow (*Salix caroliniana*), red maple (*Acer rubrum*), and buttonbush (*Cephalanthus occidentalis*). As expected, duckweed (*Lemna minor* and related species) has become naturally ubiquitous throughout the wetland.

Cattails were not planted and their anticipated colonization is being controlled. Native upland plant species have been planted on the site berms and include dahoon holly, sabal palm, saw palmetto, cocoplum, live oak, mahogany, and slash pine. Melaleuca and Brazilian pepper were removed from the site prior to construction.

Monitoring data collected by the Palm Beach County Department of Environmental Resource Management indicate that the Wakodahatchee Wetlands have attracted an abundant variety of wildlife, including turtles, frogs, otters, alligators, and birds. More than 176 different species of birds have been spotted at the site. These species thrive in the various wetland zones found at the Wakodahatchee Wetland. Open-water ponds areas attract water fowl and diving birds; emergent marsh areas are habitat for rails, moorhens, and sparrows; shallow littoral shelves are utilized by herons and egrets; islands with shrubs and snags serve as nesting, roosting, and basking sites; and the forested wetlands areas are rapidly growing into swamp forests of pond apple and bald cypress. Alligators nested in 2000 and 2001; their presence is a point of great interest to site users and an interesting proof of the self-contained, sustainable ecosystems that can develop within these types of wetland projects.

Fisheries enhancement. The Wakodahatchee Wetlands were stocked naturally and deliberately with mosquitofish to serve as a prey base for feeding wildlife and as a natural predator deterrent for mosquito larvae. Mosquitofish populations are abundant, on the order of 10 to 15 species per m^2. This provides an abundant supply of food for wading birds, such as herons and egrets, as well as aerial diving birds, such as the least tern and the belted kingfisher, while helping to maintain mosquito populations in check.

Preservation of open space. The suburban setting and the high level of public use by the Wakodahatchee Wetlands argue strongly for the value found in preserving the open space that had been conserved by the former percolation ponds. Rather than yet another dense suburban development, the neighbors of the Wakodahatchee Wetlands have a fully functional wetland wildlife park that is expected to stay as a park for as long as the public and the county deem it desirable.

Reuse of existing facilities and excavated materials. The creative reuse of the percolation ponds has proven to be one of the most popular public works projects implemented by the county. Operating costs are nominal, and public perception of this project is positive.

Public education and recreation. A mile-long boardwalk was constructed with gazebos and informative signs for easy public access and educational use (Figure I.14.15). Parking spaces are provided at the adjacent county complex. The county held a Wakodahatchee Wetlands day in 2000 and 2001, each time attracting over 1,000 people per day. Annual visitor totals to the Wakodahatchee are on the order of 2,000 people annually.

Figure I.14.15 Public-use facilities, Wakodahatchee Wetlands. A 0.75-mile-long boardwalk, three gazebos, educational signs, and 50-vehicle parking lot provide ready access to this productive wetland ecosystem.

Conclusion

Constructed wetland parks are valuable and rewarding endeavors to those who participate in the planning, design, construction, monitoring, and use. If properly designed, public safety and welfare are maintained while the natural environment is enhanced. By reviewing the examples provided and by following the general guidance provided here, more constructed wetland parks should appear feasible to landscape architects, engineers, scientists, and sponsoring agencies. If past experience holds true, public use of these facilities will continue to grow and exceed expectations.

c

Figure I.14.15 (continued)

Literature cited

Bays, J. et al., The Wakodahatchee Wetlands, Palm Beach County, Florida, *Proc. 73rd Meet. Water Environ. Fed.*, Alexandria, VA, 2000.
CH2M HILL, The Green River natural resource area: A study of harmony in diversity, Seattle, WA, 1999.
City of Kent, WA, http://www.ci.kent.wa.us/Public Works/Special Programs/grnra.htm, 2000.
Cole, J. et al., Wetland health plan, *Indus. Wastewater*, July/Aug., 35, 2001.
DuPont, Victoria Wetlands Web site, http://www.dupont.com, 2001.
Hammer, D., *Creating Freshwater Wetlands*, Lewis Publishers, Boca Raton, FL, 1992.
Kadlec, R.L. and Knight, R.H., *Treatment Wetlands*, Lewis Publishers, Boca Raton, FL, 1996.
Karpiscak, M.M. et al., Treating municipal effluent using constructed wetlands technology in the Sonoran Desert, in K.D. Schmidt, Ed., *Proc. Symp. Effluent Use Manage.*, Tucson, AZ, 1993.
Knight, R.H., 1992. Ancillary benefits and potential problems with the use of wetlands for nonpoint source pollution control, *Ecol. Eng.*, 1, 97, 1992.
Knight, R.H., 1997. Wildlife habitat and public use benefits of treatment wetlands, *Water Sci. Tech.*, 35, 35, 1997.
Kusler, J. and Kentula, M., Eds., *Wetland Creation and Restoration*, Island Press, Washington, DC, 1990.
Marble, A.D., *A Guide to Wetland Functional Design*, Lewis Publishers, Boca Raton, FL, 1991.
Mitsch, W.J. and Gosselink, J., *Wetlands*, 3rd ed., John Wiley & Sons, New York, 2000.
National Council on Highway Research and Planning (NCHRP), Guidelines for the development of wetland replacement areas, NR379, Washington, DC, 1996.
Schueler, T., Design of stormwater wetland systems: Guidelines for creating diverse and effective stormwater wetland systems in the Mid-Atlantic region, Dept. Environ. Prog., Metropolitan Washington Coun. Gov., Washington, DC.
U.S. Dept. Agric. (USDA), Soil Conservation Service, Wetland restoration, enhancement, or creation, Part 650, in *Engineering Field Handbook*, Washington, DC, 1992.
U.S. EPA, Guiding principles for constructed treatment wetlands: Providing for water quality and wildlife habitat, EPA 843-B-00-003.
http://epa.gov/owow/wetlands/constructed/guide.html, 2000.
Water Pollution Control Fed. (WPCF), *Natural Treatment Systems, Manual of Practice 16*, Alexandria, VA, 1990.
White, T., Planning, design, and construction of the Kent Green River Natural Resources Area, Kent, WA, submitted to *Urban Wetlands*, 2000.

Response

Designing wetlands for multiple benefits

A personal anecdote described in the book *Designing Wetlands: Principles and Practices for Landscape Architects and Land-Use Planners* (France, 2002) aptly describes what James Bays advocates in this chapter:

> Recently, I attended a conference at which one of the speakers (who was in the process of updating a major guidebook about wetland construction) dodged several questions from the audience concerning the possibility of designing for multiple purposes. Defending himself, this individual stressed that the guidebook he was working on for the EPA was concerned with "blue-collar" wetlands, those for which, because of obvious space limitations (i.e., suburban settings), it was therefore impossible to create functions (presumably aesthetics and wildlife benefits) other than those intended engineering solutions to either waste- or storm-water management.

As I continued, and as Bays clearly demonstrates in his chapter, "The difference between a utilitarian blue-collar wetland and a multipurpose white-collar wetland has really much less to do about absence of space than it has to do about absence of *imagination*."

Bays' chapter sets the stage for a major paradigm shift wherein created wetlands become created wetland *parks* such as those case studies he outlines as well as those described by Diana Balmori in Chapter I.8, Catherine Berris in Chapter I.9, Robert France and Kaki Martin in Chapter I.12, and Glenn Allen in Chapter I.13. As Bays identifies, not only do such wetland parks meet their primary functional objectives such as flood control, water quality improvement, and water storage or aquifer recharge, they also provide fisheries enhancement, wildlife habitat creation, and open space parkland. These latter attributes, once regarded as mere ancillary benefits, are now becoming recognized as key elements in wetland water sensitive design, as important in their own way to judgment of overall project success. In other words, because the ecological and aesthetic acumen of the public has increased so markedly over the last decade, the old-fashioned square-box, single-purpose constructed wetland will no longer be acceptable.

Bays stresses the need for wetland projects to be carefully planned in terms of hydrology, soils, and vegetation, all in relation to a series of "site selection criteria." Importantly, and in support of Wendi Goldsmith in Chapter I.17, and Robert France and Philip Craul in Chapter I.7, for example, Bays calls for the need of water sensitive designs to be brought to fore through the combined efforts of teams of individuals rather than by single egos. In his mind, successful wetland park creation requires the expertise of civil, water resource, geotechnical and environmental engineers, in addition to computer and drafting professionals, wetland biologists, and landscape architects.

chapter I.15

Applications of low-impact development techniques (Maryland)

Michael L. Clar

Abstract

Low-impact development (LID) is an innovative technology to control stormwater quantity/quality impacts at the source using microscale management practices distributed and integrated throughout the landscape. This technology, developed by Prince George's County, MD, to address perceived problems in conventional approaches to stormwater management (SWM) within the county and the state, makes multifunctional use of the urban landscape allowing one to design a hydrologically functional site (PGC, 1997). This approach results in an ecologically based approach to stormwater management that is usually more aesthetically pleasing, precludes impacts to receiving waters, and is generally less costly to construct and maintain than conventional end-of-pipe systems

This chapter provides a chronological overview of a number of projects that have been instrumental in the development of LID technology. The focus of the chapter is on case studies and demonstration projects. Although the emphasis of these projects is the state of Maryland, projects and case studies from other regions of the country are also included.

Introduction

The range of projects and case studies included in this chapter help to demonstrate the tremendous range and versatility of LID technology in addressing environmental issues resulting from land development activities. Chapter I.5 introduces the fundamental concepts and principles associated with LID. This chapter complements that earlier chapter by documenting a number of LID applications and case studies, which include

1. LID as a water quality control technique for infill development
2. LID as a water quality retrofit for existing urban areas
3. LID as a comprehensive strategy to replicate predevelopment hydrologic functions for a developed site
4. LID as a win–win strategy to provide improved environmental performance and reduced site development costs
5. LID as a volume control method to provide downstream peak discharge protection for major storm events

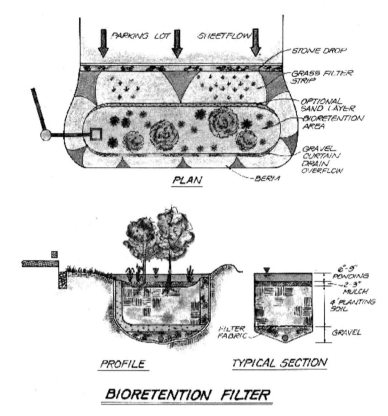

Figure I.15.1 Bioretention schematic representation (MDE, 2000). (Color version available at www.gsd.harvard.edu/watercolors. Password: lentic-lotic.)

6. LID as an improved approach to protect water supply water reservoirs
7. LID as an approach to address total impervious area (TIA) limitations

Maryland case studies

Case study 1: First steps — Prince George's County bioretention design manual

One of the first steps in the development and application of LID techniques in Maryland consisted of the development of the Prince George's County Design manual (ETA, 1993). The bioretention design manual provided one of the first integrated, landscape-based micromanagement tools for stormwater management, as shown in Figures I.15.1 and I.15.2, that made the development of the LID concept possible.

This project was conceived by Larry Coffman of Prince George's County, MD, and directed by Michael Clar of Ecosite, Inc., and began with modest objectives. These included a feasibility study to evaluate if an upland living filter could be used to achieve the water quality goals of SWM (i.e., control the first flow of surface runoff) and be used on small sites to replace the very costly and ineffective best management practices (BMPs) such as the oil-grit separator and other similar devices.

The feasibility study was conducted by an interdisciplinary team with training and expertise in water resources engineering, hydraulics, hydrology, geohydrology, soil science, landscape architecture, and biology. It demonstrated that bioretention could be used

Figure I.15.2 Bioretention cell located in a cul-de-sac illustrates multifunctional and landscape integrated features of this microscale practice. (Color version available at www.gsd.harvard.edu/watercolors. Password: lentic-lotic.)

to meet the existing water quality control criteria. In addition, it also revealed that many additional benefits could be anticipated, including improved aesthetics, significantly lower construction and maintenance costs, as well as the potential to reduce peak discharges by increasing the time of concentration and reducing the effective curve number, CN.

Based on the positive results of the feasibility study, a number of case studies were developed and documented for various uses of the bioretention concept and a "Bioretention Design Manual" was prepared.

Case study 2: Infill development with LID — Beltway Plaza Expansion

The Beltway Plaza Expansion project demonstrated the application of bioretention techniques for an infill development project. Infill development projects are very commonplace in existing urban areas. Often, these sites have an existing storm drainage infrastructure designed to convey the 2- or 10-year storms but typically lack either stormwater management or water quality control. This was the case at the Beltway Plaza located in Prince George's County, MD. It was determined that this site would receive a waiver for peak discharge control due to the existing storm drainage system but would require water quality control.

A schematic of the site is shown in Figure I.15.3. The entire parking lot for the expansion received water quality control through the use of bioretention cells, as shown in Figure I.15.4. This site proved to be a significant milestone for the use of LID technology for a number of reasons. First, it demonstrated that the bioretention concept could be used to control a large parking lot area by breaking the area up into a number of small drainage catchments: the microscale management concept. Second, it provided crucial information on the durability of this practice. The bioretention cells have been in operation since 1993 and are holding up very well. Third, it provided a site that could be monitored to document the pollutant removal performance of bioretention cells as described in the next paragraph. Fourth, it provided actual data that verify the very low and simple maintenance requirements and costs of these practices, which are approximately $200 per site/year.

Figure I.15.3 Beltway Plaza Bioretention Site illustrates how the impervious areas from this shopping center expansion were effectively disconnected and treated by designing the landscape islands as bioretention cells. (Color version available at www.gsd.harvard.edu/watercolors. Password: lentic-lotic.)

Figure I.15.4 Beltway Plaza bioretention cells. The reported maintenance costs of bioretention cells is low, averaging approximately $200/site. (Color version available at www.gsd.harvard.edu/watercolors. Password: lentic-lotic.)

The pollutant removal performance of the bioretention cells at the Beltway Plaza location was field monitored and, as shown in Table I.15.1, verified that it not only met the water quality control criteria, but actually ranked as the most effective pollutant

Table I.15.1 Pollutant Removal Performance of Bioretention Practices (% Removal Rates)

	Cu	Pb	Zn	P	TKN	NH_4	NO_3	TN
Upper zone	90	93	87	0	37	54	−97	−29
Middle zone	93	99	98	73	60	86	−194	0
Lower zone	93	99	99	81	68	79	23	43

Note: The monitoring data for the Beltway Plaza site showed that bioretention cells exhibit very high pollutant removal rates for most parameters.

Source: Davis et al. (1998).

Figure I.15.5 Schematic layout of bioretention cells at Largo Office parking lot illustrates the use of bioretention practice as an urban retrofit for an existing parking lot. (Color version available at www.gsd.harvard.edu/watercolors. Password: lentic-lotic.)

removal BMP available. In addition, the success of this site suggested that bioretention could be an effective retrofit BMP for existing urban areas.

Case study 3: Parking lot retrofit with bioretention — Prince George's County Offices

The parking lot of the Prince George's County Office Complex in Largo, MD had been designed with conventional end-of-pipe SWM controls. The Prince George's County government conducted a bioretention retrofit project to demonstrate that the bioretention cell was an effective way to retrofit existing urban areas to provide water quality benefits. Figure I.15.5 provides a site schematic map, while Figure I.15.6 shows the bioretention cell in operation shortly after completion of construction.

Figure I.15.6 Operation of bioretention cell during a storm — curb cuts intercept and direct runoff into the bioretention cell for storage and treatment. (Color version available at www.gsd.harvard.edu/watercolors. Password: lentic-lotic.)

This demonstration site also served as an important milestone in the development of LID technology in Maryland for a number of reasons. First, it demonstrated that existing impervious areas could be cost-effectively retrofitted with bioretention cells to provide water quality control. Second, it demonstrated and documented the use of various materials for the construction of the cells. Third, it provided another site that was monitored to document the pollutant removal performance of bioretention cells in a retrofit setting.

Case study 4: LID comes of age — Prince George's County LID Design Manual

The bioretention case studies described above together with considerable technical analysis of improved approaches for addressing the stormwater management impacts of development activities culminated in the development of the Prince George's LID design manual in 1997 (PGC, 1997). This manual is a milestone in LID technology for several reasons. First, it provided a complete and systemic approach for integrating the major elements of LID design. These include (1) using conservation, avoidance, and minimization, (2) reducing impervious areas, (3) disconnecting unavoidable impervious areas, (4) maintaining the predevelopment time of concentration, (5) selecting applicable control measures to approximate predevelopment hydrology, and (6) implementing public outreach and institutional activities.

The LID design manual describes and documents each step in the computational process for maintaining predevelopment hydrology. In addition a series of detailed case studies is provided in the appendices of the manual; the case studies demonstrate the application of the LID process for single-family residential, townhouse residential, and a commercial project.

Case study 5: Detailed LID comparisons with conventional SWM design

In 1997 and 1998, Prince George's County conducted a series of studies to compare the costs of using LID design with conventional SWM design practices (Greenhorne & O'Mara, Inc., 1998). Three case studies that compare LID with conventional design were developed with

Table I.15.2 Construction Cost Comparisons for Patuxent Riding

Element	Costs		Percent of Total Cost	
	Conventional	Low Impact	Conventional	Low Impact
Grading, roadway R/W	$569,698	$426,575	52	52
Storm drain	$255,721	$132,558	24	16
SWM*	$260, 858	$10,530	24	1
Bioretention, rain barrels	$0	$252,124	—	31
Total cost	**$1,086,277**	**$821,787**	**100**	**100**
Units		74		81
Unit cost	$14,679	$10,146		

* The SWM cost for LID is the SWM fee-in-lieu.

Note: The use of LID compared with conventional subdivision development provided an increase in lot yield and a corresponding decrease in unit costs.

an emphasis on mitigating hydrologic impacts. The case study approach is valuable because of the highly site-specific and interrelated outcomes of LID techniques, which cannot be sufficiently evaluated if they have been applied only to idealized or theoretical circumstances.

The three case studies included the following:

1. Patuxent Riding (residential lots)
2. Pennsylvania Riding (residential townhouses)
3. Great Eastern Shopping Center (commercial)

The results of the Patuxent Riding case study are presented in Table I.15.2 and summarized next.

Costs

The LID design provided seven additional lots while achieving a significant cost decrease in the development of the Patuxent Riding site. Cost comparisons are summarized in Table I.15.2. These cost figures demonstrate significant savings in grading and roadway costs and dramatic reductions in the costs for storm drains and SWM. Although the total roadway length was increased due to the addition of a cul-de-sac, savings in roadway costs resulted from decreases in both road width and road thickness. The storm drain costs were reduced by nearly 50% on account of a greater proportion of the conveyance made available on the surface in grassy swales instead of in underground pipes. With the elimination of the need for two SWM ponds, the SWM cost was reduced to the fee-in-lieu. The savings in SWM costs were approximately equal to the cost of bioretention and rain barrels.

The amount of woods preserved in both the conventional development and LID met the Tree Conservation Plan (TCP) requirements of Prince George's County. Because disturbance of the wooded area at Patuxent riding is slight, no cost difference was reported between LID and conventional development in terms of meeting TCP requirements.

Benefits

In comparison with conventional development, the LID for Patuxent Riding retained more water on-site, reducing runoff volume and increasing base flow in receiving streams. LID eliminated the need for an off-site easement for the SWM Pond #1 outfall, and the cost of maintaining SWM facilities was likewise eliminated. A major benefit to the developer was the significantly reduced cost per unit. The seven additional lots were added while the total construction cost was reduced by approximately $200,000. This case study was significant because it clearly demonstrated that LID can be a win–win strategy for land

304   *Handbook of water sensitive planning and design*

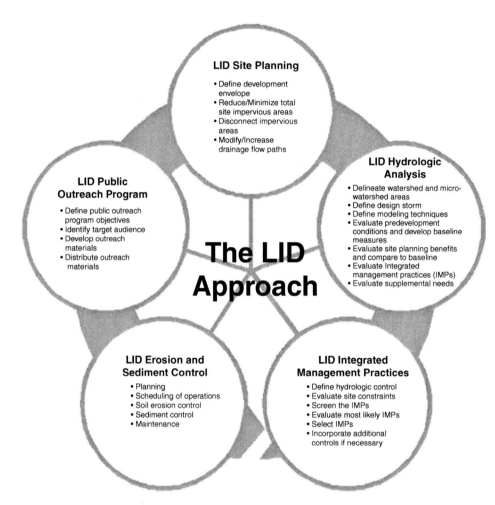

Figure I.15.7 Major components of the LID approach include 10-site planning, LID hydrologic analysis, selection, and design of LID integrated management practices, erosion, and sediment control for LID, as well as public outreach initiatives.

developers and environmental protection. Previously, it had always been presumed that providing more or higher environmental protection was associated with higher development costs. LID, however, demonstrates that better technology can provide a win–win strategy, better environmental protection, and reduced development costs.

Case study 6: LID goes national — National design manual

Under a research grant from the U.S. EPA, the contents of the Prince George's LID design manual were expanded and modified into a national LID guidance document titled, "Low-impact development design strategies: An integrated design approach" (U.S. EPA, 2000). The national manual organized the LID design process into five major groupings, which summarized in Figure I.15.7: (1) site planning, (2) hydrologic analysis, (3) selection and design of integrated management practices (IMPs), (4) erosion and sediment control requirements for LID, and (5) public outreach activities. The manual provides more detailed guidance on the incorporation of LID technology into the site planning process than is found in the Prince George's County design manual. Also, it provides more detail on a range of microscale landscape-based practices known as IMPs.

Case study 7: Pembroke subdivision — Total LID site design also provides downstream peak discharge control

The Pembroke Subdivision is a residential development located in northern Frederick County, MD, and is the first LID subdivision permitted in this county. Designed as an LID subdivision from the start, it exhibits and benefits from the win–win attributes of the LID process, which include:

1. The use of LID allowed the site design to eliminate the use of two SWM ponds, which had been envisioned in an earlier concept plan for the site. This represents a roughly $200,000 reduction in infrastructure costs.
2. In place of the SWM ponds, 2.5 acres of undisturbed open space and wetlands have been preserved. Again, a considerable savings in wetlands mitigation impacts was realized.
3. Also, the site plan gained two additional lots, increasing the 43-acre site yield from 68 to 70 lots. This added roughly $100,000 in additional value to the project.
4. Extensive use of site fingerprinting techniques allowed the site design to preserve approximately 50% of the site in undisturbed wooded condition. This design feature was very beneficial to maintaining the predevelopment CN.
5. Approximately 3000 linear ft of roads were converted from an urban road section to a rural road. This design feature replaced curb and gutter with grass swales, a $60,000 savings in construction costs. Also, the rural road section reduces the paving width from 36 to 30 ft, a 17% reduction in paving costs.

In order to satisfy county criteria for adequate downstream conveyance, a downstream impact analysis was conducted. The analysis examined the ability of an LID site design to maintain predevelopment peak discharge conditions for a range of storms, including the 1-, 2-, 10-, 50-, and 100-year storms. Many public works personnel perceive innovative SWM techniques such as LID that are capable of addressing water quality issues but insufficient to provide downstream peak discharge control for the larger flood flows associated with the 10-, 50-, or 100-year storms.

Initially, the site LID hydrologic analysis was based on the 1-year storm (2.5-in. rainfall), which is Frederick County's criteria for water quality control. The downstream analysis revealed that the 1-year storm design was not sufficient to maintain predevelopment peak discharges for the 10-, 50-, and 100-year storms. An incremental iterative procedure was then used to determine the additional control requirements that would be required to provide the required downstream control. The analysis showed that by increasing the design storm to the 2-year storm (3.0 in. of rainfall), the required downstream protection for the complete range of flood events (10-, 50-, and 100-year storms) was achieved.

These results have great significance for future SWM policy and design criteria. They clearly illustrate the tremendous advantages achieved by the volume control approach incorporated in the LID technology. Volume control allows us to replicate the predevelopment rainfall runoff processes and helps to demonstrate that the peak discharge detention approaches provided with SWM ponds in conventional SWM designs are really hydrologically flawed concepts. The flaws associated with the peak discharge detention approach are numerous and include the following problems:

1. Peak discharge control does not typically address the maintenance of groundwater recharge.
2. Peak discharge approaches alter the frequency and duration of flood flows, resulting in stream channel degradation.
3. Peak discharge approaches can actually aggravate downstream flooding conditions due to the superpositioning of runoff hydrographs.
4. Peak discharge approaches, particularly the use of regional facilities, provide no protection for streams above the regional facilities.

Case study 8: Maryland 2000 SWM design manual

The State of Maryland Department of the Environment (MDE) is in the process of updating its SWM regulations and requirements. One of the key elements of the update has been the development of a new design manual, referred to as the MDE 2000 SWM design manual (MDE, 2000). Although this statewide manual does not entirely embrace the total LID design process, it does incorporate some elements of the LID process. The manual encourages the use of the better site planning techniques as described by the Center for Watershed Protection (CWP, 1998). Many of the better site design techniques are similar to the measures described in the LID design manual to reduce the total amount of impervious area (TIA) and to disconnect as much of the remaining TIA as possible. The manual also includes a requirement for maintaining the groundwater recharge, which is consistent with the LID focus on replicating predevelopment hydrology.

Another feature of the MDE manual that is very similar to the LID design manual is the incorporation of a series of stormwater credits. The stormwater credits consist of a series of nonstructural practices that can play a significant role in reducing water quality impacts, as well as reducing the generation of stormwater runoff from the site. Six of these credits are described in the manual and are summarized in Table I.15.3. Figure I.15.8 provides a schematic representation of the natural area conservation credit, while Figure I.15.9 provides a view of a subdivision built using this credit.

LID case studies outside Maryland

The focus of this paper is on LID-related activities in Maryland, but it is important to note that the LID concept is being recognized as a viable alternative to the traditional approaches to SWM management and is being rapidly implemented in many jurisdictions throughout the United States. In addition, some applications are beginning to surface in other countries. A few of these developments are briefly identified here.

Case study 9: LID for optimum water quality protection of water supply reservoir — High Point, NC

Due to its proximity to a proposed regional water supply reservoir, the city of High Point, NC, is faced with the implementation of very stringent water quality controls related to nutrients control (i.e., phosphorus) and limitations on TIA. As part of a watershed-wide assessment and the development of a comprehensive stormwater management plan (Tetra Tech, Inc., 2000), an evaluation of the benefits of using LID technology was conducted.

The evaluation revealed that the use of LID, particularly the incorporation of bioretention techniques, could optimize the removal of phosphorus by approximately 50% over conventional pond-based BMPs. The bioretention cells can achieve phosphorus removal

Table I.15.3 Summary of MDE (2000) Stormwater Credits

Stormwater Credit	Water Quality Volume	Groundwater (gw) Recharge Volume	Peak Discharge
Natural area conservation	Area draining to natural area can be subtracted from water quality comps	Can be used w/percent area method to meet gw recharge criteria	A predevelopment forest/meadow value can be used for CN
Disconnection of rooftop runoff	Reduces runoff factor (Rv) used in water quality comps	Can be used w/percent area method to meet gw recharge criteria	Increased flow path = longer T_c; can reduce CN
Disconnection of nonrooftop runoff	Reduces runoff factor (Rv) used in water quality comps	Can be used w/percent area method to meet gw recharge criteria	Increased flow path = longer T_c; can reduce CN
Sheet flow to buffers	Contributing site area can be subtracted from water quality comps	Can be used to reduce gw recharge criteria	Can reduce CN
Open channel use	Can be used to satisfy water quality volume criteria	Can be used individually to fully satisfy gw recharge criteria	Increased flow path = longer T_c; can reduce CN
Environmentally sensitive development	Automatically meets the water quality volume criteria	Fully satisfies gw recharge criteria	Increased flow path = longer T_c; no CN credit

Note: The new Maryland stormwater management design manual incorporates a number of stormwater credits for using LID techniques.

levels ranging from 75 to 90% compared with the reported levels for SWM ponds, which range from 40 to 50%.

The LID evaluation also reinforced another advantage of the LID technology with respect to the TIA limitation requirement: a number of jurisdictions have begun to place TIA limitations on a watershed scale as a surrogate for water quality control. This approach is based on the TIA threshold concept reported in a number of publications (CWP, 1994). For a specific site, however, the LID concept can provide a win–win strategy that optimizes water quality objectives while allowing higher impervious cover for a given site. This dual strategy is accomplished in two ways. First, the LID design methodology provides procedures and techniques to hydraulically disconnect impervious areas so that, for example, a site with 70% impervious cover will be hydrologically equivalent to a site with 40 to 50% impervious cover. The second part of this strategy results from the fact that the LID micromanagement practices can be incorporated into elements of the landscape, providing a dual function for site features and thus preclude the need to dedicate and disturb (clear, grub, etc.) 8 to 10% of the total site for an SWM pond. Figures I.15.10 and I.15.11 compare the site design features for a commercial site designed with conventional controls and LID techniques.

Case study 10: LID for commercial applications — Florida Aquarium, Tampa Bay, FL

The Florida Aquarium site is an 11.5-acre asphalt and concrete parking area that serves approximately 700,000 visitors per year. An innovative stormwater management system

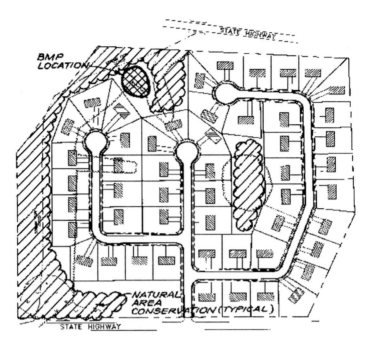

Figure I.15.8 Schematic representation of natural area conservation credit. The MDE 2000 manual provides guidance on how to incorporate natural area conservation into site design and obtain a stormwater management credit (MDE, 2000).

Figure I.15. 9 Photograph of a subdivision using the natural area conservation credit. (Color version available at www.gsd.harvard.edu/watercolors. Password: lentic-lotic.)

Chapter I.15: Applications of low-impact development techniques (Maryland)

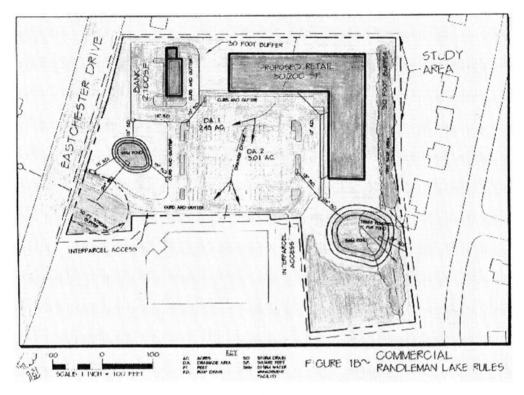

Figure I.15.10 Randleman Lake Study commercial site using conventional stormwater management to meet design criteria.

was designed using an SWM treatment train approach that incorporates many elements of LID design (Rushton, 1999). Runoff is controlled using the following BMPs:

1. End-of-island bioretention cells
2. Bioretention swales located around the parking perimeter
3. Permeable paving
4. Bioretention strips between parking stalls

This project has incorporated a long-term monitoring program that is providing valuable performance data on the hydrologic and water quality performance of this facility. Figure I.15.12 provides a site schematic plan, while Figure I.15.13 shows some of the monitoring equipment installation.

A total of 16 storm events were monitored at the Florida Aquarium site during 1998. The Southwest Florida Water Management District measured rainfall and flow from each of the subcatchments in the parking area and collected water quality samples on a flow-weighted basis. Comparisons between pavement areas controlled by the BMPs and uncontrolled asphalt areas were made for peak runoff rate, runoff volume, runoff coefficients, and water quality. Sediment cores from swales also were collected and analyzed.

The parking areas controlled by the BMPs showed a significant reduction in runoff volume and peak runoff rate. Table I.15.4 shows pollutant reductions for three pavement types; reduction is compared to pollutant concentrations in runoff from a basin without a swale.

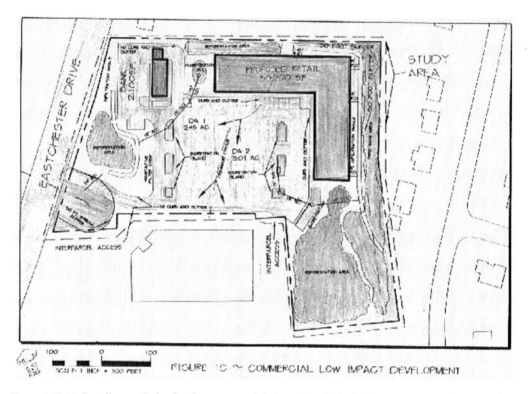

Figure I.15.11 Randleman Lake Study commercial site using LID design conventional approach to meet design criteria. Note that bioretention islands and bioswales replace ponds, and large blocks of undisturbed forest are retained on-site.

Case study 11: University of Virginia bioretention demonstration project

The University of Virginia, Charlottesville, VA, has initiated a long-term study of the performance of a bioretention cell, as shown in Figure I.15.14. This study differs from the two bioretention studies conducted in Maryland that monitored a single storm event (3 in. of rainfall). The UVA study is providing performance data based on an annual hydrologic budget. Initial, first-year results indicate that, based on an annual budget, analysis of the performance of the bioretention cells will exceed all expectations. First-year results are as follows (Yu et al., 1999): 86% for TSS, 90% for TP, 97% for COD, and 67% for oil and grease.

Case study 12: Start at the source manual — San Francisco Bay Area

The San Francisco Bay Area Stormwater Management Agencies Association has published a guidance manual titled "Start at the source: Residential site planning and design guidance manual for stormwater quality protection" (BASMAA, 1997). This document provides guidance for incorporating landscaped-based micromanagement BMPs at the site level. Much of the guidance provided in this manual is very similar to the LID design manual developed for Prince George's County, MD.

Other case studies

A number of additional case studies are being conducted throughout the country. Space and time limitations preclude more detailed treatment in this paper and will therefore be reported at a future date. Some of these projects are briefly listed here:

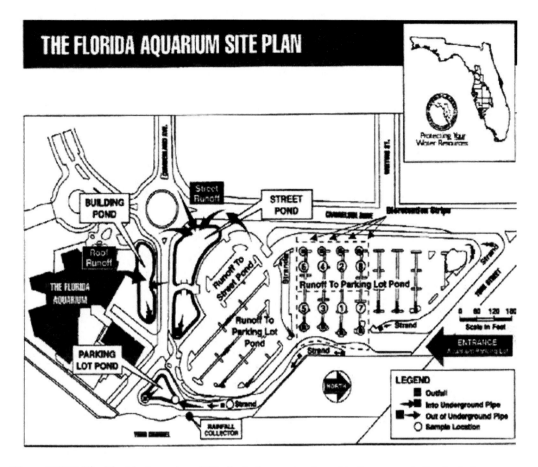

Figure I.15.12 The Florida Aquarium LID site features a treatment train approach that includes porous pavement, bioretention strips, and ponds. (Color version available at www.gsd.harvard.edu/watercolors. Password: lentic-lotic.)

1. The Maryland State Highway Administration (SHA) has recently begun an initiative that is evaluating the use of LID practices for urban and rural roads applications. The ultimate goals of this project is to develop standards and specifications for these applications, as well as the installation and monitoring of selected practices.
2. The U.S. EPA is nearing completion of a feasibility study that examines the potential for applying LID technology to retrofit existing urban areas.
3. Rockdale County, GA, has begun a series of demonstrations that will evaluate the application of LID technology in Rockdale County.
4. Jordan Cove Project, CT: a paired watershed study is being conducted in a joint effort between the U.S. EPA and the University of CT. The objective of this study is to compare the impacts at a watershed scale of conventional vs. innovative SWM control practices. The innovative control practices include many of the LID techniques.
5. LID demonstration projects in the early stages of development are being reported in MA, MN, SD, ID, DC, and Seattle, WA.
6. At the international level, LID initiatives are being reported in Canada, France, Germany, Japan, New Zealand, and Taiwan.

Figure I.15.13 Florida Aquarium site monitoring stations. Porous pavement and bioretention strips help achieve a 40% reduction in runoff. (Color version available at www.gsd.harvard.edu/ watercolors. Password: lentic-lotic.)

Table I.15.4 Load Efficiency of Pollutants Expressed as Percent Reduction for Three Types of Pavement at the Florida Aquarium Site

Constituents	Percent Pollutant Reduction*		
	Asphalt w/Swale	Cement/Swale	Permeable w/Swale
Ammonia	45	73	85
Nitrate	44	41	66
Total nitrogen	9	9	42
Ortho-phosphorus	−180	−180	−74
Total phosphorus	−94	−62	3
Suspended solids	46	78	91
Copper	23	72	8
Iron	52	84	92
Lead	59	78	85
Manganese	40	68	92
Zinc	46	62	75

* The basins with swales were compared with a basin without a swale to determine the amount of reduction in pollutant loads possible using these small alterations. Notice that the efficiencies or phosphorus are negative, indicating an increase in phosphorus load in the basins with a swale. This emphasized the importance of source control as well as runoff volume reduction.

Summary

This chapter provides a brief summary of a number of projects that illustrate the development and application of the LID technology for stormwater management. The LID technology as documented in the Prince George's County and the U.S. EPA national manuals provides a significant advance in the art and technology of stormwater management. This technological breakthrough is derived from a number of sources. The first is the attempt to define and replicate essential predevelopment hydrologic functions. The

Figure I.15.14 The UVA Bioretention Demonstration Project is providing data on the performance of bioretention cells on an annual basis. (Color version available at www.gsd.harvard.edu/watercolors. Password: lentic-lotic.)

second is the use of an integrated interdisciplinary approach that differs considerably from the single-parameter, hydraulic-based approaches of the past. The third is derived from the microscale or at the source control concept.

Stormwater management issues and alternatives can differ substantially on a watershed-by-watershed basis. Most watersheds exhibit a wide variety of interrelated stormwater management problems and issues. Some watersheds, or portions of watersheds, are dominated by older, established urban areas that were developed prior to the advent of stormwater management technology. The needs of these watersheds have largely not been adequately addressed, if they have been addressed at all in the past. Other watersheds, or portions thereof, are predominantly suburban and may be subject to various control policies that have developed over time. Still other watersheds may be predominantly rural and can present a mix of agricultural and developing watershed issues. This wide variety of conditions suggests the need for stormwater management policies and tools that can address the wide range of issues at the watershed level.

The projects and case studies presented in this paper try to provide a representative cross-section of these types of issues. Case studies are presented that address the following conditions:

1. Water quality control for infill development
2. Water quality retrofits for existing urban areas
3. LID technology to replicate predevelopment hydrologic functions
4. Volume-based LID technology to provide real downstream peak discharge control for flood events
5. LID control for optimum water quality control in water supply reservoir watersheds
6. LID technology to address the TIA issue

The author recognizes that the projects and case studies provided in this paper represent only a small sample of all the LID projects that exist. The paper does not make any representation that these are the best available examples of LID. They are included primarily on the basis of two criteria: (1) the author is familiar with and has some documentation for these projects; and (2) the projects are helpful in describing the range of applications that can be achieved with LID technology. Finally, the author invites contact from readers who are knowledgeable of other LID projects, so information can be shared and the LID database improved.

Literature cited

Bay Area Stormwater Management Agencies Association (BASMAA), *Start at the source: Residential site planning and design guidance manual for stormwater quality protection*, prepared by Tom Richman and Associates, Palo Alto, CA, 1997.

Davis, A. et al., Optimization of bioretention design for water quality and hydrologic characteristics, report 01-04-31032, Final report to Prince George's County, MD, 1998.

Center for Watershed Protection (CWP), The importance of imperviousness, *Watershed Protection Tech.*, 1, 1994.

Center for Watershed Protection (CWP), *Better Site Design: A Handbook for Changing Development Rules in Your Community*, Center for Watershed Protection, Ellicott City, MD, 1998.

Engineering Technologies Associates, Inc. (ETA), *Design manual for use of bioretention in stormwater management*, prepared for the Dept. Environ. Resour. (DER), Prince George's County, MD, 1993.

Greenhorne & O'Mara, Inc. Case studies in low-impact development: stormwater management comparison and economics evaluation, prepared by Greenhorne & O'Mara, Inc, Greenbelt, MD, for the DER, Prince George's County, MD, 1998.

Maryland Department of the Environment (MDE), *2000 Maryland stormwater design manual*, Vols. I & II, prepared by the Center for Watershed Protection and the MDE, Water Management Administration, Baltimore, MD, 2000.

Prince George's County (PGC), MD, *Low-impact development design manual*, DER, Prince George's County, MD, 1997.

Rushton, B., Low-impact parking lot design reduces runoff and pollutant loads, annual report #1, Southwest FL Watershed Manage. Dist., Brooksville, FL, 1999.

Tetra Tech, Inc., City of High Point, NC, Deep River 1 watershed assessment and stormwater plan, prepared by Tetra Tech, Inc., Research Triangle Park, NC, 2000.

U.S. EPA, Low-impact development design strategies: An integrated design approach, prepared by Tetra Tech, Inc., Fairfax, VA, for DER, Prince George's County, MD, funding provided by the U.S. EPA, Washington, DC, 2000.

Yu, S.L., Zhang, X., Earles, A., and Sievers, M., Field testing of ultra-urban BMPs, *Proc. 26th Ann. Water Resour. Plan. Manage. Conf.*, Wilson, E., Ed., ASCE, Tempe, AZ, 1999.

Response

Values of demonstration projects and case studies of stormwater source management

This chapter by Michael Clar clearly demonstrates that low-impact development represents a significant advance in the technology and art of stormwater management. As introduced in Chapter I.5 by Larry Coffman and adapted by Richard Pinkham and Timothy Collins in Chapter I.4, Robert France and Philip Craul in Chapter I.7, and Thomas Liptan and Robert Murase in Chapter I.6, low-impact development presents an integrated and interdisciplinary approach to water sensitive design that differs considerably from the single-parameter hydraulic approaches of previous stormwater management.

Clar makes the point that because the stormwater management issues differ from site to site due to watershed specificity, low-impact development approaches must be flexible in order for them to be adapted to any situation. This is why demonstration projects such as those described in this chapter offer convincing evidence toward the eventual general acceptance of these new stormwater technologies. The fact that one of the case studies presented by Clar was a retrofit located in the parking lot right beneath an environmental director's office window goes far to limiting charges of hypocrisy that developers often (and justifiably) voice toward environmental regulators; i.e., the wisdom and strength of "practicing what you preach" cannot be overestimated. Another interesting case study shows the strategy of breaking up the flow paths of large parking lots into small, manageable drainage sub-catchments where bioretention swales can be easily fitted.

This chapter adds to the discussion by presenting evidence obtained from comparative economic analyses that low-impact development approaches can actually have lower construction and maintenance costs while, at the same time, offering improved aesthetics and other multifunctional uses of the urban landscape. The concept of stormwater credits for uncoupling runoff from impervious surfaces is a direction worth pursuing further.

chapter I.16

Restoring and protecting a small, urban lake (Boston, Massachusetts)

Nicholas Pouder and Robert France

Abstract

Chandler Pond, a small urban lake in Boston, has experienced cultural eutrophication due to watershed inputs of phosphorus. By the late 1990s, the problem had become so severe that the bottom sediments of the lake, where much of the phosphorus was stored and frequently resuspended to promote algal growth, were removed by dredging. Although such dredging immediately improved water quality, it was determined that an underlying cause of the problem — the continued inputs of new aqueous phosphorus and nutrient-laden sediments to the lake through stormwater runoff and shoreline erosion — had not been fully addressed. The present project focused on raising public awareness about the possibilities of mitigating future eutrophication through on-site methods. Specifically, a demonstration project restored a section of the eroding shoreline through bioengineering techniques, in and around which were placed wetland plants to stabilize the soil, cleanse inflowing stormwater, and provide aesthetic and recreational opportunities. A plan to address inputs from the watershed is also being prepared.

Introduction

Urban lakes face a host of environmental problems arising from their location within densely populated cities (Tourbier and Westmacott, 1992). Fertilizer application to residential lawns and community golf courses, as well as frequent deposition of dog feces, ensure that a large pool of nutrients is available within (sub)urban watersheds. Given the high degree of impervious surfaces in such settings (Chapter I.3), stormwater runoff is a serious problem in terms of water quantity (ASLA, 1996; Chapter I.1). When this runoff moves across vegetated surfaces such as lawns and golf courses, phosphorus from fertilizer and dog waste become mobilized, and together with suspended sediments, can be transported to urban waters where they pose a serious threat to water quality (Terrene Institute, 1994). The resulting increased algae and vascular macrophyte growth contribute to the bottom sediments of these lakes upon plant death, eventually decreasing water depth as the nutrient-laden sediments accumulate. Wind-driven resuspension of bottom sediments over shallow-water lakes as well as extensive shoreline erosion can both continually supply nutrients to the water column and support yet further primary production, thereby exacerbating an already serious water quality problem. The problem of eutrophication of urban lakes has three obvious solutions:

1. One might think that the most obvious and successful procedure for restoring eutrophied urban lakes would be to simply reduce or stop the input of all anthropogenic phosphorus. As an ultimate strategy, however, such an option is limited due to the large source of nutrients contained within the sediments now situated at the bottom of such lakes. Due to the accumulation of nutrients previously transported from the watershed to the lakes, and the potential of those nutrients to serve as a permanent and renewable source of enrichment due to wind resuspension, it is often never enough to simply halt the input of new sediments or nutrients. This becomes very much a case of the "sins" of fathers and mothers being exacted upon the offspring.
2. In such circumstances, it becomes necessary to restore urban lakes by actively removing past nutrient inputs. Deepening lakes — in a sense setting back the clock of their degradation caused by urbanization — by dredging bottom sediments is a very costly, yet becoming all-too-necessary, technique to rejuvenate lakes to a more pristine past. Many examples of dredging rehabilitation exist in urban settings such as Boston (e.g., Barnett, 1999).
3. It would be wrong, however, to think that once the symptom of lake eutrophication is temporarily "fixed," that the work of environmental managers is over and the problem(s) alleviated or mitigated. Dredging urban lakes with no attention paid to watershed causality or implementation of an effective plan to reduce future nutrient inputs is like treating cancer with a bandage (France, 1999). Only through the use of retrofitted BMPs (or best management practices) (e.g., Schueler et al., 1992; Chapters I.4, I.5, and I.15) and altered community development rules (Center for Watershed Protection, 1999), as part of a program of comprehensive watershed planning and management (Center for Watershed Protection, 1998; Chapters II.8 and II.9), can one really say that any urban lake is truly "restored" (see Chapters II.4 and I.9 for descriptions of two other Boston-area urban lakes).

Due to the presence of large neighborhood populations and obvious shortages of green spaces, urban lakes are frequently characterized by extensive shoreline erosion brought about by shoreline strolling and dogs running in and out of the water (Chapter II.4). These shorelines are not only aesthetically unappealing, but they can also contribute to lake eutrophication in two important ways:

- Upon eroding into the water, shoreline soils break apart and release their burdens of nutrients for uptake by planktonic algae and nuisance aquatic plants.
- Due to bankside instability, nearshore aquatic plants are often unable to become established, thereby reducing the capability of this first line of eutrophication defense in the phyto-uptake of high nutrient stormwater runoff before the algae can assimilate it.

The purpose of our intervention in an urban lake was to create a demonstration project that would raise neighborhood awareness about the opportunities present for cost-effective in-lake management of eutrophication. By restoring a section of the severely degraded shoreline, we hoped to reduce the potential for future, much more expensive, approaches to lake restoration such as periodic dredging.

Figure I.16.1 Chandler Pond before lake restoration. Note abundant aquatic plants.

Project history

Chandler Pond (Figure I.16.1), located at Gallagher Memorial Park in the Brighton section of Boston, presents many of the attractions and problems typical to contemporary urban wetland-ponds in microcosm. It is a 10.8-acre man-made pond, constructed in the 19th century for ice production; over the years it has become heavily contaminated as a result of sediment input from its highly urbanized 185-acre watershed, 45% of which is residential and 55% recreational (parks and golf course).

The pond's original average depth was approximately 3 ft, yet over subsequent decades sedimentation reduced it to 1.9 ft. As a result, poor water quality and excessive growth of nuisance aquatic macrophytes degraded the pond's habitat, recreational, and aesthetic values. As a first step to correct the problem, the Boston Parks and Recreation Department (BPRD) implemented a project to restore the pond to its original depth. In 1999 the pond was drained (Figure I.16.2) and subsequently dredged (Figure I.16.3; Heron, 1998). The removal of 25,000 yd^3 of nutrient-rich sediment provided immediate improvement of overall pond conditions by increasing water depth and reducing the availability of nutrients to aquatic macrophytes, which had become a nuisance prior to dredging. Yet the underlying cause of sedimentation had not been remedied.

A special condition of the Boston Conservation Commission's (BCC) approval for the dredging project encouraged the BPRD to "address potential upstream sources of sedimentation, and nutrients which may lead to future siltation and eutrophication of Chandler Pond" (BCC, 1998). In response, the BPRD conducted a preliminary assessment of the site, which identified shoreline erosion as "a significant problem contributing to siltation of the pond, excessive nutrient growth, and degraded wildlife habitat" (McLaughlin, 2000) (Figure I.16.4). Excessive pedestrian use and overgrazing by resident waterfowl were determined to be potential causes of this shoreline erosion.

To address the on-site erosion, the BPRD, under the guidance of Timothy Smith, applied for and received grant funding from the Massachusetts Department of Environmental Management, Lake and Pond Grant Program and retained the Pouder Design Group to prepare a shoreline master plan and demonstration shoreline stabilization project. Stated goals of the project were to:

Figure I.16.2 Chandler Pond during draining in summer 1999.

- Improve water quality
- Enhance habitat for wildlife
- Reduce on-site inputs of sediment, urban runoff, nutrients, and other contaminants into the pond
- Improve passive recreational opportunities
- Recreate a naturalized riparian wetland in an urban setting
- Develop a project that would serve as a model for other urban wilds restoration projects in the city

The project team began their work by assessing the severity of shoreline degradation. Native and naturalized vegetation were observed on approximately one-third of the 5000-ft. shoreline. A dense monotypic stand of cattails 10 to 50 ft wide existed along the pond's western edge, and several small patches of cattails were scattered around the

Figure I.16.3 Dredging operations at Chandler Pond in summer 1999.

remaining perimeter. Smaller patches of purple loosestrife and Japanese knotweed were distributed around the pond. A portion of the northeastern and northwestern shoreline was densely vegetated with woody species.

The remainder of the shoreline was either covered with turf or unvegetated. With the exception of several clumps of vegetation and areas of turf, virtually the entire southern shoreline was barren of vegetation (Figure I.16.5). These vegetated clumps typically contained a single tree with a shrub layer and/or emergent or submergent herbaceous species. Lawn areas bordering the pond were mown to the edge of water, limiting riparian habitat and reducing buffering capacity.

Some erosion was also evident along the steep slopes of the pond's privately controlled northern shoreline. With varying success, homeowners had implemented haphazard methods of slope stabilization including walls and vegetation. Yet it was immediately apparent that the most unstable and severely degraded portion of the shoreline was located along the pond's publicly accessible southern side. Areas of bank erosion coincided with the areas of highest observed pedestrian use. Park visitors in this portion of the site were frequently seen feeding waterfowl or fishing, and erosion was most severe in favorite fishing and waterfowl feeding areas. The erosion caused unstable slopes and an undercut shoreline (Figure I.16.4). Gullies formed in some areas.

Pedestrian circulation at the site consisted of a wide asphalt sidewalk directly adjacent to Lake Shore Road and an eroded "desire path" parallel to the shoreline approximately midway between the sidewalk and the pond's southern shoreline. A pedestrian/maintenance vehicle path was located within the park to the west of the pond.

Despite its degraded condition, the pond provided habitat for numerous species of birds, reptiles, amphibians, fish, and small mammals (Figures I.16.6 and I.16.7). BPRD staff and consultants, as well as area residents, reported sightings of numerous species. Canada geese are a dominant species with dozens of resident birds, well above BPRD's estimate of an acceptable population. The high population of geese at the site causes excessive erosion as noted previously. Nutrient loading from goose droppings is also a concern. Furthermore, goose droppings diminished the passive recreational potential of lawn areas.

Figure I.16.4 Shoreline regions of Chandler Pond exhibiting substantial erosion.

Planning and design recommendations

To address the on-site causes of erosion and sedimentation, a long-range Shoreline Master Plan was prepared by the Pouder Design Group with input from BPRD, W.D.N.R.G. Limnetics, and the Chandler Pond Preservation Society (CPPS), a nonprofit community action group comprised of interested citizens and area residents. To guide future park improvements, conceptual and generalized management guidelines, infrastructure improvements, bank stabilization, and vegetative enhancements were proposed (Figure I.16.8). One such recommendation was the immediate implementation of a demonstration project

Figure I.16.5 Shoreline habitat suffering from excessive pedestrian use and waterfowl grazing.

Figure I.16.6 Searching for turtles following draining.

(discussed next), which could address the most serious areas of bank degradation, as funding permitted.

Other recommendations of the long-range plan included:

- Reduce the frequency of mowing of a band 10 ft wide along the pond edge while continuing the frequency of mowing for the remainder of turf to (a) provide enhanced filtration of runoff, (b) discourage geese from using turf areas, thereby reducing the amount of goose droppings on lawn areas, (c) reduce nutrient loading (d) lower maintenance costs, and (e) allow for the self-sowing of riparian species.
- Retain areas of mown turf for passive recreation.
- Stabilize all portions of the eroded bank using biological (plant) controls; provide coir fiber rolls in areas of severe erosion or bank degradation.

Figure I.16.7 One of the dozens of turtles rescued following draining and preceding dredging.

- Create new wetland areas at stormwater inlets to act as sediment traps.
- Control access to the water, thereby limiting destruction of riparian areas by excessive foot traffic.
- Remove the existing sidewalk and desire path, replacing them with a single Americans with Disabilities Act (ADA)-compliant walkway.
- Link the new walkway to two ADA-compliant fishing structures.
- Restrict and discourage access to prime bird breeding habitat.
- Create a boardwalk and bird-viewing stand to allow continued bird-watching opportunities in this area while at the same time limiting disturbances of nesting birds.
- Retain existing and provide additional subsurface structure for wildlife habitat.
- Develop, in conjunction with the CPPS, ecological educational and signage programs directed at watershed residents and park uses.
- Monitor and photo-document treated and untreated areas both pre- and post-project for chemical, physical, and biological indicators of sediment and water quality.

Shoreline restoration demonstration project

Installation of the pilot/demonstration program was intended to assess the effectiveness of low-cost shoreline stabilization methods. The most severely degraded areas within the park along the pond's publicly accessible southern shoreline were selected for this purpose. A primary goal of the demonstration project was to identify which of these cost-effective restoration treatments could be used to stabilize other degraded shoreline areas, both at Chandler Pond and at other locations.

The demonstration project proposed the preservation of native existing vegetation; hand removal of invasive vegetation; and introduction of native species in deep emergent, shallow emergent, wet meadow/pond edge, and upland habitats. Selected plant species were typical of natural freshwater marshes in eastern Massachusetts and were acquired from a Massachusetts source.

Chapter I.16: Restoring and protecting a small, urban lake (Boston, Massachusetts) 325

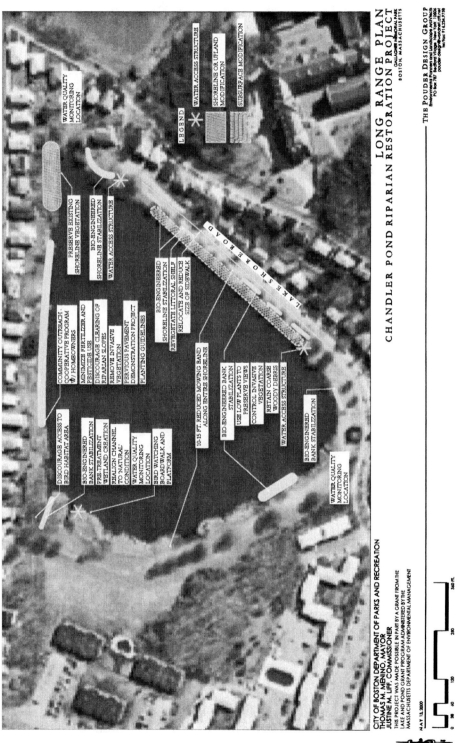

Figure I.16.8 Management options for the restoration of Chandler Pond.

Other goals of the demonstration project were to:

- Enhance needed feeding, brooding, and protective habitat for a range of native insects, birds, herpitiles, and fish.
- Stabilize bank soils and physically trap sediment and nutrients, making these unavailable to nonnative, invasive, nuisance species.
- Chemically retain sediment, nutrients, and other pollutants within plant tissue, further limiting their availability to nonnative, invasive, nuisance species; and promote the self-sowing of other native wetland plants.
- Preserve water and locate new plantings to retain important view corridors. Use of low plantings in foreground areas would frame views of and across the pond.

Community outreach

Community involvement was considered a key factor in the successful outcome of the project. The CPPS sponsored an evening information session at which the consultant presented plans of the demonstration project, explained the process, and solicited feedback. CPPS members were strongly committed to the success of the project and provided volunteer labor for watering of newly installed plants, removal of invasive vegetation, and informal "surveillance" of the project to minimize vandalism. Additionally, the CPPS including regular articles about the project in its newsletter.

Implementation

To evaluate alternative methods of bank stabilization, three combinations of plants and biologs were proposed. These were:

- A combination of biologs and vegetation
- Biologs only
- Vegetation only

Eleven restoration areas (Figure I.16.9) of native vegetation consisting of 26 different species (Table I.16.1) were proposed along unvegetated portions of the southern shoreline. Plantings (Table I.16.2) were proposed to begin about 2 ft above the waterline and extend into the pond to a water depth of 2 ft in a zone 6 to 10 ft wide. Planting was also designed to deter travel by waterfowl between the land and water, thus reducing the impacts of waterfowl excrement. Equally important was the desire to provide an attractive visual display of colorful flowers and foliage throughout the growing season and to both enhance and preserve views of the lake from the adjacent shoreline (Figure I.16.10).

Planting depth zones (Figure I.16.11) were based on inundation tolerance. Deep emergent marsh species were installed in average pool depths of –12 to –24 in. Shallow emergent marsh species were planted in average pool depths of –3 to –12 in. Wet meadow edge plants were planted directly in the saturated biolog and from an elevation of +6 to –3 in. Shoreline fringe species were planted at elevations greater than +6 in. The shoreline fringe plants required regular watering during their grow-in period. Approximately 2750 plants were installed.

Installation was conducted by a habitat restoration specialist. Planting commenced in early June and was completed in 2 weeks. Plants and biologs were delivered to the site in the early morning. Plants were stored in a cool, moist location to reduce desiccation. Installation began in late morning. Eleven areas were selected for intervention (Figure I.16.9); these

Chapter I.16: Restoring and protecting a small, urban lake (Boston, Massachusetts) 327

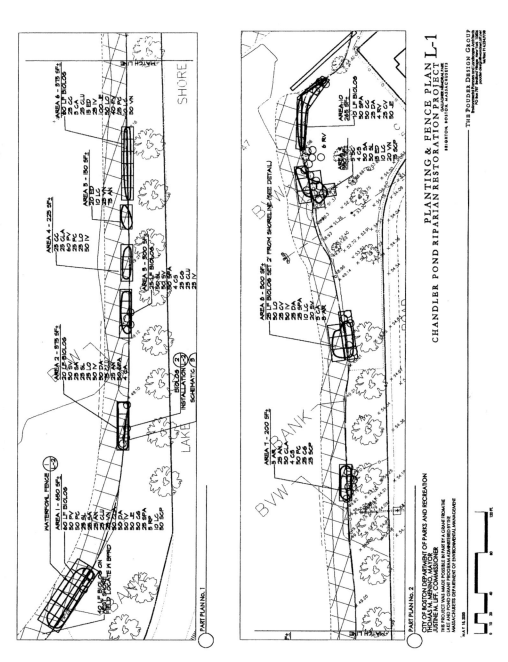

Figure I.16.9 Planting plans for shoreline restoration of Chandler Pond. See Table I.16.2 for planting instructions to installation personnel. Plant abbreviations are described in Table I.16.1.

Table I.16.1 Planting Schedule, Chandler Pond Riparian Restoration Project

Code	Scientific Name	Common Name	Zone	Wetland Indicator	Spacing	Quantity
AN	*Aster novae-angliae*	New England aster	Wet meadow/edge	FACW−	12" o.c.	150
AR	*Alnus rugosa*	Alder	Shoreline	FACW+	As noted	8
CA	*Cornus amomum*	Silky dogwood	Shoreline	FACW	As noted	10
CC	*Calamagrostis canadense*	Blue joint grass	Shallow emergent	FACW+	12" o.c.	100
CG	*Chelone glabra*	Turtlehead	Wet meadow/edge	OBL	12" o.c.	100
CLA	*Carex lacustris*	Lake sedge	Shallow emergent	OBL	12" o.c.	150
CLU	*Carex lurida*	Lurid sedge	Wet meadow/edge	OBL	12" o.c.	150
CS	*Cornus sericea*	Red twig dogwood	Shoreline	FACW+	As noted	12
CV	*Carex vulpinoidea*	Fox sedge	Wet meadow/edge	FACW+	12" o.c.	50
DA	*Dulichium arundinaceum*	Three-way sedge	Shallow emergent	OBL	12" o.c.	150
ED	*Eupatorium dubium*	Joe pye weed	Wet meadow/edge	FAC	12" o.c.	50
IV	*Iris versicolor*	Blue flag iris	Shallow emergent	OBL	12" o.c.	250
JE	*Juncus effusus*	Soft rush	Shallow emergent	OBL	12" o.c.	200
LC	*Lobelia cardinalis*	Cardinal flower	Wet meadow/edge	FACW+	12" o.c.	50
LO	*Leersia oryzoides*	Rice cutgrass	Shallow emergent	OBL	12" o.c.	150
PC	*Pontederia cordata*	Pickerel weed	Deep emergent	OBL	12" o.c.	150
PV	*Peltandra virginica*	Arrow arum	Deep emergent	OBL	12" o.c.	150
RP	*Rosa palustris*	Swamp rose	Shoreline	OBL	As noted	5
RV	*Rhododendron viscosum*	Swamp azalea	Shoreline	OBL	As noted	10
SA	*Scirpus acutus*	Hard stem bulrush	Deep emergent	OBL	12" o.c.	100
SC	*Sambucus canadensis*	Elderberry	Shoreline	FACW−	As noted	5
SL	*Sagittaria latifolia*	Duck potato	Deep emergent	OBL	12" o.c.	150
SP	*Scirpus pungens*	Three square bulrush	Shallow emergent	OBL	12" o.c.	150
SPA	*Sparganium americanum*	Burreed	Shallow emergent	OBL	12" o.c.	250
SV	*Scirpus validus*	Soft stem bulrush	Deep emergent	OBL	12" o.c.	100
VN	*Veronica noveboracensis*	New York ironweed	Wet meadow/edge	FACW+	12" o.c.	100

Table I.16.2 Notes on Planting for Construction Personnel

- Remove all purple loosestrife and Japanese knotweed, including all roots, encountered in planting areas.
- Grade eroded and undercut shoreline slopes to provide smooth transition between upland and subsurface areas or as directed by BPRD.
- Use of power equipment for excavation and grading is prohibited.
- Work areas are shown in their approximate locations. Verify work areas with BPRD prior to planting.
- Use no fertilizers unless directed to do so by BPRD.
- Plant materials provided by BPRD shall be planted on day of delivery or shall be stored and maintained in healthy condition by the contractor. Store plants in a secure, shaded, out of wind, and well-ventilated site.
- The contractor shall ensure that soil conditions at time of planting are sufficiently moist to maintain plant health.
- Plants shall be carefully handled to prevent injury and desiccation. Plants installed above the waterline shall be watered by the contractor.
- Prior to planting, contractor shall stake out all planting areas for approval by BPRD. Plants shall be installed upright. All planting procedures, tools, and methods subject to approval by BPRD.
- Plugs shall be planted in the soil at least one (1) inch deeper than grown in the nursery and to a depth which ensures that the root structure is at least one inch but not more than two in. Below the soil surface.
- Fencing shall be installed at each planting area immediately following planting.
- The wetland plant material shall be evaluated for acceptance thirty (30) days after the following have been completed or satisfied:
  1. All specified fencing has been installed and accepted adjacent to the planting areas.
  2. Slopes have been graded to the satisfaction of the BPRD.
  3. The plant material has been installed to the satisfaction of the BPRD.

consisted of one area stabilized only with a biolog, six areas that received a combination of biolog and plants, and four areas at which only plants were used.

The contractor installed biologs by excavating a shallow trench parallel to the shoreline (Figure I.16.12). Excavated soil was used to backfill on the upland side of the biolog. Some biologs were set directly against bank; others were set as much as 24 in. out from bank, thereby creating a shallow ponded area on the shore side of the biolog (Figure I.16.13). Hardwood stakes were driven into the substrate on both the waterside and the shore side of each biolog.

Fencing

Installation of temporary fencing (Figure I.16.14) was also proposed to enclose newly planted areas and reduce herbivory from waterfowl and muskrats. The fence consisted of 2-in. square wood stakes and biodegradable twine strung in 10-ft "cells" at several levels both below and above the water surface (Figure I.16.15). The fence was intended to remain in place for a 1-year period, until plantings had sufficiently rooted and were no longer vulnerable to herbivory. Within days of installation, however, it became apparent that unleashed dogs were breaking through the goose exclusion fence into at least one planting area. Waterfowl subsequently entered the area through the broken fence and damaged a high percentage of newly installed plants. To educate dog owners about the impact of their actions, laminated color signs describing the project were subsequently installed (Figure I.16.16). The signs contained a brief description of the project and included color "before" and "after" images of a stable shoreline.

Figure I.16.10 Vegetation a few months after shoreline planting.

Monitoring

Although the long-term success of the project cannot be determined for several years, some preliminary observations were apparent. First, the use of biodegradable twine for the fence echoed the naturalistic approach of the project; however, a more durable fence should be used in urban areas where dogs may be present, especially on the upland side of planting areas. Similarly, signage should be installed concurrently with plantings to alert dog owners to the potential damage that their pets might cause. Flexibility with plant

Chapter I.16: Restoring and protecting a small, urban lake (Boston, Massachusetts) 331

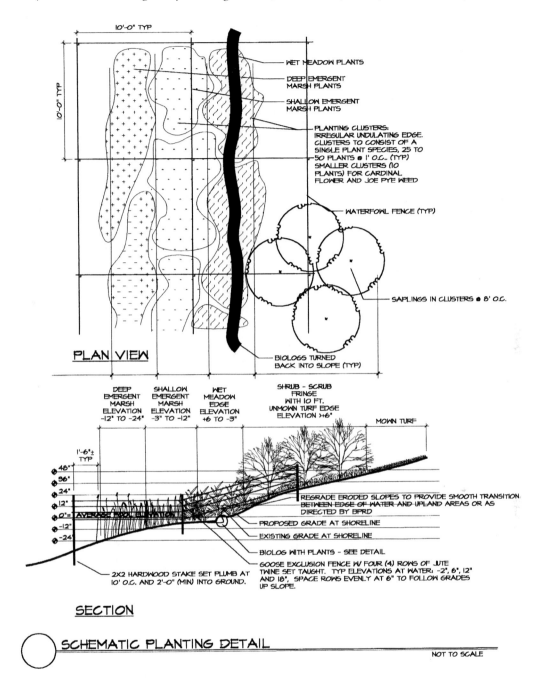

Figure I.16.11 Schematic representations of planting depth zones.

species selection was also required. A late-season crop failure at the supply nursery required the substitution of several species of plants. Last, arrangements for secure storage and watering of plants must be provided if they are not installed on the day of delivery.

In terms of the lake itself, a year following shoreline treatment, most of the restoration areas exhibit strong growth of the species that were planted. One localized problem that developed is the aggressive colonization of purple loosestrife along most of the shoreline, possibly a result of the water drawdown that preceded bottom sediment dredging (Smith, 1999). However, a more serious and systematic problem not expected to be anywhere near

Figure I.16.12 Installation of biologs.

completely addressed by our small-scale demonstration project concerns the bloom of algae that has returned to the pond to a degree comparable to what it was prior to dredging. This has led to feelings of anxiety and confusion by pond-side residents (Smith, 1999), leading them to question the possible "waste" of hundreds of thousands of dollars spent on the dredging, especially in light of the consequent inconvenience that ensued during the entire process. This result underscores the need for a much more comprehensive watershed approach to ensure true sustainability of restoration projects.

Future — scaling up to the watershed

The primary aim of the present project was to "protect the investment" of recent dredging in Chandler Pond. The polluted stormwater continues to flow into Chandler Pond unaltered, so the expectation must be that the sedimentation process, augmented by continued shoreline erosion from those areas outside of the present demonstration project, is already under way again. Unless steps are taken to reduce the nutrient loading and sedimentation of the pond, the dredging will prove to be no more than a temporary solution. With current understanding of shoreline bioengineering and wetland treatment techniques in a framework of watershed management (Chapters I.17, II.4, and I.14), there is no excuse for letting Chandler Pond's water quality deteriorate again. At the start, it is recognized that the steps needed to achieve a stable and acceptable water quality in the pond are manifold, and that a comprehensive review of the Chandler Pond watershed and a community liaison/education program would be an important part of the strategy.

Chapter I.16: Restoring and protecting a small, urban lake (Boston, Massachusetts)

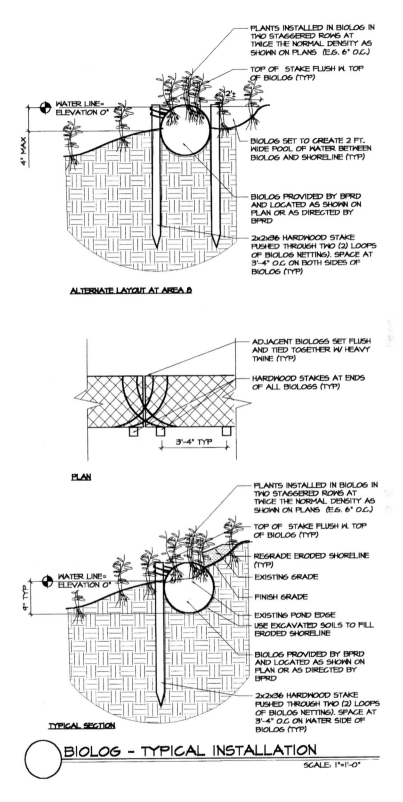

Figure I.16.13 Schematic representation of biolog installation.

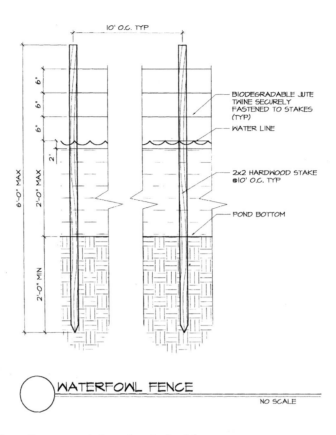

Figure I.16.14 Schematic representation of waterfowl fence.

In this respect, it is always important to place site-specific restorative designs in both a larger landscape and a larger management context, in this case dealing with where the two meet — the watershed. Watersheds are important to urban communities because they help to foster an embodiment of sense of place in a developed landscape and because their waters aid in providing much-needed solace in daily lives. Importantly from an environmental perspective, watersheds are the most appropriate units in which to effectively manage water resources in the urban landscape (Chapters II.8 and II.9). After years of deleterious human activity in its watershed, Chandler Pond has undergone a vastly expensive dredging operation to rejuvenate its health. It is egregiously naïve, however, to think that once the symptom is thus "fixed," the work is over and the problem is mitigated or alleviated. (The recently returned blooms of algae indicate that the lake is already becoming stressed again.) A cardinal need exists to put Chandler Pond back into its landscape and to develop an effective and comprehensive watershed management plan that will prevent, or at least reduce, the severity of the problem(s) that necessitated the dredging in the first place (Table I.16.3). Chandler Pond desperately needs a watershed plan similar to those recently completed for nearby Hall's Pond in Brookline (Chapter I.11) or Fresh Pond in Cambridge (Chapter II.4). This is the only working solution to protect the environmental integrity of the temporarily restored Chandler Pond, lest we face another, by then perhaps prohibitively expensive, redredging several decades in the future. Toward this end, a preliminary report (France, 1999) provides a first attempt at itemizing the requisite elements needed to be addressed in development of such a master watershed plan for Chandler Pond (Table I.16.3), some of the dangers to be avoided in the process, and a preliminary estimated budget needed to accomplish this task.

Chapter I.16: *Restoring and protecting a small, urban lake (Boston, Massachusetts)* 335

Figure I.16.15 Waterfowl exclusion fencing and netting in restoration areas.

To tackle issues of urban watershed renewal and repair necessitates a comprehensive vision that is best approached in small steps in order to ensure the greatest likelihood of success (EPA, 1997). The BPRD is also working with W.D.N.R.G. Limnetics to compile a long-term, watershed-based management plan, a part of which is intended to reduce and mitigate sediment and nutrient sources before stormwater enters Chandler Pond.

Major elements of the post-dredging watershed plan include both structural and nonstructural corrective elements, many of which are addressed through the shoreline

CHANDLER POND SHORELINE RESTORATION

Before **After**

This project will reduce shoreline erosion and provide habitat for fish and wildlife. The temporary fence is needed to prevent geese from damaging young plants. Please help us by keeping your dog out of the planting areas and on a leash.

Thomas M. Menino, Mayor
Justine M. Liff, Commissioner

THIS PROJECT WAS MADE POSSIBLE IN PART BY A GRANT FROM THE LAKE AND POND GRANT PROGRAM ADMINISTERED BY THE MASSACHUSETTS DEPARTMENT OF ENVIRONMENTAL MANAGEMENT.

Figure I.16.16 Instructional signage.

restoration demonstration project described in this chapter. The present project involving shoreline bioengineering and littoral wetland replanting is an integral part of the larger watershed plan, particularly addressing the elements A.1.a and B.1.a outlined in the master plan proposal (France, 1999) shown in Table I.16.4. Other components in the restoration of Chandler Pond that have been previously investigated include model estimates for the hydrologic water budget (Fugro, 1996; Element A.3.c in Table I.16.4), design of a wetland created for the inflow embayment (France et al., 2000; Element A.1.b in Table I.16.4), design and implementation of a long-term chemical monitoring program (Smith, 1999; Element A.3.b in Table I.16.4), design of an upstream created wetland and the potential for daylighting the inflow stream from the golf course (France et al., 2000; Elements C.1.b and c in Table I.16.4), and redesign of the riparian park (Boston Parks and Recreation, 1997, 1999; Elements B.3.a and b in Table I.16.4). This latter redesign is important, for it reinforces that the shoreline restoration undertaken for Chandler Pond, similar to that planned for nearby Fresh Pond (Chapter II.4), is part of a larger effort to help reconnect people with urban lakes through establishment of water sensitive riparian parks.

Table I.16.3 Overall General Goals for Developing a Watershed Plan for Chandler Pond

Goals for Natural Resources
- Restore the pond's one-time natural character and water quality
- Benefit wildlife in the urban setting
- Preserve or replace riparian (shoreline) tree cover
- Foster a new, more natural assemblage of native fauna through active stocking and maintenance
- Reduce or eliminate invasive, nonnative plants

Goals for Resident and Visitor Uses
- Enhance enjoyment and opportunity to observe nature
- Foster a sense of place for the community
- Limit recreational uses to those that are compatible with natural resource objectives
- Entail minimal maintenance

Goals for Education
- Promote educational uses of Chandler Pond
- Foster stewardship
- Contribute to general environmental advocacy

Source: Adapted from Center for Watershed Protection (1998).

Table I.16.4 Specific Elements Needed to Be Addressed in a Master Plan for Chandler Pond

A. Within-Pond
1. Floral habitat
   a. Littoral fringe wetland restoration design and planting scheme
   b. Inflow embayment wetland creation design, construction plan, and planting scheme
2. Faunal biology
   a. Species assemblage desirability study and stocking program implementation
   b. Long-term monitoring plan development
3. Physical-chemical
   a. Assessment of requisite elements to be surveyed
   b. Long-term monitoring plan development
   c. Design outline for developing a water budget model for the pond

B. Around-Pond
1. Shoreline vegetation
   a. Bankside bioengineering feasibility study, restoration design, and installation plan
   b. Tree-bush planting scheme design
   c. Nearshore grass maintenance management plan
2. Shoreline hard structures
   a. Fishing platform installation design and construction plans
   b. Examination of possibilities for stormwater inflow pipes repair and maintenance
3. Park
   a. Pedestrian pathway restoration design and construction plans
   b. Naturalized park vegetation planting scheme design

C. Upstream and Inflows
1. Point source
   a. Long-term monitoring plan development from the golf course to the pond
   b. Examination of potential for daylighting buried inflow stream with design and construction plan
   c. Inflow wetland creation design, construction plan and planting scheme
2. Nonpoint sources
   a. Development of a stormwater management plan for the pond based on a comprehensive monitoring plan

Table I.16.4 Specific Elements Needed to Be Addressed in a Master Plan for Chandler Pond

D. Surrounding Watershed
1. Land-use planning
   a. Determination of impervious cover model
   b. Examination of land use management techniques and future development scenario predictions
2. Land conservation
   a. Prioritization of future water sensitive development and protection/infiltration sites
3. Aquatic Buffers
   a. Examination of protective buffer strip creation potential, design, and planting schemes
4. Better site design
   a. Development of guidance directives for reducing deleterious effects on stormwater tailored to site specifics in Chandler Pond watershed
5. Erosion and sediment control
   a. Development of guidance directives for reducing erosion effects tailored to site specifics in Chandler Pond watershed
   b. Prioritized site exploration assessment of potential for erosion control BMP (best management practices) retrofits to existing urban framework with detailed budget
6. Stormwater BMPs
   a. Development of guidance directives for reducing stormwater inputs tailored to site specifics in Chandler Pond watershed
   b. Prioritized site exploration assessment of potential for stormwater bioretention BMP retrofits to existing urban framework with detailed budget
7. Nonstormwater discharges
   a. Investigation of structural and nonstructural controls for limiting lawn and wastewater discharges and uncoupling illicit connections with accompanying management plan and detailed implementation strategy and budget
8. Watershed Stewardship Programs
   a. Outline for program of fostering choices for public and private stewardship to sustain watershed management

Source: Adapted from France (1999).

Acknowledgments

This project could not have been undertaken without the dedicated effort and insightful wisdom of Timothy Smith of the Boston Parks Department, and the concerned citizenry who are dedicated to Chandler Pond. This chapter is dedicated to the environmental vigilance of Dr. Mark Chandler. The authors also thank Tim Smith for providing an update on the present status of the pond.

Literature cited

American Society of Landscape Artists (ASLA), Integrating stormwater into the urban fabric, Portland State Univ., 1996.
Barnett, D., Dredging and landscaping at Auburn Lake. *Mount Auburn Cemetery Newsletter*, 1999.
Boston Conservation Commission (BCC), Order of Conditions for the Conventional Dredging of Chandler Pond, Brighton, MA, 1998.
Boston Parks and Recreation Department (BPRD), Chandler Pond, Brighton Park improvement plan, 1997.
BPRD, Urban Wilds Program, Chandler Pond shoreline restoration pilot project, 1999.
Center for Watershed Protection (CWP), *Rapid Watershed Planning Handbook. A Comprehensive Guide for Managing Urbanizing Watersheds*, Ellicott City, MD, 1998.

CWP, *Better Site Design: A Handbook for Changing Development Rules in Your Community*, Ellicott City, MD, 1999.
EPA, Top 10 watershed lessons learned. United States Environmental Protection Agency Report, 1997.
France, R., Proposal for developing a master plan for Chandler Pond. W.D.N.R.G. Limnetics. Unpubl. Report., 1999.
France, R. et al., Proposal for environmental and recreational improvement to Chandler Pond through removal of nutrients from golf course and urban runoff by creation of an inflow-forebay treatment wetland complex. Harvard Design School Student Project, 2000.
Fugro Associates, Hydrologic evaluation and preliminary field investigation of Chandler Pond, Boston, MA, 1996.
Heron, Let the digging begin, Chandler Pond Preserv. Soc. Newsletter, no. 7, 1998.
McLaughlin, B.J., Request to amend order of conditions, DEP File No. 6-790, 2000.
Pouder Design Group, Chandler pond riparian restoration project, 2000.
Schueler, T., Kumble, P., and Heraty, M., A current assessment of urban best management practices, techniques for reducing non-point source pollution in the coastal zone, U.S. EPA, Washington, DC, 1992.
Smith, T., Chem monitoring plan for Chandler Pond, personal communication, 1999.
Terrene Institute, Urbanization and water quality, a guide to protecting the urban environment, Washington, DC, 1994.
Tourbier, J.T. and Westmacott, R., *Lakes and Ponds*, The Urban Land Inst., Washington, DC, 1992.

Response

Buying time by bioengineering

Sometimes the efforts of environmentalists can seem frustratingly like the labors of Sisyphus when it comes mitigating, much less getting ahead of, the degradation surrounding us. One of the most important messages in the EPA's *Top 10 Watershed Lessons Learned* is the need for, and reassurance in, small iterative steps forward. In other words, if we wait to raise all the expertise and funding necessary to completely restore or recover a body of water, it may be too late to do so in any sensible fashion given the relentless rate at which degradation proceeds. This chapter by Nicholas Pouder and Robert France demonstrates the utility in undertaking a small-scale bioengineering project in a severely polluted urban lake.

A major purpose of the shoreline restoration project described in this chapter was to raise public consciousness about proactive, and largely inexpensive, means that can, and indeed should, be engaged in order to help sustain the quality of the water in the lake following the expensive draining and dredging of contaminated bottom sediments. Great educational power exists in such demonstration projects, and it goes far beyond whatever physical benefits might ensue to the particular sites upon which they are enacted. Water sensitive design projects have the greatest likelihood of success with attention directed to sociological issues such as community outreach. This may be especially true when using techniques that are a little "out of the box" in terms of current practice, as, for example, the various integrated management options involved in low-impact development (Michael Clar in Chapter I.15) or the use of permeable pavement (Bruce Ferguson in Chapter I.1) and multifunction treatment wetlands (James Bays in Chapter I.14).

Another important issue that Pouder and France raise concerns the need to scale up site-specific projects to the watershed scale. Specifically, in terms of bioengineering, this is something that Wendi Goldsmith conceptually elaborates upon in Chapter I.17 and Thomas Benjamin practically demonstrates in Chapter II.4. Pouder and France's listing of salient elements that need to be addressed in formulating a master plan for the lake in this study are not specific to only that location; instead, they offer a useful set of guidelines to aid water sensitive planning and design at any urban location.

chapter I.17

Integrating ecology, geomorphology, and bioengineering for watershed-friendly design

Wendi Goldsmith

Abstract

Within watersheds, cumulative impacts from development have caused widespread physical instability, degradation of water quality, and deterioration of habitats. Research findings support that the protection, enhancement, and restoration of many natural landforms and ecosystem functions can provide the most cost-effective solutions for watershed management. Although reasonable consensus has been reached regarding measures to protect existing intact resources, current examples of project work reveal that engineers, biologists, landscape architects, geologists, and other professionals adopt divergent approaches and standards for resource enhancement and restoration. This chapter describes a process of project goal setting (based on ecological assessment), site analysis (based on geomorphology, hydrology, and pedology), and treatment design (based on bioengineering principles and rigorous review from a geotechnical and/or hydraulic engineering perspective as appropriate) that can be applied on a variety of scales to achieve practical watershed-friendly design. A brief overview of the underlying science and engineering concepts offers a rationale for adopting the procedure. Briefly presented case studies offer examples of how the interdisciplinary process plays out in various situations. Although integration with large-scale watershed management problems are addressed, smaller site-scale design topics are emphasized.

The case for sustainability

Engineers, architects, landscape architects, and other professionals are called upon to plan and design projects connected to new development or in response to problems stemming from old development. Standards of professional practice and responsiveness to clients' needs dictate that cost-effectiveness, safety and usability, constructability, and predictability of implementation and performance are the chief guiding principles. Another pervasive influence is "defensive design," which often takes the shape of overdesign as a substitute for careful analysis and planning, or mechanistic design, which seeks to wrest a site or structure into one final immutable condition instead of addressing inevitable and sometimes desirable processes of change and evolution. For project types ranging from suburban residential subdivisions to urban infill commercial structures, from transportation corridors to flood control facilities, work products generated by design professionals answer to influences such as professional liability, scope of work under contract, schedule and budget, and conventions that are often outdated and, by definition, not forward-looking. One topic that gets lip service but seldom enters meaningfully into the design or maintenance of projects is sustainability. While the more mundane and practical factors contributing to the design process clearly cannot be neglected, society has an overriding, though infrequently pursued, interest in ensuring the sustainability of our natural resources, infrastructure, and quality of life. Designers have an interest in embracing sustainability as an outcome (e.g., Chapters II.3 and II.11), as it can position them to be leaders in their fields, proactively identifying and addressing needs beyond the bounds of a site or construction schedule, generally saving money and creating goodwill for their clients.

It has become evident, moreover, that healthy communities are those that recognize the need for a balance between socioeconomic vitality and stewardship of the environment. This phenomenon is not recent, because cultures ranging from the Mayans to ancient Rome rose and fell on the productivity and ultimate demise of natural systems. A metaphor that is occasionally used to illustrate the importance of a balanced approach is the three-legged stool. The legs represent strong social, economic, and ecological elements. When each is in place and providing full support, the stool is stable. Failure to maintain one or more of the elements and the stool falls over. Since World War II, numerous failures have been made relating to community development in the U.S. Sprawl, for example, serves few of our needs for social interaction, is wasteful of economic resources, and is environmentally destructive. Sustainability of this pattern of growth is highly unlikely, particularly because of its lack of consideration of the value of natural systems and processes. One blatant example is the water supply for the city of New York, located 100 miles away in the Catskill Mountains, which for nearly a century provided clean water for millions of city residents. In recent years, the suburban fringe has extended to the Catskills, where water quality is beginning to decline in proportion to the new shopping malls and residential subdivisions that are replacing local forest cover. The EPA estimated the cost of compensating for the change in land use with a water treatment facility at $6 billion (Hu, 2000). Sustainability is cheaper than symptom treatment in a reactive mode and also is a better cause for the community to rally around and participate in, plus it creates quality of life that can be equitably shared.

Various helpful efforts have been made to define sustainability, though none have achieved widespread recognition and adoption in the United States. One system that a growing number of countries in Europe have adopted is the "Natural Step" (Robert, 1997). Karl-Henrik Robert, a Swedish oncologist, developed this method of evaluating human impact on the environment. He recognizes the finite nature of the system interconnecting the earth's crust and the atmosphere, with only sunlight as an input. All human activities, he believes, must serve to maintain or manage the equilibrium relationships of the system in order to achieve sustainability and essential human health. Everything that we produce or construct must conform to four guidelines that Robert refers to as *system conditions*: (1) substances from the earth's crust must not systematically increase in nature; (2) substances produced by society must not systematically increase in nature; (3) the physical basis for the productivity and diversity of nature must not be systematically deteriorated; and (4) we must be efficient enough to meet basic human needs. The Natural Step involves taking one radical step toward reorienting society's priorities to meet the four system conditions. Robert's seminal paper expounding on the principals of the Natural Step received extensive peer review and wide acclaim among numerous leaders in the scientific community. Achieving sustainability requires an all-pervasive effort: while many of the aspects outlined in the Natural Step require national policy initiatives, restructuring of markets, and changes in personal consumption habits, many issues apply to site-scale design choices. No place or action is separate from the pursuit of sustainability.

The elements of the Natural Step have as a foundation the basic characteristics of a stable landscape ecosystem, including (1) water is retained and processed via vegetation and soils, (2) nutrients are recycled, (3) landforms and soils are not eroded or degraded, and (4) biological populations are diverse and able to reproduce successfully. Without stability in the form and function of our landscape-scale systems, sustainability cannot be achieved. Historically, all human land uses from agriculture to urbanization have caused significant modification and related impacts to the landscape and its ecological functions. The nature, extent, and timing of the response by the natural system may vary due to a number of factors. In some cases, the effects may be tolerable over an extended time period (the decline in air quality since the dawn of industrialization), though in many, they can be catastrophic (erosion and flooding caused by increased peak flow of stormwater runoff from paved/roofed areas). In all cases, decline and degradation result, and sooner or later we recognize the need to manage the system to compensate, mitigate, or repair. In those cases where the solutions to the perceived problems are based on a comprehensive evaluation of the causes, effects, and relationships of the system as a whole, instead of focusing on the symptom, sustainability is best achieved. For instance, improving the efficiency of manufacturing technology and using cleaner energy sources have a more beneficial effect than refining combustion methods and scrubbing emissions from smokestacks. Similarly, treatments that allow rainfall to be captured on-site, as opposed to directed on an accelerated basis toward rivers, is more effective than efforts to deepen and widen rivers for flood conveyance. All design teams would be well served by the inclusion of scientists and natural resource managers trained and experienced at identifying effective means of achieving sustainability, instead of allowing the design process to be driven by the prevalent set of influences.

Interdisciplinary design

In most professional design scenarios, architects or engineers are placed in charge of the project design process. The work they perform closely follows the original scope prepared by the client, generally represented by a person of fundamentally the same training and background. Thus, any essential bias or limitation in the original understanding or approach to performing project work is not likely to be noticed or comfortably broached by the designer in charge. Refinements to the schedule, exact methods employed, and even budget are addressed as needed, but the basic premise is seldom reevaluated (which is most unfortunate when sustainability is not the basic premise). Various scientists are often engaged to address the matter of obtaining necessary environmental permits. Rarely are these scientists included with the timing or collaborative effort needed to help guide the design itself. In fact, most scientists whose experience dwells in preparation of permit applications and supporting studies do not possess expertise in planning, design, or resource management that would help them to play a meaningful role in guiding project design. Whereas most environmental permitting processes are reasonably effective at obstructing or correcting gravely problematic proposals, they seldom serve much practical use toward promoting optimal stewardship practices. Often the scientific studies produced are not reflected or incorporated in any way in the final design, nor are they necessarily studying the most important issues in terms of guiding the design toward sustainability. In short, the presence of scientists, engineers, and other designers on a project team does not in itself signify that the design process incorporates interdisciplinary collaboration.

When well executed, interdisciplinary design can be effective at achieving richly sustainable land-use practices, as demonstrated by a growing number of case studies (see also Chapter I.7). True interdisciplinary design is characterized by a creative, inclusive, objective, and iterative process. The nature and scale of the process can be adapted based on the size and schedule of the project. The cost of interdisciplinary design is frequently higher than its conventional counterpart due to the inclusion of suitable team members of varied disciplines, and a review and coordination process that allows iterative evaluation, incorporation, and refinement of appropriate design elements; however, this approach need not be considered an expendable luxury that is only suited for special projects with generous schedules and budgets. Time and time again, interdisciplinary design produces efficiencies in meeting permitting and implementation schedules, improvements in construction and maintenance costs, reductions in off-site impacts, and enhancements in aesthetics and perceived quality in general. After having completed a few successful projects, no other approach will be acceptable; as testimony are the numerous design firms, municipalities, and even entire countries that have wholeheartedly embraced interdisciplinary design in support of sustainability.

Though there is not one single recipe for success at interdisciplinary design, several features can aid in the formulation of such an approach:

- Tailors assessment methods and design approach to site conditions
- Responds to community participation
- Emphasizes collaboration and facilitation rather than hierarchy and direction
- Welcomes brainstorming and out-of-the-box ideas
- Considers factors beyond the construction phase
- Evaluates system-wide issues including past and future potential conditions
- Views problems within their broad environmental, social, and economic context

The three-step process

Once the interdisciplinary approach has been accepted for use on a project and appropriate team members have been included, a procedure must be adopted. The procedure we favor at The Bioengineering Group, Inc. (TBG), is a three-step process: identify ecological constraints and opportunities for every project; base design on stable landforms, healthy soils, and balanced hydrology; ensure excellence in design and execution through sound engineering and landscape architectural practice, emphasizing green solutions.

The following outline is intended to provide a framework for demystifying the science and the art of applying interdisciplinary design. The entire method is an evolving tool drawing on well-documented scientific principles applied in complex, multifunctional applications. Depending on the scale, the context, and the intent, this approach can be followed for achieving excellent results with some fairly common design scenarios, but it has also been useful when applied in a highly experimental mode. Each point is worthy of its own in-depth article, and many exist in the literature, though few focus on interdisciplinary restoration applications (Barrett, 1997; Larson and Goldsmith, 1997; Goldsmith, 1998; U.S. Army Corps of Engineers, 1995).

Setting project goals and objectives based on ecological assessment

Very often, ecology and biology are seen as sciences that pertain to rural, wild, or generally nonurban areas, and little attention is paid to these subjects in the context of urban settings. Also, they are seen as extraneous themes that can be addressed only when time, money, and vocal supporters are abundant. Little effort has been made to recognize the practical values connected with protecting, maintaining, and improving the conditions and functions of our flora and fauna. Landowners have lacked both motivation and guiding principles for identifying and prioritizing sensible measures and have failed to integrate them into common development activities at the site scale. For the architect, engineer, or builder, the challenge of interdisciplinary design is to address the big picture view of a client's problems and their potential solutions, even when the client has not specifically asked for them. In practice, the evaluation of the biological and ecological components is generally broken down into the following topics.

Site potential

- Prioritize restoring to past conditions, either partially or fully.
- Consider mitigating upstream or regional problems.
- Evaluate how site design could contribute partially to a potential future condition (e.g., creating salmon spawning habitat even though fish passage problems will not be solved by this project).

Limiting factors

- The usual suspects include water quality impacts, lack of suitable in-stream conditions, overcompacted soils, invasive species, etc.
- The list should include those factors that are intractable as well as those that can be managed or corrected.
- Account for both natural and anthropogenic factors.

Food web

- Recognize linkage between terrestrial and aquatic habitat.
- Provide for suitable primary production (plants).
- Address needs of consumers, both invertebrates and vertebrates.
- Address needs of predators, both invertebrates and vertebrates.
- Establish realistic plans for the needed range of species, not only charismatic megafauna or game species.

Physical complexity

- Generate microtopography and suitably varied soil conditions.
- Favor natural drainage patterns including complex planform and section at all scales.
- Enhance edge habitat through integration of complex shapes at all scales, strive to create numerous and varied nooks and crannies as physical niches.

Biodiversity

- Plan a diverse seeding, planting, and maintenance approach.
- Do not envision a "stocking program," but plan to have numerous critters tend to their own needs if the physical/chemical basics are there.
- Focus on early successional plant species in most cases.
- Aim for the widest number of species possible, instead of targeting just a few.

Biogeochemical cycles

- Address the interrelationships between plants, soils, and water to provide appropriate redox conditions.
- Establish the potential for robust N, P, and K cycles.
- Use the landscape for managing soil contamination and/or water quality impacts whenever possible.

Guiding design alternatives based on sound application of earth science

Every site will have its own unique land-use history, and each region or landform will possess characteristic conditions. However, one thing remains constant: landscapes have evolved their features in response to climate over time to achieve a state of dynamic equilibrium and/or slow, controlled change. Almost equally constant is the fact that forestry, agriculture, and urbanization practices have altered and destabilized these powerful yet sensitive equilibrium relationships. Failing to recognize, respect, and/or harness landforming processes will inevitably lead to project design that does not achieve sustainability. The chief repercussions include causing offsite impacts, requiring repair or reconstruction, and demanding high ongoing maintenance efforts. Careful selection of site design elements can reduce initial construction costs in some, though not all, situations, but this approach consistently offers improved long-term costs. A broadly useful list of considerations follows:

Watershed hydrology

- Use equilibrium hydrology (i.e., undeveloped, precolonial conditions) as a benchmark for achieving sustainable function.
- Balance the water budget in terms of evapotranspiration, infiltration, and runoff.
- Recognize that analysis of existing versus post-construction conditions will not yield a sustainable outcome, because the site is likely already degraded.
- This method allows watershed-scale problems to be addressed remarkably well at the site scale and on a piecemeal basis.

Stream pattern and process

- Perform at least a basic fluvial geomorphic assessment in order to ascertain the suitability of the existing conditions.
- Endeavor to protect, enhance, or radically restore stable stream cross-section, profile, and planform geometry whenever possible.
- Respect stream form and function as it pertains to low flow, or typical daily conditions, at least as much as for less frequent recurrence intervals.
- Do not inadvertently or purposefully introduce measures that cause disturbance of stream pattern and process.
- Do not assume that all stream/river channels should be treated as sacred spaces not to be disturbed. Many have been trashed and can be much improved.
- Apply principles of fluvial geomorphology to guide the design of suitable channel restoration measures.
- Use qualitative methods to develop and/or refine restoration schemes, but never omit quantitative analysis for perennial streams.

Sediment transport

- Recognize that all streams and rivers serve to carry sediment, not just water, generated within their watersheds.
- Carry out design and construction in recognition that in-stream conditions are not fixed or rigid and will be reworked by sediment transport processes.
- Identify sources of sediment and characterize sediment yield of the basin.
- Perform, at minimum, a rough characterization of bed load, suspended load, and wash load for all perennial streams.
- Analyze critical values for sediment transport hydraulics for larger streams.

Flow conveyance

- Seek to promote appropriate base flow conditions.
- Recognize and/or manage channel-forming discharges.
- Be prepared to address regulatory needs to prevent adverse flood impacts.
- Explore nontraditional measures to address flow conveyance such as increasing channel roughness to retard downstream flooding.
- Consider how channel modification or instability may have exacerbated flooding problems (such as downcutting reducing access to floodplain storage volume and translocating problems downstream).

Role of soils and vegetation

- Assess the contribution of existing soils and/or vegetative cover or reinforcing roots on physical stability of slopes and streambanks.
- Recognize that runoff and erosion are highly sensitive to variation in soil character and plant cover.
- Assess the actual or proposed site-specific soil and plant cover conditions in maintaining infiltration and/or evapotranspiration.
- Understand that soils and structures will require maintenance for ongoing infiltration functions unless vegetation promotes permeability through perturbation by roots and burrowing organisms.
- Seek to preserve or reestablish healthy soil conditions and self-maintaining plant cover.

Formulating and implementing elements of planning and design for sustainability

The most elegant design solutions are successful at achieving multifunctionality with efficiency and value. Not unlike poetry, they condense multiple meanings and great significance into compact formats. In pursuing sustainability, planning must be applied to protect existing resources; engineering, to validate and ensure performance; and landscape architecture, to deliver functionality rather than decoration. Having set project goals and objectives based on sustainable ecological function, and steered the selection and development of features and treatments based on earth science, the design process will likely be aimed in a wise direction. The actual design process may be either simple and abbreviated for smaller projects, or quite elaborate with many iterative steps and repeated review and coordination by various disciplines. We have found that the following guidelines apply to a wide array of situations and designers' styles.

Bioengineering

- Incorporate vegetation into site design for its functional values.
- Apply woody and herbaceous vegetation for streambank protection and slope stabilization.
- Apply vegetation for management of contaminated soil and water via treatment wetlands, phytoremediation, and riparian buffer enhancements.
- View bioengineering as an extension of, and a complement to, conventional engineering.
- Combine bioengineering with conventional engineering where appropriate.

Geotechnical engineering

- Rely on landform analysis to drive site features and use bioengineering first, then validate or refine design based on geotechnical analysis.
- Use conventional measures such as soil benching to address soil stability below the zone of influence of vegetation.
- Avoid over-compaction of soils, relying on assessment of root contributions to soil shear resistance instead.
- Incorporate geofabrics to complement, not hinder plant and soil health.

Hydraulic engineering

- Develop channel form and bank treatment characteristics first, and validate or refine using quantitative hydraulic analysis.
- Consider hydraulics at current, built-out, and managed conditions (such as with feasible upstream detention features).
- Adapt the hydraulics to allow geomorphically sound, vegetatively rich treatments, rather than ruling out vegetation if the first analysis does not validate it.
- Use upstream/downstream or reference reach channel characteristics and roughness to calibrate any modeling.

Civil/environmental engineering

- Incorporate a wide and uncommon array of stormwater best management practices in all site designs toward achieving sustainable hydrology.
- Avoid the use of conventional catch-basin-and-culvert systems whenever possible.
- Do not assume that end-of-pipe water treatments can adequately deliver suitable management of water quantity and quality.
- Consider the use of phytoremediation for the cleanup or containment of soil and groundwater contamination.

Landscape architecture

- Select native species, including grass seed or sod, with rare and purposeful exceptions.
- Incorporate plant communities, instead of isolated species.

- Address self-propagation instead of planned obsolescence.
- Stockpile, store with care, and reapply topsoil to conserve nutrients, organic matter, and propagules.
- Minimize reliance on mowing and mulching, favoring rougher and looser styles of plant cover that are compatible with self-propagating plant communities.
- Accommodate human access to open space and water elements, even if they did not receive attention before restorative work.

Regional/urban planning

- Prevent sprawl.
- Reuse brownfields.
- Protect and manage greenway corridors connected with sensitive and functional resource areas.
- Demand the greenest of green techniques in the most densely developed settings instead of considering them "too far gone."

Sound construction practices

- Ensure that public safety is addressed in all designs, but refer to natural conditions as an ideal standard for site design.
- Ensure that access and staging and realistic scheduling are possible for any proposed sitework.
- Specify if access shall be conducted in a sensitive manner, such as by barge or using low ground-pressure equipment, or even only pedestrians and hand tools.
- Accommodate special planting forms or quantities of unusual species by arranging to contract grow in advance.
- Assume that the contractor will be unfamiliar — even reluctant — to execute the design as intended.
- Provide adequate quality assurance procedures including construction oversight.
- Anticipate use of common materials and equipment, even if you specify uncommon applications.
- Account for required maintenance during the establishment period, and beyond, through contractor clauses or separate arrangements.

This three-step design approach has been used on more than 100 projects conducted by TBG, ranging from residential-scale projects to river restoration efforts involving 20 miles of heavily used large channels. Three sample case studies are briefly presented to outline the primary ecological goals, the basic earth science principles applied, and the particular engineering approach or construction measures selected.

Merrick Brook, Scotland, CT

Project highlights

- Streambank stabilization using bioengineering features
- Application of stable channel geometry principles
- Vernal pool creation
- Enhancement of habitat for brook trout and brown trout

Figure I.17.1 Merrick Brook, Scotland, CT. Newly realigned channel with bank stabilization using root wads and native plantings, plus in-stream enhancements using boulder weirs.

Merrick Brook is part of the Talbot Wildlife Management Area and provides an important fisheries habitat, which includes reproducing populations of native brook trout *Salvelinus fontinalis*, introduced brown trout *Salmo trutta*, and other aquatic and terrestrial wildlife. The persistent erosion along one extreme channel bend caused degradation of the fisheries habitat directly at and immediately downstream of the bend. The resulting downstream sediment loading adversely impacted the benthic aquatic organisms. The CT DEP retained TBG to design a streambank stabilization project using bioengineering techniques and fluvial geomorphic assessment principles to protect and enhance the aquatic habitat.

Restoration design

The restoration design (Figure I.17.1) established a channel with a more suitable radius of curvature for the given channel bankfull width and discharge than what presently existed, thus reducing concentrated energy losses against the eroding meander bank. The design curve radius is consistent with measurements of stable meander bends upstream and downstream of the project area.

A riffle pool sequence was also established throughout the constructed channel to create the suitable energy gradient and efficient flow and sediment transport conveyance. Weir structures (cross weir vanes) were used to provide a stable transition at the head and tail of the newly constructed meander bend, as well as long-term grade control and energy dissipation. The weirs provide additional aeration of low flows to improve water quality, and an excavated area downstream of the weir provides scour pools for aquatic organisms in the transition to the riffle.

The meander bend was stabilized with root wads, live fascines, and transplants of native shrubs and sods containing rooted brush species to provide immediate erosion protection and long-term vegetative component that can shade the water and lessen water temperature rises during the summer months.

Hearthstone Quarry Brook, Chicopee, MA

Project highlights

- Integration of bioengineering and conventional engineering structures
- Stable channel geometry principles
- Detailed pre- and post-project monitoring
- Riparian function restoration
- Urban stream resource enhancement

The Hearthstone Quarry Brook (HQB) runs through a dense residential area and receives stormwater runoff from area roads, highways, and houses. With a planned road expansion that would route even more stormwater flow to the brook, the city had to address the problems of severe channel incision, erosion, and loss of wetland resources. With grant-writing assistance from TBG, the city received an EPA s. 319 grant to fund a project that would demonstrate the utilization of bioengineering techniques for channel stabilization, nonpoint source pollution abatement, and habitat enhancement.

Site assessment and bioengineering design

TBG conducted a channel assessment to measure critical morphologic parameters and determine the suitable design channel dimension and channel material sizes. Staff calculated shear stress conditions using field data and used modeling to help determine the location of bioengineering applications and suitable materials. Due to the high degree of shade and steep banks at the site, vegetation selection was critical. A historical analysis of channel conditions was conducted as well, to assess rates of degradation, evaluate potential resource functions, and refine design objectives.

Monitoring and construction oversight

TBG oversaw installation of the bioengineering applications, planting, and placement of coarse woody debris. In addition, TBG staff provided technical assistance in accessing the site and designing dewatering techniques during construction. Due to the poor access, the confined incised valley, adjacent residential property, and saturated clay soils TBG had to closely supervise construction and make on-site modifications to the engineering designs. TBG staff designed several natural log dam structures to provide grade control and aquatic habitat and integrated vegetation into several gabion structures to increase riparian functions (Figure I.17.2). In order to evaluate the unique combination of techniques used at the site, TBG designed a monitoring scheme to assess pre- and post-construction channel stability, habitat values, and vegetation performance.

Salem Salt Marsh, Salem, MA

Project highlights

- Design/build responsibility complex for degraded saltmarsh
- Phytostabilization of residually contaminated organic soils
- Brownfields design and remediation
- Reduction of invasive species (*Phragmites australis*)
- Enhanced tidal flushing process and created stable landforms

Figure I.17.2 Hearthstone Quarry Brook, Chicopee, MA. After several years of development, a dense cover of riparian vegetation borders a stable channel supporting appropriate organisms.

Water quality and habitat improvement

Background

The historic stockpiling of industrial waste within the saltmarsh resulted in serious degradation to the otherwise healthy tidal system. At the time of property transfer, the former owner was required to perform a cleanup of the marsh and restore it to its original condition. Following the removal of the waste, TBG was selected to restore the saltmarsh.

Objectives and constraints

The project's objectives and methodology were tightly framed to restore the site's original saltmarsh vegetation and to reduce the vitality of a number of dense pockets of Common Reed (*Phragmites australis*). Residual levels of lead and other metals necessitated an aggressive and effective stabilization strategy to address public concern. Phytostabilization measures using native saltmarsh plant species allowed engineering controls and habitat improvements to be achieved simultaneously. The main responsibilities of TBG were (1) to survey the site to determine appropriate elevations and locations for each of the specified saltmarsh species, (2) to develop a planting plan based on the survey, (3) to direct and supervise regrading activities intended to improve tidal circulation throughout the restoration zone, (4) to procure and propagate plant stock within the narrow geographic parameters specified, and (5) to coordinate and perform plant installation and phragmites reduction activities. In all, nearly 70,000 individual saltmarsh plants were planted. Despite the practical matter of being able to proceed only during low-mid-tide cycles, this 2+acre site was planted on schedule and in compliance with project specifications.

Figure I.17.3 Salem Salt Marsh, Salem, MA. Even the most degraded and impacted urban sites can be restored using sound design and implementation strategies to achieve beauty and function.

Results

Monitoring visits performed by TBG in 1997 showed a near-100% success rate in growth for the saltmarsh species. Phragmites reduction and phytostabilization of contaminated soils has also been found successful. The beautiful landscape now serves as an educational resource for local college and high school students (Figure I.17.3).

Literature cited

Barrett, K.R., Introduction to ecological engineering for water resources: The benefits of collaborating with nature, presented at the Ann. Conf. New England Water Environ. Assoc, 1997.

Goldsmith, W., CPESC, Soil reinforcement by river plants: Progress results, *Proc. ASCE Conf. Wetlands Eng. River Restoration*, 1998.

Hu, W., U.S. says New York City may have to spend $6 billion on filtration, *New York Times*, June 1, 2000.

Larson, M. and Goldsmith, W., Incised channel stabilization and enhancement integrating geomorphology and bioengineering, *Proc. Conf. Manage. Landscapes Disturbed by Channel Incision*, Wang, S.S.Y., Ed., 1997.

Robert, K.H., *The Natural Step: A Framework for Achieving Sustainability in Our Organizations*, Pegasus Communications, Cambridge, MA, 1997.

U.S. Army Corps of Engineers (CECW-E/CECW-O), Environmental engineering initiatives for water management, *Civil Works Eng. Tech. Lett.*, TL 1110-2-362, http://www.usace.army.mil/inet/usace-docs/eng-tech-ltrs/etl1110-2-362/entire.pdf, 1995.

Response

Sustainability through interdisciplinarity

This chapter by Wendi Goldsmith provides an extremely useful mix of hopefully quixotic aspirations with grounded reality therapy. Goldsmith's likening of the multifunctionality of the best water sensitive design projects to the multiple meanings residing in the best of poetry is perceptive. Her clarion call for interdiscipline plurality in the effective design and lasting existence of projects is especially apt if those projects occur at high-visibility sites such as those described by Robert France and Philip Craul in Chapter I.7, Diana Balmori in Chapter I.8, Catherine Berris in Chapter I.9, and Glenn Allen in Chapter I.13. A strong point made in this chapter is the need to include scientists in the actual design process, not merely having them only participating in the environmental permitting. This is a topic demonstrated in the work described by France and Craul in Chapter I.7. As Goldsmith notes, "True interdisciplinary design is characterized by a creative, inclusive, objective, and iterative process." Given the rapidly rising acumen and coincident expectations of the public for multiple functions, Goldsmith is correct in stating that "no other approach will be acceptable." The "three-step" process described in this chapter to achieve this end should prove valuable to many undertaking water sensitive designs and planning.

The daunting challenge to accomplishing all this is correctly identified by Goldsmith: educating clients to think in the long term and to understand that initial project development costs may be higher in the short term. Nicholas Pouder and Robert France experienced this challenge with their own project described in Chapter I.16, for example.

By moving from what Goldsmith calls "defensive design" toward sustainable designs, in the spirit of Daniel Williams' objectives for watershed management described in Chapter II.11, an important lesson becomes clear: the need to consider factors beyond the construction phase — "beyond" in both space and time. The emphasis for sustainable design in a "watershed-friendly" format is to move from running around and stomping out individual fires to considering what is starting those fires in the first place. Or, to use an expression popular in the environmental movement, to move from rearranging the deck chairs on the Titanic for a better view of our collective demise, to grabbing the wheel and steering us away from the danger all together.

Constraints, challenges, and opportunities in implementing innovative stormwater management techniques

Participants: Richard Pinkham, Larry Coffman, Michael Clar, Thomas Liptan, Herbert Dreiseitl, Wendi Goldsmith, Dennis Haag, Desheng Wang, Catherine Berris, Bill Wenk, Alex Felson, and Robert France

Past

- Desire for and love of tightly regulated, top-down control of an element — water — with a physical nature that resists such an approach
- Rapid drainage paradigm; i.e., sending water down the pipes and not taking responsibility
- Imbalance between understanding of hydraulic engineering sciences (e.g., how to design a trapezoidal channel) and hydrologic engineering sciences (e.g., what the proper design storm is and how it may affect the entire watershed), leaving us with fragmented, partial approaches that do not really solve the problem
- Regulatory control restricted to a few professional groups with a narrow, institutionally enforced knowledge base
- Flooding and water pollution are accepted standards of urbanization
- Perspective of stormwater issues as a relocative problem that someone else at another location will deal with later
- Undervaluing water's cultural importance
- Even if we knew today what ecological design was, we could not do it; i.e., often the greatest limitations are due to institutional barriers and antiquated development codes that codify the status quo

Present

- Acceptance of limits to urban growth, water resources, and human ingenuity with respect to solving environmental problems
- Early attempts at multiple use of water as it moves down the pipe
- Recognition that alterations in stormwater management are multigenerational and that solutions need to be self-sustaining through time
- Understanding of the roles of natural, predevelopment processes, and how stormwater behaves there
- Gradual evolution of institutional changes to circumvent previously codified and unimaginative regulations
- Narrowing the gaps between clients, regulators, and general society

- Intermingling of academic knowledge with real-world experience to facilitate and speed up acceptance of innovative solutions
- Beginning awareness of appropriate technologies with associated attempts at implementing innovative design solutions
- Fusion of technologies and legislative actions for wetland protection, groundwater recharge, and public recreation
- Acknowledgment of the importance of cumulative effects

Future

- Search for means to take new technologies and redesign them to deal with ecological issues
- An approach for assessing performance that is more like medicine in terms of requirements for quantifiable data, threshold criteria, and checks of implemented solutions
- Further education of the comprehensive role water plays in the functional infrastructure of cities
- More knowledge needed about the hydrologic importance of groundwater recharge in a watershed context
- Implementation of a true hydrologically functional approach supported by a corpus of systematic and vigorous scientific investigation and engineering analysis
- Desire for many more high-visibility demonstration projects ably demonstrating creativity and revealing processes; i.e., need to try and prove techniques
- Requirement for shared leadership of new, fresh voices along with a corresponding movement toward embracing a plurality of visions in innovative stormwater management; i.e., adoption of the tradeoff leadership strategy exemplified by geese in flight
- Greater inclusion of ethics in professional practice; e.g., movement away from large engineering, end-of-pipe solutions that generate large fees, toward small, dispersed, and less lucrative micromanaged solutions
- Realization of the limitations of technological solutions alone

Moving from single-purpose treatment wetlands toward multifunction designed wetland parks

Participants: James Bays, Jake Cormier, Nicholas Pouder, Bryan Bear, and Robert France

Past

- Single-purpose functionality rules in wetland design
- Shortage of awareness and absence of guidelines in how to bring people into contact with wetlands, especially by those who fund projects
- Fear and ignorance about how some treatment wetlands, especially those designed for sanitary effluents, can be made multifunctional
- Erroneous belief that this is just some new, untested, techno-fix temporary solution; i.e., "don't innovate with me!"
- Poor inter-/transdisciplinary communication among professionals designing and constructing wetlands; e.g., between landscape architects and civil engineers
- Shortage of empirical data on how designed wetland sites are used by the public
- Absence of community involvement
- Arrogance of some landscape architects in thinking that they can and should design everything
- Prevalence of the two extremes; i.e., barren and ugly, square-box wetlands and overgardened, exotic-planted wetlands
- Lack of long-term maintenance
- Overwhelming majority of wetland creation and restoration projects are undertaken because they have to be; i.e., reactive instead of being proactive
- Unrealistic belief in mitigation as a tried and proven strategy for replacement of natural wetlands lost during development

Present

- Recent perception of wetlands designed to achieve multiple objectives; i.e., treatment wetlands are now recognized to have an ancillary role of attracting wildlife, and with them, humans
- Development of wetlands for environmental education
- Establishment of a good corpus of physical design guidelines for wetland creation; e.g., importance and influence of size on biodiversity
- Most are public projects located in the suburbs
- Start of recognition of a "wetland premium" in terms of real estate values for proximal property owners

- Involvement of landscape architects in the presentation of design plans to the public as an accepted and integral step in raising public interest in wetland creation projects
- Gradual realization that nature is the final and best architect
- Wetlands created in suburban locations are becoming appreciated for the feelings of natural experience they engender; i.e., realization of Olmsted's grand vision of extracting nature to make it convenient

Future

- Multidisciplinary team right from the start; i.e., all created wetland parks that have public-use applications must involve a landscape architect, a hydraulic engineer, a wetland designer, a site civil engineer, and a watershed process engineer or hydrologist, all sharing a strong belief in "osmosis learning"
- Need to explore new and varied ways of allowing people to become enriched by their wetland experiences while at the same time not disturbing wildlife
- Opportunities for multiple uses need to be brought into planning and design discussions much earlier in the process, and both the public and public works officials need to be included
- Turn multifunctionality on its head; i.e., "ancillary" functions such as enhancing biodiversity need to be made premier with other uses, such as stormwater detention, pitched as "by-the-way" extra benefits
- Need for much more documentation about the challenges, successes, and lessons learned from established wetland creation projects
- Cardinal requirement for more rigorous post-construction monitoring in order to persuade others that these projects have value
- More attention needs to be paid to safety concerns of wetlands as part of public works projects; i.e., the liability issue in the U. S. compared with such places as Canada and Europe may be the single greatest limitation to the design of aesthetically beautiful wetland parks that encourage unfettered visitor access
- Avoidance of hubris; i.e., we should not lose sight that what we are really designing (at least in the short term) is an interpretation or surrogate of nature and not necessarily real nature (though well-designed projects may evolve toward such over time)
- Honest acknowledgment of the ethical and practical weaknesses and limitations involved in wetland mitigation
- Further education of clients toward moving from site-specific design issues to larger regional planning objectives; i.e., water does end in the wetland
- Embrace the concepts of, and find ways to further promote, diffusion innovation in terms of education
- Always remember that no matter who is paying the bills, the ultimate client is the place — the landscape

part II

Water sensitive planning

Overview: New interpretations in the management of watersheds and riparian buffers and corridors

Part II comprises 17 chapters and responses that deal with two of the most important issues associated with water sensitive planning: management of watersheds and riparian buffer-corridors. Together, these chapters go far toward advancing the concept that, in order to be sustainable and therefore effective, water sensitive planning has to be concerned just as much about managing human interactions as about simply managing water resources alone. In other words, as the physical scale of projects increases, so too must the sociological scale.

Four chapters focus on the management of riparian buffers and corridors. Frank Mitchell provides an introduction to the complex issues and variable management options for protecting water quality and preserving shoreline biodiversity. The chapter by landscape architect Charles Flink expands this vision by addressing the important role that such regions play in supporting human recreation and societal infrastructure. Two of the chapters — those by James MacBroom and by Leslie Zucker, Anne Weekes, Mark Vian, and Jay Dorsey — integrate these objectives and discuss how rivers should be managed and restored by viewing them as landscape systems rather than as independent aquatic entities.

Five chapters focus on implementation of watershed management planning in urban settings. Four chapters — those by landscape architect Thomas Benjamin, by landscape architect David Blau, by Dennis Haag, Stephen Hurst and Bryan Bear, and by Kelly Cave — demonstrate the various concepts that need to be addressed in order to protect urban water resources in the face of existing degradation and future development, both often taking place in an environment of complex sociopolitical interactions. The chapter by Gail Krantzberg and Judi Barnes grapples with the intriguing and important issue of measuring the recovery of impaired urban waters.

Eight chapters address water sensitive land-use planning in rural areas. Three of these — by landscape architect Margot Cantwell, by Robert France, and by Neil Hutchinson — deal with protecting the integrity of boreal lakes from watershed development, riparian forestry, and recreational use. In his chapter, landscape architect Daniel Williams expands the physical sphere of water sensitive planning, arguing that in order for water management to be truly sustainable in certain situations, it should be approached on the scale of entire regions instead of specific watersheds. In a similar spirit, landscape architect Amir Mueller, Robert France and landscape architect Carl Steinitz expand water sensitive planning temporally in their presentation of a methodology based on alternate futures modeling. Jeffrey Schloss's chapter introduces the benefits and pitfalls ensuing from use of geographic information systems in water sensitive planning. The remaining two chapters — those by John Felkner and Michael Binford and by Robert France, John Felkner, Michael Flaxman, and Robert Rempel — provide examples of how GIS analysis can aid decision making about both overt and subtle land-use management concerns.

chapter II.1

Shoreland buffers: Protecting water quality and biological diversity (New Hampshire)

Frank Mitchell

Abstract

Most interest in shoreland buffers has been related to water quality protection. Protecting biological diversity, however, requires much larger buffers. This chapter presents several concepts that apply to shoreland buffers whether water quality or wildlife habitat is the primary concern:

- Buffers are most effective when they are part of an overall protection strategy that includes whole watersheds or habitat areas.
- Connectivity is crucial. A continuous buffer along a stream, for example, will have more integrity and effectiveness than a fragmented one.
- Core protected areas will add a level of protection beyond what routine buffers can provide.
- Even a fragmented buffer provides some benefits regarding water quality and wildlife. Some gaps can be filled over time through restoration efforts.
- Citizens appear to support the water resources and other features that shoreland buffers can protect. They also appear to support actions needed to conserve ecological and social functions and values of buffers.
- Shoreland buffers can be established through voluntary or regulatory means.
- Buffers can be designed and established to adequately protect both water quality and wildlife habitats.

Introduction

"Buffer" is used with different meanings in different contexts. For the purposes of this discussion, a *shoreland buffer* is defined as a naturally vegetated upland area adjacent to a surface water or wetland. This often means uncut or undisturbed forest, minimally disturbed or managed forest, and abandoned pasture or fields reverting to forest (Figure II.1.1). This is different from a *setback*, in which there is a distance requirement between surface waters or wetlands and activity that might affect them but not a specification of the type of land cover in the setback zone. A *filter strip* usually refers to an area

Figure II.1.1 Profile of stream buffer. The shoreland buffer in this profile consists of the forested areas on either side of the stream, with farmland to the left and residential land to the right. The natural roughness of the land supports retention and filtering of surface runoff and groundwater infiltration. Tree roots help hold soil in place, and debris the trees produce nourishes stream ecosystems. The trees can also sequester nutrients flowing across the buffer aboveground and in groundwater. (Drawing by Linda Isaacson.)

of herbaceous vegetation between agricultural or similar land uses and surface waters or wetlands. The term *riparian buffer* is sometimes used to describe streamside, but not necessarily wetland or lakeshore areas, and some who use the term apply it primarily to wetlands adjacent to a stream or river, or to the immediate shoreline. Therefore, the term "shoreland buffer," as used here, differs from these other terms in that it implies:

- The topography of the land adjacent to water or wetlands is natural
- An expectation of natural vegetation
- Inclusion of buffers for wetlands as well as surface waters
- Inclusion of upland areas beyond a stream bank, lake shore, or wetland edge
- Application to a variety of possible land uses beyond the buffer
- A minimum but flexible width based on site-specific conditions

Some current views of buffers

Most historical interest in shoreland buffers has been related to water quality protection and to some degree aesthetic concerns, but more recently the wildlife habitat functions of shoreland areas have received increasing attention. If landscape design that incorporates shoreland buffers is to be ecological, it must consider these habitat functions and values as well as water quality and visual concerns. Chapter II.5 presents considerations for meeting objectives of multiple-function stream corridor planning. Chapter II.3 also discusses how water-related greenways can serve to meet water quality, hydrologic, transportation, recreational, and other needs.

Research on buffer effectiveness has been the basis for recommendations about buffer width and other aspects of design and maintenance.[1–7] For example, in New Hampshire,

Table II.1.1 Wildlife within a 100-ft Shoreland Buffer: Habitat Needs of Representative Species

Species	Habitat Needs
Eastern (red-spotted) newt *Notophthalmus v. viridescens*	Woodland habitat for terrestrial juveniles (efts) for the 2–7 years they spend on land
Northern 2-lined salamander *Eurycea b. bislineata*	Foraging area — adults may wander 330 ft on rainy nights; dispersal of juveniles (only 25% return to natal streams)
Green frog *Rana clamitans melanota*	Dispersal habitat
Wood frog *Rana sylvatica*	Habitat for most terrestrial activity, often well away from water
Spotted turtle *Clemmys guttata*	Habitat for most terrestrial activity — will travel up to 2600 ft from water to find temporary food sources
Wood turtle *Clemmys insculpta*	Habitat for most activities; spend most of their time within 1,000 ft of water, but will travel up to 1 mile to search for food; will nest up to 330 ft away
Northern water snake *Nerodia s. sipedon*	Habitat for dispersal and hibernation
Bats *Myotis* and other *spp.*	Roosting sites — prefer to roost within 1,300 ft of water
Beaver *Castor canadensis*	Enough foraging habitat — most foraging within 330 ft of water; dispersal routes
Mink *Mustela vison*	Hunt up to 600 ft from water; den sites may be up to 330 ft from water
Black bear *Ursus americanus*	Den sites; enough area for travel — adult males require up to 19 square miles depending on habitat and food sources
Bald eagle *Haliaeetus leucocephalus*	Nest sites — most eagle nests are within 1,300 ft of shorelines; protection from human disturbance
Red-shouldered hawk *Buteo lineatus*	Nest sites — this species is found only where buffers are 330 ft or more
Area-sensitive forest birds	Sufficient breeding habitat for species that need buffers wider than 330 ft

Source: Adapted from Chase, V.P., Deming, L.S., and Lataviec, F., *Buffers for Wetlands and Surface Waters: A Guidebook for New Hampshire Municipalities*, 2nd ed., Audubon Society of NH, Concord, NH, 1997.

an interagency project that produced the publication *Buffers for Wetlands and Surface Waters, A Guidebook for New Hampshire Municipalities*, used research results to recommend a "reasonable minimum" 100-ft, naturally vegetated upland buffer around streams, rivers, ponds, lakes, wetlands and estuaries. The basis for the recommendation was the recognition that a 100-ft-wide buffer can be expected to typically remove at least 60% of pollutants such as eroded soil and phosphorus.[1]

Protecting biological diversity, however, requires much larger buffers. A 100-ft buffer will provide food sources, cover, and other habitat components for many species, particularly smaller ones, but would only partially provide these needs for others. Buffers recommended by wildlife biologists can extend to over 1,000 ft from a water body or wetland.[1,8] The rationale for buffers of this size is based on the travel and dispersal needs of certain species, especially those with life cycles that utilize both aquatic and upland habitats, and protection from nest predation and parasitism for animals such as interior forest nesting birds.[9] Table II.1.1 presents examples of the habitat needs of representative species that use shoreland buffer areas.

Shoreland buffers alone will not fully protect water quality or wildlife. In both cases, buffers will be most effective if they are part of a larger overall protection strategy that includes whole watersheds or habitat areas.

Figure II.1.2 Woody debris in a stream. Leaves, branches, and other debris falling into streams from the surrounding forest provide important habitat structure and food for stream fauna. In smaller streams, such detritus can account for the majority of food energy in the system. (Photo by Frank Mitchell.)

What buffers do (buffer functions)

Buffer functions and their benefits are ecological and social. They include:

- Hydrologic effects such as promoting groundwater recharge, moderating flooding by intercepting rain and snow melt, reducing sunlight penetration (thereby slowing the rate of spring snow melt), and storing flood waters that overflow stream banks, lakes, and wetlands.
- Water quality maintenance through at least partial removal of sediment, phosphorous, and nitrogen from runoff passing through a naturally vegetated buffer. Forested buffers also contribute to neutralization of acid precipitation, due to ion exchange that occurs as precipitation filters through the forest canopy.[10] The percentage of pollutants removed depends on the pollutant load, the nature of the material, the amount of runoff, and the character of the buffer area. The pollutant removal rate is not generally a linear relationship with buffer width, but decreases with increasing buffer width. More pollutant removal occurs in the first 100 ft than the second hundred, for example.[2]
- Wildlife habitat opportunities, such as providing foraging and nesting habitat as well as cover for a mix of upland, aquatic, and wetland species. Buffer areas can also serve as travel routes for migratory and nomadic, as well as resident, species. They also support plant diversity in the ecotones represented. Buffers protect surface waters and wetlands from temperature increases, which reduces water's capacity to hold oxygen. Leaf litter and woody debris from buffers along smaller streams supply most of the energy processed by the stream. The woody debris also traps the leaf litter, making it available to organisms over a long period of time (Figure II.1.2). These streamside buffers help stabilize banks as well, and naturally undercut areas beneath tree roots offer cover for fish, turtles, and other creatures.[1,11]
- Recreation and aesthetics benefit from the visual screen buffers provide along surface waters, and they frame wetlands and surface waters in the landscape, particularly in hilly or mountainous terrain.

Factors affecting buffer functions

All buffers are not the same. Each, as part of a living system, functions differently. The characteristics of a shoreland buffer affect its functional capacity. Several factors other than buffer width influence the effectiveness of shoreland buffers. Some of these are described here.

For water quality

- Soil type — Less erodible soils within a buffer pose less risk of buffer degradation from runoff passing through it (channeling). More permeable soils allow greater infiltration of rain and melted snow into the ground.
- Vegetative type and density — Woody plants at a natural density generally do the best job of holding soil in place within buffers. They also contribute to buffer stability by intercepting rain and snow (in the range of 20% in the northeastern United States, depending on species) and reducing sunlight penetration, thereby moderating snow melt that often precedes peak runoff times.
- Surface roughness of the land — Undisturbed terrain retains its capacity to trap surface water through ponding and groundwater infiltration.
- Season — In northern areas, buffer effectiveness can be reduced substantially during times of the year when the ground is frozen and plants are dormant.
- The nature of the land beyond the buffer — Buffers have their limits as to how much water or pollution they can absorb. The more intense the land use (more likely to generate pollution such as eroded soil) above a buffer, the more the buffer is in jeopardy of having its assimilative capacity exceeded.[1]

For wildlife habitat

- What the buffer connects to (landscape position) along its length — Buffers can function both as resident ("in-place") habitat and as travel routes for wildlife. As resident habitat, a buffer's value is supplemented by other habitats with which it physically connects. This is important because larger habitat blocks are known to support greater biodiversity than smaller ones. Two blocks of habitat, for example, can function as a larger unit if connected by an area of adequate width. Animals using a buffer as a travel route require the route to lead somewhere productive, i.e., to a viable habitat block. In either case, then, resident habitat or travel route, a buffer functions best as habitat when it is connected to other habitats, including other buffers (Figure II.1.3).
- Integrity of the connectivity of the buffer — One of the greatest challenges in establishing shoreland buffers is avoiding or minimizing the effects of fragmentation of the buffer. The network of roads in most parts of the country makes it difficult to avoid fragmentation altogether, and the ecological effects of roads can be substantial. These effects can include direct mortality (roadkill), modification of animal behavior, alteration of the physical and chemical environments, and the introduction of exotic species. These effects occur in terrestrial habitats along roadsides as well as extending downstream in aquatic and wetland systems with moving water (Figure II.1.4). This can fragment the aquatic and wetland habitats by altering hydrology, increasing sedimentation, and introducing pollutants. These effects can, in turn, lead to discontinuous floral and faunal assemblages in such aquatic systems.[12,13]

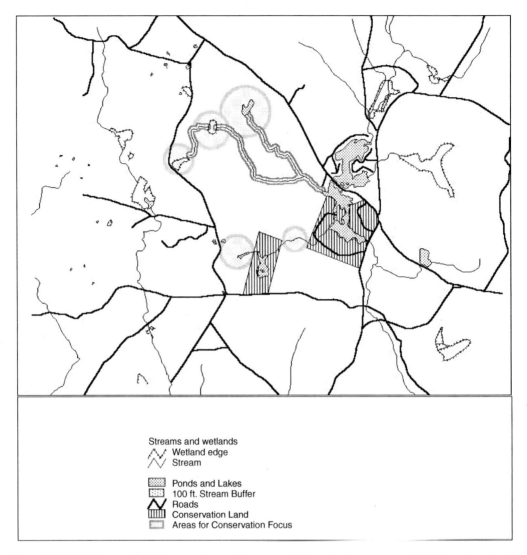

Figure II.1.3 Buffers as habitat connections. The most effective buffers connect to blocks of protected land and connect with other wetland and surface-water buffers. This map shows how riparian buffers can connect existing conservation areas and suggest areas for focusing on riparian buffer protection. In the example, the circles depict wetlands, stream segments, and their surrounding uplands that, if protected, would add significantly to the habitat value of the protected stream buffer in the watershed. Such protection would have water quality, habitat, and aesthetic benefits.

- Abruptness of the edge between the buffer and the land use beyond it — When buffers become fragmented strips between water and land, they may be subject to negative "edge effects" of predation and parasitism as well as physical effects such as wind, drying, temperature increase, and blowdown of trees. Edge habitats tend to harbor disproportionate populations of predators such as blue jays (*Cyanacitta cristata*), crows (*Corvus brachyrhynchos*), raccoons (*Procyon lotor*), skunks (*Mephitis mephitis*), red foxes (*Vulpes vulpes*), and dogs and cats. Animals like these use the edge habitats and penetrate from it into woodlands.[1,8]

Chapter II.1: Shoreland buffers: Protecting water quality and biological diversity

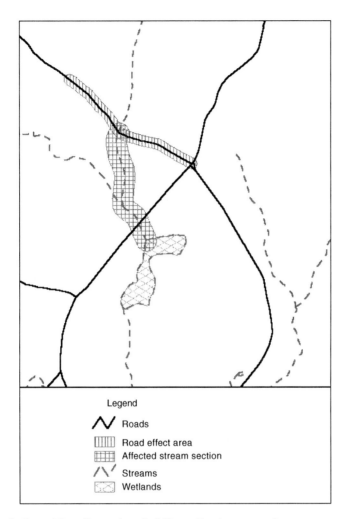

Figure II.1.4 Road effects. The effects of roads follows flowing water downstream, sometimes changing physical, chemical, and biological conditions for a substantial distance downstream.

In the northeast, the brown-headed cowbird (*Molothrus ater*) is the nest parasite of concern in relation to migrant songbirds. Cowbirds lay their eggs in nests of other species, which then usually raise them as their own. Cowbird young hatch quickly and tend to be larger than their host species, giving them a competitive advantage in feeding. They may also physically remove their host species' offspring from the nest. Cowbirds are native to the midwest and have expanded their range throughout the northeast in response to land clearing for agriculture and other uses.[1]

- Some biologists feel that negative edge effects can be reduced by managing buffer areas to have gradual or patchy ("soft") edges on the upland side, rather than abrupt, or "hard" ones. Another suggested technique is establishing or maintaining dense vegetation along the upland buffer edge.[8,9] In effect, this means a transitional upland buffer to support the wetland/surface water buffer habitat functions more fully (a buffer of sorts for the shoreland buffer).
- Habitat structure (layering of vegetation, dead trees with cavities, etc.) — The structure provided by a riparian forest determines which species can use the habitat. Habitat structure includes:

- Horizontal diversity (the horizontal arrangement of plant communities in an area)
- Vertical diversity (the degree to which plants are layered within an area)
- Soil qualities
- Dead standing trees
- Downed logs
- Boulders, cliffs, etc.[14]

- Vegetation type — The species of plants in an area generally determine the animals that will occupy the area. Dense stands of evergreen trees, for example, are well know for their value as deer wintering areas in New England. "Hard mast" (nut producing) trees, such as oaks and beech, provide food for a number of species (such as bear, deer, squirrels, and jays).[15]

Establishing and maintaining buffers

Assuming the vegetation is already present, protected shoreland buffers can be established through voluntary or regulatory means. The author's state of New Hampshire offers examples of both of these approaches, as described in the following sections.

Voluntary measures

Publications

Education about buffers has become widespread in the United States in recent years, resulting in a number of publications and other educational materials on the topic. The northern New England state of New Hampshire is an example of such efforts. The publication *Buffers for Wetlands and Surface Waters: A Guidebook for New Hampshire Municipalities*[1] was produced cooperatively by four public and private organizations and has been distributed widely in the state and used as the basis for dozens of educational presentations. It was produced primarily to assist public officials in making decisions about regulatory buffers. It has influenced public policy decisions from the local to the state level. *Buffers for Wetlands and Surface Waters* recommends a reasonable minimum 100-ft buffer for wetlands and surface waters to protect water quality and documents the rationale for this recommendation. Though some habitat value is maintained with the 100-ft recommendation, the publication acknowledges that this is generally an insufficient size buffer to protect biodiversity and provides guidance on using larger, voluntary buffers for this purpose.

A related publication, *A Guide to Developing and Redeveloping Shoreland Property in New Hampshire, A Blueprint to Help You Live by the Water*,[16] was also an interagency effort involving 13 public and private organizations. The booklet's purpose is to provide landowners "with the understanding and tools necessary to develop your shoreland property in a manner that meets [the landowners'] goals while at the same time maintaining the quality of the water body for everyone."

Good Forestry in the Granite State: Recommended Voluntary Forest Management Practices for New Hampshire[17] is yet another collaborative project, this time involving 15 groups. *Good Forestry in the Granite State* presents recommendations for protecting water quality, wetlands, and riparian areas in a forest management context. Its purpose is to "provide New Hampshire landowners, and the professionals work with them, practical recommendations on sustainable management practices for individual forest ownerships." Similar to *Buffers for Wetlands and Surface Waters*, this publication has been widely distributed and used in educational programs throughout the state. Because it emphasizes voluntary

Table II.1.2 Legally Required and Recommended Voluntary Shoreland Buffers for Forestry in New Hampshire

Feature	Legally Required		Recommended	
	Management Zone	No-Harvest Zone	Management Zone	No-Harvest Zone
Intermittent streams	None	None	100 ft	None
First- and second-order streams	50 ft	None	100 ft	None
Third-order streams	50 ft	None	300 ft	25 ft
Fourth-order and larger streams	150 ft	None	600 ft	25 ft
Pond (<10 acres)	50 ft	None	100 ft	None
Lake (>10 acres)	150 ft	None	300 ft	25 ft
Nonforested wetland (<10 acres)	None	None	100 ft	None
Nonforested wetland (>10 acres)	None	None	300 ft	25 ft

Note: Recommended Voluntary Best Management Practices (RVMPs) for shoreland buffers on forestry lands in New Hampshire, presented in *Good Forestry in the Granite State*, advise substantially larger buffers than regulations provide. The regulations are focused on water quality protection, whereas the RVMPs consider habitat needs as well.

Source: Adapted from NH Forestry Sustainability Standards Work Team, *Good Forestry in the Granite State: Recommended Voluntary Practices for New Hampshire,* Soc. Protection of NH Forests, Concord, NH, 1997.

adoption of buffer recommendations and a strong consideration of habitat requirements, this document makes recommendations for substantial buffer widths. For example, *Good Forestry in the Granite State* suggests buffers from 100 ft (from intermittent streams to second-order streams, ponds, and open wetlands) to 600 ft for fourth-order and larger streams (Table I.1.2)[17]

Landowner stewardship, or voluntarily maintaining sufficient shoreland buffers in forest, residential, and other landscape settings is happening all the time.

Current use of property tax assessment can provide some incentive for landowners to keep land in forestry, agriculture, or other open-space uses.

Landscape design for development projects can incorporate adequate or even optimal protection for water resources by including shoreland buffers.

Land conservation can protect shorelands over the long term. Conservation easements and land acquisition for conservation purposes are the most common tools employed.

Regulatory measures

State regulations

An example of state regulation of shoreland buffers can be found in New Hampshire's Comprehensive Shoreland Protection Act, which requires shoreland buffers (150 ft, except a building may be built within 50 ft of the shore) along the states lakes (>10 acres), fourth-order streams, and tidal waters.

State forestry and agriculture best management practices can be legally enforceable and focus mainly on reducing erosion and sedimentation from these land uses. They may require buffers ("filter strips") of varying width depending on slope and the type of water resource being protected.

Municipal zoning

Many communities have municipal zoning ordinances that offer some degree of protection for shorelands.

Public attitudes toward buffers

Results of public opinion surveys

Awareness of the functions and values of shoreland buffers among scientists, landscape planners, and regulators is relatively recent, so one would not expect that the public would be particularly well versed about the workings and importance of buffers in providing critical habitats and maintaining water quality. However, there is evidence of public support for the resources buffers can protect and for some types of buffer implementation methods. Three examples from New Hampshire illustrate this.

In 1996 a Community Vision Committee surveyed town residents in Deerfield, NH (pop. of 3124 in 1990), asking questions on a variety of topics related to the character people wanted their town to have in the future. Questions included open-ended, multiple-choice, and rating-scale (agree/disagree, very important, important, not important, etc.) types. The response rate was about one-third.[18]

In 1998 a private conservation advocacy group, the Concerned Citizens for a Safe Environment, completed a survey of landowners along the Lamprey River in Raymond, NH (estimated pop. of 9398 in 1999). The group got a 25% response rate to the survey, which was done as part of a planned nomination of the river into the state's Rivers Management and Protection Program. Seventy-five percent of survey respondents reside in the town and 72% use the property as a full-time residence.[19]

The Friends of the Cold River, a private conservation group, completed three landowner surveys in 1999, in cooperation with the University of New Hampshire Cooperative Extension and the Upper Valley Regional Planning Commission. One survey questioned landowners within the river's watershed, the second targeted those within ¼ mile of the river, and the third questioned those within ¼ mile of a lake that is the river's headwaters. The discussion below covers the first two surveys. Of the landowners in the river corridor survey, one third owned land that was actually on the river. The Friends of the Cold River received a response rate of 32% from the 1984 surveys it sent. Six rural communities are in the Cold River Watershed.[20,21]

These surveys, though not scientifically conducted, give valuable insight into landowner attitudes and opinions, at least for the northeastern state in which they were done. The consistency between the Lamprey River and Cold River surveys, in different parts of the same state, supports an interpretation of credibility of the results despite the fact that the surveys were conducted by laypeople.

Attitudes toward the value of surface waters, wetlands, and shoreland property

Deerfield community vision survey

- The first question of the survey was open-ended, asking, "What do you think is (are) the most desirable features of Deerfield?" Eighteen of 299 respondents (6%) wrote "lakes" or "rivers" in a section on physical character of the town.
- Ninety-seven percent of survey respondents answered that "clean water" was very important or important to "protecting the town's environment and natural resources." Ninety-five percent indicated that "water resources" were this important.[18]

Concerned citizens for a safe environment landowner survey

- Landowners responding who indicated that the river played a part in their decision to purchase their property: 60%

- Respondents who answered that they "think the Lamprey River contributes to the quality of life" in the community: 95%
- Top reasons for this view, in order of frequency, were

Fishing	100%
Wildlife and waterfowl habitat	95%
Scenic value	93%
Open space	75%

- Landowners who cited "shoreline development" (the least cited category) as the quality of life contribution: 15%[19]

Cold River area surveys

	Watershed landowners	River corridor landowners
Live on the land that was the subject of the survey (primary residence)	60%	69%
Plan to continue the current land use of their property	84%	83%
Plan to keep the property in the family	50%	41%

Rated the following "quality of life" features associated with the Cold River as "high" on an importance scale of low–medium–high–no opinion:

	Watershed landowners	River corridor landowners
Good water quality	90%	87%
Diverse wildlife habitat/wildlife and waterfowl habitat	76%	81%
Forest lands	77%	71%
Scenic qualities	75%	80%
Fisheries habitat	Not included	73%
Wetlands	Not included	64%
Agricultural lands	68%	55%

- Cited as a top concern (when asked to list top three concerns):

Water pollution	Not included	47%
Building too close to the river	Not included	28%
Clearing too close to the river	Not included	24%

- Landowners observed:

Water pollution	Not included	19%
Building too close to the river	Not included	14%
Clearing too close to the river	Not included	22%[20,21]

Attitudes toward resource protection

Deerfield community vision survey

In response to the question, "Would you favor an increase, a decrease or no change in the current levels of each of the following?" (a list of land uses followed), 65% of respondents chose "town conservation land" as a land use for which they wanted an increase. This was the answer getting the greatest response for an increase. Private undeveloped land and open space were the third and fourth most desired future land uses.[18]

Concerned citizens for a safe environment landowner survey

- Landowners who intend to keep their land in its present use — 90%
- Those who plan to build on the land — 2%
- Those who plan to subdivide — 2%

In answer to the question, "How important is it that each of the following characteristics associated with the Lamprey River and its corridor are conserved?" responses of "very important" or "important" were given as follows:

Open space	93%
Scenic quality	83%
Water quality	82%
Wildlife and waterfowl habitat	75%
Fishing access	66%
Fisheries habitat	63%
Wetland ecosystems	63%
Residential development opportunity	18%
Commercial or industrial development	2%

- Respondents who reported noticing "development too close to the river": 40%
- In response to the question, "Do you believe that any of the following measures should be taken to protect the Lamprey River and the special opportunities it offers to the area?" landowners answered:

Limit shoreline commercial development	83%
Protect water quality	80%
Limit shoreline industrial development	75%
Protect the scenic character of the river corridor	75%
Protect wildlife and waterfowl habitat	73%
Protect fisheries habitat	60%
Limit shoreline residential development	50%
Protect the free flowing nature of the river	50%
No additional protection needed	8%[19]

Cold River area landowner survey

	Watershed landowners	River corridor landowners
Landowners who indicated intent to protect some or all of their land from development	37%	19%
Landowners who plan to sell their land at some point without subdividing	18%	13%
Those who plan to subdivide	5%	1%
Respondents who have a written management plan for their farm or forest land	19%	19%
Those interested in developing one	25%	25%

- Landowners who were not aware of published best management practices for farm and forest land management (water quality BMPs)   58%   58%

For the question, "Which of the following would be of interest to you for your lands?" responses were as follows for "high" interest:

	Watershed landowners	River corridor landowners
Maintain the land's natural beauty	84%	Not included
Protect streams and wetlands	80%	Not included
Protect from development	73%	Not included
Identify and maintain wildlife habitat	69%	Not included
Improve timber/firewood production	44%	Not included
Improve production of forest products	40%	Not included
Improve soils	34%	Not included
Leave it alone (no management)	30%	Not included
Grow horticultural products	24%	Not included
Improve animal or crop production	23%	Not included
Not sure/not applicable	13%	Not included[20,21]

Opinions on appropriate protective methods

Deerfield community vision survey

The Deerfield survey did not deal with issues of specific protective measures.[18]

Concerned citizens for a safe environment landowner survey

In response to a question about which specific steps would be "appropriate for river and river corridor protection," landowners answered:

Minimum setbacks for new construction	65%
Floodplain protection regulations	48%
No additional protection needed	48%
Purchase property in the river corridor (from willing seller)	40%
Purchase of development rights in the river corridor	35%
Voluntary easement donation program	25%[19]

Cold River area surveys

The Cold River surveys included the similar questions, "Which of the following options are appropriate for conserving land and the natural resources associated with it?" and "Which of the following do you feel are appropriate for protecting the Cold River and the resources associated with it?" Responses of a "high" level of appropriateness were as follows:

	Watershed landowners	River corridor landowners
Incentives for landowners to manage land appropriately	68%	60%
Landowner education	60%	61%
Education of municipal officials	58%	53%

Coordinated planning among towns	58%	46%
Conservation planning in utility and road projects	55%	49%
Voluntary protection through conservation easements	53%	40%
Voluntary protection through public acquisition	37%	34%
Stricter enforcement of state regulations	34%	34%
Stricter enforcement of existing local regulations	35%	37%
Stronger local regulations	20%	20%
Stronger state regulations	18%	17%[20,21]

Conclusions from public opinion surveys

These survey results indicate that citizens and landowners in these geographic locations generally support protection of the features and resources associated with water resources in general and shorelands in particular. Many of the values respondents cite as important to them are dependent to some degree on healthy shoreland ecosystems for their integrity. Values intrinsic to aquatic and wetland systems are of primary importance to most survey respondents. These include good water quality, wildlife habitats, scenic values, and "open-space" values. Survey respondents reported interest in nonconsumptive uses at a greater level than interest in resource extraction activities.

Similarly, people are concerned about threats to these resources that shoreland buffers can help ameliorate — water pollution, land clearing and building near rivers, and loss of wildlife habitat.

A high proportion of landowners report their intent to keep their land in its current use, an attitude that bodes well for voluntary shoreland protection, at least until the land changes ownership.

When asked what to do to protect resources related to rivers and shoreland areas, people strongly approved of limiting shoreland development, protecting water quality, wildlife habitats, and scenic and recreational values. They generally favored voluntary over regulatory approaches to accomplish this, with the exception of Lamprey River respondents, a majority of whom favored minimum setbacks for new construction (buffers were not a choice in the question). Voluntary measures most preferred included landowner incentives, education, and improved local planning, but permanent land protection through conservation easements or public acquisition was favored by one-third to one-half of the Lamprey and Cold River landowners.

An interesting variation from the consistency seen through much of the surveys appeared in the Cold River survey. Thirty-seven percent of landowners in the watershed said they intend to protect some or all of their land from development, but only 19% in the river corridor said this. However, 5% of watershed landowners claimed to intend to subdivide, as opposed to only 1% of river corridor landowners.

It is also interesting to note that the level of concern expressed by landowners in the Cold River survey is apparently greater than their preparedness to act accordingly. A majority of the Cold River landowners had no written management plan for their land and were not aware of published best management practices (BMPs) to protect water resources. Almost half (45%) of the Cold River Watershed landowners, however, stated that they know "where to find appropriate personal assistance in managing their lands."

Summary of buffer concepts

Several concepts apply to shoreland buffers for both water quality protection and wildlife habitat conservation:

- Buffers are most effective as part of an overall watershed protection strategy.
- Connectivity is crucial. A continuous buffer will have more integrity and effectiveness than a fragmented one.
- "Core areas" can add a level of protection beyond what routine buffers can provide. For water quality, this might mean a more substantial buffer around a wetland that has a particularly high stormwater storage and pollution removal capacity. For wildlife, the core area might be an ecological reserve of significant size to which shoreland buffers (as travel routes for wildlife) connect.
- Even a fragmented buffer provides some benefits for water quality and wildlife. Some gaps can be filled through restoration efforts. (Chapter II.2 describes considerations for river restoration planning, including riparian greenways.)
- The character of the land influences the functional quality of the buffer. Soil type, vegetation type, and slope are the factors that most influence buffer function.

Conclusions

- Buffers can protect both habitat and water quality functions and values, but protecting habitats requires wider, connected buffers.
- Voluntary and regulatory methods may be used to protect buffers, preferably in combination. Voluntary measures have the potential to provide optimal protection, but rely on landowner initiative. Regulatory measures apply broadly but typically provide less than ideal, and sometimes less than adequate, protection. Thus, a strategy that encompasses both approaches can result in minimum standards being met widely, with more desirable protection occurring in areas where landowners have chosen to provide it. Such voluntary protection can be assured over long time periods through conservation easements and land acquisition by public agencies or private conservation groups.
- Citizens appear to support the water resources and other features that shoreland buffers can protect. They also appear to support actions needed to conserve ecological and social functions and values of buffers, particularly voluntary ones.
- A landscape design and management approach that embraces water quality protection and conservation of biodiversity will give consideration to shoreland buffers as a means to support both functions. Though buffers alone cannot fully protect water quality and biodiversity, they are a key piece in a larger landscape view that would encompass these goals. A landscape design that follows this path will complement regulatory efforts by others to protect the same features, resources, functions, and values. Aesthetics and other social values will also be supported by such an approach.

Literature cited

1. Chase, V.P., Deming, L.S., and Lataviec, F., *Buffers for Wetlands and Surface Waters: A Guidebook for New Hampshire Municipalities*, 2nd ed., Audubon Society of NH, Concord, NH, 1997.
2. Desbonnet, A. et al., Vegetated buffers in the coastal zone, a summary review and bibliography. Coastal Resources Center Technical Report No. 2064, Univ. RI Grad. Sch. Oceanography, Naragansett, RI, 1994.
3. Castelle, A.J. et al., *Wetland Buffers: Use and Effectiveness*, Adofson Associates, Inc., Shorelands and Coastal Zone Management Program, WA Dept. of Ecol., Olympia, WA, 1992.
4. Groffman, P.M. et al., *An Investigation into Multiple Uses of Vegetated Buffer Strips*, Univ. RI, Dept. Nat. Resour., Kingston, RI, undated.

5. Palmstrom, N., *Vegetated Buffer Strip Designation Method Guidance Manual*, I.E.P., Inc., Consulting Environmental Scientists, Northborough, MA, 1991.
6. Rogers, Golden & Halpern, *Wetland Buffer Delineation Method*, NJ Dept. of Environ. Protection, Trenton, NJ, 1988.
7. Roman, C.T. and Good, R.E., *Buffer Delineation Model for New Jersey Pinelands Wetlands*, Rutgers, State Univ. NJ, New Brunswick, NJ, 1985.
8. Smith, D.S. and Hellmund, P.C., Ed., *Ecology of Greenways*, Univ. MN Press, Minneapolis, 1993.
9. Noss, R.F. and Cooperrider, A.Y., *Saving Nature's Legacy, Protecting and Restoring Biodiversity*, Island Press, Washington, DC and Covelo, CO, 1994.
10. Hornbeck, J.W., Likens, G.E., and Eaton, J.S., Seasonal patterns in acidity of precipitation and their implication for forest stream ecosystems, *Water, Air and Soil Pollution*, 7, 355, 1977.
11. Carroll, D.M., Deadfalls, turtles and trout, *Wild Earth*, 11, 57, 2001.
12. Trombulak, S.C. and Frissell, C.A., Review of ecological effects of roads on terrestrial and aquatic communities, *Conserv. Biol.*, 14, 18, 2000.
13. Forman, R.T. and Deblinger, R.D., The ecological road-effect zone of a Massachusetts (U.S.A.) suburban highway, *Conserv. Biol.*, 14, 2000.
14. DeGraff, R. et al., New England wildlife: Management of forested habitats, General Technical Report NE-144, USDA Forest Service Northeast Forest Experiment Station, 1992.
15. DeGraff, R.M. and Yamasaki, M., Bird and mammal habitat in riparian areas, in Verry, E.S., Hornbeck, J.W., and Dolloff, C.A., Eds., *Riparian Management of Forests of the Continental Eastern United States*, Lewis Publishers, Boca Raton, FL, 1999.
16. Lobdell, R., *A Guide to Developing and Re-developing Shoreland Property in New Hampshire: A Blueprint to Help You Live by the Water*, 2nd ed., North Country Resource Conservation and Development Area, Inc., Meredith, NH, 1995.
17. NH Forestry Sustainability Standards Work Team, *Good Forestry in the Granite State: Recommended Voluntary Practices for New Hampshire*, Soc. Protection of NH Forests, Concord, NH, 1997.
18. Deerfield Community Vision Committee, Results of the Deerfield vision survey, Town of Deerfield, NH, 1996.
19. Concerned Citizens for a Safe Environment, Survey of landowners along the Lamprey River, Raymond, NH, 1998.
20. Friends of the Cold River, Cold River area landowners survey, Acworth, NH, 1999a.
21. Friends of the Cold River. 1999b. Cold River corridor survey, Acworth, NH, 1999b.

Response

Buffer strips: More than green eyelashes?

The region where land meets water — the riparian *ecotone* — is known to demand special attention in environmental management. Shoreline buffers can function protectively as the first line of defense for water bodies from land development, as, for example, discussed by Robert France in Chapter II.16 with respect to soil erosion, or for human occupants of floodplains, as, for example, described by Leslie Zucker, Anne Weekes, Mark Vian, and Jay Dorsey in Chapter II.5 in terms of water management. Because shoreline buffers provide not only these functions but also serve as human amenities in green recreational corridors (Charles Flink in Chapter II.3), much attention has been devoted to measuring their size and extent on a landscape scale (Jeffrey Schloss in Chapter II.12 and Robert France, John Felkner, Michael Flaxman, and Robert Rempel in Chapter II.14). This chapter by Frank Mitchell is important in emphasizing the need for bridging, conceptually as well as physically, across spatial scales.

Mitchell repeatedly makes the point that to be truly effective, shoreline buffers must be coupled to the larger landscape of whole watersheds and continuous habitat areas. The importance of this simple precept for the profession of water sensitive planning cannot be overstated. This is particularly the case for concerns about wildlife biodiversity. As Mitchell mentions, though a buffer strip of only 100 ft or so may be sufficient to filter out sediments and entrap contaminants (as, for example, shown by France's work in Chapter II.16), such widths are woefully inadequate for protecting riparian wildlife.

Mitchell's chapter is most informative in its demonstration of how the concepts of landscape ecology such as fragmentation, connectivity, edge effects, patches, nodes, etc. should affect shoreline planning. As he correctly states, one of the greatest challenges in establishing networks of shoreline buffers is in avoiding or minimizing gaps, which can later be filled in with localized restoration efforts. Water sensitive planners must therefore learn how to place their regions of interest and study back into the larger landscape in order to effectively buffer deleterious land-use changes. For example, roads, which Mitchell touches upon briefly in this chapter, may have a negative influence upon riparian wildlife (e.g., directly through roadkills or indirectly through breeding migration interference) even though they may be located many kilometers distant.

Finally, Mitchell's chapter becomes even more significant in his encompassing of the sociology implicit, yet frequently ignored, in shoreline protection. The role of education in fostering initial interest and then in sustaining surveillance is of critical importance. The results from the described public surveys demonstrate the support professed by landowners for both human and environmental (wildlife) benefits accruing from shoreline protection. Water sensitive planners desperately need this sort of information at an early stage in order to justify to the skeptics (who may hold the purse strings) that their intents are obviously in the best interests of both the wildlife *and* the human populations.

chapter II.2

River restoration planning (Connecticut)

James G. MacBroom

Abstract

River restoration refers to the modification of a previously disturbed river's physical, chemical, or biological properties to simulate those of a healthy natural river. This chapter describes several types of river restoration projects with emphasis on multidisciplinary collaboration and public participation.

Introduction

Restoration of degraded rivers is a complex process involving expertise in a broad range of physical and social sciences. Historically, flood control, water supply, water quality, urban waterfronts, hydroelectric power generation, and navigation improvements were treated as independent programs to facilitate the nation's growth and economic development; however, this has resulted in adverse environmental and social impacts. Today, we seek to integrate the separate programs into unified efforts to balance ecological and human needs and to create a sustainable environment. This creates a need for design professionals and regulators to have cross training between the traditional scientific and planning disciplines while maintaining the skills of a specialist. The classic academic programs need to introduce students to the role of interdisciplinary teams that span a broad range of expertise. Specific project types include dechannelization, removal of old dams, creation of linear greenways and trails, instream habitats, daylighting enclosed channels, and the cleanup of abandoned industrial properties.

River restoration projects require a thorough knowledge of watershed processes and their interactions, including hydrology, hydraulics, water quality, fluvial morphology, plus the biological sciences. In addition, developed areas must focus on community-sensitive designs that address land use, demographics, aesthetics, recreational opportunities, public access, historic sites, and linkages to cultural features.

Several trends have reinforced the interest in river restoration in recent years. The first is the water quality improvements since passage of the federal Water Pollution Control Act in 1972. Few people are interested in recreation in polluted streams. As water quality improves, so does public demand for water access, recreation, and aesthetics. A second major factor is the demise of many older 19th-century, water-dependent industries. Many dams built for waterpower or water intakes are no longer actively used and can be removed, and urban riverfronts with crowded industrial facilities that were formerly water-dependent are now available for adaptive reuse.

Human impacts on rivers

Many early communities developed along rivers that provided transportation, water supply, fisheries, waterpower, and irrigation. The clearing of forestland, filling of wetlands, and drainage of agricultural lands decreased groundwater recharge and increased surface runoff even in colonial periods. More recently, urbanization has had a tremendous impact on both surface-water quality and quantity. Impervious areas increase the percentage of precipitation that becomes surface runoff, while paved gutters, channels, and storm drains accelerate the rate at which water moves downstream, causing larger flood flow rates.

The higher peak runoff rates found in heavily disturbed watersheds and urban areas increase water velocities in rivers, leading to increased erosion of the channel bed and banks with a general enlargement of the waterway area. The erosion process disturbs the channel substrate and aquatic habitat, reducing biological diversity and productivity. Downstream areas may receive excessive sediment loads, leading to poor water quality and habitat impacts.

Other human activities have impacted riverine systems by directly altering the natural rivers. Many communities dredged or realigned rivers to channelize their flow for navigation, flood control, or land reclamation purposes. Channels that have been straightened will have steeper slopes with higher velocities that encourage erosion, and the shorter length reduces the size of aquatic habitats. Widening channels for flood control reduce flow velocities, often leading to very slow velocities and shallow base flow depths with subsequent sediment deposition. Channelization may also increase downstream peak flood flows by reducing water storage on floodplains and in wetlands.

Filling channels or floodplains and stabilizing riverbanks with stone, concrete, and retaining walls reduce riparian vegetation and habitat diversity. In addition, artificial linings limit public access and affect aesthetics. Many smaller tributary streams have been enclosed underground, resulting in a loss of all ecological functions (Figure II.2.1).

Watershed management

A river's flow rate and water quality are dependent on the land-use practices within their entire watershed. Consequently, watershed management is an essential part of maintaining healthy productive rivers.

Comprehensive watershed planning involves three separate but interrelated planning levels that must be coordinated (MacBroom, 1998). The watershed includes all the land that naturally drains to the specific point of interest. Its perimeter, called the watershed boundary, marks the divide between one basin and another. Watersheds may range in size from a few acres to thousands of square miles. Small watersheds are generally more sensitive to land-use conditions and changes while large watersheds typically have broad floodplains and channels that attenuate much of the land-use influence. The hydrologic cycle and pollution sources need to be addressed for the entire drainage basin, which includes all lands that contribute runoff to the project site. The influences of topography, soils, vegetation, land uses, and runoff need to be considered.

The intermediate planning level consists of the stream corridors, including natural resources such as floodplains, wetlands, most aquifers, and riparian buffer zones. These areas are essential transitions between upland areas and waterbodies. Riparian corridors carry flood flows, recharge aquifers, filter pollutants from overland flows, reduce water velocities and peak flow rates, and provide wildlife habitat areas. Many riparian corridors have been disturbed by agricultural and urban land uses, plus linear features such as highways, railroads, and dikes that tend to follow rivers and thus isolate riverfront areas.

Chapter II.2: River restoration planning (Connecticut)   381

PROGRESSIVE URBANIZATION OF A STREAM

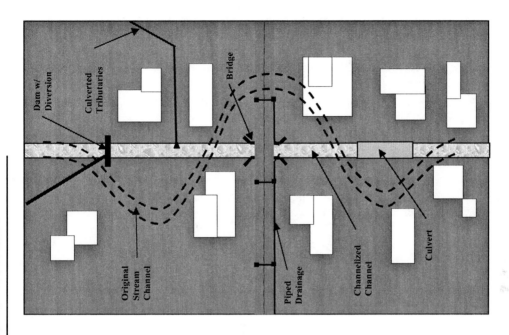

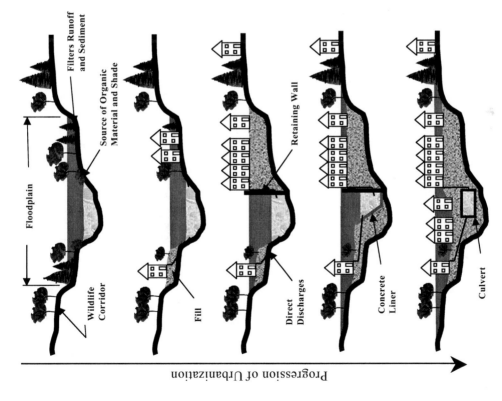

Figure II.2.1

Important elements of urban river corridor planning include their relationship with adjacent neighborhoods, active recreation needs, environmental education, and habitat linkages.

The final planning level focuses on the actual streams and river channels, plus ponds and lakes. It addresses aquatic habitat, channel capacity and stability, fish passage, and water quality. In urban areas, water use, public access, and recreation are also important factors.

Intense land-use activities in urban, industrial, and agricultural areas can produce significant point and nonpoint sources of pollution. The need to address the water quality impacts of combined point and nonpoint sources of pollution is forcing the U.S. EPA and individual states to develop total maximum daily (pollutant) loads for impaired rivers. This places more emphasis on the overall watercourse quality rather than setting effluent discharge limits or specifying the best management practices. Nonpoint sources originate from a broad range of human activities, therefore, a watershed management approach is necessary to evaluate and manage cumulative water quality impacts.

Watershed management is also adaptable for controlling peak flow rates and runoff volumes. This involves land-use and site design criteria to minimize impervious cover, increase infiltration to groundwater, preserve wetlands, and avoid accelerating runoff movement. Low-impact site developments include use of porous paving, grass swale drainage systems, curbless roads to encourage sheet flow, and drywells for roof runoff. The emphasis is on minimizing stormwater runoff at its source rather than trying to collect it and rapidly pass it downstream.

The concept of river restoration has evolved in response to an interest in improving degraded resources. River restoration ranges from urban areas that seek to revive community values, to rural areas that seek to improve aquatic habitats. Even small restoration projects require careful attention to the basic water resource-related sciences, as noted in the examples that follow.

Dam removal

Many of the nation's rivers have dams that were constructed for public, or industrial water supply, waterpower, navigation, and recreation. The United States has an estimated 75,000 significant dams (Maslin and Sicchio, 1999). There is nationwide interest in the removal of old, obsolete dams that are no longer functional. Some have been removed due to physical deterioration of aging structures, others due to the abandonment of 19th-century, water-power-operated mills.

Whereas many dams continue to provide essential public benefits such as storing water supplies or creating hydroelectric power, other dams that are no longer functional can actually have negative impacts. Specific negative impacts due to dams include altering the river's sediment transport, trapping contaminated sediment, reducing flow velocities, decreasing dissolved oxygen levels, inundating riparian wetlands or floodplains, obstructing canoes or kayaks, and, most important, blocking fish movement, which creates fragmented habitats. Dams also affect downstream areas because their discharges have reduced sediment loads that encourage downstream erosion as the river seeks to balance sediment loads versus transport capacity (ASCE, 1997). In addition, older dams that are not properly maintained can create hazardous conditions in the event of a failure.

The impact of dams on fish movements and populations is a major issue in coastal areas with anadromous species, as well as in inland waters. Anadromous fish such as salmon, shad, alewife, and herring spend their adult lives in the ocean but return to swift freshwater rivers to reproduce. Passage facilities, including fish ladders, elevators, and trap and haul operations have had limited success.

At first glance, some may think of dam removal as a simple demolition project, but the actual removal of the structure is often the easy part. The key issue is to create a subsequent channel suitable for fish passage and to control excess sediments. Each species of fish has unique needs in terms of allowable and optimum flow depths and velocities required for passage. The post-dam channel design requires a careful hydrology analysis of seasonal flow rates to identify criteria for mean flows during the migratory runs in addition to flood flows. A hydraulic analysis is required to determine the final channel dimensions and slope for fish passage and flood conveyance, integrated with a geomorphic assessment of river patterns and stable conditions (Figure II.2.2).

One of the critical issues related to dam removal projects is the management of sediments often found in the impoundments. The impoundments trap most of the river's bedload sediments and a portion of the suspended sediments, depending on the ratio of the watershed size to the pond volume and area. Sediments within the impoundments must be explored and decisions made with regard to leaving the sediments in place or removing it. In some cases, sediments may be contaminated and could harm downstream water quality. Sediment management options include removal to off-site disposal areas, stabilization in place, on-site relocation, or allowing gradual erosion to occur.

After dams are breached or removed, the unvegetated soils in their former impoundments may need to be reshaped or landscaped to minimize erosion. The reclaimed land may be used for wetland habitat or parks.

The Naugatuck River restoration project in western Connecticut is an example of a complex, multifaceted program to rehabilitate a formerly polluted urban river. For 150 years, the Naugatuck River Valley supported copper and brass industries dependent on the river for mechanical power, cooling water, rinse water, steam boiler water, and waste assimilation. Most of the industries have either closed or become less water-dependent, enabling communities to reclaim former industrial properties and restore the river. Key elements include an advanced wastewater treatment plant in Waterbury to improve water quality, new sanitary sewer interceptors to prevent overflows, and fish passage at eight abandoned industrial dams for species that can take advantage of improved water quality. The fish passage component of the project includes removal of the Union City, Freight Street, Platts Mill, Chase Brass Dams, and Anaconda Dams, a fish ladder at Brays Buckle Dam, a ramp at the Plume and Atwood Dam, and a bypass channel at the Tingue Dam. Additional project features include improved public access, fish stocking, riverbank clean upstream, tree planting, and modification of sediment bars at tributary stream confluences. The program stakeowners include federal and state agencies, local communities, private businesses and industries, and nonprofit conservation groups (Wildman and MacBroom, 2000). The latter includes Trout Unlimited, the Naugatuck River Watershed Association, and The National Fish and Wildlife Foundation.

Dechannelization

Dechannelization refers to efforts that reverse previous channelization processes. It may include removing deteriorated retaining walls or linings, replanting the river banks, reshaping the channel, providing low-flow channels, etc. All these activities require a thorough evaluation of the river's hydrology and hydraulics. In addition, the channel shape and alignment should conform with geomorphic principles. The inner channel should be proportional to annual peak flow rates, with a supplemental vegetated floodplain to convey excess flows (Brookes, 1988). Figure II.2.3 depicts some of the available design concepts.

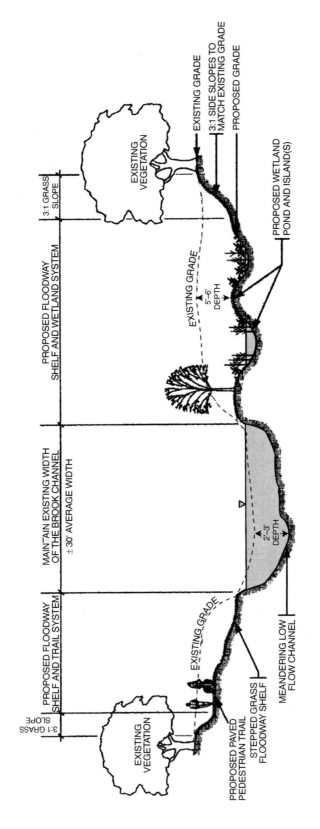

Figure II.2.2

Chapter II.2: River restoration planning (Connecticut)

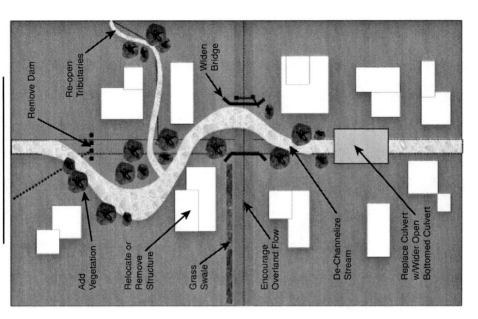

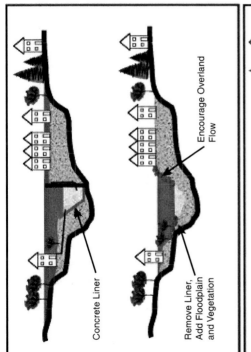

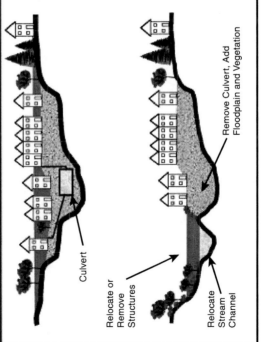

Figure II.2.3

Traditionally, manmade channels have relatively straight alignments with compact, steep-sided cross-sections of a trapezoidal or rectangular shape designed for large flood flows. They often have high flow velocities that require concrete or stone riprap linings for erosion protection. As a result, they have low environmental value due to limited vegetation, poor diversity, lack of in-stream shelter, wide depth and velocity fluctuations, warm-water temperatures, and minimal aeration. Large flood control channels tend to be too wide to concentrate dry weather base flows and provide poor sediment transport.

A strong interaction exists between restored rivers and their communities. This demands that the project team include urban planners and landscape designers with community involvement in the planning process.

Channel daylighting

In extreme cases, communities have totally enclosed watercourses and placed them underground, typically to reclaim floodplain land for development or roads and to control flooding. Enclosed watercourses vary from small streams placed in storm drains to larger rivers that are in conduits or beneath buildings and parking lots. In Hartford, CT, two miles of the Park River is totally underground as part of flood control efforts. An added incentive to bury rivers occurred along polluted areas with obnoxious odors or unsanitary conditions. Enclosing rivers obviously prevents recreation, destroys aquatic habitat, and disconnects people from water resources (Pinkham, 2000).

Among the benefits to opening up previously enclosed rivers are enlarged habitat, water quality, hydraulic capacity, recreation, fish passage, boat passage, and reduction of expensive maintenance of underground structures. Efforts to daylight rivers are often related to urban redevelopment and the creation of open spaces and linear parks that emphasize water resources and public access. Daylighting projects requires extensive community planning participation and a flexible, multidiscipline project team.

The new channel should be designed to provide a viable aquatic habitat and to replicate natural channels. Opportunities are created for urban parks and linear trail systems. Although most documented daylighting projects have been small, several larger urban projects are currently in the planning and design stage.

Riverfront access

Riverfront access in urban areas is often restricted by water-dependent industrial uses or highways and railroads built along river corridors. Other riverfronts have been isolated by flood control dikes or walls. As a result, communities are left with little or no access to rivers that stimulated the community's early growth. Poor access limits aesthetic opportunities, recreation, and reduces a community's interest in environmental conditions.

The Cherry Creek project in Denver, San Antonio River Walk, Baltimore Harbor, and Hartford's Riverfront Recapture are fine examples of adaptive reuse of previously isolated or underused waterfronts in urban areas. Successful projects combine urban planning and landscape design to create public space and areas for special events. Urban waterfronts stimulate economic and recreational activity and provide public access.

The Riverfront Recapture Inc. program was initiated in 1981 to reunite the Connecticut River with its adjacent communities. Flood control dikes, railroad tracks, and an interstate highway had isolated downtown Hartford from its waterfront, blocking physical access and the view of the river. A riverfront master plan was created in 1982, including an ambitious vision of the future that included greenways, parks, trails, boat launches, excursion boats, an amphitheater, a tall-ships wharf, and a plaza. The riverfront project is credited for stimulating other programs, including a new University of Connecticut football stadium (under construction), and the $500 million multiuse Adrian's Landing project with a hotel in addition to residential and retail space. Recent riverfront activities include bass tournaments, triathlon races, festivals, fireworks, concerts, and rowing clinics.

Habitat improvements

Rivers that have been channelized or simply subject to increased flooding due to watershed changes often have degraded aquatic habitat. Specific adverse conditions include reduced diversity, wide, shallow channels, loss of vegetation, limited shelter, poor water quality, few benthic species, rapid flow variations, high sediment loads, and excessive velocities.

The U.S. EPA has recognized that water quality improvements alone cannot restore all streams to the desired swimmable and fishable level without additional habitat and ecosystem help (EPA, 1996).

Many of these problems can be addressed by physical alteration of the channel's shape and form. Among the available techniques and features available to improve aquatic habitat are erosion controls, stormwater quality management, variable channel depths, pools and riffles, flow deflectors, boulder clusters, undercut banks, coarse substrate, and use of an irregular alignment. Aquatic habitats also benefit from replanted riverbanks that shade the water, provide cover, and help to attenuate runoff pollutants.

Flow management

Human activities have often resulted in altering the natural river flow rates due to both land-use activities and the diversion of water for consumptive or irrigation uses. It is not uncommon for urban watersheds to have higher flood flows and lower base flows than natural watersheds. Consumptive water uses that reduce river flow include public water supply, irrigation, and evaporative cooling water. Interbasin water diversions also reduce the river flows necessary to protect aquatic systems, provide recreation, and assimilate wastewater. Short-term river flows can be altered by nonconsumptive uses such as hydroelectric plants and industry.

Flow management is an important part of restoring rivers with regulated discharge rates. There must be sufficient flow to preserve the ecological integrity of riverine systems and maintain water quality. Flow rates must be sufficient to transport fine sediment, avoid excessive deposition, and periodically scour coarse substrates. The area and volume of aquatic habitat are directly proportional to critical flow rates. For most of the country, low flows occur during the summer and often limit aquatic habitats. In-stream flow needs vary on a seasonal basis to correspond with the normal life stages of aquatic species. The management of water quantity and water quality is an integral part in achieving the federal Clean Water Act goal for all rivers to be both fishable and swimmable.

In some watersheds, human activities may intentionally or inadvertently increase dry weather streamflow, helping aquatic habitat, recreation, and aesthetics. For example, in Connecticut's Farmington River watershed, large flood control and water supply reservoirs retain excess wet-weather runoff in the winter and during storms, then slowly release water in the summer to augment natural runoff. The flow release pattern was developed via a detailed instream flow study involving the interested parties, led by the National Park Service, Connecticut Department of Environmental Protection, Metropolitan District Commission, and Farmington River Watershed Association. The augmented flows enhance recreational activities including white-water boating, fishing, and aesthetics.

Geomorphic considerations

The naturalistic approach being applied in many river restoration projects would not be possible without key scientific advances over the past 20 years. The evolution of river restoration sciences is shown on Figure II.2.4. It is recognized that the alignment, slope, and cross-sectional shape of natural rivers have evolved in response to their flow rates and sediment loads and that aquatic ecosystems of plants and animals that live in or adjacent to rivers have adjusted to these conditions. Our improved ability to restore rivers in recent years is due to the gradual merger of hydraulic engineering and fluvial morphology. These disciplines historically evolved independent of each other along two separate tracks: hydraulic engineering focused on numeric analysis of hydrodynamic systems and structures, while fluvial morphology developed from geologists observing natural systems. This allows for increased quantification of river mechanics, which can be applied to forecast fluvial behavior and responses.

The geomorphic design process is similar for dam removal, daylighting, dechannelization, and habitat improvement projects. It begins with a thorough assessment of existing hydrologic and geologic conditions, including the characteristics in nearby stable channels. One then has to identify the preferred natural channel slope, alignment, and cross-section parameters for the project site. If the preferred natural conditions cannot be provided, then other variables must be altered (slope, alignment, sediment, etc.) to create alternate conditions.

Conventional hydraulic analysis of open channels defines the flow capacity for a given size, or conversely determines a channel size for a stipulated flow rate. For any given flow rate, however, many different combinations of channel width, depth, slope, roughness, and shape are possible. Consequently, supplemental techniques are necessary to find the optimal geometric proportions to simulate natural channels.

Several approaches are used to assess stable conditions for unlined channels. The empirical regime relationships provide general guidance but have been criticized for their incomplete consideration of physical processes. The hydraulic geometry equations of stable channel width, depth, and slope are based on plots of empirical field data that are best used on a regional basis. The theoretical tractive stress approach and sediment transport analysis are available for technical evaluation of stream stability and equilibrium conditions. They provide guidance on channel slopes versus cross-sectional shapes, but do not address channel shape or patterns. The author has applied a combination of the empirical regional regime data and theoretical analysis in the design of channels that simulate alluvial rivers, in order to minimize the use of rigid channel linings and provide habitat opportunities.

The first technique consists of locating a stable channel, upstream or downstream of the subject area, that can be used as a model to determine channel widths, depths, and slopes for the restored channel segment. It requires that the duplicate channel have similar discharge rates and geology and that the surrogate channel be in equilibrium. For these reasons, it is difficult to apply where rapid watershed changes are occurring.

The second technique is based on the empirical field data on channel geometry measurements collected by the U.S. Geological Survey and others beginning in the 1950s (Leopold, 1994). The studies found that within a specific climatic zone, the width, depth, and slope of stable natural channels were proportional to their bankfull flow capacity and that the bankfull capacity was equal to flood flow with an average recurrence frequency of approximately 1.5 years. Consequently, by determining a channel's flow, one could then determine optimum channel proportions to simulate natural conditions. This is a powerful tool for use with alluvial channels in sedimentary floodplains but is less valid for degrading or aggrading channels. It is also unsuitable for watersheds undergoing rapid hydrologic changes resulting from urbanization.

Chapter II.2: River restoration planning (Connecticut)

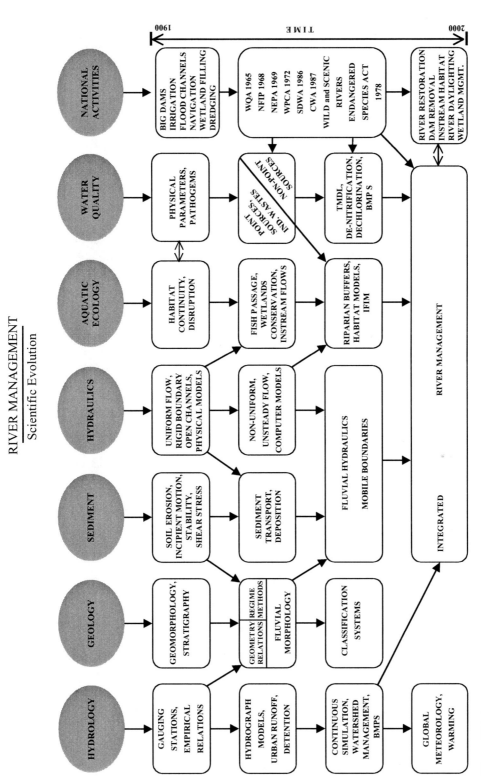

Figure II.2.4

Whereas the first two techniques are empirical, the final approach to channel restoration is based on engineering theories for substrate stability and sediment transport. The basic concept for providing a channel in equilibrium is that a river's ability to transport sediment will be equal to the sediment inflow rate to a given segment. Mathematical techniques vary from simple allowable velocity and shear stress analysis to one or two dimensional computer models (Yang, 1996).

Using the previously described techniques, we are learning how to recreate naturalistic channels and to forecast the impact of future watershed conditions.

Public participation

River restoration has broad public support and is often initiated by local community groups or nonprofit conservation organizations working in conjunction with government agencies. Public participation is a key element in creating successful projects that form a linkage between community needs and our environment.

Public awareness of environmental and community needs plus local knowledge of restoration opportunities help to generate political support for funding, land acquisition, and regulatory permitting. In some cases, private funds are raised to prepare master plans for grant applications and cost estimates, and private funds can be used as a matching contribution to generate more funds.

Other areas of public interest that can be part of river restoration programs in urban areas include preservation of historic sites and buildings, environmental education, job creation, and youth programs.

A typical project flow chart (Figure II.2.5) would include public participation at several levels. Early participation helps to scope the project and establish local goals and objectives, while later participation helps to define project alternatives and priorities.

Harbor Brook restoration project — A river restoration example

Meriden, CT has embarked upon an ambitious program to restore 4 miles of Harbor Brook. Harbor Brook is a post-industrial river with an urban watershed of 12.3 square miles. It has been dammed, diked, diverted, polluted, partially filled, and partially enclosed over the past 200 years. It is prone to frequent damaging floods, chronic bank erosion, and variable water quality and provides no recreational uses.

The proposed project was stimulated by severe downtown flooding in 1982, 1992, and 1996, causing millions of dollars of damages, plus a strong desire for removal of derelict riverfront factory buildings and reviving the central city. Although initiated due to flood control concerns, the project has evolved into a multifaceted river restoration program. Key components of the $30 million project include:

- Acquire open space.
- Detain and treat stormwater at four sites.
- Replace nine undersized, deteriorated bridges over the river.
- Remove seven bridges without replacement.
- Open up and "daylight" two enclosed conduits, creating a downtown park.
- Restore open channel, with floodway and instream habitat.
- Create a linear park and trail system along the river.
- Improve instream habitat with pools, riffles, coarse substrate.
- Remove old channel retaining walls, reduce bank slopes.
- Reduce bank slopes.
- Replant groundcover and overhead canopy trees.

Chapter II.2: River restoration planning (Connecticut)

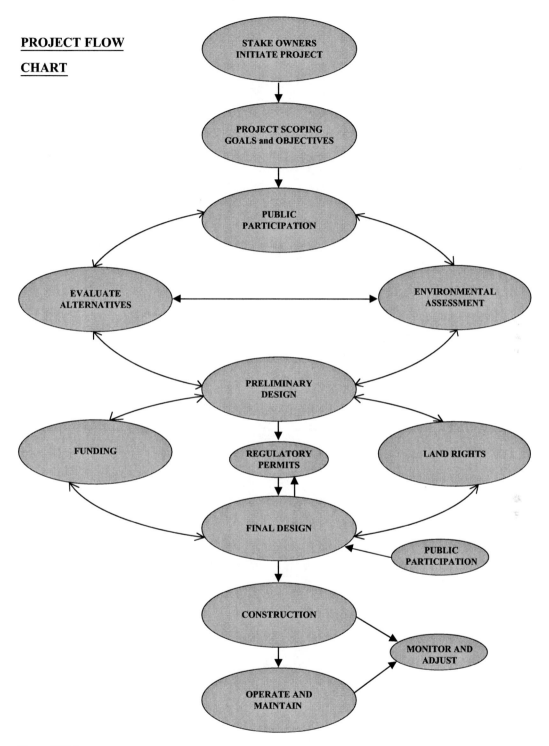

Figure II.2.5

As part of the project, a partially vacant shopping plaza and parking lot that were built over the river in the 1960s will be demolished to allow "daylighting" over half a mile of enclosed river. The site will become a city park in the downtown area, creating a village green (Figure II.2.6).

Figure II.2.6

The upstream stormwater detention sites are being proposed along tributary streams to reduce peak flows and redistribute runoff, preventing the project from having impact upon downstream areas. In addition, land-use policies will reemphasize stormwater infiltration and runoff quality controls plus stormwater treatment to improve water quality. Wetlands will be created in previously filled floodplain areas.

The project planning was guided by an appointed Flood Control Implementation Agency that included the City Engineer, City Council officials, and members of the public. The agency coordinated with city planners, the City Planning Commission, Wetlands Agency, Conservation Commission, nonprofit watershed association, downstream city of Wallingford, plus state and federal agencies.

Literature cited

ASCE Task Committee, *Guidelines for the Retirement of Dams and Hydroelectric Facilities*, American Society of Civil Engineers, New York, 1997.

Brookes, A., *Channelized Rivers*, John Wiley & Sons, New York, 1988.

Environmental Protection Agency (EPA), *Ecological Restoration: A Tool to Manage Stream Quality*, Office of Water, Washington, DC, 1996.

EPA, *Stream Channel Restoration: Principals, Processes, and Practices*, Federal Interagency Stream Restoration Working Group, Washington, DC, 1998.

Leopold, L.B., *A View of the River*, Harvard Univ. Press, Cambridge, MA, 1994.

MacBroom, J.G., *The River Book*, CT Dept. Environ. Protection Bulletin #28, Hartford, CT, 1998.

Maslin, E. and Sicchio, M., Ed., *Dam Removal Success Stories*, Friends of the Earth American Rivers and Trout Unlimited, Washington, DC, 1999.

Pinkham, R., *Daylighting, New Life for Buried Streams*, Rocky Mountain Institute, Old Snowmass, CO, 2000.

Wildman, L. and MacBroom, J.G., Dam removal, a tool for river restoration on the Naugatuck River, *Proc. ASCE Water Resourc. Eng. Conf.*, Minneapolis, 2000.

Yang, C.T., *Sediment Transport, Theory and Practice*, McGraw-Hill, New York, 1996.

Response

Water quality improvements are not enough

River restoration, James MacBroom considers, is a complex process involving physical and social sciences directed toward seeking balanced approaches to ecological and human needs in a sustainable environment. Water quality improvements alone are not enough to restore rivers due to the host of physical degradations that have been wrought upon them. Dechannelization, dam removal, daylighting, riparian brownfield cleanup and redevelopment, in-stream habitat creation, and greenway establishment are all factors that should be addressed in engaging in water sensitive planning along river corridors.

MacBroom believes that the planning and implementation of river restoration projects will not succeed unless four tenets are met:

1. All aquatic elements in the watershed are considered such as floodplains, wetlands, and aquifers, in addition to rivers and streams.
2. Public participation is sought at an early stage to form a conceptual linkage between the needs of riparian communities and those of the environment, such that, through time and the planning process, this will develop into physical linkages including riverfront access and greenway trail networks, as discussed in Chapter II.3 by Charles Flink.
3. River management professionals require cross training between the traditional scientific and the planning disciplines while maintaining the skills of the specialist.
4. All participants in the process, from the public to the planning professionals, need to have, to varying degrees, education and understanding about fluvial processes and morphology, as discussed in Chapter II.5 by Leslie Zucker, Anne Weekes, Mark Vian, and Jay Dorsey.

chapter II.3

Greenways as green infrastructure in the new millennium

Charles A. Flink

Abstract

Greenways are not a new land-use concept in the United States; they have, in fact, been in existence since the mid-1800s (Little, 1990). The modern American greenway landscape has taken on a new and important function. Recognized for years as a recreational amenity, greenways have emerged in the 1990s as a multiobjective land use that can satisfy a variety of concerns facing urbanizing America. Greenways are now viewed as an important element of "green infrastructure." The term "infrastructure" is fairly well known throughout the United States and has historically been used to describe things that humans have created to "improve the qualify of life," such as sewers, roads, highways, and utilities. Green infrastructure identifies the natural elements of the land that also serve to define the quality of life. The streams, lakes, native vegetation, landforms, animal life, and geologic formations of an area define a sense of place, community character, and livability of American towns and cities. Greenways have successfully been used as a method for cataloging these unique community resources and setting in motion the strategies necessary to preserve, enhance, and manage these resources.

This chapter profiles a variety of greenway projects and programs from throughout the nation and define how greenways serve as resources for floodplain management, water quality protection, alternative transportation, recreational resource, and open-space stewardship.

Introduction

As we begin a new millennium, the United States finds itself at a critical juncture point. More than 100 years of sustained growth and development across a vast, hospitable continent has resulted in a nation that is the envy of most modern societies. This growth has been made possible by virtue of a comprehensive and integrated system of infrastructure. When judged by its size, level of service, delivery to even the most remote locations, and accessibility, infrastructure within the United States has served to define our standard of living as one of the highest in the world; however, there is a flipside to our manmade infrastructure superiority. The resulting damage that has occurred to both our natural and social infrastructure has severely impaired numerous ecological systems and destroyed the character of our communities.

Figure II.3.1 Raleigh greenway. This natural trail winds along Crabtree Creek in Raleigh, NC, and is an example of a greenway corridor that serves multiple benefits as green infrastructure. (Photo by Charles Flink.)

One of the greatest challenges facing the United States in the next 25 years will be to conserve, manage, and modernize our nation's infrastructure to meet the future needs of a growing society. In order for this to be accomplished in a realistic and cost-effective manner, fundamental changes will need to occur in the way that we view the purpose, function, and operation of infrastructure. As a nation, we need to redefine infrastructure more broadly to include the critical interrelationships that exist between man and nature.

As part of this redefinition, we need to promote land uses and systems that enable us to best utilize our limited resources. Our society is in need of exemplary models of land use that extol the virtues of preservation and conservation, promote energy efficiency, encourage community action, and provide multiple benefits. Greenways offer us such a model and can become an integral part of our natural, manmade, and social infrastructure (Figure II.3.1).

Infrastructure, by definition

According to *The American Heritage Dictionary of the English Language* (1981), infrastructure is defined as "the basic installations and facilities on which the continued growth of a community, state, etc. depend, (such) as roads, schools, power plants, transportation and communication systems." In order to more clearly define the full impact of infrastructure, and the essential interrelationship between man and nature that impacts infrastructure, I suggest that three important types or divisions of infrastructure deserve consideration: manmade, natural, and social.

> *Manmade infrastructure* is the type that most people are familiar with and includes utility systems, such as electrical generation, manufacturing systems, potable water, graywater treatment, solid waste disposal, stormwater discharge, and telecommunications transmission; transportation systems, such as roadways, highways, bridges, mass transit service, and all other land- and water-based travel ways.
> *Natural infrastructure* is the naturally occurring physical foundation of the world, which includes major components of our planet's ecological systems, such as soil, water, air, and vegetation.

Social infrastructure consists of the political systems, religious beliefs, moral values, laws, conduct, and livelihood that we abide by as individuals and that collectively forms the spiritual foundation of our communities and societies.

If you accept the definition for these three types of infrastructure, then you might also agree that there has been conflict among these divisions for thousands of years. It is the relationship between these divisions that ultimately defines man's quality of life and shapes our common future. It has been proven by Lynch (1962) and McHarg (1969) that achieving balance and harmony between these divisions offers man an opportunity to enjoy a higher standard of living. But balance is difficult to achieve even under controlled conditions.

Gilbert Grosvenor, past president of the National Geographic Society, concluded (1990) that "the forces of man are now of such immense power, (that) they can threaten the very forces of nature, the very physical forces that make up the Earth itself." Perhaps for the first time in the history of humankind, it is fully understood that collectively, our species of 6 billion is capable of tipping the balance between the three divisions of infrastructure. As worldwide population continues to grow at an exponential rate, the conflict between these divisions becomes more evident. It is very possible that the world's entire standard of living could be dramatically altered and degraded if we fail to change the methods by which we develop and manage the interrelationships of infrastructure.

Infrastructure in the United States

It is worthwhile to briefly consider the plight of the three types of infrastructure within the context of the United States, partly because many societies of the world measure their standard of living by the quality of life Americans have enjoyed for more than 100 years. To a large degree, the success of this nation has been built around the world's most comprehensive, accessible, and integrated system of infrastructure.

Our human-made infrastructure provides us with a network of roadways and highways that span the continent, linking urban and rural areas from coast to coast; fiber optic lines that transmit voice and data from the most remote areas of New Mexico to the burgeoning metropolis of New York City in a matter of seconds; water management systems that reroute the natural drainage of the Rocky Mountains to desert areas of California and Arizona, where abundant supplies of fruits and vegetables are grown to feed our population. Our green infrastructure has supported this nation's steady growth and development, providing ample timber to build homes, minerals and raw materials to supply factories and manufacturing centers, abundant fertile soil to grow crops, and a seemingly endless supply of fresh water to quench our thirst. Our social infrastructure, built on democratic principles, has supported the land of the free for more than 200 years. We offer safe haven to the oppressed, religious freedom to the disenfranchised, equal opportunity to the minority, and the ability for an individual to participate in the governing of our communities.

Despite these heralded successes, the United States faces daunting infrastructure problems that threaten our standard of living and economic future. For example, Americans lose more than 1.5 billion hours of productive time each year stuck in traffic jams because our urban mobility has been severely impaired (Surface Transportation Policy Project, 1991). Drinking water systems in major metropolitan areas, such as Milwaukee and New York City, are aging and over capacity, such that the delivery of potable water to millions of consumers is no longer taken for granted. Thousands of miles of rivers are classified as degraded and severely threatened by urbanization, poor agricultural practices, and illegal dumping of waste products. The below-cost deforestation of our National

Forests not only destroys a significant national resource, but also threatens thousands of species of plants and animals. Our nation's health care system is spiraling out of control, suffering as much from an unhealthy populous as from the greed of the medical industry. Civil and ethnic dissatisfaction has degraded the quality of life for inner-city neighborhoods in Los Angeles, Detroit, and Miami. Finally, the American family, pillar of social structure, has been weakened during the past 40 years by a variety of competing interests, not the least of which has been a lack of quality-oriented time to enjoy with each other the pursuits of life.

How did the United States arrive at this juncture?

For the most part, we have developed our human-made infrastructure with little respect or concern for the natural infrastructure. We have changed the function of entire ecological systems to lessen the impact of *potential* natural hazards. For example, expensive flood control projects have dramatically altered vast, complex upstream and downstream ecosystems, while benefiting only a few local residents. We have used a manufacturing-based philosophy to exploit natural resources without first understanding the full function of these resources. As one example, the stormwater management systems of experimental superfarms in eastern North Carolina significantly altered the salinity of the second-largest estuary on the continent. Combined with poorly managed fishing practices, this indiscriminate change to the infrastructure of a land-based ecological system has severely crippled an entire water-based food chain and abundant fishery.

Our human-made infrastructure systems have historically been developed as individual units, with little regard for *multiple use* of land and water resources. Because we have always viewed this continent as bountiful and capable of supporting individual pursuits, little effort has been made to coordinate the development of transcontinental, regional, or local infrastructure systems. These systems are now so vast, uncoordinated, and old that it takes supercomputers to manage their operation and delivery. Today, minor problems in these systems now cause major interruptions of service to increasing numbers of consumers.

Furthermore, we have historically destroyed the social fabric of local communities in order to satisfy straight-line, high-speed, human-made infrastructure systems. This has been accomplished under the heading of progress. In other cases, to improve the efficiency of an infrastructure system, we have rerouted its path of travel around areas of congestion, eradicating the unique charm and indigenous qualities of local communities; thus, we have created a monotonous landscape whose character can be found repeatedly within every town and city in the nation.

As the American landscape continues to change from rural to urban, we realize that land uses have been encouraged during the past 50 years that are not compatible with an optimal and efficiently managed infrastructure system. Looking to the future, we are forced to encourage the redevelopment of what is now low-intensity land-use areas into higher densities, in order to make more efficient use of existing infrastructure.

Additionally, in a capitalistic society, the following question must always be asked: How much will it cost to modernize and maintain our existing human-made infrastructure, as well as develop new systems to meet the demands of urbanizing America? Some infrastructure specialists place the cost somewhere in the hundreds of billions of dollars, a price tag that may be out of reach, irrespective of a healthy world economy.

Have we reached the point of no return? Are there opportunities to change course? I suggest that one emerging land-use concept offers Americans hope for our future. Greenways, often regarded as recreation based corridors or linear parks, is a land use concept that has been evolving during the past 20 years and is presently being implemented to

Greenways as infrastructure

Greenways are multiple-use corridors that respect the inherent qualities of natural systems and accommodate manmade systems in a way that is compatible with nature. Greenways conserve open space, protecting riverine systems, vegetation, native soil, and geomorphology that is essential to the health and vitality of the local landscape. They soften patterns of urban growth, offering green vegetated buffers between land uses that can effectively cool the air mass, deflect and absorb objectionable noises, and block unsightly views (Flink and Searns, 1993).

Well-planned greenway systems have been proven to offer efficient means of alternative transportation for community residents. An integrated trail system provides choice in the mode of travel between popular points of origin and destination. Greenways offer a way to protect our nation's natural and cultural heritage, promoting eco-tourism, protecting historic settlements, buildings and landscapes, and offering opportunity for self-interpretation and education. They have also been shown to promote economic development by increasing the value of adjacent property, offering improved access to community resources, and improving the quality of life for local residents.

By protecting the basic foundation of natural infrastructure, vegetation within greenways can cleanse the air of pollution, and absorb overland runoff to minimize pollution to our creeks, streams, and lakes. They can mitigate the effects of urban flooding by protecting flood-prone areas from encroachment and development. Greenways serve as important corridors for wildlife, protecting habitat, offering sanctuary during migration, and promoting biological diversity.

Successful greenways promote goodwill throughout a community. They offer access to landscape resources close to where people live and work. They provide an outlet for community activism and have become the new "main street" for many towns and cities throughout the United States.

Greenways offer communities with an opportunity to establish infrastructure that is based on the local abilities of ecological systems. This linear system, when planned and developed properly, provides community residents with safe, efficient access and service to a variety of natural resources, much like a roadway network, water supply lines, or electrical system. Under ideal circumstances, communities would plan for greenways in the same manner.

Greenways have been shown to have a positive impact on the different types of infrastructure. A few specific examples are described next.

Transportation

Traditionally, our transportation system, a component of human-made infrastructure, has been measured by mode of travel, number of highway miles, automobile carrying capacity, and speed of travel. If transportation systems were instead measured on the basis of their impact on nonrenewable resources, air quality, energy efficiency, and urban mobility, then the resulting end products — the travel way, route, and location — would have a different look and function.

Greenways offer an opportunity to transform traditional transportation design, development, and management. They do so first by accepting the fact that mobility is of prime importance in defining urban transportation systems and by offering a choice in the mode of travel. Second, energy efficiency is absolutely critical to our future survival. George

Figure II.3.2 Greenway cyclists. A grandfather and grandson enjoy the Trolley Line Greenway in downtown Kansas City, MO. Thousands of miles of greenway trails serve the transportation needs of Americans daily. (Photo by Charles Flink.)

Woodwell, director of the Woods Hole Research Center in Woods Hole, MA, observed that "from the perspective of a biologist, it appears urgent to recognize that the era of fossil fuel has passed" (Hay, 1990). Yet, as we evaluate our nation's current transportation system, we realize that nearly 50% of all oil consumed is for automobile travel (Surface Transportation Policy Project, 1991).

In fact, greenways are a reasonable choice of travel for trips under 5 miles and might be the preferred choice for many trips under 2 miles if the infrastructure were in place today (Flink and Musser, 1992) (Figure II.3.2). Several communities are enjoying the benefit of having comprehensive and interconnected greenways as one component of the local transportation network. Seattle, WA, Portland, OR, Davis, CA, Denver, CO, Madison, WI, and Raleigh, NC, have come to rely on their greenway systems as one method for providing a choice in travel. Offering a choice in travel may be the most effective method for reducing traffic congestion, improving air quality, and promoting a more healthy lifestyle. The Transportation Efficiency Act for the Twenty First Century (TEA-21) emphasizes multimodalism as the future goal for our nation's transportation infrastructure. Because greenways are built to be fully accessible to all persons regardless of ability, they satisfy the concern regarding mobility. Greenways offer nonpolluting, nonoil-consumptive, clean, and green travel ways throughout our communities.

Water quality management

Protecting our potable water supply is dependent on mitigating the effects of nonpoint source pollution — which in urban areas is discharged through our stormwater management systems. For the past 75 years, stormwater management has largely been associated with street system drainage and the detention of post-development exit flows from intensive land-use development. The solutions to these drainage concerns have been based on calculating how much water is capable of flowing through a circular pipe — otherwise defined as "hard engineering." As part of revisions to the Clean Water Act in 1990, the

Figure II.3.3 Denver flood channel/greenway, Denver, CO, has transformed several of its urban streams into multiobjective stormwater management, transportation, and recreational corridors. (Photo by Charles Flink.)

EPA's new national water quality program is aimed at rethinking the hard engineering approach to solving stormwater management. Promoting a "soft engineering" approach, EPA's Nonpoint Discharge Elimination System guidelines radically change the way exit flows are discharged and accommodated within downstream sites (Flink and Musser, 1993). In the future, stormwater will have to be absorbed within floodplain lands, or through Best Management Practices, rather than being discharged directly into flowing streams (Figure II.3.3).

Greenways as a component of stormwater management infrastructure provide a mechanism for protecting floodplain lands and establish areas where floodwaters can be absorbed naturally. Denver's South Platte River Greenway system and its tributary streams are an early example of the soft engineering approach to stormwater management. The greenway is an inexpensive alternative to a proposed $600 million concrete-lined channel flood control project originally offered by the Army Corps of Engineers. The greenway project cost $18 million to construct, resulting in a savings of $582 million from the original Corps proposal. It not only provides flood protection for lands along the South Platte River and its tributaries, but also an interconnected 40-mile-long system of multiuse trails that serve as access points for maintenance of the riverine system. The greenway provides residents with access to the flowing waters of the Platte and safeguards the city's urban core from flooding. In addition, Denver earns bonus points from the Federal Emergency Management Agency, National Flood Insurance Program, under its Community Rating System, which awards communities with lower insurance rates for protection of floodplain lands.

Other communities are following Denver's lead and beginning to develop comprehensive greenway systems where the primary function is flood control and water quality management. The Louisville–Jefferson County Metropolitan Sewer District is one such community where a comprehensive effort is underway to revise the traditional approach to channelization of natural streams. The community expects to implement a multiobjective approach to stream corridor management, which will serve to protect natural habitat, improve the capacity of native streams, and offer an interconnected system of off-road trails for recreation and alternative transportation.

Figure II.3.4 The Swift Creek Recycled Greenway, Cary, NC, is a public–private partnership project that demonstrates the creative use of waste materials and trash in the construction of public-use facilities. (Photo by Charles Flink.)

Recycling of solid waste

The town of Cary, NC, became one of the first communities in the nation to develop a greenway facility using the principles and products of recycling. Faced with a diminishing capacity in the local landfill, and a successful community-wide recycling program, the town accepted an invitation from Greenways Incorporated, a local environmental and landscape architecture consulting firm, to develop the Swift Creek Greenway using recycled materials. A unique public–private partnership was established between the town and the firm, which included more than 22 other public and private-sector organizations. A one-mile pilot project was envisioned that would demonstrate how typical household solid waste could be reused to create essential facilities for the greenway, including bridges, benches, trash receptacles, signage, and safety railing. The hard-surfaced trail was developed using bottom ash from a local coal-fired electrical generation plant. Recycled asphalt and shredded rubber tires were used to develop the surface of the trail tread (Figure II.3.4).

The greenway demonstrates that trash and debris can no longer be thought of as disposable items. Instead, they are new resources for communities all across the United States who face similar problems regarding waste management. Further, a slogan developed for the project, "Buy Recycled. It's Second Nature," offers a needed boost to the recycling industry, which has suffered in recent years due to a lack of appropriate product marketing. The project clearly demonstrates that recycled products can achieve equal or greater performance than similar products made from natural resources. In addition to the stated project goals, the greenway has an environmental education component and serves to link local neighborhoods with a large office park complex, providing an efficient alternative route for travel from home to work. The project received a Take Pride in America Award for its accomplishments and is serving to enhance Cary's reputation as a progressive and livable community.

Figure II.3.5 The Stones River Greenway, Murfreesboro, TN, is an example of a greenway that offers educational opportunities. The greenway trail links together Civil War battlefield sites and documents one of the most important battles in American history. (Photo by Charles Flink.)

Environmental education

The city of Murfreesboro, TN, has a storied if not dubious place in American history. Site of the one of the bloodiest battles in the Civil War, the town has struggled with its identity and future as a bedroom community for metropolitan Nashville. With assistance from Congressman Bart Gordon and the National Park Service, the community decided to link together the resources associated with the Stones River National Battlefield and develop a multipurpose greenway that will highlight the community's unique history. One of the primary motivations behind the project is to increase tourism and economic development for the community through the development of a riverside greenway. Education is the other principal pursuit of the project (Figure II.3.5).

Community leaders developed a series of wayside exhibits that provide greenway users with opportunities to learn more about the Battle for Stones River. Additionally, the local landscape has been interpreted so that residents and visitors can better understand the unique natural features of Stones River and middle Tennessee. The project links the downtown area to suburban neighborhoods and has become the first off-road bicycle facility within the community, offering an alternative route of transportation. Local greenway advocates firmly believe that the greenway has improved the quality of life and standard of living for residents of the community.

Improving social conditions

In the classic work *Design with Nature*, Ian McHarg states (1969), "We need not only a better view of man and nature, but a working method by which the least of us can ensure

Figure II.3.6 Greenways can be a deterrent to crime. In downtown Nashville, TN, the Shelby Safewalk has helped transform an urban neighborhood. Similar projects can now be found in Chattanooga, TN, and Miami, FL. (Photo by Charles Flink.)

that the product of his works is not more despoliation." Several greenway projects throughout the United States have served to reverse poor land-use practices and enhance local community pride. The Brooklyn–Queens Greenway is a remarkable effort to remove concrete and asphalt in downtown New York City so that trees can be installed, and urban residents will be encouraged to walk and bike from neighborhood to neighborhood. In Nashville, TN, and Greensboro, NC, two economically depressed neighborhoods have embraced the greenway concept as a way of restoring social order and quality of life to degraded urban conditions. Devoid of significant natural features, these communities have rallied around the concept of multiple-use greenways as a method for instilling pride and ownership in the neighborhood landscape (Figure II.3.6).

The typical urban trappings — concrete sidewalks, telephone poles, overhead utility wires, vacant lots, and abused vegetation — have been reorganized and replaced with decorative paving, human-scale street lights, trees, pocket parks, and flowering plants. Project implementation involves local participation in the demolition and construction of the greenway. From this involvement, local residents have become stakeholders in the future success of the greenway and will hopefully pass along their pride and accomplishment to future residents of the community. In addition, restoration of the designated greenway corridor offers local residents with a viable choice in travel and improves the efficiency of utility systems throughout the community.

Conclusion

The future development of our infrastructure will determine to the greatest extent the impact that civilization has on our planet's natural environment. We have clear and relatively easy choices to make. We can choose to modify the way in which we define, plan, develop, and manage our infrastructure, or we can continue to promote infrastructure development that will eventually degrade our social and economic future and destroy the ecological systems necessary to sustain life on the planet.

Since the release of the President's Commission on Americans Outdoors Report in 1987, our nation has narrowly defined greenways principally as a recreational amenity for our communities (Americans Outdoors, 1987). I suggest that the most significant attribute of greenways is not their recreational value, but their potential as a land use that successfully accommodates multiple environmental and social functions. As a component of infrastructure, greenways can provide our communities with a land resource that successfully balances the needs of man with the abilities of the nature.

As we face the next 25, 50, and even 100 years, it is very appropriate that we take time to rethink traditional forms of land use, especially with regard to infrastructure development. I believe that multiple use of linear land corridors will be one key to accommodating our future physical growth. At the heart of this multiple-use concept, I feel that greenways will serve as the foundation within which other infrastructure components can and should be developed.

Literature cited

Acting in the national interest: The transportation agenda, Surface Transportation Policy Project, Washington, DC, 1991.

Americans Outdoors: The Legacy, The Challenge, The Report of the President's Commission, Island Press, Washington, DC, 1987.

Flink, C. and Searns, R., *Greenways, a Practical Guide to Planning, Design and Management*, Island Press, Washington, DC, 1993.

Flink, C. and Musser, T., Current planning guidelines and design standards being used by state and local agencies for bicycle and pedestrian facilities, working paper for the U.S. Dept. Trans., Fed. Hwy. Admin., Washington, DC, 1992.

Flink, C. and Musser, T., A nationwide inventory of stream corridor planning programs, working paper for the Louisville–Jefferson County metropolitan sewer district, Louisville, KY, 1993.

Grove, N., Greenways paths to the future, *National Geographic*, 177, 76, 1990.

Hay, K.G., Using natural gas rights-of-way as greenways: A feasibility study, The Conservation Fund, Washington, DC, 1990.

Little, C.E., *Greenways for America*, The Johns Hopkins Univ. Press, Baltimore, MD, 1990.

Lynch, K., *Site Planning*, MIT, Cambridge, MA, 1962.

McHarg, I.L., *Design with Nature*, Natural History Press, Garden City, NY, 1969.

Response

Corridors that integrate natural, societal, and social elements

As Charles Flink discusses in this chapter, water sensitive planning should transcend consideration of mere physical spaces. In particular, greenways can very much be utilized to help define and instill a quality of life at the same time as shaping a sustainable future. This is similar to the broad aspirations characterizing comprehensive, regional planning as Daniel Williams describes in Chapter II.11.

Greenways, once regarded as only recreational amenities, are now recognized to provide many more benefits to communities, functioning as sort of green "main streets." Such systems, Flink argues, represent models for integrating human and environmental objectives in land-use planning that can come together in concepts of open-space stewardship. Not only does modern greenway planning address the capability of these linear parks to ably serve their original resource purpose for humans in terms of tourism, protection of historic settlements, alternative transportation areas, and increased nearby property values, such planning also considers these river borders with respect to their roles in floodplain management, water quality protection, and wildlife habitat preservation, as discussed elsewhere in this book (Frank Mitchell in Chapter II.1; James MacBroom in Chapter II.2; and Leslie Zucker, Anne Weekes, Mark Vian, and Jay Dorsey in Chapter II.5).

chapter II.4

Natural resource stewardship planning and design: Fresh Pond Reservation (Massachusetts)

Thomas S. Benjamin

Abstract

Fresh Pond Reservation represents one of metropolitan Boston's premier public open spaces. The reservation surrounds Fresh Pond Reservoir, the terminus of the city of Cambridge's water supply. As such, Cambridge places a high value on protecting the reservoir's water quality. Originally designed by Olmsted and Eliot in the 1890s, the reservation is now aging under heavy recreational use, with erosion from hillsides and pathways threatening the reservoir's water quality. Further, ecological values, public safety, and aesthetics are now also threatened as diverse vegetation types and zones become overrun by a few invasive vine and shrub species, their proliferation related to ongoing soil disturbance.

The *Fresh Pond Reservation master plan* (adopted January 2001) provides long-term guidance for resource management and enhancement as the reservation moves into its second century. The plan's goals are now being realized through a number of projects, the most significant of which is a comprehensive Landscape Management Plan to control soil erosion and invasive species. Two other major infrastructure projects currently under construction at the also reflect the goals of the plan: (1) Cambridge's new water purification facility; and (2) a sewer separation/bikeway project. Additional projects to manifest the plan's vision include a master plan for landscape and drainage improvements at the reservation's golf course, landscape improvements surrounding a new assisted-living facility on the premises, and possible naturalization of a former soccer field currently in use as a construction staging area. The last two projects have been incorporated into one Northeast Sector Landscape Improvements Project covering approximately 25 acres at Fresh Pond. Finally, two pilot projects have emerged from the plan as well — the Perimeter Road Alternative Paving Pilot Study and the Reservoir Fencing Alternatives Pilot Study.

Introduction

Many of our nation's parklands and prized urban open spaces face a common challenge — how to balance water quality and natural resource protection with the diverse needs of the public. This challenge is being faced internationally, as well, as population pressures increase on ever more scarce open space and water resources. Fresh Pond Reservation in

Cambridge, MA, is no exception. Here, the connection between water quality protection and wise stewardship of the landscape in which water resources reside, or the natural interaction between land and water resources, has been broadly realized in a major forward-looking planning effort that is now being implemented. With a little luck, the anticipated success of this effort will provide an international model for natural resource stewardship planning and design.

After a century of increasingly heavy recreational use, the 162-acre Fresh Pond Reservation is showing its age, especially with soil compaction and erosion from hillsides. Invasive plant species are overrunning natural areas. Moreover, shoreline erosion is contributing to the degradation of numerous water bodies on the reservation, including Fresh Pond Reservoir — the terminus of the city of Cambridge's water supply. Mitigating these environmental stressors is critical if Fresh Pond is to remain a natural haven in the city for a second century (Figure II.4.1).

Thankfully, the city of Cambridge and its residents have recognized the need for a long-term natural resource management program for Fresh Pond. They are moving forward with an ambitious master plan (Fresh Pond Reservation master plan) to preserve the water quality of Fresh Pond Reservoir while implementing much-needed landscape enhancements. This master plan is a model for water sensitive landscape design and restoration. Its lessons are poignant and applicable to the preservation of our country's urban parks for generations to come (Figure II.4.2).

History

Fresh Pond Reservation is one of metropolitan Boston's premier public open spaces. Originally designed by the Olmsted firm and Charles Eliot in the 1890s, the 162-acre reservation has long served as a natural haven in the city. It is an interesting and complex ecosystem, loved no less fervently by its diverse users than Central Park by New Yorkers.

The reservation's natural features are a challenge to manage: wetlands and upland woods; stands of large evergreen and deciduous trees; shrub borders; steep hillsides; ravines and small ponds; seasonal marshes, meadows, grassy areas, dense thickets, irregular vegetated shorelines; and various soil types and hydrologic conditions. Adding to this complexity is the fact that the reservation serves as a protective buffer for the 163-acre Fresh Pond Reservoir, the terminus of the city of Cambridge's water supply (Figure II.4.3).

Throughout much of the reservation's history, municipal management and maintenance have been minimal. City of Cambridge residents, long characterized by their strong tendency toward activism, have taken the initiative to preserve the reservation's natural resources for more than four decades. In 1997, however, the city of Cambridge and the Fresh Pond Master Plan Advisory Committee (FPMPAC) joined forces to develop a natural resources stewardship plan (*Fresh Pond Reservation Natural Resource Stewardship Plan*) — one focused primarily on watershed management (Benjamin, 1999).

Natural resource inventory and stewardship plan

The resulting stewardship plan process for Fresh Pond consisted of two major components: the natural resource inventory (Fresh Pond Reservation Natural Resource Inventory) and the plan itself. The inventory phase comprehensively investigated and recorded natural resource conditions at the reservation and around the reservoir. This systematic, ecological study of the reservation's open spaces focused on water quality issues and the identification of surface conditions such as erosion and compaction, vegetation cover types, and shoreline and stream conditions. It also identified many areas of notable natural or cultural character, including healthy stands of old trees and historic water treatment structures.

Chapter II.4: Natural resource stewardship planning and design: Fresh Pond Reservation 409

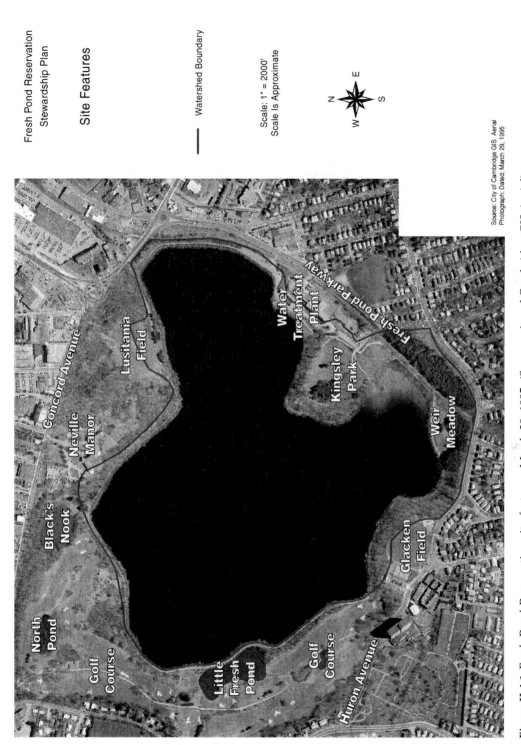

Figure II.4.1 Fresh Pond Reservation site features, March 29, 1995. (Source: city of Cambridge GIS Aerial).

Figure II.4.2 Fresh Pond Reservation master plan, prepared by the Fresh Pond Master Plan Advisory Committee, adopted 2001.

Stewardship plan recommendations

With strong community support, the inventory findings and the stewardship plan recommendations for Fresh Pond were accepted in early 1999. The plan provided recommendations for general topics reservation-wide, such as slope stabilization, wetland management, and trail management. It also provided extensive recommendations for 11 subareas of the reservation including several water bodies and a municipal golf course. As a reflection of the reservation's diverse needs, the recommendations themselves ranged widely in character:

- Erosion control and shoreline/upland stabilization
- Reduction of invasive plant species
- Enhancement of meadow areas
- Control of trail access
- Improvements to trail surface and drainage
- Reuse of on-site materials, such as downed trees, for trail resurfacing
- Composting of leaves for widespread reuse in soil improvement efforts

In January 2001, the Cambridge City Council took these recommendations to heart. It gave the green light for work to proceed on seven water sensitive design projects at the reservation, as part of the comprehensive Fresh Pond Reservation master plan into which the stewardship plan's recommendations were incorporated (Figures II.4.4, II.4.5, and II.4.6).

A key player in taking the stewardship plan and, later, the master plan from conception to reality has been the city of Cambridge's Water Department. In particular, the city's

Chapter II.4: Natural resource stewardship planning and design: Fresh Pond Reservation 411

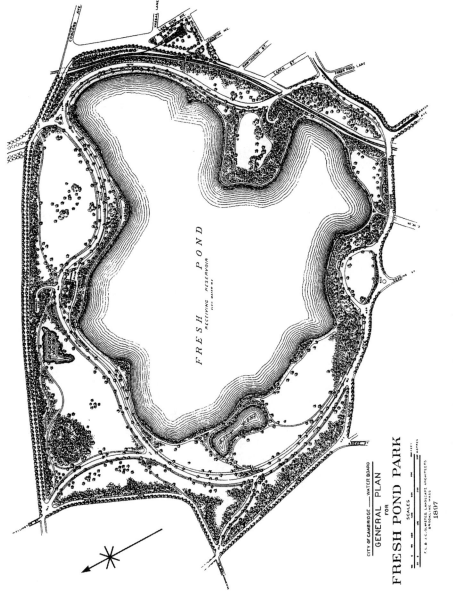

Figure II.4.3 Original plan for Fresh Pond by the Olmsted Firm and Charles Eliot, (1897).

412 Handbook of water sensitive planning and design

Figure II.4.4 Fresh Pond vegetation types map prepared in geographic information systems (GIS) format for the Fresh Pond Reservation Natural Resource Inventory. (Source: Rizzo Associates, Inc., 1998.)

Table 1. Developed Land, Open Water and Vegetated Land in Fresh Pond Reservation

Major Land Cover Type	Area (acres)
Developed Land	10.3
Open Water	162.6
Fresh Pond	156.6
Little Fresh Pond	3.5
Black's Nook	2.0
North Pond	0.6
Vegetated Land (see Table 2)	152.0
Total Reservation Area	324.9

Table 2. Vegetative Cover Types at Fresh Pond Reservation

Vegetative Cover Type	Area (acres)
Upland Forest	52.1
Softwood forest	8.1
Hardwood forest	38.5
Mixed forest	5.5
Scrub/Shrub Upland	2.1
Meadow/Open Field	0.5
Wetland	21.4
Forested wetland	17.4
Scrub/shrub wetland	3.3
Emergent wetland	0.7
Landscaped/Maintained	75.9
Golf course	50.4
Other	25.5
Total Vegetated Land	152.0

Source: Fresh Pond Natural Resource Inventory, Table 1

Figure II.4.5 Tables from the Inventory.

watershed manager, Chip Norton, has been a vital link in coordinating the multiple interests represented by the committee and others with the city's ongoing work, especially with regards to large, preexisting infrastructure improvement projects. With both a strong understanding of the larger context in which the reservation and reservoir sits and the city's fiscal constraints and day-to-day operational needs, Mr. Norton has helped mold the master plan's grand vision into tangible and feasible capital projects. It has been crucial to the process to have a proactive public official located on-site who possesses the broad view of how all the pieces fit together and who also has the ability, and level of commitment, to get things done in the city.

Master plan implementation

Of the seven projects currently under way or about to begin that incorporate Fresh Pond master plan goals, the Landscape Management and Maintenance Plan represents the first step toward manifesting the goals of the natural resource stewardship plan: to preserve water quality through improvement of soil conditions; and to increase biodiversity and overall aesthetic interest through vegetation management. The Landscape Management Plan's scope exemplifies and mirrors the many recent and current efforts elsewhere in the emerging, multidisciplinary field of restoration ecology. The key components of this project are:

- Vegetation management through control of invasive tree, shrub, vine, and herbaceous species to enhance biodiversity and native species growth
- Erosion control and stabilization of the most severely eroded and highly visible upland slopes and nonreservoir shoreline section(s) to protect the reservoir from silt-laden surface runoff

Priority areas have been selected based on the severity and nature of invasive plants and/or erosion conditions, the visibility to the public, the proximity to other preexisting construction projects and their impacts (e.g., recently disturbed soils), potential ecological value (e.g., edge zones considered highly valuable), and the size and manageability of the

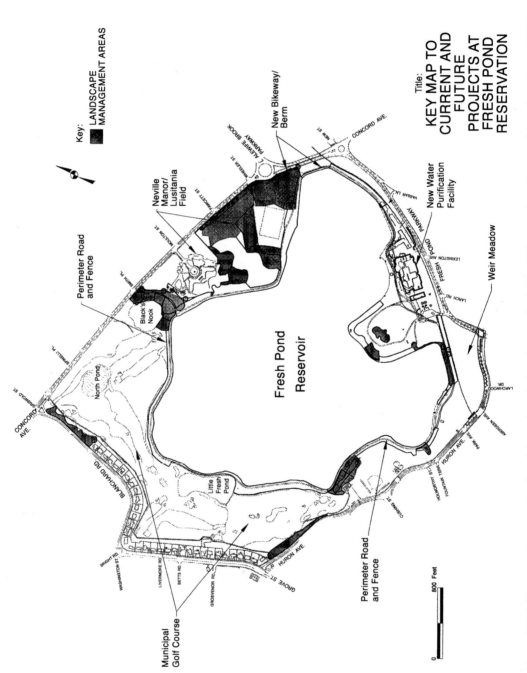

Figure II.4.6 Key map to current and anticipated capital projects at Fresh Pond Reservation. (Source: Rizzo Associates, 2001.)

Figure II.4.7 Oriental bittersweet (*Celastrus orbiculatus*) vine, an invasive species, choking a hardwood forest and edge zone at Fresh Pond.

unit. An area in excess of 25 acres has been proposed for vegetation management, slope stabilization, or both. This represents approximately half of the reservation's "natural" area, typically woodlands or overgrowing old fields (Figure II.4.7).

Landscape Management Plan — vegetation and soils management

In terms of the level of effort, complexity, and cost, the vegetation management component is the larger piece of the Landscape Management and Maintenance Plan. Its primary objective is to control invasive species in high-priority areas so that biodiversity with native species may be increased. These areas include forested wetland and upland areas, as well as meadows and edge zones.

Methods under consideration for invasive species control include both mechanical and chemical techniques. The mechanical techniques being considered are cutting, hand-pulling, mowing, smothering, and prescribed burning or spot burning. Due to their reliance on abundant labor, mechanical techniques may only being considered for limited areas, such as highly environmentally sensitive areas near the reservoir and open-water bodies; however, mechanical or manual removal techniques will be emphasized to the greatest possible extent.

For larger, less sensitive areas, environmentally sound chemical techniques, specifically the use of herbicides, may represent the most feasible and cost-effective method for controlling invasives. This approach is endorsed by leading experts in invasives control. In fact, a strong argument in favor of chemical treatment over mechanical methods is that mechanical removal of established plants is more disturbing to soils. This is due to the trampling by workers engaged in the act of pulling plants out of the ground, especially where invasive stands are dense (e.g., buckthorn in Lusitania Woods) (Figures II.4.8, II.4.9, and II.4.10). Further, state and federal level environmental regulations strongly discourage soil disturbance of any kind within wetlands. Approximately half of the area currently proposed for landscape management work at Fresh Pond falls within regulated wetland resource areas.

Environmentally sound chemical techniques emphasize direct herbicide application into the individual plant stems or trunks to be treated, rather than area or blanket spraying. These techniques include injection of herbicide into invasive tree species (e.g., tree-of-heaven,

Figure II.4.8 Buckthorn (*Rhamnus spp.*), an invasive shrub species that thrives on disturbed soils, dominates as much as 95% of the forest understory in parts of Fresh Pond Reservation.

Figure II.4.9 Mechanical removal of invasive species along Fresh Pond Reservoir's shoreline. The Fresh Pond Reservation Landscape Management Contract currently being prepared will address invasive species control.

Chapter II.4: Natural resource stewardship planning and design: Fresh Pond Reservation

Fresh Pond Vegetation Management Project
[Sample] Monitoring Form

Monitoring Location: Water Purification Facility
Monitoring Station Number: 1
Monitor's Name: Jane Doe
Monitoring Date: 1 June 2003
Date of Last Monitoring: 27 October 1999

Monitoring Data

Vegetation Layer	Botanical Name	Common Name	Percent (%) Cover	Percent (%) Cover Last Monitoring	Percent (%) Change (±) Since Last Monitoring	Invasive	Planted (P)/ Volunteer (V)	Vegetation Health (Poor, Fair, Good, Excellent)
Tree	*Acer saccharinum*	Silver Maple	35	30	+5			G
Tree	*Ulmus americanus*	American Elm	20	20	0			F
Tree	*Fraxinus pennsylvanica*	Green Ash	15	10	+5			G
Tree	*Betula nigra*	River Birch	20	10	+10			G
Tree		OPEN	10	30	−20			
Shrub	*Rhaninus frangula*	Glossy Buckthorn	10	90	−80	x		
Shrub	*Cornus racentosa*	Gray Dogwood	50	10	+40			E
Shrub	*Viburnum denatum*	Arrowwood	10	0	+10		V	E
Shrub	*Kalmia angustifolia*	Sheep's Laurel	10	0	+10		V	G
Shrub	*Azalea viscosum*	Swamp Azalea	5	0	+5		P	G
Shrub	*Vaccinium corymbosum*	Highbush Blueberry	10	0	+10		P	G
Shrub		OPEN	5	0	+15			
Groundcover	*Vincetoxicum nigrum*	Black Swallow-wort	5	25	−20	x		F
Groundcover	*Celastrus scandens*	Oriental Bittersweet	0	5	−5	x		P
Groundcover	*Athyrium felix-femina*	Lady Fern	10	0	+10		P	E
Groundcover	*Arctostaphylos uva-ursi*	Bearberry	10	0	+10		V	E
Groundcover	*Polygonatum biflorum*	Solomon's Seal	15	0	+15		P	G
Groundcover		OPEN/BARE	40	70	−30			

Comments: Fairly dense native groundcover developing throughout plot area; remaining buckthorn appears stressed; tree canopy almost fully covering area

Figure II.4.10 Vegetation management monitoring form. Monitoring will be an integral component to landscape management strategies being planned for the reservation.

Norway maple) and cutting and dabbing herbicide onto the stumps of invasive shrubs and vines (e.g., buckthorn, oriental bittersweet) called the "cut-stump" method. The only exception to these direct and highly controlled applications would be to herbaceous invasives (e.g., garlic mustard), which can only be chemically controlled through foliar spraying. Generally speaking, herbaceous invasives have not yet become a major problem at Fresh Pond.

Two widely available "general-use" herbicides that have been endorsed by numerous experts in natural areas management include Round-Up® (registered trademark of Monsanto Company, St. Louis, MO) (active ingredient: glyphosate) and Garlon® (registered trademark of The Dow AgroSciences Company, Indianapolis, IN) (active ingredient: triclopyr). Both products have been used extensively around the country for invasives control. Round-Up has been particularly endorsed for its rapid chemical breakdown after application, although Round-Up is a nonspecific herbicide that, if improperly used or accidentally overthrown, can kill both woody and herbaceous species. By contrast, Garlon can have a longer chemical breakdown time but is specific to woody species only and will not harm herbaceous species if spilled or overthrown. These chemicals have been successfully used to control invasives in Massachusetts by the New England Wildflower Society, Trustees for Reservations, and the Middlesex County (MA) Mosquito Control Unit, among others. Nationally, controlled herbicide use has proven effective in combatting invasives to restore native species diversity by the Nature Conservancy, Minnesota Department of Natural Resources in state parks, the McHenry County (IL) Conservation District, and the University of Wisconsin's Curtis Arboretum, among others (http://www.newfs.org/ publication.html, http://tncweeds.ucdavis.edu/handbook.html, http://www.nps.gov/plants/alien/).

Following chemical or mechanical treatment, selected areas will receive soil amendments (mechanically treated areas only) and/or seeding with a noninvasive or less invasive cover species such as annual ryegrass, sheep fescue, white clover, buckwheat, and/or switchgrass. The cover species will serve a number of purposes, including:

- Immediate soil cover providing erosion control and protection from trampling impacts in newly "opened" areas
- Immediate competition for invasive species germination
- Initial decompacting of compacted soils
- Organic mulch and nutrient source for impoverished soils
- Improved stormwater runoff uptake
- Improved aesthetics for areas with extensive vegetation removal

The Management Plan will provide the first phase of natural resource improvements by focusing on invasives removal and selective thinning of the tree canopy (in forested sections) to increase light penetration to lower forest strata. Restoration plans will be prepared and executed under the closely associated Northeast Sector Landscape Improvement Plan. The restoration plans will focus upon rebuilding the dense, diverse forest understory strata through intensive replanting with native species. The restoration strategy will also diversify edge zones/hedgerows with native berrying shrub plantings. The old field and meadow areas restoration will focus upon periodic mowing to maintain the ecologically productive open character, overseeding/planting with native wild grasses and wildflowers, and establishing new meadows in previously disturbed areas. All vegetation improvements are intended to increase stormwater filtration and uptake while better protecting soils from excessive erosion. Improvements will also increase wildlife habitat opportunities and year-round aesthetic interest for reservation users.

Chapter II.4: Natural resource stewardship planning and design: Fresh Pond Reservation

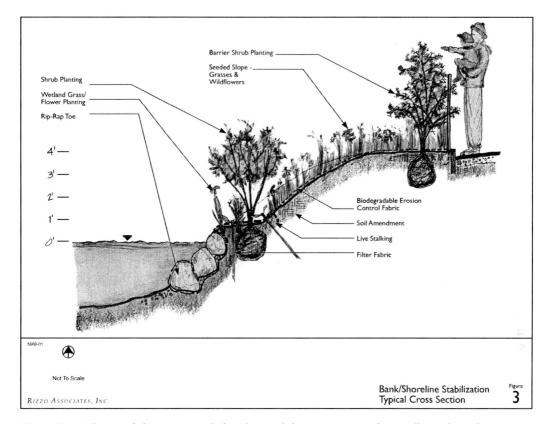

Figure II.4.11 Low-tech bioengineered shoreline stabilization concept for small ponds in the reservation. (Source: Fresh Pond Reservation Natural Resource Stewardship Plan, Rizzo Associates, Inc., 1999.)

Vegetation management monitoring

An important component to the Landscape Management and Maintenance Plan will be to explore different management techniques in different areas and to closely monitor the effectiveness of the various techniques in general and relative to one another. The monitoring program will provide the basis to assess future management needs for the selected areas. It is important to note that the initially treated areas will likely require additional follow-up management. Management activities, including reapplication of herbicides and follow-up cutting or hand-pulling, may be necessary for a number of years subsequent to the initial treatment to achieve an acceptable level of invasives control and to begin to restore vegetative diversity (Figure II.4.11).

Erosion control and slope stabilization

The second component of the Landscape Management and Maintenance Plan is erosion control and slope stabilization. These efforts will be directed toward the steepest, most eroded, and most visible slopes and shorelines, as well as those that directly threaten the water quality of Fresh Pond Reservoir. Low-tech, low-cost bioengineering approaches will be used to achieve initial stabilization. Techniques may include some or all of the following:

- Securing base of slope or bank with biodegradable coconut roll (coir fascine) and porous gravel base
- Intercepting gully-bound runoff in gravel based dry wells or french drains

- Spreading composted leaf litter over the selected slope areas
- Creating organic pockets with vertically driven wood stakes and compost
- Harvesting acorns and nuts at the reservation and spreading them over the organic layer (composted leaves)
- Spreading topsoil over the organic layer
- Securing the organic layer with biodegradable erosion control fabrics
- Seeding the areas with a noninvasive cover crop (e.g., white clover, switchgrass) and/or meadow wildflowers and wild grasses in sunny exposures
- Planting native sapling trees, shrubs, and herbaceous plants
- Placing wind-fallen branches, branch bundles, and other coarse-textured woody debris over the treated slopes to discourage trampling

In addition, the placement of cut logs and live brush bundles (fascines) or the like buried and staked in place across slope faces may also be considered where necessary and feasible. Slope stabilization will also involve the removal of invasive trees (e.g., Norway maples) to improve light penetration to the ground level. The goals of the upland slope work are soil stabilization in the short term and full forest structure restoration of all vegetation strata in the long term. Final design methodology will be informed by extensive soil sampling of the areas in question (Figures II.4.12, II.4.13, and II.4.14).

Water purification facility/Weir Meadow landscape improvements

The Landscape Management and Maintenance Plan is an independent capital project. However, other notable projects in the Fresh Pond Master Plan involve bringing existing landscape plans into compliance with the goals of the natural resource stewardship plan, now incorporated into the master plan. indeed, it was the public's call for a well-coordinated approach to landscape planning at the reservation that helped to galvanize the community-based master planning process in 1997. One such project is the improvement plan for the city's new state-of-the-art water purification facility (which went online in summer 2001) in the reservation and an associated meadow area (http://www.ci.cambridge.ma.us/~Water/startup.html).

The Weir Meadow is a popular, ±5-acre sloping lawn area that has historically been home to the reservoir's weir structure. The structure regulates piped flow into Fresh Pond from Cambridge's "upcountry" reservoirs located approximately 8 miles to the west. Landscape improvements to the water purification facility and the Weir Meadow area have been accomplished through restoration-oriented (naturalized) planting strategies. Such strategies have emphasized the use of native species to protect water quality and improve aesthetics and ecological values. The strategies have included:

- Replacement of lawn areas with lower maintenance groundcovers and meadows
- Use of low-maintenance lawn species where lawn remains
- Restoration of forest strata with addition of groundcovers, shrubs, understory trees, and canopy trees

In addition, heavy revegetation of the site is intended to improve overall soil health by protecting the surface from erosion, breaking up the soil through root growth, increasing infiltration opportunities, and adding organic matter as plant materials are annually cycled back into the earth. Improved soils equate to reduced surface runoff into the reservoir.

Another important design strategy is the inclusion of substantial wetland-oriented bioswales, at both the water purification facility and Weir Meadow sites, to capture and

Chapter II.4: Natural resource stewardship planning and design: Fresh Pond Reservation

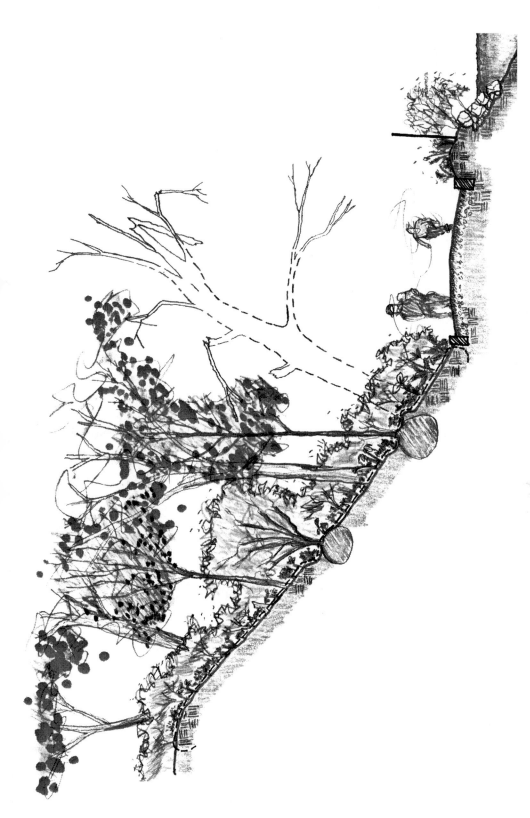

Figure II.4.12 Upland slope stabilization and forest restoration concept. (From the stewardship plan.)

Figure II.4.13 Deep shade of invasive Norway maple (*Acer platanoides*), at left, prevents lower forest strata, growth leaving soil surface on steep slopes open to erosion, which, in turn, increases silt load of surface runoff into Fresh Pond Reservoir (directly behind fence).

Figure II.4.14 Plans for the reservation's Weir Meadow include heavily planted bioswales, designed for the primary purpose of surface runoff filtration, to be located approximately where pools form on the existing lawn and overflow into the reservoir. (Right) Newly installed bioswale.

filter runoff prior to its entry, primarily as groundwater, into the reservoir. The bioswales will be heavily seeded and planted with wetland herbs and encircled by dense native wetland shrub borders. One of the most valuable contributions of the overall design will be a substantial increase of high-quality edge habitat and hedgerows — a great benefit to wildlife (Figures II.4.15 and II.4.16).

A public art piece has also been included in the facility's design, near a major entrance, in the hopes of raising public awareness about the city's watershed and water delivery system (http://www.harriesheder.com/www.camb.htm).

Reservoir fence and Perimeter Road resurfacing alternatives

Two exciting studies are emerging from the Fresh Pond master plan — both of which examine alternatives to better aesthetically frame the reservoir while protecting it from

Figure II.4.15 Cambridge's new state-of-the-art water purification facility, to go online in 2002, is complemented by a water-friendly landscape around it. (Left) View of new facility from Kingsley Park. (Right) View of newly landscaped facility with new bikeway in foreground and Fresh Pond Parkway (vehicular) at left.

Figure II.4.16 Persistent flooding and drainage problems at Fresh Pond's municipal golf course are being addressed by a master plan for the course now under way.

surface runoff contamination. The Reservoir Fence pilot study will determine the most aesthetically complimentary and economically feasible fence type or types to replace a deteriorating existing chain-link fence. An existing fence encircles the entire ±2.5-mile perimeter of Fresh Pond Reservoir, with the notable exception of one short unfenced section of shoreline immediately adjacent to the new water purification facility. Due to the highly urban context in which the reservation and reservoir sit, the city has determined it necessary to entirely fence its drinking water supply. In the city's view, unlimited water access by people and dogs would represent an uncontrollable threat to drinking water quality. Further, the reservation has a number of other significant water bodies, most notably Little Fresh Pond and Black's Nook, which are easily accessible to reservation users. However, due to their accessibility these ponds tend to be highly eroded along their unprotected shorelines. Current and future projects will address reinforcement of these shorelines.

Figure II.4.17 Pilot section of stabilized aggregate Perimeter Road resurfacing. Stabilized aggregate surface is porous, natural in appearance, and easy on the feet. Stabilized aggregate acts as a "sponge" when wet preventing runoff into the reservoir.

In the meantime, the fence pilot project implementation is under way with a new 4-ft-high, black polyvinyl-clad chain-link fence installed along the shoreline in front of the new purification facility. Related to the fence pilot study are ongoing efforts to improve views to the reservoir from the reservation, primarily through the selective thinning of vegetation along the reservoir's shoreline. Some consideration is also being given to the creation of additional controlled access openings in the reservoir's fence to get reservation users closer to or over the water, though this is only in the conceptual stage.

More directly related to surface runoff management, the Perimeter Road resurfacing pilot study is investigating the viability of the Stabilizer (brand) product (http://www.stabilizersolutions.com), a porous, nonchemical aggregate binder material. The goal of the study is to repave the entire ±2.5-miles of deteriorating asphalt road that encircles the Fresh Pond Reservoir with an aggregate bound with Stabilizer.

Stabilizer's use here is expected to allow for much increased infiltration of surface runoff prior to its reaching the reservoir. This durable, low-maintenance material has proven effective at Battle Road National Historic Park in Lexington, MA. An approximately 1500-ft-long "pilot" section of stabilized aggregate was installed in 2001 and is under study. (Figure II.4.17).

Northeast sector landscape improvements

A recently begun project at Fresh Pond Reservation is the design of landscape and site improvements to Neville Manor, which is being converted into an assisted-living center, and nearby Lusitania Field, a former soccer field being used as the construction staging area for the new water purification facility. The project area also encompasses a large upland and wetland forest and old field edge zones.

Although in its early planning stages, current components of this project are envisioned to include construction of a new youth soccer field and a meadow to maximize songbird habitat with an environmental art concept closely integrated into the overall design. As discussed earlier, the 25-acre project area will significantly overlap with high-priority areas for vegetation management, as specified in the Landscape Management and

Maintenance Plan. As such, the project will take the management strategies specified therein to the next level by preparing full restoration plans for the areas to be managed for invasive species. Design work is expected to be completed by the end of 2002 with site work expected to begin in 2003.

Golf course master plan

In addition to providing open space around Fresh Pond Reservoir, the reservation is home to the city's only municipal golf course. The Thomas P. O'Neill Municipal Golf Course, which covers approximately one third of the reservation's total area (±51 acres maintained), has been in continuous play since the 1920s. Although it has been upgraded at various times since its inception, the course has been plagued with significant drainage problems. Moreover, its playability has not been examined comprehensively for years. Further, the complex stormwater drainage system which traverses the course, and involves the neighboring community of Belmont, has never before been comprehensively studied, to the city's knowledge. However, the city has recently hired a consultant to do exactly that.

Master planning for the golf course is being approached from the dual perspectives of playability and environmental soundness. The planning is underway and conceptual recommendations for playability improvements will be reviewed in the near future. With the city conducting a comprehensive stormwater and drainage study of the golf course, the master planning for its use will soon be informed by critical drainage data. The drainage study will identify and quantify stormwater infrastructure problems contributing to flooding at the course and recommend solutions for related improvements. Such a study will be critical to improving drainage problems at the course and for protecting water quality in the adjacent reservoir (Figure II.4.18).

Bikeway corridor landscape improvements

Yet another project currently under way is the Fresh Pond Bikeway Corridor project. This project focuses on landscape improvements to a 1,000-ft-long section of new bikeway corridor associated with a much larger sewer separation project occurring within and along existing Fresh Pond Parkway, a major arterial route in Cambridge. The landscape improvement area covers about two acres. About half of the bikeway, within the project area, runs adjacent to the reservoir's highly used Perimeter Road within the narrowest section of the reservation. The reservation's width here is only about 60 ft from the reservoir's shoreline to the edge of curb at Fresh Pond Parkway. The project may be broken into two distinct sections. The first covers the narrow section between the reservoir and the Parkway, the second extends the bike path along the Parkway edge of a lowland forest subject to seasonal flooding.

Due to the project's immediate proximity and groundwater connections to the reservoir, water quality protection was a central goal of the landscape design work, which was completed in 1999 and 2000. Other important design goals included visual and noise buffering of Perimeter Road from adjacent Parkway traffic (and ongoing roadway reconstruction), strong separation between bike path and Perimeter Road to avoid user conflicts between bicyclists and pedestrians, and enhancement of views toward the immediately adjacent reservoir. Further, in keeping with the master plan goals, the design work also aimed to maximize habitat values and forest restoration opportunities.

In response to these goals, a number of design strategies were employed here and a variety of water-conscious features included. Perhaps, the most visually notable feature was the creation of a 3- to 5-ft high, 30-ft-wide linear berm running between the Perimeter Road (reservoir) and the bike path itself. The berm's north side, closest to the Parkway

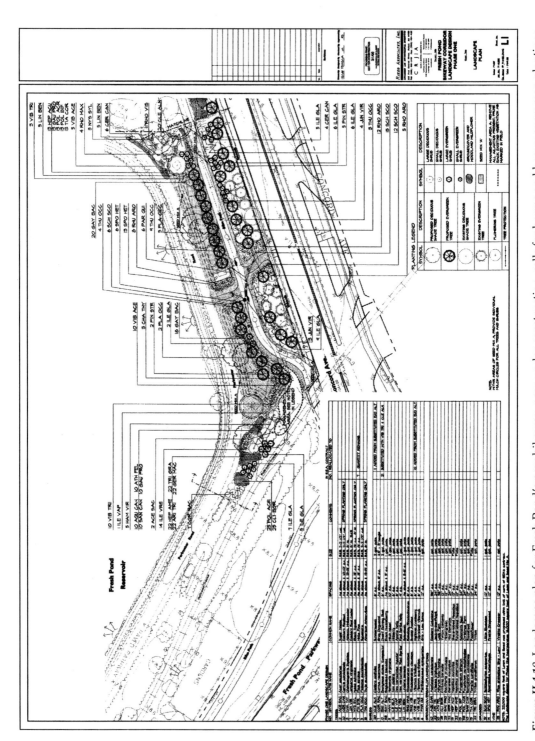

Figure II.4.18 Landscape plan for Fresh Pond's new bikeway, now under construction, calls for berms and heavy evergreen plantings to buffer the reservation and reservoir from adjacent parkway traffic impacts. (From Fresh Pond Bikeway Corridor landscape design, Pizzo Associates, Inc. and Carol R. Johnson Associates, Inc.)

Chapter II.4: Natural resource stewardship planning and design: Fresh Pond Reservation 427

Figure II.4.19 Berm construction along new bikeway, April 2001. Perimeter Road on right along reservoir fence, bikeway on left (behind berm), and Fresh Pond Parkway on far left (behind trees and Jersey barriers).

and adjacent to the bike path, will be heavily planted with native evergreen trees and shrubs, while the sunny south side will be planted with a low-maintenance lawn mixture to provide an informal grassy sitting area with opened views to the reservoir. The berm will serve multiple purposes. The primary purpose will be to buffer noise and visual impacts from the Parkway. The berm will also help keep stormwater runoff and airborne dust/garbage (particulates) out of the reservoir. To complement the view to the reservoir from the berm, selective tree thinning along the shoreline will occur. The thinning will also remove invasive species and increase the presence of favored desirable natives species (e.g., river birch), which will further enhance aesthetic and habitat values.

Another important strategy used in the woodland section adjacent to the Parkway was the creation of densely vegetated bioswales. The bioswales will capture and filter runoff from areas adjacent to the Parkway and slowly filter it into the groundwater table. This will serve to protect water quality in both the adjacent lowland woods and in the nearby reservoir. Other planting strategies included dense plantings of a diverse variety of native trees, shrubs, perennial grasses and flowers, and groundcovers to increase edge habitat values and restore complex forest strata.

As with all landscape improvement projects at the reservation guided by the master plan's environmentally conscious vision, the Bikeway Corridor design work was the product of extensive public involvement at all phases, but particularly during the critical design development phase (Figure II.4.19).

Conclusion

The extensive and complex landscape improvements occurring at Fresh Pond Reservation today, and in the pipeline for the near future, are very much shaped by the singular goal of water quality protection. In the design and execution of these projects one is constantly reminded of the critical connection of the landscape to water bodies via surface and groundwater drainage and the need for water-minded stewardship. As the Reservation enters its second century, armed with the strong vision outlined in the master plan, the future for water quality preservation and the restoration of highly functioning natural

Figure II.4.20 The master plan's goals seek to protect the reservoir's water quality through wise stewardship of landscape that surrounds it.

Figure II.4.21 Soil stabilization and improvement through intensive vegetation management, emphasizing restoration of biodiversity, is a central concept in the plan's approach.

systems looks bright, indeed. Hopefully, the forward-looking strategies used at Fresh Pond will succeed in preserving water quality long term and serve as an excellent example for stewardship of aging, yet much-loved public open spaces everywhere (Figures II.4.20 and II.4.21).

Acknowledgments

Special thanks to Karen Maki and Sandra Fischer of Rizzo Associates for editorial review and graphic support, respectively. Further thanks to Amy Green, Jane Taylor, Audrey Sall, Richard Grant, Cynthia Baumann, Dave George, and Ken Deshais, all of Rizzo Associates, and to Joseph Ingoldsby of J.E. Ingoldsby & Associates for their respective roles, large and small, in the preparation of the Fresh Pond natural resource stewardship plan and related planning and design efforts. From the city of Cambridge, the significant efforts of the following individuals and committees must be recognized: Chip Norton; Fresh Pond Master Plan Advisory Committee (FPMPAC 1999–2001); Fresh Pond Stewardship Plan Oversight Committee members (1998–1999), including James Barton, Deborah Howe, Patricia Pratt, Janice Snow, and Louise Weed; and Julia Bowdoin, Richard Rossi, Lisa Peterson, Paul Ryder, Robert Carey, and J.C. Girouard. Additional thanks are extended to others who have significantly participated in the implementation phases of the projects at Fresh Pond Reservation, including John Amodeo, Jennifer Jones, Randy Sorenson, Ben Dieterle, and Howard Snyder, Carol R. Johnson Associates, Inc.; John Kissida, John Olcott, and Nate Sanford, Camp, Dresser & McKee, Inc.; George Sargent, Tom Devane, and Debby Sargent, Wogan & Sargent Golf Course Architects, Inc.; Julie Mair Messervy.

Literature cited

Benjamin, T.S., Building consensus for natural-resource and water-quality preservation at the Fresh Pond Reservation, *Erosion Control*, Sept.–Oct. 1999.

City of Cambridge (Fresh Pond Master Plan Advisory Committee), Fresh Pond Reservation master plan, prepared May 2000, adopted January 2001.

City of Cambridge Water Department Web site:
http://www.ci.cambridge.ma.us/~Water/startup.html.

Harries, Mags, Web site: http://www.harriesheder.com/www.camb.htm.

New England Wildflower Society, Conservation Notes, Invaders, 2, 1998, http://www.newfs.org/publication.html.

Rizzo Associates, Inc. and J.E. Ingoldsby Associates, Fresh Pond Reservation natural resource stewardship plan, February 1999.

Rizzo Associates, Inc. and J.E. Ingoldsby Associates, Fresh Pond Reservation natural resource inventory, October 1998.

Rizzo Associates Web site: http://www.rizzo.com/portfolio_cambridge_fresh_pond.htm.

Stabilizer Solutions, Inc. Web site: http://www.stabilizersolutions.com/.

Tu, M., Hurd, C, and Randall, J.M., *Weed Control Methods Handbook: Tools and Techniques for Use in Natural Areas*, The Nature Conservancy, April 2001, http://tncweeds.ucdavis.edu/handbook.html.

U.S. National Park Service/Plant Conservation Alliance, Weeds gone wild project, Washington, DC, jil_swearingen@nps.gov and/or http://www.nps.gov/plants/alien/.

Response

Protecting and restoring treasured landscapes: Complexity and integration

Landscapes are not only situated in physical space, they can also be layered within the same space as determined by variable uses. The case study described by Thomas Benjamin is particularly interesting due to the high visibility of the site. Not only does the Fresh Pond Reservation protect the city of Cambridge's drinking water supply, it is also the largest green space in the city and is used by golfers, pedestrians, cyclists, dog walkers, joggers, etc. On top of that, the landscape has historic significance, as, for example, being designed and planted by the master landscape architects, Olmsted and Eliot. Dealing with the concerns raised by each of these interest groups, some mutually opposed, is an onerous task indeed.

Certainly, the most straightforward message from Benjamin's study is one of recognizing of the false dichotomy often existing between what is a water sensitive "design" project and what is a water sensitive "planning" project. This chapter, more than most in this book, is very much a straddler between the two thematic sections because Benjamin describes a series of site-specific designs whose conceptual origin and strategy of implementation are very much rooted in approaches of traditional planning. The approach taken of moving from a natural resource inventory as part of a "stewardship plan" used to identify problem areas and provide recommendations to a water sensitive management plan is a strategy to be admired. The watershed assessment planning process described by Dennis Haag, Stephen Hurst, and Bryan Bear in Chapter II.8 is patterned in a similar vein.

This chapter also addresses an array of topics considered in more detail elsewhere in this book, such as shoreline bioengineering covered by Wendi Goldsmith in Chapter I.17 and Nicholas Pouder and Robert France in Chapter I.16, recreational planning of riparian trails introduced by Charles Flink in Chapter II.3, and biofiltration swales described by Robert France and Philip Craul in Chapter I.7, and Thomas Liptan and Robert Murase in Chapter I.6. Benjamin's chapter assumes real significance in his presentation of a case study showing how all these elements, in addition to landscape aesthetic enhancements, can be effectively integrated into a single, comprehensive solution.

Several other elements of this study offer important lessons to water sensitive planning professionals that are worth reiterating:

1. The extensive public participation process at all phases of planning and acknowledgment that such bottom-up concerns were instrumental to motivating the entire project from the start
2. The piggy-backing of the resource plan with ongoing public utility improvements at the water purification facility, including art installations
3. The use of pilot projects as an incremental approach to both educate the public and affect the watershed alterations required
4. Understanding that nonstructural planning solutions involving tending and management may be just as important as structural designs
5. Recognition of the need to experiment with different management techniques for different locations, with the consequent requirement for monitoring or surveillance through time

chapter II.5

Treating rivers as systems to meet multiple objectives

Leslie Zucker, Anne Weekes, Mark Vian, and Jay D. Dorsey

Abstract

Planners, local decision makers, and landowners are increasingly asked to meet multiple objectives for stream resources. The ability of the stream system to meet multiple objectives depends on complex relationships between system components (e.g., the channel, riparian area, and floodplain). Similar to other systems with multiple variables, rivers appear to exhibit an emergent order that transcends the properties of component parts; however, management approaches have tended to ignore the interconnectedness of river processes. This chapter examines aspects of river system management that could be improved through the application of complex systems theory. The application of systems theory to river planning and management can help to meet multifunctional goals.

> After three years
> I no longer saw this mass.
> I saw the distinctions.
>
> But now, I see nothing
> With the eye. My whole being
> Apprehends.
> My senses are idle. The spirit
> Free to work without plan
> Follows its own instinct
> Guided by natural line,
> By the secret opening, the hidden space,
> My cleaver finds its own way.
> I cut through no joint, chop no bone.
>
> — from *Cutting Up an Ox*, Chuang-Tzu (approx. 300 B.C.E.)

Introduction

This enigmatic poem presents a simple meat cutter as artful master of his trade. The meat-cutter is able to move his blade to the place where the least effort is required to meet his

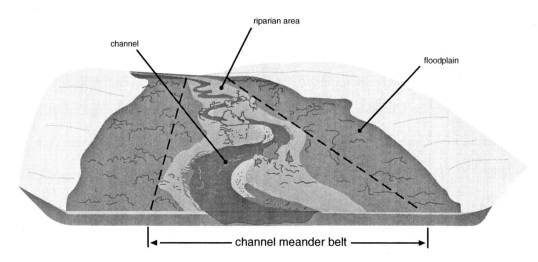

Figure II.5.1 A river includes a channel, riparian area, and floodplain among other natural features. The river uses the riparian area and the floodplain to dissipate an immense quantity of hydraulic energy. The channel meander belt might shift over the entire floodplain in a period of several hundred years; however, within management time frames, the meander belt might be confined to a relatively fixed position. (Adapted from Palmer, L., River management criteria for Oregon and Washington, in *Geomorphology and Engineering*, D.R. Coates, Ed. Dowden, Hutchinson & Ross, Inc., Stroudsburg, PA., 1976.)

goal. Rivers are the artful masters of their own existence. They balance the forces of moving water and sediment against the landscape, assuming a form that minimizes the energy required to do the necessary work.

In this chapter, we examine rivers from the perspective of systems science. The elements that make up a river (such as water flow, nutrients, plant and animal communities, and bedload) have a tendency to organize into an interconnected and complex system. One assumption of systems theory is that systems self-organize into emergent states that cannot be inferred from their parts (Waldrop, 1992).

This assumption suggests the potential to expand river science beyond traditional approaches. We explore the management implications of ignoring the tendency of rivers to behave as systems, and suggest that application of systems theory to river planning and management might improve our ability to meet multifunctional goals. The word "river" is used in reference to systems of all sizes: river, stream, creek, or ditch.

In 1976, the geologist Leonard Palmer wrote that "given reasonable freedom to continue its normal functions, the river will provide for the discharge of water and sediment, and will sustain an environment of high quality for wildlife habitat and for humans" (Palmer, 1976). He goes on to say, "Management must provide a sufficient corridor to allow continuation of meander belt progression [downstream]." Meander belt progression refers to the downstream migration of meander bends as the stream erodes the outside of banks and deposits sediment on the inside of bends and on adjacent floodplains (Figure II.5.1).

Palmer's management philosophy recognizes that such dynamic processes can cause a river to adjust its course and perform a number of other actions, or functions. This philosophy recognizes a certain kind of stability — that a river "given reasonable freedom to continue its normal function" will discharge water and sediment while sustaining a high-quality environment. Although stream management goals may include stability, we have generally failed to recognize that the river is often able to provide the outcomes we desire without human manipulation. Some engineers and scientists have begun to describe the river as "the carpenter of its own edifice" (Leopold, 1994). One well-known practitioner of applied river morphology has said that "river form and fluvial processes evolve simultaneously and operate through mutual adjustments toward self-stabilization" (Rosgen, 1996).

Generations of hydraulic engineers have used mathematical equations and rigid structural controls to define and refine the operations of a river's component parts. Although reductionist science, statistical variability, and empirical field techniques are powerful tools with which to examine both the universal and landscape-specific processes of rivers, these tools necessarily ignore the collective behaviors that arise from the properties of parts.

Traditional stream management has mirrored the tendency of science to break systems into parts. Separate resource agencies emerged to address management goals related to individual stream functions. Examples of stream function include runoff and stormwater conveyance, decreased storm flow peaks, groundwater recharge, sediment and debris sorting, nutrient cycling, pollutant retention and transformation, water cooling and oxygenation, channel maintenance, fish and wildlife habitat, and recreation and aesthetics. Planners, engineers, and designers within separate organizations are increasingly asked to sustain one or several stream functions simultaneously and, to that effect, choose designs that manage the relevant properties of component parts. Often, however, these designs fall short of meeting multifunctional goals (NRC, 1992). Greater success might be achieved if multifunctionality is viewed as a desirable property of dynamic river *systems*.

New management approaches will emphasize the complexity of river systems, addressing characteristics such as channel morphology and the integrity of the biological community, instead of isolated attributes such as channel stabilization or simplification of the stream fringe. This might require a significant change of attitude for the designer, the resource manager, and the public at large. Historically, the activities of managers (including resource agencies, planners, and individual landowners) have not been well coordinated. As competing interests in individual functions increasingly become apparent, stream managers and planners must find ways to coordinate their management goals and practices. The required changes can be viewed as properties of a social system that is as dynamic and complex as the river itself.

The following sections explore application of systems science to aspects of river management and planning.

Interactions

According to systems theory, systems are controlled by interactions among component parts. Like other systems, a river's ability to maintain multiple functions is dependent upon the magnitude and dynamics of such interactions. Loss of connectivity between component parts can alter the flow regime, degrade water quality, and affect the health and distribution of aquatic communities.

Connectivity in a stream is expressed in several dimensions: longitudinally (upstream/downstream), laterally (channel/floodplain), between scales (microhabitat/macrohabitat), and temporally (timing of flows). One of the most significant examples of connectivity is the interaction between channel flows and the floodplain. Filling of floodplains, channelization, berming, and riparian vegetation removal are common practices that limit interactions between stream system components. A number of processes are potentially affected, including vegetative growth, groundwater recharge, stream flow regime, erosion and deposition, and nutrient cycling.

For example, some of these management practices result in lowering of the water table below the riparian rooting zone, which alters nitrogen cycling within riparian areas.

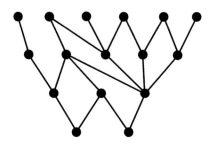

Figure II.5.2 A schematic representation of a food web showing few species with many links and many with only one or two links. The food web is presented as a graph. Highly connected species can exist at different trophic levels depending on the particular food web. (Adapted from Sole, R.V. and Montoya, J.M., Complexity and fragility in ecological networks, Santa Fe Institute Working Paper 00-11-060, 2000.)

Nitrogen removal within the riparian area is biologically mediated through plant uptake and bacterial processes. The biophysical structure necessary for maximum nitrogen removal includes a dense and deep plant root structure, and organic matter such as dead roots and fallen leaves. By lowering the water table, management practices can prevent nitrate-laden shallow groundwater from interacting with the rooting zone where these conditions exist.

Nitrogen removal is also an example of how organisms help to mediate local and regional flows of energy and materials (Naeem et al. 1999). Many stream ecosystem processes are sensitive to declines in the diversity of aquatic organisms. Biological diversity stems not only from the numbers of species involved, but also from the richness and variety of interactions organisms have with their environment and with other organisms.

It appears that successful management of aquatic communities requires the maintenance of characteristic types and levels of connectivity to which biotic communities are adapted. Some aquatic food webs are characterized by numerous weak connections between species and very few strong interactions (Montoya and Sole 2000) (Figure II.5.2). Although food webs show resilience to random removal of species, disturbances that result in selective removal of highly connected species may be enough to promote ecosystem collapse (Sole and Montoya 2000).

Thus, the type and rate of disturbance, and in addition, the degree to which its spatial effect is either amplified or muted by connectivity, are important factors in maintaining biodiversity and multiple ecosystem functions. Within stream systems that exhibit moderate levels of connectivity, aquatic communities can be locally unstable, yet globally stable. In contrast, landscapes that are highly connected can be globally unstable, because disturbances are able to spread rapidly through a homogeneous system (Green 1994).

Periodic ecosystem disturbances, such as flooding or fragmentation, provide critical input to the structure and dynamics of stream communities. Some evidence suggests aquatic communities require a moderate regime of disturbance fluctuation over time and in space in order to remain healthy (Reice et al. 1990). In contrast, as a part of watershed development and water management, we tend to drastically alter disturbance regimes by preventing flooding, clearing floodplain vegetation, or by directly modifying the channel. These changes are particularly harmful to gravel-spawning fish and benthic invertebrates when alteration of natural channel morphology and hydrology changes sediment deposition dynamics (Kondolf et al. 1987). In rivers with natural channel morphology (e.g., pool/riffle development), channel-forming flows at or near bankfull tend to remove fine sediments deposited on stream beds while leaving beneficial coarse gravels (Kondolf and Wilcock 1996; Wilcock et al. 1996).

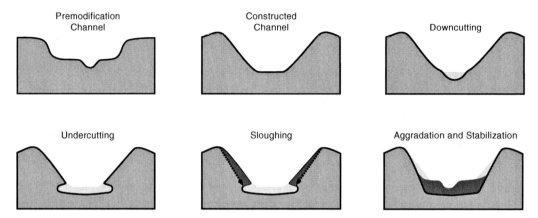

Figure II.5.3 A representation of channel evolution following channelization observed in West Tennessee. The modified cross-section is shown superimposed over the original natural channel. The dashed line represents the original modified cross-section. After modification, the channel begins to downcut and then undercut banks, which eventually slump into the channel. A new, low-flow channel cuts a path through sediment deposits. (Adapted from Simon, A. and Hupp, C.R., The recovery of alluvial systems in response to imposed channel modifications, West Tennessee, U.S.A., in *Vegetation and Erosion*, Thornes, J.B., Ed., John Wiley & Sons, New York, 1990.)

In order to maintain high levels of biodiversity, planners, designers and managers should strive to maintain natural ranges of the type, rate, and extent of disturbance, and to avoid creating overly homogeneous or fragmented landscapes that limit interactions.

System adjustments

The functionality of a system depends on the nature and arrangement of its component parts. From this perspective, channel form at the scale of the stream reach might be understood as an emergent property of dynamic interactions between variables. Within the field of fluvial geomorphology (the field that studies how running water and the landscape shape each other), stream form is considered a function of the following interrelated variables: stream discharge, sediment load, sediment size, slope, velocity, width, and depth — as these adjust to resistance created by vegetation and geology (Leopold et al., 1992). A change in any one of the variables causes the other variables to adjust.

Systems theory proposes that a preferred arrangement for the system exists and that systems evolve toward the preferred state. The flux of energy and matter through dissipative systems is a driving force generating order. Of the many possible channel forms that could result from the driving variables, the channel tends toward a most probable form related to the dissipation and distribution of energy within the system (Leopold, 1994). Researchers in fluvial geomorphology have noted that alluvial channels appear to travel through different stages of "channel evolution." One widely cited example of this theory is shown in Figure II.5.3, which documents evolution of channel morphology toward a new state following disturbance (Simon, 1989).

The culturally accepted stream-planning and management approaches in many areas of the country do not allow river systems to self-organize. This is particularly true in urbanized or settled areas, where infrastructure has been built within the floodplain, and in areas that require drainage of wet soils for development. In these situations, stream channels are typically maintained in a modified form.

The history of the midwestern United States is closely associated with the conversion of wetlands and wet soils to productive agricultural lands. As settlement of wetlands proceeded,

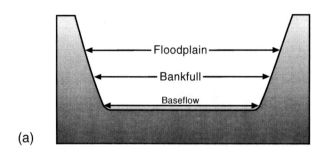

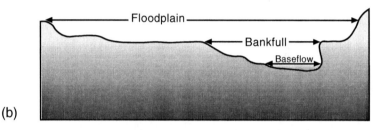

Figure II.5.4 Cross-sectional views of (a) a trapezoidal-shaped drainage ditch, and (b) a natural channel cross-section. The drainage ditch has been deepened to increase water-carrying capacity and to lower the water table in a timely fashion following storm events. Notice the dimensions of the channel have been changed so that the bankfull flow is now spread over a wider area, slowed and made shallower. This promotes deposition. In addition, flows that once accessed the floodplain are now contained within the channel. This increases erosive force on the ditch banks. (Adapted from Rosgen, D., *Applied River Morphology*, Wildland Hydrology, Pagosa Springs, CO, 1996.)

the first drainage efforts were ditching programs, introduced primarily to remove standing surface water. Today, a mixture of surface and subsurface drainage measures are used to keep basements and yards dry; remove water from parking lots and roads; lower the water table for agricultural production; and treat septic system effluent in leach fields.

The traditional drainage ditch was historically designed for, and highly effective at, achieving one function: rapid downstream conveyance of unwanted rainwater. Drainage improvement has been seen as a way to help rivers and promote stewardship of the land. A straightened, trapezoid-shaped channel was typically accepted as the ideal design, as evidenced by the comments of one Wisconsin engineer who wrote that "the streams running through [swamps and bottoms] are not swift enough, meandering about and taking double, or even more, the necessary time to cross a certain distance. This meandering, crooked or irregular course of a stream is also the course [sic] of its unequal width and depth, and of the wear and tear of its banks" (Kessenger, 1890).

A schematic of a typical drainage channel cross-section compared to a natural stream cross-section is shown in Figure II.5.4. In an attempt to achieve channel stability and control flooding, natural channels are often widened and straightened. This design is intended to expand channel capacity by increasing the width of the channel. Channel straightening increases the channel slope, which accelerates mean flow velocity during peak runoff events and facilitates the timely removal of water.

Over time, however, adjustments within the channel can threaten these functions, necessitating frequent repair of the modified channel. Flow distributed across the uniform

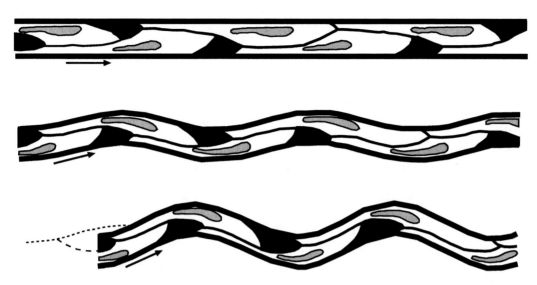

Figure II.5.5 Some straightened channels begin to develop deposition and erosion patterns that force meander redevelopment shortly after modification. Water rarely flows in a straight line.

bottom of the ditch promotes sedimentation and channel aggradation. Flows begin to accelerate around depositional areas as the channel regains meander geometry and slope (Figure II.5.5). Meander formation acts against ditch walls, which erode and slump, adding additional sediment to the channel (this process is also illustrated in Figure II.5.3).

In some systems, straightening and piping of headwater streams to facilitate development result in faster delivery of larger flows downstream. Increased discharge leads to downstream channel instability and adjustment to new and often larger channel configurations. If adequate channel capacity is not available downstream, flooding occurs.

A typical response to flooding is construction of levees or other containment structures. This strategy can fail to prevent flooding if aggradation continues to cause flood elevations to rise. For example, for hundreds of years upstream land management practices added large amounts of silt to the Yellow River of China. Levees were continuously built higher to contain flood flows. As a result of this centuries-long cycle, the river bed is now at a higher elevation than the surrounding land in some areas.

Urbanization exacerbates the effects of channelization by changing the water cycle within watersheds (Ferguson, 2001). Water drained from suburban housing developments and urban roofs, roads, and parking lots is typically directed to the existing stream network. Water that would have seeped slowly through soils and back to the stream channel now reaches the channel as runoff. The same storm event produces higher flood peaks and larger, faster flows. As a result, erosive forces on stream banks and beds increase.

A typical management response to bank erosion is hardening of the banks with concrete slabs, tires, or other means in order to hold soil in place. This prevents the channel from adjusting to a new size that accommodates increased stormflows. The river system responds by carving a new pattern in reaches that have not been hardened, and the resulting erosion can lead to bed aggradation in another part of the system.

Both urban and agricultural drainage systems must be maintained in the desired state at considerable expense. In 47 western-Ohio counties alone, almost $3 million was spent on drainage ditch maintenance in 1996 (Atherton, 1998). These costs and others must be balanced against the benefits of traditional approaches that maintain a single objective.

Planning and management approaches that recognize typical river system responses should be investigated for their ability to lower long-term costs and maintain a full range of natural function. Today's river scientists increasingly view meandering channels and connected floodplains as mechanisms by which potentially destructive energy is dissipated from the system.

Planning approaches

Although a great deal is known about river system function (i.e., the importance of floodplains, nutrient cycling, riparian corridors, channel processes, etc.), society still continues to make behavioral choices — decisions regarding stream management — that have severe unwanted consequences, such as ecological degradation and risk to life and property. Why do we continue to make these choices? Similar to river ecosystems, the human community is a "self-organizing system" whose dynamics may be understood through the lens of complexity theory.

In any particular watershed, the number of players with a stake in managing the river is startlingly large (Figure II.5.6). Interactions between stakeholders are fluid and complex. Each of the stakeholders holds different meanings and values for the river. In the terms of Luhman (Luhman, 1982), each stakeholder group approaches the dialogue around stream management from the perspective of its own "meaning world." These "meaning worlds" are related but functionally differentiated — for example, the worlds of property rights, of economics, of governmental bureaucratic culture, of environmental law, etc.

Within particular meaning worlds, individuals engage in conversations that manifest the social system, articulate the individuals' perceived interests, and define the boundaries of stakeholder behaviors. In the current socioeconomic climate, the discourse between meaning worlds tends to result in one or several meaning worlds trumping all others as the ultimate determinant of resource management acts. As a result, the dialogue produces actions that can privilege one function of the river system at the expense of many others.

The fact that stakeholders, including watershed residents, policy shapers, and educators, generally share a limited vision of the stream system may contribute to this outcome. Stakeholders who view their streamside property or function of interest as separate from the entire system may not recognize the full consequences of individual actions. As a result, the most politically expedient or profitable solution is often chosen, although the consequences include costly system adjustments and loss of overall stream function.

Systemic changes are often difficult to predict within a complex system. To continue basing actions on ignorance of system responses, however, puts land managers in the position of constantly reacting to the negative consequences of past decisions. Approaching the planning process with a vision of the river as a complex system could change the ultimate valuation of design and management approaches.

One strategy for transforming the shared vision of stakeholders is to introduce a new meaning world in the guise of a new institution or organization that holds a vision of the stream as a complex system and explicitly makes this vision a subject of its discourse. Two organizational formats that possess this capacity are (1) the stream, river, or watershed association, and (2) the stream, river, or watershed management advisory group, both containing members with diverse interests in stream function. These groups tend to have development of multiobjective management plans or related planning processes as a primary objective.

Participation in a stream, river, or watershed association ostensibly expands the member's frame of reference to a larger scale of system function (i.e., from the reach traversing the individual's property, to the entire system). Participants realize that the function of a particular reach (for example, stability of a stream bank near one's house) is contingent on the way other actors manage their respective properties. Where individual interest transcends the limits of a particular meaning world, a new state of stream corridor

Chapter II.5: Treating rivers as systems to meet multiple objectives 439

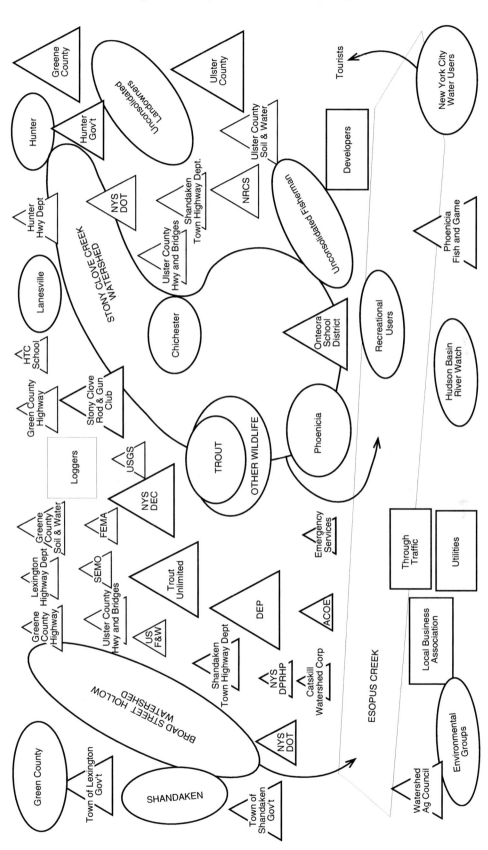

Figure II.5.6 Map of stakeholders in two adjacent subwatersheds in the Catskill Region of the New York City Water Supply Watershed, produced by representatives of landowners, municipal officials, sport fishing groups, and area resource agencies during a participatory stream management planning session initiated by the NYC Dept. of Environ. Protection, Stream Management Program. Each triangle, circle, or rectangle represents a distinct management interest.

management may emerge that emphasizes long-sighted values such as ecology, community accord, and quality of life.

In the case of a stream or watershed management advisory group, previously uncoordinated, disparate management goals and practices are considered in a unified context, and contradictions in management approaches can be made explicit. The group is a forum for a new discourse that addresses the consequences of practices aimed at preserving a single stream function. Process linkages in the physical system can become the subject of conversation within and between management organizations.

Although these organizational forms have the capacity to generate new meaning worlds that reflect the complexity of rivers, the emergence of unifying dialogues is not a foregone conclusion. Because stakeholders may have historically antagonistic discourses (and practices), it can be useful to employ professional facilitators of small-group process, whose primary role is to mitigate the tendencies of participants to regress into purely self-referencing discourses. Such processes might reveal coupled interests that were not previously apparent, allowing management responses to emerge that support multiple human interests and natural system functions.

In addition to facilitation, simple frameworks for understanding both river and human behavior might be key to shifting the dialogue between diverse interest groups. One example is use of the Rosgen stream classification system (Rosgen, 1994) (Figure II.5.7) to impart some understanding of channel processes to a lay audience. While fluvial geomorphology has been the subject of academic study for decades, Rosgen's typology has enabled discussion of the relationship between stream form and function, and the effects of various management practices on system function, with local officials, planning boards, and interest groups throughout the country (Benoit and Wilcox, 1997; Azary, 1999; Lovegreen, 2001).

Summary

Rivers of any size, whether a tiny rivulet or the mighty Mississippi, are systems that follow physical laws and exhibit similar patterns of behavior that can be described roughly with mathematical functions. This allows us to predict their behaviors reasonably well. For instance, we know that streams will flood, and we can predict the frequency with which certain sized floods will return.

What we cannot predict is the exact date or year that floods will occur. We know that rivers within certain valley types and within certain climates tend to form patterns that can be classified. We cannot predict the exact course a river will take, or the exact way an aquatic ecosystem will respond to a flood event. The reason we cannot make these exact predictions is that rivers, at all scales, are multivariate, complex systems. Rivers exhibit properties that cannot be predicted from their component parts. Furthermore, rivers are subsystems nested within and connected to larger global systems.

Although we can predict that rivers will flood (albeit, without knowing exactly when), we continue to build structures within floodplains. As a result, floods produce predictable outcomes such as loss of property that we unreasonably continue to see as unmanageable tragedies. If our conversations begin to reflect an awareness and appreciation of the stream as a system, it is likely we will develop organizational structures that consider our collective interests in all of the functions that rivers manifest.

As river managers and planners, perhaps we too can become as artful as the meat cutter in the poem at the beginning of this chapter. We suggest that ecologically sensitive water planning, design, and management can meet multiple goals for the river by applying the strategies in the following section.

Chapter II.5: *Treating rivers as systems to meet multiple objectives* 441

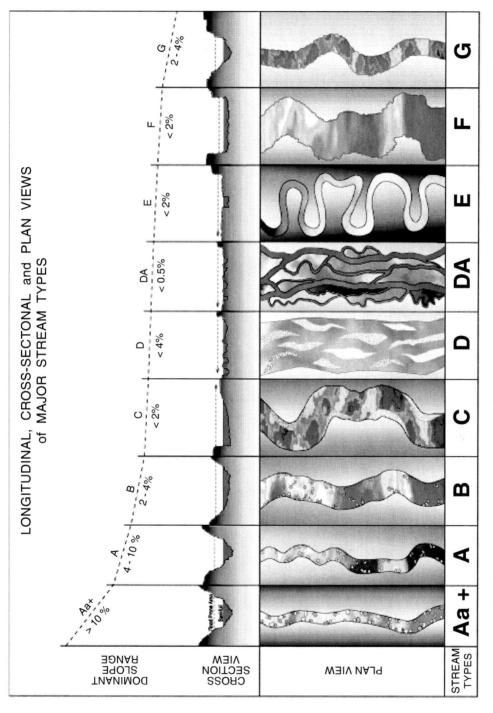

Figure II.5.7 Level I of Rosgen's stream delineation and classification process provides a general characterization of valley type and landforms and identifies the major stream types, A through G. (From Rosgen, D., *Applied River Morphology*, Wildland Hydrology, Pagosa Springs, CO, 1996. With permission.)

Planning and management

- Allowing the stream to utilize its floodplain — this requires limiting floodplain uses to low-impact activities without altering or narrowing the floodplain
- Recognizing rivers and streams as systems with emergent properties at all scales of organization
- Creating organizational structures to deal with management of streams as systems

Site design and engineering

- Allowing the channel to manage itself, except in extreme cases (such as logjam removals that threaten existing infrastructure) or for initial restoration of stream processes
- Designing culverts and bridge crossings to allow maximum meander belt width
- Maintaining important connections between and within physical and biological components (subsystems) of the river system

Biological

- Maintaining and restoring biologically healthy stream corridors
- Maintaining important connections between and within physical and biological components (subsystems) of the river system

How does the meat cutter become a master? By transcending, without losing, the components of the system:

> But now, I see nothing
> With the eye. My whole being
> Apprehends.

Literature cited

Atherton, B., Drainage improvement assessment methods and subsurface drainage practices in Ohio, Ph.D. thesis, Ohio State Univ., Columbus, 1998.

Azary, I., Using the Rosgen stream classification system as a pedagogical tool, presented at the California Geographical Society annual meeting, San Diego, CA, May 1, 1999.

Benoit, T. and Wilcox, J., Applying a fluvial geomorphic classification system to watershed restoration, *Stream Notes*, Fort Collins, CO, July, 1997.

Ferguson, B., Stormwater management and stormwater restoration, in *Handbook of Water Sensitive Planning and Design*, France, R.L., Ed., CRC/Lewis Publishers, Boca Raton, FL, 2002.

Green, D.G., Connectivity and complexity in landscapes and ecosystems, *Pacific Conserv. Biol.*, 1, 194, 1994.

Kessenger, H., Draining swampy districts, *The Drainage J.*, 12, Sept., 1890.

Kondolf, G.M., Cada, G.F., and Sale, M.J., 1987. Assessment of flushing flow requirements for brown trout spawning gravels in steep streams. *Water Resour. Bull.*, 23:927–935.

Kondolf, G.M. and Wilcock, P.R., 1996. The flushing flow problem: defining and evaluating objectives. *Water Resour. Res.*, 32(8):2589–2599.

Leopold, L.B., *A View of the River*, Harvard Univ. Press, Cambridge, MA, 1994.

Leopold, L.B., Wolman, M.G., and Miller, J.P., *Fluvial Processes in Geomorphology*, Dover Publications, New York, 1992.

Lovegreen, M., Bradford County, Pennsylvania Conservation District Manager, personal communication, 2001.
Luhman, N., *The Differentiation of Society*, John Wiley & Sons, New York, 1982.
Montoya, J.M. and Sole, R.V., Small world patterns in food webs, Santa Fe Institute, working paper 00-10-059, 2000.
Naeem, S. et al., Biodiversity and ecosystem functioning: Maintaining natural life support processes, *Issues in Ecology*, 4, 1999.
National Resource Council (NRC), *Restoration of Aquatic Ecosystems: Science, Technology, and Public Policy*, National Academy Press, Washington, DC, 1992.
Palmer, L., River management criteria for Oregon and Washington, in *Geomorphology and Engineering*, Coates, D.R., Ed., Dowden, Hutchinson & Ross, Inc., Stroudsburg, PA, 1976.
Reice, S.R., Wissmar, R.C., and Naiman, R.J., Disturbance regimes, resilience, and recovery of animal communities and habitats in lotic ecosystems, *Environ. Manage.*, 14, 647, 1990.
Rosgen, D., A classification of natural rivers, *Catena*, 22, 169, 1994.
Rosgen, D., *Applied River Morphology*, Wildland Hydrology, Pagosa Springs, CO, 1996.
Simon, A., The discharge of sediment in channelized alluvial streams, *Water Resourc. Bull.*, 25, 1177, 1989.
Simon, A. and Hupp, C.R., The recovery of alluvial systems in response to imposed channel modifications, West Tennessee, in *Vegetation and Erosion*, Thornes, J.B., Ed., John Wiley & Sons, New York, 1990.
Sole, R.V. and Montoya, J.M., Complexity and fragility in ecological networks, Santa Fe Institute, working paper 00-11-060, 2000.
Waldrop, M.M., *Complexity*, Simon & Schuster, New York, 1992.
Wilcock, P.R. et al. 1996. Specification of sediment maintenance flows for a large gravel-bed river. *Water Resour. Res.*, 32(9):2911–2921.

Response

Beyond the banks:
Holistic planning of rivers as more than the sum of their parts

If there is a single and strong, take-away message from this chapter by Leslie Zucker, Anne Weekes, Mark Vian, and Jay Dorsey, it is that rivers are not isolated entities in the landscape, but behave as complex systems whose management requires consideration of the interconnectedness of both elements and processes. Traditional approaches to river management have been limited due to tendencies to break up and study the individual influencing variables such as stormwater runoff, groundwater recharge, sedimentation, nutrient cycling, etc. as if they were independent, ignoring the simple fact that water, more than anything else, is a landscape integrator. This belief is echoed in Chapter II.2 by James MacBroom.

For water sensitive planning of river systems to occur in an effective way, we need to dramatically shift our focus in planning for rivers to planning for river *systems*. "Systems," Zucker and her co-authors argue, are not static systems but are constantly adjusting (in the short term) and even evolving (in the long term) in a dynamic equilibrium of self-organization.

As Zucker et al. state, for this process to be truly successful, water sensitive planners must not forget that the complex river systems also support and are supported by complex social systems. Such planners must search for more effective means of technology transfer to help educate the lay public and riparian stakeholders. Adjustments to current institutional frameworks might also have to be reexamined to help implement planning decisions. Finally, homeowners must be constantly and reiteratively educated about the importance in regarding river systems from a collective watershed perspective that extends beyond the bounds of their individual property lines.

chapter II.6

What progress has been made in the Remedial Action Plan program after ten years of effort? (Ontario, Canada)

Gail Krantzberg and Judi Barnes

Abstract

This chapter provides a brief history and the current status of the Canada/Ontario Great Lakes Remedial Action Plan (RAP) program. It highlights the progress that has been achieved in developing and implementing actions to rehabilitate the Great Lakes since the inception of the RAP Program in 1987, and argues for new methods of documenting and communicating accomplishments and impediments. Strategies that address the impairments related to micropollutants are progressing. Governments have been focusing on reducing loadings of chemicals from industrial sources and extensive abatement activities have advanced point source controls. This has resulted in declining levels of contamination in water, sediment, and biota, including the elimination of several fish consumption restrictions in a number of areas of concern (AOCs). Habitat rehabilitation is also proceeding well, due to the ability of the RAP process to engage volunteers and partners in local action that is both tangible and visible. Conversely, the diffuse nature of nonpoint source inputs, including stormwater and combined sewer overflows, continues to require considerable financing. The current climate of constraints in government spending means that limited resources are available to implement storm and wastewater controls at a rate anticipated when RAPs began. Nevertheless, noteworthy accomplishments include new, cost-effective technologies that are effective at controlling nutrients and bacterial loadings. To date, more than half the actions predicted to be necessary to restore beneficial uses in Ontario have been implemented. Our environmental quality index features the number of beneficial uses impaired and the extent of impairment. The higher the value of the index, the greater the degree of impairment. In many locations, the index values have declined appreciably.

The Remedial Action Plan program: Why and how?

Industrial, municipal, and recreational uses of the Great Lakes have imposed great stress on the basin ecosystem. Insults include toxic substances in air, water, and sediment, bioaccumulation and food web magnification, bacterial contamination, and eutrophication. Native fish, bird, mammal, and plant species have been lost due to inputs of pollutants, the introduction of exotic species, and the profound destruction of habitat.

The Canada–U.S. Great Lakes Water Quality Agreement (GLWQA), first signed in 1972 and amended in 1978 and 1987, is the primary mechanism for ensuring a coordinated, binational approach to management of environmental quality in the Great Lakes basin. Its purpose is to "restore and maintain the chemical, physical, and biological integrity of the Great Lakes basin ecosystem" (IJC, 1988). In 1985, the Great Lakes Water Quality Board of the International Joint Commission identified 42 degraded areas of concern around the Great Lakes. Areas were characterized by some or all of the following conditions:

- Restrictions on fish and wildlife consumption
- Tainting of fish and wildlife flavor
- Degradation of fish and wildlife populations
- Fish tumors or other deformities
- Bird or animal deformities or reproduction problems
- Degradation of benthos
- Restrictions on dredging activities
- Eutrophication or undesirable algae
- Restrictions on drinking water consumption, or taste and odor problems
- Beach closings
- Degradation of aesthetics
- Added costs to agriculture or industry
- Degradation of phytoplankton and zooplankton populations
- Loss of fish and wildlife habitat

According to the language of the Great Lakes Water Quality Agreement, these are referred to as impaired "beneficial uses." With the signing of the 1987 Great Lakes Water Quality Agreement, Canada and the United States agreed to restore these locations by developing and implementing remedial action plans (RAPs). Figure II.6.1 depicts the locations of the Canadian areas of concern.

Restoration of beneficial uses within the Areas of Concern is the primary goal of RAPs, which are characterized as proceeding in three stages. Stage 1 involves identifying impaired beneficial uses and sources of environmental degradation. In stage 2, the restoration goals are determined by the local communities with help from agency experts. Actions required to restore the impaired uses are identified along with a timetable for implementation. Stage 3 records that the restoration goals and targets have been met and impaired beneficial uses have been restored. Once beneficial uses have been restored to the satisfaction of the public and the governments, the site is stated to no longer have the attributes of an Area of Concern and is removed from the list. This is termed "delisting." In Canada, one area has been delisted (Collingwood Harbour) and by the end of 1998, the Areas of Concern were beyond the planning phase of stage 2. All are implementing actions to restore the areas of concern. RAP success depends on accomplishing visible elements. This is partly due to the tenet that informed individuals, groups, and clubs are interested in getting involved and making a difference. Public involvement in RAPs has been a major breakthrough for ecosystem management and recovery (Hartig and Zarull, 1992).

Methods for tracking progress

Canada and Ontario reaffirmed their commitments to joint development and implementation of Remedial Action Plans for the Canadian Areas of Concern through the 1994 Canada–Ontario Agreement Respecting the Great Lakes Basin Ecosystem (COA, 1994). The 1994 COA committed Canada and Ontario to restore 60% of impaired beneficial uses across all 17 areas of concern, leading to the delisting of 9 areas by the year 2000. All AOCs

Chapter II.6: *What progress has been made in the Remedial Action Plan program?* 447

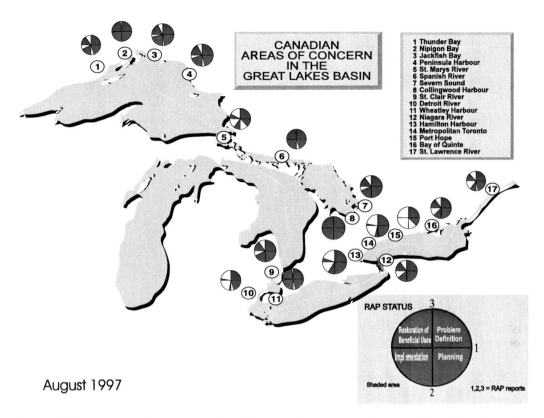

Figure II.6.1 Location and status of the Canadian Great Lakes areas of concern. (From Krantzberg, G., *Aquatic Restoration in Canada*, Backhuys Publications, 1999. With permission.)

and impairments were considered to be equally important, and efforts were directed at all locations. It is becoming increasingly apparent, however, that while this is was challenging objective, it does not account well for cumulative and step-wise progress in environmental recovery. It was also conceived that if all implementation was complete by 2000, then an area of concern would be delisted. This, in fact, is not the case, as we point out in the following paragraphs.

Progress on RAP development and implementation has been described by the parties, jurisdictions, and the International Joint Commission in a fairly bureaucratic context, that is, the completion and submission of stage 1, 2, and 3 reports to the IJC. Progress has been criticized as slow when measured against the finding that only one of all Canadian and U.S. Areas of Concern has been delisted. This is a very limited evaluation of the current status and achievements in the areas of concern.

What are needed are measurements of incremental progress toward fully achieving the goals and restoration targets at the Areas of Concern. RAPs require a long-term commitment in order to restore beneficial uses (Water Quality Board, 1991). Along the way, intermediate indicators of progress could be reductions in the number and extent of stressors such as declining chemical discharges or additive initiatives to rehabilitate degraded habitat (Environment Canada and U.S. EPA, 1994) By recognizing that progress is achieved, and should therefore be documented and celebrated in a step-wise fashion, RAP participants will be better able to maintain and broaden partnerships and momentum for Great Lakes rehabilitation and protection for the long term (Water Quality Board, 1996).

To address the shortcoming to date in reporting progress, Canada and Ontario have been making use of a status pie diagram that depicts the extent to which planning,

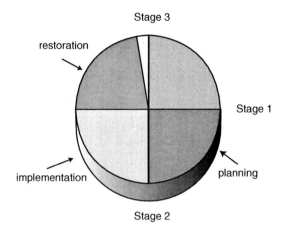

Figure II.6.2 Status pie diagram showing RAP development, implementation, and restoration of beneficial uses. (From Krantzberg, G., *Aquatic Restoration in Canada*, Backhuys Publications, 1999. With permission.)

implementation, and restoration are complete (Figure II.6.2). In each Remedial Action Plan status pie diagram, the first three quadrants represent problem identification (stage 1), plan preparation (stage 2), and plan implementation (component of stage 3). The fourth quadrant represents the degree to which the beneficial uses have been restored, that is, the progress made toward meeting the Area of Concern's delisting targets. The shaded portion of each quadrant shows the degree to which the RAP has accomplished its task. The numbers around the edge of the pie diagram are points at which reports are submitted to the International Joint Commission. Of note in the pie diagrams, progress in stage 2 is depicted as occurring in two quadrants in planning and implementation. This demonstrates that implementation is ongoing and not contingent on completion of the overall plan.

The pie diagram as shown in Figure II.6.2 portrays that problem definition has been completed (stage 1), a plan for restoring beneficial uses, along with an implementation framework, has been developed (stage 2), and all actions proposed for restoring beneficial uses have been implemented (third quadrant). The fourth quadrant is not entirely shaded, however, reflecting that environmental recovery is not complete.

This scenario could exist for a number of reasons. For example, in the case of eutrophication, several RAPs have set phosphorus loading reduction targets from point and nonpoint sources in order to achieve receiving water quality of specific characteristics. When phosphorus load reductions have been met, residual phosphorus in the system may slow ecosystem response. The attainment of the delisting target in the receiving waters could take time. Similarly, RAPs are setting targets for rehabilitation of riparian zones and coastal wetlands. It is feasible that the necessary actions to physically rehabilitate habitat to a condition anticipated to support healthy wildlife communities have been completed. Colonization and recruitment may require time to respond, and a sufficient number of years of monitoring will be required to clearly document change. Another increasingly familiar condition is in the context of sediment management. In some Areas of Concern, the preferred option is to institute source control and allow for natural recovery. Once sources have been controlled, impairments will remain until cleaner sediment covers the more polluted deposits, or biodegradation proceeds; recovery could take several to many years.

In the instances depicted in Figure II.6.2, it is important to report to the Great Lakes community that all reasonable actions have been taken and no further active intervention is needed or planned at the present time. This in itself is a statement of victory. Of course, there should be sufficiently rigorous criteria applied to the phrase "all reasonable action," to ensure that the decision to no longer actively intervene is not misused as an excuse for inaction. Also paramount is that a formal monitoring program is in place to track natural recovery and attainment of the RAP delisting targets. We suggest that until the beneficial uses are restored to the satisfaction of the community and government agencies, delisting is inappropriate.

We propose the following criteria be used to verify that all reasonable intervention has been taken, and natural recovery is required prior to delisting. We also recommend this be used as a starting point for discussion among RAP participants, the parties, the jurisdictions, and the IJC:

- All reasonable and practical implementation has been completed, with the tools available in the present. What is reasonable and practical will vary in time, and future opportunities may exist to further intervene and speed recovery.
- The severity of the impairments will influence the rate of recovery. The time scale for natural recovery, however, should be mutually agreed upon by all RAP participants.
- The local public are to be satisfied that current conditions will respond to actions implemented and that natural recovery will be monitored.
- A pollution prevention or other maintenance plan is in place to reduce the risk of future degradation and to ensure that natural recovery can proceed.
- A process is in place to respond to future development pressures such that environmental recovery is sustainable and to implement new actions as technology advances. As a corollary, contingency plans have been developed to further intervene if monitoring results warrant a more aggressive strategy.
- Commitments to a monitoring plan and program are in place to measure progress toward environmental restoration and a mechanism is established to report periodically to the public.

Delisting would occur at such time as experts and the public concur that the delisting targets detailed by the AOC have been met. This places significant pressure on the parties to design and commit to assessing ecological recovery and intervening further if predicted recovery is not responding in the manner projected.

Status of the Canadian Areas of Concern

Across the Basin, federal, state, and provincial agencies are discussing policy and process around delisting the areas of concern. No RAP has been completed yet in the United States, and several RAPs are coming to completion in Canada. Completing RAP implementation, however, is not the same as restoring environmental quality. The focus on delisting has, in some instances, detracted from the objective of incrementally measuring gains in environmental recovery. Most recently, we have been examining methods to more precisely quantify and illustrate the last two quadrants of the status pie diagram, that is, estimating the extent to which recommended remedial actions have been implemented (third quadrant) and the degree to which the restoration targets for the beneficial uses have been attained (fourth quadrant). The point is to highlight obstacles and celebrate progress at many levels in order to sustain long-term rehabilitation and recovery (WQB, 1996). To continue to restore the Great Lakes will require a clear articulation of the accomplishments to date and the challenges remaining.

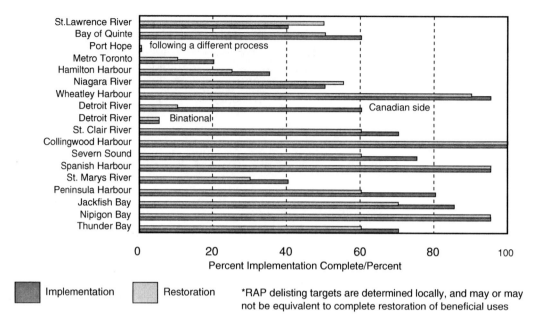

Figure II.6.3 Progress on RAP implementation and achievement of restoration targets. (From Krantzberg, G., *Aquatic Restoration in Canada*, Backhuys Publications, 1999. With permission.)

Figure II.6.3 presents a synopsis of progress for each Area of Concern. The mean degree of implementation across all AOCs by 1997 was 57% ± 35%, with the mean degree of attainment of the restoration targets being 51% ± 33%. The values portrayed in Figure II.6.3 are based on an evaluation of progress on individual impairments at each AOC. This evaluation is reported in the Remedial Action Plan Update (COA RAP Steering Committee, 1997) and by Krantzberg et al. (1997).

Progress varies with the nature of the impairment

It is instructive to examine the variability associated with progress in restoring the beneficial uses. Table II.6.1 presents the number of Canadian AOCs that have particular beneficial uses designated as either impaired or requiring further assessment. Those impairments associated with contaminated sediment and habitat destruction are pervasive across the AOCs. Of particular note is that eutrophication remains to be a problem, despite the present perception that conventional pollutants are no longer an issue and that toxic chemicals must be the central focus for environmental researchers and managers. The nearly singular discussion on toxic chemicals and Great Lakes environmental quality minimizes the real threats posed by excessive nutrient and bacterial contamination and neglects the overarching force of habitat despoilment in restructuring aquatic and nearshore terrestrial communities.

Table II.6.2 illustrates progress in restoring the impaired beneficial uses in the Canadian AOCs. In summarizing the current status for each impairment, Table II.6.2 notes in parentheses the number of AOCs that, at the onset of the program, considered the beneficial use to be either impaired or potentially impaired. In several cases, it was uncertain as to whether a particular beneficial use was impaired and further assessment was required. The actual values we used to estimate progress in restoring beneficial uses only considered cases where a change in the status of an impairment was due to implementing a remedial action. The values, then, do not include uses that were considered not impaired

Chapter II.6: What progress has been made in the Remedial Action Plan program?

Table II.6.1 Number of AOCs Reporting a Beneficial Use as Impaired or Requiring Further Assessment as of 1997

Beneficial Use Impairment	Number of AOCs
Fish and wildlife consumption advisories	9
Tainting of fish and wildlife	2
Degraded fish and wildlife populations	9
Fish tumors or other deformities	5
Bird/animal deformities or reproduction problems	1
Degraded benthos	12
Restrictions on dredging	11
Eutrophication	10
Restrictions on drinking water	3
Beach closings	10
Degraded aesthetics	9
Added costs to agriculture and/or industry	2
Degraded plankton communities	4
Loss of fish and wildlife habitat	12

Table II.6.2 Overall Progress on Restoring Restoration of Beneficial Users

Implementation and Restoration <50%	No. AOCs Represented	Implementation and Restoration >50%	No. AOCs Represented	Completely Restored
Fish tumors or other deformities	(11), n = 5	Restrictions on fish and wildlife consumption	(15), n = 9	Added cost to agriculture or industry (4), n = 2
Eutrophication of undesirable algae	(10), n = 10	Tainting of fish and wildlife flavor	(5), n = 2	
Beach closures	(11), n = 10	Degradation of fish and wildlife populations	(13), n = 9	
Degradation of aesthetics	(11), n = 0	Loss of fish and wildlife habitat	(16), n = 12	
Degradation of phytoplankton and zooplankton populations	(8), n = 4	Bird or animal deformities, reproductive problems	(8), n = 1	
		Degradation of benthos	(14), n = 12	
		Restrictions on dredging activities	(17), n = 11	
		Restrictions on drinking water consumption; taste/odor problems	(5), n = 3	

Note: Values in parentheses refer to the number of areas of concern that have addressed the beneficial use at some point within the past 10 years. Where the value "n" used for statistical analysis is less than the value in parentheses, it is due to clarifying whether or not a beneficial use was impairment and not a result of remedial measures being taken.

based on a more in-depth assessment, instead of recovery in response to remedial actions. As well, in some instances, some beneficial uses were originally designated as impaired due to a misinterpretation of the intent of the GLWQA. This was the case, for example, where RAPs assumed that exceeding a sediment chemical guideline meant that there were restrictions on dredging, yet no dredging was necessary or predicted to take place at that location. (Contaminants in sediment are considered in the context of impairments such as degraded benthos, consumption advisories, and others.)

As illustrated in Table II.6.2, RAPs have made substantial progress in habitat rehabilitation. This is partly due to the tenet that informed individuals, groups, and clubs are interested in getting involved and making a difference. Public involvement in RAPs has

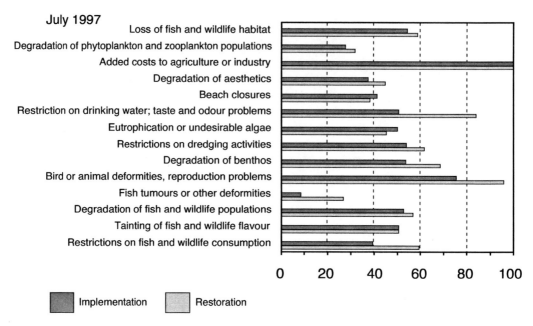

Figure II.6.4 Mean degree of progress on restoring beneficial uses in the Canadian areas of concern. (From Krantzberg, G., *Aquatic Restoration in Canada*, Backhuys Publications, 1999. With permission.)

been a major breakthrough for ecosystem management and recovery (Hartig and Zarull, 1992). A lesson learned is that RAP success depends on accomplishing visible elements, such as relatively short-term, focused projects that contribute to habitat rehabilitation (Figure II.6.4). A step-wise approach to restoring habitat achieves progressive gains in ecological integrity.

With governments focusing on reducing loadings of chemicals from industrial sources, extensive abatement activities have advanced point source control, resulting in declining levels of contamination in water, sediment, and biota, as well as concurrent improvements in benthic community structure, declines in deformities in wildlife, and fewer restrictions on fish consumption in several AOCs (MOE, 1997).

Conversely, the diffuse nature of nutrient and bacterial inputs, from nonpoint source inputs and combined sewer overflows, continues to require considerable effort. Without exception, funding is the major concern of agencies and the public involved in implementing RAP recommendations (MacKenzie, 1996). The availability of adequate financial resources to address infrastructure problems is a common concern across the RAPs. The costs to municipalities of remediating and preventing further adverse stormwater quantity and quality impacts are substantial. Although no province-wide expenditure estimates are available, clean-up costs of stormwater-induced impacts at Ontario's 16 RAP sites were estimated to be $2.5 billion. Local stormwater management programs can have annual costs in excess of $5 million (Cameron et al., 2001). The current climate of provincial and federal resource restraints means that limited resources are available to implement storm and wastewater controls at a rate anticipated when RAPs began. This obstacle to progress points to the need for research into new financing methodologies such as reported by Apogee Research (1991, 1997) and Water Environment Federation (1994). Not withstanding the above, considerable progress has been made in demonstrating new, cost-effective technologies that are effective at controlling nutrients and bacterial loadings (COA, 1997).

Development of an environmental quality index to measure change

As an alternative measure of environmental quality at the AOCs, we adapted a water quality index based on an approach developed in British Columbia and described by Rocchini and Swain (1995). The water quality index has features that quantify the number of parameters that exceed a standard and the extent to which the standard is exceeded. We substituted "the number of parameters exceeded" by "the number of beneficial uses impaired," and "the extent" by a qualitative measure of the degree to which the use is impaired, based on a consensus among RAP practitioners for each of the Areas of Concern. The higher the value of the index, the greater the degree of impairment. The index is calculated as follows:

$$EQI = \sqrt{(N \times 10/14)^2 + \Sigma(D_{i \times n}/N)^2}$$

where EQI is the environmental quality index; N is the number of beneficial uses impaired, normalized to a scale of 1 to 10; D is the degree to which each of the uses is impaired, on a scale of 1 to 10, with 10 being completely impaired. The maximum value for the index is 14. Where an impairment is present but not due to local sources (such as fish consumption advisories), a full value of 10 (fully impaired) was assigned. Similarly, where further assessment is still required to determine whether a beneficial use is impaired, we made the assumption that the use was fully impaired.

For the impaired beneficial uses, the extent of recovery was based on our interview data and consensus-based system. We used the assumption that at the point of beginning the RAP in 1987, the average degree of impairment was 10 (fully impaired). If by 1997, the degree of impairment was 40%, then the value for D was 4. With no clear rationale for ranking the environmental or social importance of the beneficial uses, all were weighed equally. Similarly, we considered the number of impairments to be equally important as the extent to which they are impaired.

As illustrated in Figure II.6.5, considerable improvements in ecosystem quality have been observed across the Basin. The mean environmental quality index in 1987 was 11.7, with an improvement to 7.7 by 1997. Based on our knowledge of environmental conditions at the AOCs, values less than 6.0 represent considerable recovery. Some extensive improvements are reflected by the index as noted at a number of AOCs.

For example, in Nipigon Bay, source control at Dow Chemical has virtually restored benthic communities. Industrial upgrades have removed tainting problems, and the habitat strategy has dramatically improved fish populations (Figure II.6.6). The Environmental Quality Index reflects these improvements, declining from 12 to 4.5. Due to substantial investments and improvements in industrial effluents discharged by Omstead Foods to Wheatley Harbour, beneficial uses are responding and the Environmental Quality Index has declined from 10.5 to 2. Similarly, to infrastructure improvements, control of industrial point sources and habitat enhancement in Spanish Harbour is mirrored by a decline in the index from 11.2 to less than 2.

The Environmental Quality Index is just one further tool that can be used to portray progress in restoring the AOCs. It diverts the focus from the singular milestone of delisting to a continuous scale where actions that result in real environmental change can be documented.

Summary and recommendations

Measuring, recording, and reporting on progress in restoring beneficial use impairments are a dynamic process, requiring continual reassessment as implementation status

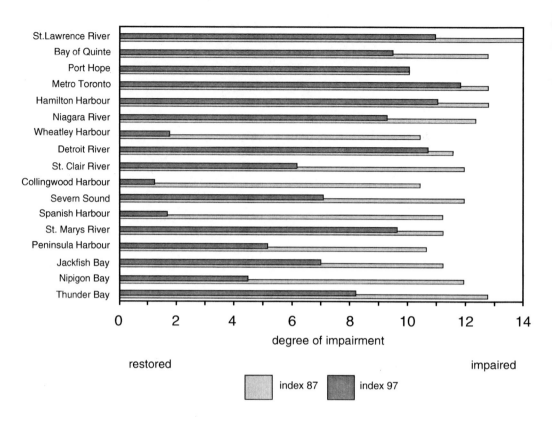

Figure II.6.5 Environmental quality index for the areas of concern showing improvements from 1987 to 1997. Please refer to the text for an explanation of the calculation. (From Krantzberg, G., *Aquatic Restoration in Canada*, Backhuys Publications, 1999. With permission.)

changes. For example, in some of the AOCs, major projects are just getting under way now that the environmental planning, assessment, or approval process has been completed. These activities were only marginally accounted for in this analysis, because actual implementation had not yet proceeded and the substantial improvements to environmental quality have yet to be realized; however, the fact that a RAP has reached the point at which a major project can be launched is itself a clear success. Other forms of progress include the establishment of local, sustainable implementation teams to continue rehabilitation and protection initiatives for the long term. Multiagency and binational cooperative arrangements are also markers of achievement that the RAP process enables.

This has been the first attempt to quantify environmental change across the Canadian AOCs. It is intended in part to gives credit to thousands of individuals for their accomplishments. A record of success is an important element in maintaining momentum. For many AOCs, the long-term strategic planning necessary to tackle complex problems, such as infrastructure and sediment remediation, builds on the record of success. It also provides a view to where obstacles to progress lie.

It is important to provide an opportunity to motivate all partners to overcome the challenges that remain to reach the goals and targets for delisting. The ability to demonstrate progress sustains public and political confidence and support (IJC, 1989). We recommend the public, the agencies, and the IJC press for such accounting.

NIPIGON BAY

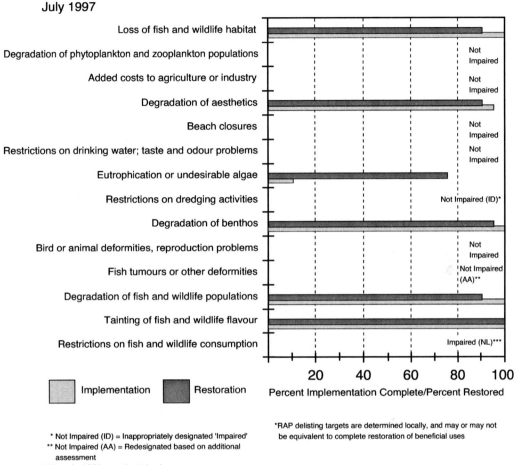

Figure II.6.6 Progress on restoring beneficial uses in Nipigon Bay. (From Krantzberg, G., *Aquatic Restoration in Canada*, Backhuys Publications, 1999. With permission.)

Acknowledgments

The authors gratefully acknowledge the willingness of all those contacted to be candid in their perspectives and for the wealth of information they provided. Their expert opinions were invaluable in attempting to measure progress toward restoration of environmental quality and identifying future challenges to restore and protect the Great Lakes Basin. RAP Teams and Public Advisory Committee members all assisted in evaluating how they had advanced over the past 10 years. The values assigned are theirs and not those of the authors.

Literature cited

Apogee Research, The user pay approach to stormwater management and its potential application in Ontario, prepared for Environment Canada, 1991.

Apogee Research, Financing options for stormwater quantity and quality management in the RMOC, prepared for the RMOC, Environment Canada, and the Ontario Min. Environ. Energy, 1997.

Canada–Ontario Agreement (COA) respecting the Great Lakes Basin ecosystem. 1994 Cleanup Fund, Great Lakes 2000 Cleanup Fund Project summaries report, Environment Canada, Downsview, ON, 1997.

COA RAP Steering Committee, The remedial action plan update, available from Ontario Ministry of Environment, Great Lakes Programs, 40 St. Clair Ave. W., Toronto, ON, www.cciw.ca/glimr/raps/intro.html, 1997.

Cameron, J. et al., User pay financing of stormwater management: A case study in Ottawa — Carleton, Ontario, submitted for publication in *J. Environ. Manage.*, 2001.

Environment Canada and U.S. EPA, Progress in Great Lakes remedial action plans: Implementing the ecosystem approach in Great Lakes areas of concern, Toronto and Chicago, 1994.

Hartig, J. and Zarull, M., *Under RAPs, Towards Grassroots Ecological Democracy in the Great Lakes Basin*, Univ. Michigan Press, 1992.

International Joint Comm. (IJC), Revised Great Lakes Water Quality Agreement of 1978, as amended by protocol signed in Ottawa, ON, Nov. 18, 1987, consolidated by the IJC, Jan. 1988.

IJC, Report on Great Lakes water quality, Great Lakes Water Quality Board, Windsor, ON, 1989.

Krantzberg, G., Ali, H., and Barnes, J., Status report on incremental progress in restoring beneficial uses at the Canadian areas of concern, a site specific analysis, Ontario Min. Environ., Toronto, ON, 1987.

MacKenzie, S.H., *Integrated Resource Planning and Management*, Island Press, 1996.

MOE, Guide to eating Ontario sport fish, Ontario Min. Environ. and Energy, and Ontario Min. Nat. Resour., PIBS 590B07, 1997.

Rocchini, R. and Swain, L.G., The British Columbia water quality index, Water Quality Branch, Environ. Protection Dept., B.C. Mini. Environ., Lands and Parks, 1995.

Water Environment Federation, *User-Fee-Funded Stormwater Utilities*, Alexandria, Va, 1994.

Water Quality Board (WQB), *Review and Evaluation of the Great Lakes Remedial Action Plan Program*, IJC, Windsor, ON, 1991.

WQB, *Position Statement on the Future of Great Lakes Remedial Action Plans*, IJC, Windsor, ON, 1996.

Response

Measuring recovery of impaired waters

The Great Lakes Basin is the largest freshwater body in the world and is one of the most highly populated areas on the North American continent. As a result, the specific locations where the two meet — the industrial harbors — have become degraded. This study demonstrates that innovative solutions are needed and are possible in even the most complicated of situations; "complicated" in terms of jurisdictional complexity involving multiple agencies across binational borders.

One of the take-away lessons from this chapter by Gail Krantzberg and Judi Barnes is in identifying the benefits ensuing from prioritizing sites to be targeted for recovery efforts — the identified "areas of concern." By concentrating efforts on those particular sites of greatest impaired "beneficial use," a step-wise approach can be adopted that can lead to progressive gains in ecological integrity. These improvements, first felt for the immediate sites, will, with time, spread to the entire system due to the ability of water to carry benefits (as well as, of course, pollutants) far. The importance of this message cannot be overemphasized. Given the magnitude of ailments plaguing our waterbodies, it is all too easy to become disheartened and dismayed at the restorative jobs that need to be done. This chapter by Krantzberg and Barnes provides a vivid demonstration supporting one of the cardinal tenets of the EPA's *Top 10 Watershed Lessons Learned* — namely, watershed management works best when undertaken in incremental steps (e.g., see Nicholas Pouder and Robert France in Chapter I.16).

Similar to many developed watersheds, those of the Great Lakes Basin faces problems of waterfront habitat rehabilitation, aquatic restoration, and pollutant source reduction. As also observed by Kelly Cave in Chapter II.9 for Detroit, Krantzberg and Barnes record that the slowest progress has been made in dealing with the elusive or "occult" — in the true sense of the word, nonpoint source pollutants. Nevertheless, the common sense approach they describe for measuring progress in their efforts toward recovery, termed RAPs, or "remedial actions plans," is commendable and could easily be adapted elsewhere: problem definition → restoration plan → implementation framework development → adoption of proposed actions → monitoring recovery. And finally, the generation of an environmental quality index is an important tool to help document and communicate accomplishments and impediments to the public.

chapter II.7

Watershed management plans: Bridging from science to policy to operations (San Francisco, California)

David Blau

Abstract

Watershed management plans are nearing completion for San Francisco's two urban watersheds — the 23,000-acre Peninsula watershed and the 40,000-acre Alameda watershed. The watershed lands surround five storage reservoirs that provide drinking water to 2.3 million Bay Area customers.

The watershed plans are the result of an in-depth, 9-year planning process that has included extensive community outreach and agency involvement. A comprehensive database was assembled for each watershed and mapped using Geographic Information System (GIS) technology. This data set was then used to create a watershed tool kit comprised of five parts:

1. Water quality vulnerability
2. Ecological sensitivity
3. Erosion sensitivity
4. Cultural resource sensitivity
5. Fire hazard

The tool kit is serving as the foundation for the long-range management plan and will also serve as a day-to-day reference base for decisions on the compatibility of land-use proposals in the future. The planning process has included cutting-edge work on such subjects as water quality vulnerability analysis and modeling; fire hazard analysis and modeling; and grazing management. The grazing management plan was a particularly challenging component of the work, considering the sensitivity of Bay Area residents to risks of exposure to *Cryptosporidium*.

Unlike many watershed management plans that stop at the policy statement level, this plan attempts to translate broad policy goals and objectives into specific management actions and then down to day-to-day operation and maintenance practices. One product of the effort, for example, is a "user's manual" for the watershed keepers in the field.

The plan document is structured in such a way that a reader can easily track from policy to actions to operations. The document is generously illustrated with photographs of watershed conditions, GIS data maps and analytical tools, planning process diagrams, and plan recommendations.

Background

More than 130 years ago, the predecessor of the San Francisco Public Utilities Commission (SFPUC), Spring Valley Water Works, had a vision of protected watershed lands that would provide a pure and reliable drinking water supply from the developing economy of San Francisco. In the last half of the 19th century, Spring Valley Water Works began purchasing the watershed lands that are now managed by the SFPUC. They first acquired the 23,000-acre San Francisco Peninsula and then the 40,000-acre Alameda Creek watershed in the East Bay. Today, these two watersheds remain largely protected and continue to serve their primary purpose — to collect and store a reliable supply of high-quality water for the homes and businesses in the San Francisco Bay Area.

The SFPUC's mission for managing their two urban watersheds — the Peninsula and Alameda watersheds — is "to provide the best environment for the production, collection, and storage of the highest quality water for the City and County of San Francisco and suburban customers." The SFPUC seeks to accomplish this by "developing, implementing and monitoring a resource management program which addresses all watershed activities." The watershed management plan "applies best management practices for the protection of water and natural resources and their conservation, enhancement restoration, and maintenance while balancing financial costs and benefits." The Peninsula and Alameda watershed management plans were prepared in response to this mission statement and because existing SFPUC policies do not address the management of watershed lands in a comprehensive or integrated manner.

The plans provide a policy framework for the SFPUC to make consistent decisions about the activities, practices, and procedures that are appropriate on SFPUC watershed lands. To aid the SFPUC in its decision making, the plans provide a comprehensive set of goals, policies, and management actions that integrate all watershed resources and reflect the unique qualities of the watersheds.

In addition to serving as a long-term regulatory framework for decision making by the SFPUC, the plans are also intended to be used as watershed implementation guides by the SFPUC's Land and Resources Management Section (LRMS) staff. The plans provide the LMRS manager and staff with management actions designed to implement and establish the goals and policies for water quality, water supply, ecological and cultural resource protection, fire and safety management, watershed activities, public awareness, and revenue enhancement. The plans also enable LRMS staff to address and plan for future management issues such as fire management, erosion control, range management, public access, security, development encroachment, and ecological resource management.

In these plans, the SFPUC has taken an approach to watershed management that considers water quality protection as the first and foremost goal. The primary goal is to:

- Maintain and improve source water quality to protect public health and safety

In addition to the primary goal, the following six secondary goals are also supported by the plans' policies and management actions:

- Maximize water supply
- Preserve and enhance the ecological and cultural resources of the watershed
- Protect the watersheds, adjacent urban areas, and the public from fire and other hazards

- Continue existing compatible uses and provide opportunities for potential compatible uses on watershed lands, including educational, recreational, and scientific uses
- Provide a fiscal framework that balances the financial resources, revenue-generating activities and overall benefits, and an administrative framework that allows implementation of the watershed management plans
- Enhance public awareness of the water quality, water supply, conservation, and watershed protection issues.

The planning process and public involvement

The watershed management planning process commenced in August 1992 and, when complete, it will have spanned nine years. The process addressed planning for both the Peninsula and Alameda watersheds simultaneously, allowing for similar goals and policies to be established for all of the SFPUC's local watershed lands. The process will culminate with the completion of the final Peninsula and Alameda watershed management plans, as well as the EIRs that will evaluate the environmental impacts of the plans in compliance with the California Environmental Quality Act (CEQA). The planning process consists of the seven stages described next, as well as an extensive, ongoing public and agency participation program.

Establish goals

One primary and six secondary goals for watershed management were established at the outset of the project by the Watershed Planning Committee (WPC), a group of SFPUC division and department representatives who assisted the planning team with the plan development and review. These goals are listed above and were used by the planning team throughout the planning process to provide direction for alternative and plan development. The goals serve as a foundation for the plan's policies and management actions and will also serve as a basis for ongoing evaluation of plan implementation.

Assemble database and prepare resource vulnerability maps

Mapping of watershed resource information was conducted on the SFPUC's GIS. Each resource type (e.g., vegetation, wildlife, etc.) was entered into the GIS system and became a separate map (or layer). Selected layers were then "sandwiched" together to provide information-rich composite maps. A set of resource vulnerability/sensitivity maps was also created for each watershed.

Formulate alternatives

The analysis of water quality, natural resources, cultural resources, and wildfire severity data gathered for the watershed was incorporated with the public comments and public survey results to form three watershed management alternatives. These alternatives applied to the management of both watersheds and were used to explore the range of options between a totally closed and totally open watershed.

Evaluate alternatives and select preferred plan

The three alternatives were evaluated against the primary and secondary goals and the requirements set by the various agencies with jurisdiction over the watershed. The alternatives were also presented at public, agency, and staff workshops. Although those in attendance at the public workshops initially preferred a more permissive approach to the management of the watersheds, the resource data, the public survey, and input from the agencies pushed the preferred plan toward a more protectionist approach that stresses controlled access to the watersheds, provides an improvement in water quality, and gives a balance among ecological resource protection, water supply needs, and watershed activities.

As a result of the alternatives evaluation process, the watershed management plan was formed and approved through and SFPUC resolution in January 1995. The preferred plan, as it was approved in January 1995, applied to both watersheds. Subsequent amendments to the Peninsula watershed management plan include the Southern Peninsula Watershed Golf Course Element (March 1997) and the Fifield/Cahill Ridge Trail Element (June 1997). Subsequent amendments to the Alameda watershed management plan included the Sunol Valley Resources Management Element (May 1996) and the Alameda Watershed Grazing Resources Management Element (July 1997).

Prepare management plans and EIRs

The general direction provided by the SFPUC in the preferred plan has been developed into two specific watershed management plans — the Peninsula watershed management plan and the Alameda watershed management plan. Each plan provides policies for decision making and actions for day-to-day management, which are specific to the character and resources of each watershed.

The environmental impacts of each plan were evaluated in a programmatic EIR, along with an evaluation of the other alternatives. The results of the EIRs necessitated some revisions to the plans and required that mitigation measures be incorporated into the final plans.

Public, agency, and staff participation program

An extensive outreach program was developed at the outset of the project and included interviews with key stakeholders; public, agency, and SFPUC staff workshops; newsletters; a public opinion survey; and development of a watershed Web page. The program was designed to elicit information and opinions from the public, SFPUC staff, and agency representatives, as well as to keep these interested parties abreast of the process as it proceeded.

Interviews

Early in the planning process, 60 individual interviews were conducted with various public agencies, local water agencies, adjacent cities and counties, neighborhood organizations, and special interest groups (e.g., golf, fishing, trails, mountain biking, and equestrian). The issues and concerns raised at these interviews were carried forward as the key issues to be addressed during the planning process.

Public, agency, and staff workshops

At six key points during the planning process, a series of workshops were held with the public. An agency workshop and a workshop with SFPUC staff were also held at these points in the planning process. These workshops allowed the planning team to present new information to the public, staff, and agencies, while receiving comments at each stage in the process. During the course of the project, meetings focused on the finding of the resource inventories and the GIS mapping, the management alternatives developed for each watershed, identification of a preferred alternative, and EIR scoping. Additional workshops were held to address issues pertaining to specific plan elements, including the management of the Sunol Valley resources, the Southern Peninsula Watershed Golf Course, the Fifield/Cahill Ridge Trail, equestrian uses, and grazing on the Alameda watershed.

Newsletters

Six newsletters were prepared to announce upcoming workshops and to summarize key issues related to watershed planning. Newsletters may be requested from the LRMS division of the PUC. Topics addressed include a summary of public opinions heard at the first series of workshops; an overview of data collected on each watershed; presentation of the watershed management alternatives; and a presentation of the preferred alternative, as well as articles on critical watershed issues, such as quality protection and fire.

Public opinion survey

A statistically valid telephone survey of 578 randomly selected households, all of which receive water from the SFPUC, was conducted in September 1993. The purpose of the survey was to ascertain the opinions of these SFPUC consumers on issues of water quality, the watershed management goals, recreational access to the watersheds, environmental protection, financing, and other issues related to watershed management. The survey results were used in conjunction with all of the other data and information collected to develop the watershed management alternatives.

The results of the customer survey tended to be more protective in terms of water quality and watershed management than the opinions voiced at the workshops. For example, 71% of the respondents stated that ensuring water quality was the most important goal of watershed management, while only 3% indicated that providing recreational access was the most important goal; 85% of the respondents wanted the same or less public access than is currently provided. Respondents were relatively restrictive in the type of public access they would allow. A majority of respondents supported natural resource studies (92%), jogging (81%), hiking (75%), and guided tours (75%). Less than half of the survey respondents would allow mountain biking (41%), fishing (28%), or golf courses (26%).

Web page

A watershed Web page (www.ci.sf.ca.us/puclrms) was developed, allowing users to review the plan online and download portions of the plans that are of interest to them. The technology also allows the Web page user to view the GIS maps. The Web page includes information on upcoming public involvement opportunities, directions for applying for access permits, and general information on the watersheds.

Technical work: Baseline and analysis

A tremendous array of resource data was compiled as input to the watershed management plans. The information served as the basis for the goals, policies, and management actions set forth in the plans.

GIS database

During the spring and summer of 1993, scientists from the EDAW team surveyed the Peninsula and Alameda watersheds to identify and map the existing natural and cultural resources. In addition to the field studies, existing data prior to watershed studies were

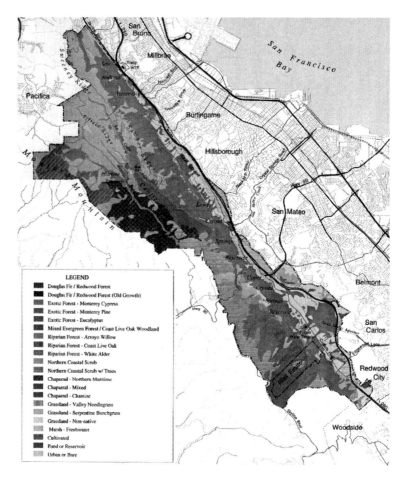

Figure II.7.1 Example of GIS database surveying the Peninsula watershed. (Color version available at www.gsd.harvard.edu/watercolors. Password: lentic-lotic.)

also reviewed. Data maps were prepared that identified the location and the extent of each resource, and data tables were created to indicate the special characteristics of each resource. GIS layers prepared as part of this project included vegetation, soils, slopes, aspect, geology and faults, mineral resources, landslide susceptibility, protected species habitat, water contamination sources, cultural resources, aquatic habitat, wildlife habitat, wildfire fuels, wildfire hazards, land use, lease areas, and protected plans. In addition, specific maps were created to analyze and convey information on specific topics such as grazing and grazing management on the Alameda watershed and trail access on the Peninsula watershed (see Figure II.7.1 as an example of the GIS database).

Land-use/water quality relationship models

Because the primary goal of the watershed management plans is to protect and enhance water quality, a tool to convey and better understand the relationship between land-use activities and water quality was developed. A series of land-use/water quality impact models was developed for each existing and anticipated activity, and the potential construction, maintenance, and operational impacts were defined. Activities that were delineated in this manner included golf courses, trails, grazing, and fishing.

Ultimate water quality impacts were divided into six general categories — turbidity and sediment; microorganisms; THM precursors; VOCs, SOC, and pesticides; nitrogen, phosphorus, and algae; and heavy metals.

One of the most important outcomes of these models was the realization that the risk to water quality associated with various watershed activities was in certain cases very high but very difficult to quantify. The outcome of this analysis was either to prohibit activities that posed too great a risk to water quality or to develop policies and management actions that reduce the impacts of a particular activity to an acceptable level of risk.

Composite vulnerability maps and the watershed tool kit

Resource vulnerability and sensitivity zones were identified using the GIS and compiled into a "tool kit" for each watershed. Each tool kit consisted of a set of five maps: the water quality vulnerability map, the composite ecological sensitivity map, the cultural resource sensitivity map, the erosion and land instability map, and the wildfire severity map. These maps were used to identify those areas of the watersheds most sensitive or vulnerable to disturbance and therefore least suitable for accommodating watershed activities. Zones of high vulnerability/sensitivity are areas in which an activity, use, or facility is most likely to have a negative impact on a resource.

The high-vulnerability/sensitivity zones from each of the individual tool kit maps were combined to create the composite high-sensitivity zones shown in Figure II.7.2 for the Peninsula watershed and Figure II.7.3 for the Alameda watershed. The composite maps illustrate that few places within either watershed are completely free of resource vulnerability/sensitivity, and in many areas two or more of the high-resource sensitivity areas overlap, indicating that activities in most locations on the watershed are likely to have an impact on at least one, if not multiple, resources.

Development of the water quality vulnerability zones

A system for identifying those areas of the watershed where activities or disturbance have the greatest potential to impact water quality was developed specifically for this project. Development of these zones, termed *water quality vulnerability zones* (WQVZs), involved extensive use of the GIS. The key criteria used to develop these zones were proximity to water varied by intensity of rainfall, wildlife concentrations, vegetation, slope, and soils. Initially, individual water quality vulnerability zone maps were prepared for microorganisms, particulates, and other groups of substances known to impact water quality. These individual water quality vulnerability zone maps were then combined to create a composite WQVZ map for each watershed (see Figure II.7.4 for an example).

Development of the resource sensitivity zones

The sensitivity maps for ecological resources, cultural resources, erosion and land instability, and wildfire severity were developed in much of the same manner as the WQVZs. The information for each individual data layer (e.g., vegetation) was assigned a ranking of high, medium, or low sensitivity. All areas of high-sensitivity were assigned a high sensitivity ranking on the composite maps, while a numeric ranking system was designed for the areas of medium and low sensitivity, which, depending on the density of resources in one area, ended up with an overall ranking of high, medium, or low sensitivity in the composite map. These composite maps indicate that in most cases, the watersheds include a significant amount of sensitive lands for at least one of the resource types (see Figure II.7.5 for the wildfire severity map).

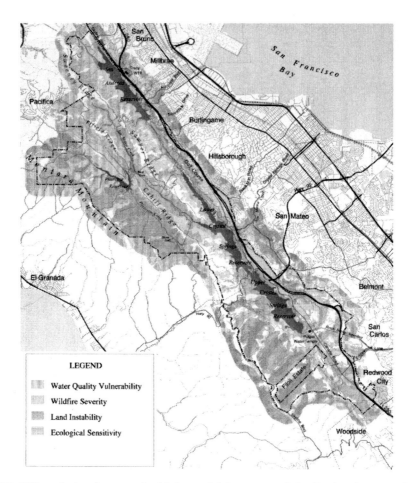

Figure II.7.2 GIS analysis of composite high-sensitivity zones of the Peninsula watershed. (Color version available at www.gsd.harvard.edu/watercolors. Password: lentic-lotic.)

Key planning issues

The key watershed management issues facing the SFPUC include fire hazard management, road management, trail management, golf course management, and control over upstream activities and urbanization. The watershed management plans strive to address these issues through the development of sound management actions, which SFPUC staff can implement.

Fire hazard management

Key issues related to fire hazard management include a significant buildup of fuels, the fear held by neighboring communities of controlled burns, the stringent air quality requirements established by the Regional Air Quality Board for controlled burns, the difficulty in balancing watershed activities with the need for fire protection, and the extensive fuel management activities that must take place to reduce the fire hazard on the watersheds.

Because fire is an extremely complex issue, fire management plans were completed for each watershed. These plans provided detailed actions developed to address the above issues and included a prioritized list of necessary treatments and treatment intervals. Recommendations for staff training and equipment purchase were also included. The GIS

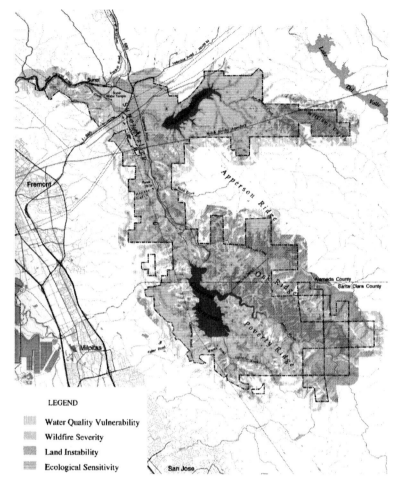

Figure II.7.3 GIS analysis of composite high-sensitivity zones of the Alameda watershed. (Color version available at www.gsd.harvard.edu/watercolors. Password: lentic-lotic.)

was used extensively to prepare models of fires and fire movement, given certain temperatures and wind conditions.

Road management

At present, roads are the single greatest, ongoing contributor of the sediment on the watersheds. Sedimentation and related erosion lead to treatment difficulties and may result in drinking water taste and odor problems. The plans include extensive management actions aimed at limiting the construction of new roads, closing and revegetating unnecessary roads, and establishing best management practices (BMPs) for road maintenance and management. Because of the impacts roads can have on water quality, the road-related management actions were given first priority in the phasing of the plan actions.

Trail management

Key trail management issues are related to the Fifield/Cahill Ridge Trail, which was recently incorporated into the Peninsula watershed management plan based on a request

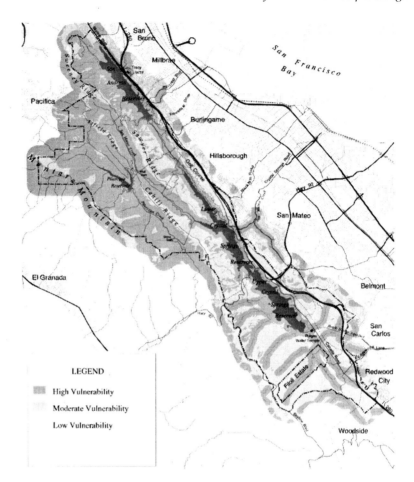

Figure II.7.4 Composite water quality vulnerability zones. (Color version available at www.gsd.harvard.edu/watercolors. Password: lentic-lotic.)

from the San Francisco Board of Supervisors. Because the trail passes through a high fire hazard area of the watershed, provides access to many linking watershed service roads, and passes through potential significant habitat areas, the trail poses significant management challenges. The trail proposal put forth by the trail proponents calls for unrestricted access for hikers, bicyclists, and equestrians. The Peninsula watershed management plan includes two alternatives to the unrestricted access proposal: docent-led access or access by annual permit, with an electronic gate tracking system. Both alternatives allow the SFPUC greater control over the number of users and greater control over off-trail use. These two alternatives also provide the SFPUC with the opportunity to close the trail or reduce the number of trail users, based on monitoring results, to protect the resources from damage. All three alternatives were analyzed for their impacts in the Peninsula watershed management plan EIR.

Golf course management

In January 1995, the SFPUC adopted the watershed management preferred alternative for the Peninsula and Alameda watersheds. The preferred alternative was based on the studies of the watersheds, their resources, and the sensitivity and vulnerability of these resources. The watershed management preferred alternative, approved in 1995, called for retention of existing golf courses, consideration of golf course expansion in areas of low sensitivity and vulnerability, and no new golf courses.

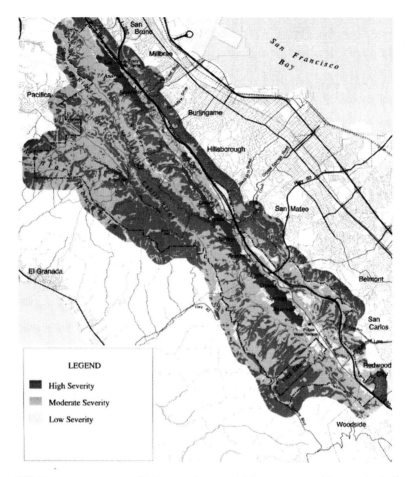

Figure II.7.5 Wildfire severity map. (Color version available at www.gsd.harvard.edu/watercolors. Password: lentic-lotic.)

On March 4, 1997, the SFPUC held a public hearing to consider revisions to the preferred alternative. After extensive discussion and review of citizen testimony, the SFPUC revised the preferred alternative to consider a golf course on the southern Peninsula watershed.

Potential issues and concerns related to the development of the golf course are its construction and operation, including the potential impacts to water quality from pesticides, herbicides, and fertilizers; the location of the site in the high- and medium-water quality vulnerability zones; increased paved and impervious surfaces, as well as extensive earthmoving, which may contribute particulates to water bodies. Of perhaps greatest concern from a water quality standpoint is the fact that human error and failure of the systems in place to protect water quality can, and do, occur and can lead to direct contamination of drinking water supplies.

Water supply concerns include the large amount of water traditionally used by golf courses, which reduces the supply available for drinking water. Ecological resource concerns include the loss of complexity of the original ecosystem due to extensive grading; the disruption of habitat that impacts migration corridors; and the impacts of golf course chemicals on aquatic species. The additional traffic generated by the golf course could generate safety, air quality, and traffic concerns.

The Peninsula watershed management plan includes a number of policies that require the golf course to adhere to strict standards in its development, construction, and operation.

These policies are intended to guide SFPUC decision making regarding the golf course and assure that if a golf course is ultimately approved, water quality and ecological resources are protected during its construction and operation. Key policies relate to the containment of runoff, the lining of water impoundments, the implementation of integrated pest management strategies, and the requirements to adhere to environmental design standards for golf courses. Other key policies require that the costs of operating and monitoring the golf course are not borne by the rate-payers, and that adequate staff and equipment be allocated to adequately monitor the construction and operation of the course. Management actions were developed and intended for use if the project is approved and related primarily to project-level studies and ongoing monitoring.

The golf course was analyzed at the programmatic level in the Peninsula watershed management plan EIR. Should the impacts at this level be found to be insignificant, the golf course may proceed to project-level design and environmental review.

Upstream activities

Upstream activities are of great concern on the Alameda watershed, as the upper two-thirds of the watershed is not owned by the SFPUC. Current uses on these lands include cattle grazing and residential development. Cattle grazing raises concerns about pathogens entering the creeks and eventually the SFPUC water bodies. The lack of streamside protection and the potential for erosion and particulates to enter the creeks and water bodies is also of concern. Residential development issues are related to septic systems and their possible failure, and the improper use and disposal of household chemicals, both of which raise water quality concerns. Domesticated pets are also a concern because of the disruption of wildlife and their habitats.

Because the SFPUC has no direct control over these lands, management actions were developed to encourage cooperation and collaboration among the various agencies involved in managing the upstream lands. Key agencies include the two counties with jurisdiction of the upper watershed and the local resource conservation district. Actions were also developed to assist in the development of educational materials to help upstream landowners understand the relationship between the activities they undertake and water quality.

Management challenges

Three major challenges face the land managers at the SFPUC as they strive to manage the watersheds in an integrated and responsible manner, described next.

Adjusting to changes in policy direction over time

Comprehensive watershed management planning is a multiyear process. At the staff level, flexibility, continuity, and endurance are needed to adjust to changing decisions and to keep various components of the plan in sync as pieces change.

As previously discussed, the development of watershed plans will span 9 years at the completion of the process. During this time, major components of the plans were reconsidered, resulting in a number of significant changes. For example, the watershed management preferred plan, approved by the SFPUC in January of 1995, was developed to provide the greatest protection to water quality and watershed resources. This plan prohibited new golf courses and unsupervised access to internal trails; however, the prohibitions on new golf courses and access to internal trails were reconsidered based on the long history of these issues, direction from the San Francisco Board of Supervisors, and additional public testimony. This resulted in amendments to the preferred alternative to include the southern Peninsula watershed golf course and the Fifield/Cahill Ridge Trail in the Peninsula watershed management plan.

A number of implications are associated with this level of change:

1. It cannot be assumed that because the players, the opinions, and the facts remain the same, the policy decisions will be stable. Instead, as the process proceeds, the arena where decisions are made changes. In one setting, technical data and public opinion surveys may carry a lot of weight, while in another, the relative political power of various parties will have more of an impact.
2. Schedules are impacted as proposed changes to the plan are analyzed and subsequently made.
3. Changes to one part of the plan have implications for other parts of the plan. Some of these are not immediately obvious and take time to consider.
4. Inevitably the process moves from stages dominated by technical data and staff to arenas where political and financial factors become more significant. Whereas presumably all changes stay consistent with the overall goals of the plan, the specifics may or may not match the recommendations of the technical staff, who support and manage the process. At different stages, staff at different levels of the organization become actively involved in the process. This also results in changes of direction. While the plan itself is changing, staff turnover also occurs. One of the challenges of managing this kind of project is to maintain staff-level commitment to the process and product when continuity is lacking in both the staffing of the project and significant aspects of the direction of the project.
5. The cycle of a utility planning project begins with the utility staff, goes through a series of stages, and returns to the staff for implementation. If, during the interim stages, the staff loses a sense of ownership in the plan, implementation will be impaired.

Short-term decision making versus long-term planning

Although it takes years to finalize and publish a watershed management plan, the planning process itself affects the way the watershed is managed. Change does not begin when the plan is published. If staff are adequately involved in the planning process, change is initiated by the planning process.

Early in the watershed planning process, it was felt that leasing decisions should be deferred until the plan was finalized. However, as the process moved forward, which involved delays to handle difficult decisions and changes in direction, it became apparent that many decisions needed to be made prior to the publication of the final plan. For example, the Crystal Springs Golf Course lease was due to expire and new lease terms and conditions were needed. The information developed through the planning process was used to create a new lease with new provisions relating to water quality and watershed protection.

Grazing on the Alameda watershed also became an issue during the planning process, with members of the public and certain decision makers requesting an end to grazing on the watershed. This request was due to concerns over the pathogen *Cryptosporidium*, which causes severe stomach distress and is of concern in San Francisco due to its high percentage of individuals with compromised immune systems.

Resource specialists and the grazing community felt that grazing was vital to the health of the ecosystem, because it reduced fire hazard on the watershed and could be managed in such a way as to significantly reduce the risk of pathogens entering the water bodies.

Working closely with the local resource conservation district and members of the community opposed to grazing, SFPUC staff and the consultant team developed a grazing management plan that retained grazing while introducing lease provisions, which provided much greater protection for water quality. Timing of calving, development of exclosures around streams and waterbodies, and limits on the number of cattle grazed were all included in the new lease provisions. Significant penalties were also included if the lease provisions were violated. The plan was especially significant in that it brought together a number of opposing groups to develop a constructive solution that was agreeable to all parties.

Implementation and funding

Watershed management is not only about what you allow others to do to your watershed; it is also about what you do to your watershed. Therefore, you need to provide the tools and guidance that will enable staff to provide a higher level of watershed protection and maintenance.

Implementation of the plans will require additional funding, additional staffing, additional training, and development procedure manuals for staff. This commitment is essential to achieve the goals of the plan. A number of steps have been taken to ensure that this occurs.

A training manual for the staff has been prepared based on the plan recommendations and will be provided to all staff involved in watershed work. The goal of the manual is to educate staff so that they may better protect water quality and the watershed's natural resources as part of their day-to-day activities. It is intended for use by nonresource specialists and will assist watershed workers in defining areas of high-sensitivity/vulnerability, locations of special status species that should not be disturbed, etc. (Figure II.7.6 is an illustration from the training manual.)

Training sessions are also planned for staff, which will familiarize watershed workers with the plans and provide them with the tools necessary to do their jobs more effectively, while protecting water quality and the natural resources.

Within each management plan, the management actions have been prioritized to assist staff with implementation. Priorities were based on staff input and the actions that were determined to be the most essential to water quality protection and improvement. The staff will be able to include several of these prioritized actions in each year's budget request, which ensures that plan implementation continues.

Certain implementation steps began early in the project. One of these was the purchase of a GIS system for the Land Resources Management Section (LRMS). The consultant team helped the LRMS select a watershed GIS system best suited to its needs, as well as interfaced well with both the consultant's GIS and other GIS systems within the city. The consultant then set up the system and trained staff to work on the GIS in order to respond to map queries and update the LRMS' internal database. This system now resides with the LRMS and continues to be used extensively to respond to data requests, to provide maps, and to periodically update the resources layers based on new information.

Conclusions

The key to a successful watershed management plan is implementation. Implementation generally occurs following plan adoption, but often the need to respond to an urgent issue or problem requires action during the planning process. This is especially true in a multiyear watershed management planning process, where new issues and concerns arise frequently.

Figure II.7.6 Illustration from the training manual based on plan recommendations. (Color version available at www.gsd.harvard.edu/watercolors. Password: lentic-lotic.)

It is important to decide which issues can and should be addressed immediately, and which should wait until completion of the plan and the environmental review process. Issues that pose a threat to the health and safety of the consumer, or another significant risk, are best addressed as they arise. Issues that do not pose a health or safety risk and that may have significant negative environmental consequences, or may be highly controversial, may not be appropriate to address prior to plan completion and environmental review. Often, these issues require additional time to build consensus, collect additional data, develop alternatives, and identify a final solution.

Implementation during the course of the planning process is most successful once data collection is complete and the key issues have been identified. This information can be used to develop interim plans to address urgent health and safety issues that must be addressed prior to plan completion.

Implementation following plan adoption requires ongoing dedication from the staff assigned to implementing the plan. Staff and managers must be provocative in securing funds, training staff in new activities and methods to execute plan recommendations, and follow-through to get the job done.

A system to evaluate plan recommendations against accomplishments on an ongoing basis is also critical to successful implementation. The plan's policies, because they provide guidance for long-term decision making, are generally fixed. The actions to implement the policies and the means and methods to implement them, however, are changing as technology, legislation, and understanding of the resource change. Ongoing evaluation of plan actions and accomplishments allows watershed staff to adjust the actions, and the means and methods to accomplish them, in order to fit the current state of the art.

Response

Sociology of implementing adaptive management

A primary goal of many watershed management plans is to protect and enhance water quality by using models that analyze the risks associated with various land-use activities (see, for example, Chapter II.8 by Dennis Haag, Stephen Hurst, and Bryan Bear, Chapter II.12 by Jeffrey Schloss, Chapter II.13 by Margot Cantwell, and Chapter II.17 by Neil Hutchinson). This chapter by David Blau identifies a critical aspect in implementing successful watershed plans — namely, that comprehensive water sensitive planning is a multiyear process that needs to accommodate potential organizational changes by factoring in flexibility through adaptive management. Although many watershed management plans stop at the policy statement level, the one that Blau describes shows how production of a "user's manual" tool kit provides one means of translating policy goals and objectives into specific management actions and operations. Two cardinal tenets are included in Blau's message: first, clear goals are essential at the get-go; second, successful watershed planning is more than just science (i.e., it is policy and implementation as well).

This watershed tool kit is spatially based, wherein water quality, ecological, and erosion sensitivities are considered along with cultural resource sites and fire hazards. All are integrated with various land uses to prioritize areas of vulnerability. The tool kit is designed to be used for both long-term forecasting as well as day-to-day decision making.

A final element of interest in this particular study includes the attention paid to arrive at balanced social goals for the watersheds. Approaches such as use of a Web site, interviews with stakeholders, newsletters, special-interest workshops, and general public surveys go far toward setting the stage for adaptively managing these watersheds through time in a truly sustainable fashion, as, for example, that outlined by Daniel Williams in Chapter II.11.

chapter II.8

Watershed assessment planning process (Johnson County, Kansas)

Dennis A. Haag, Stephen A. Hurst, and Bryan J. Bear

Abstract

The Kansas Urban Resource Assessment Project (URAMP) was established in 1997 to evaluate the effects of urban development policies and practices on riparian and wetland areas at the watershed level. URAMP was a pilot study funded by a U.S. EPA grant administered by the Kansas Water Office, with assistance from George Butler Associates, a hired consultant, and support from several state and federal agencies. The study involved a comparison of two watersheds, one urban and the other rural, located in Johnson County, KS.

Many of the URAMP conclusions and recommendations have been incorporated into new watershed planning and development policies and practices being implemented at the state and local levels of government. This paper discusses the current status of watershed planning in Johnson County. It presents a case history of a developing suburban residential community that required stormwater variances to allow for the construction of adjacent wetland, riparian, and park lands. The wetland and riparian areas are being installed as part of a residential housing development to provide multiple uses that include mitigation bank, stormwater treatment, science education, public recreation, and residential housing amenity. The project will be discussed from three planning and permitting perspectives: (1) private developer/engineer; (2) state water resource planning/Kansas Water Office; and (3) local planning/city of Overland Park, KS.

Introduction and background

In 1996 and 1997, the Kansas Water Office, the state's water policy and planning agency, completed an innovative project sponsored by the U.S. EPA Wetland Protection Grant Program. This project, the Urban Resource Assessment and Management Project (URAMP), was conducted in Johnson County, KS, a fast-growing suburban area near Kansas City, MO, (Figure II.8.1). It brought together federal, state, and city planners and private developers to explore the impacts of urbanization on wetland and riparian resources.

URAMP examined two watersheds — Tomahawk Creek, an almost fully developed urban watershed, and Wolf Creek, a rural watershed that was just beginning development (Figure II.8.2). The goal was to bring together a group of diverse stakeholders and, through a collaborative, team-building process, explore alternatives to the "we have always done it this way" approach to natural resource management.

478 Handbook of water sensitive planning and design

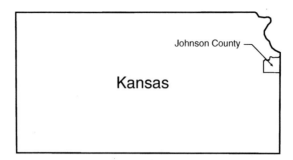

Figure II.8.1 The project is located in Johnson County, KS, one of the fastest growing suburban communities in the Kansas City, MO, metro area.

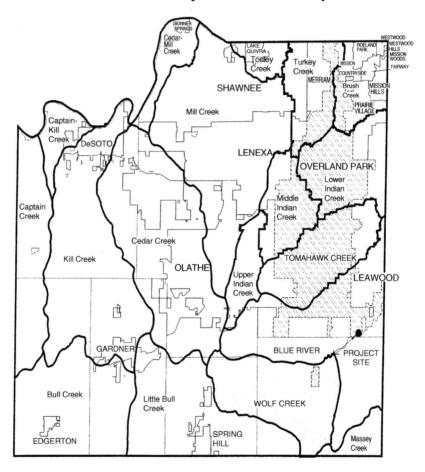

Figure II.8.2 The URAMP study compared the impacts of urbanization on wetland and riparian areas in two watersheds: Tomahawk Creek, an almost fully developed urban watershed; and Wolf Creek, a rural watershed just beginning development.

The study was divided into three phases: phase I, biological assessment; phase II, watershed assessment; and phase III, workshop and demonstration. Phase I was performed by the Kansas Dept. of Wildlife and Parks, Pratt, KS, with species sampling, identification, and enumeration; water sampling and testing; habitat and land-use evaluation, and stream flow measurements. Identification of major causes of stream degradation was also a major component of phase I. Phase II was conducted by George Butler Associates, a local planning and engineering firm located in Lenexa, KS. This phase consisted of assessing urban development impacts and their causes, reviewing urban development policies, and providing alternative methods of resource management that may reduce negative impacts. This phase also included a series of stakeholder meetings to address the above concerns. Phase III consisted of a regional conference for local decision makers on recommended alternative wetland and riparian area management strategies. Robbin Sotir, of Robbin Sotir and Associates, also conducted a soil bioengineering demonstration project with stakeholders and local planners.

The Kansas Water Office envisioned its role in URAMP as one of a cultural change agent. The project afforded diverse stakeholders such as engineering and consulting firms, homebuilders, developers, and local and state planners the opportunity to explore their attitudes toward riparian and wetland resources. They were able to examine the values and benefits, perception versus reality, of these resources and educate themselves on innovative management alternatives. Planning and resource agencies within Johnson County, and generally throughout the region, as well as developers are initiating many of these management alternatives.

The Wilderness Valley and Mitigation Bank project

One of the post-project outcomes was a project that designed and constructed a wetlands and riparian mitigation development plan, which used many of the concepts and recommendations of URAMP (KS Water Office, 1997). In particular, this project implemented two concepts and one recommendation: using wetlands and riparian areas to convey and treat urban stormwater runoff and (2) creating a wetland mitigation bank. With the support of a local school district, county park district, and the city of Overland Park, the project has integrated a wetlands park and science center into a new, upscale residential community. The project is called Wilderness Valley and Johnson County Wetlands Mitigation Bank (Figure II.8.3). It is located in the fringe of the floodplain of the Blue River watershed, Overland Park, KS (Figure II.8.4).

Johnson County, KS, is the state's most populated county. It currently has 21 cities and a population in excess of 460,000. Significant demand exists for new development of all types, including residential communities, which is a primary cause of urban sprawl. The demand for land has led to numerous conflicts with environmental concerns such as the use of floodplains, destruction of wetlands and riparian zones, and impacts on the quality of surface waters. A comparison of 1935 and 1991 USGS maps for Lenexa, KS, which is located adjacent to Overland Park, indicates that many of the wooded riparian areas have been eliminated by development. In addition, over 50% of Kansas' wetlands have been lost due to development (Kansas Wetlands and Riparian Areas Alliance, undated). Figures II.8.5 and II.8.6 provide examples of the major impact to streams and riparian areas typically caused by stormwater channel improvements, which most cities require. URAMP addressed this type of problem.

The original scope of Wilderness Valley was to construct a typical "urban-sprawl" residential community with concrete stormwater channels. The ultimate goal was to maximize the use of the Blue River floodplain by excavating and filling for development of lots (Figure II.8.7). After a review of the needs URAMP identified, however, the scope

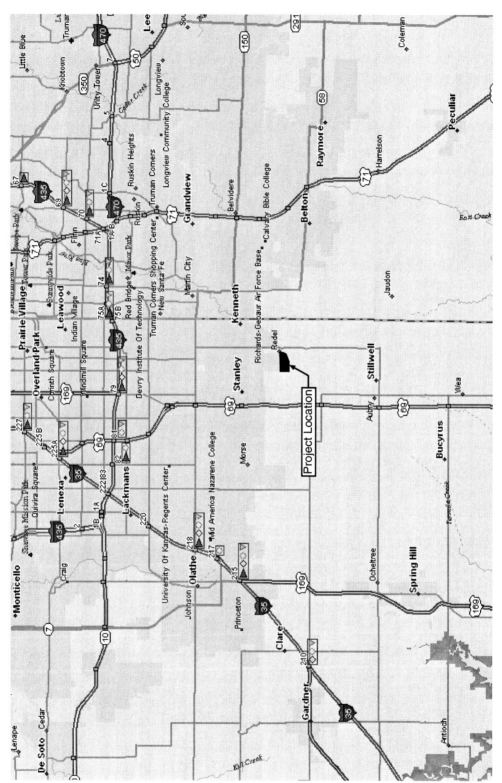

Figure II.8.3 The Wilderness Valley and Johnson County Wetlands Mitigation Bank Project is located just west of the Missouri–Kansas state line in Overland Park, KS.

Blue River Watershed Map

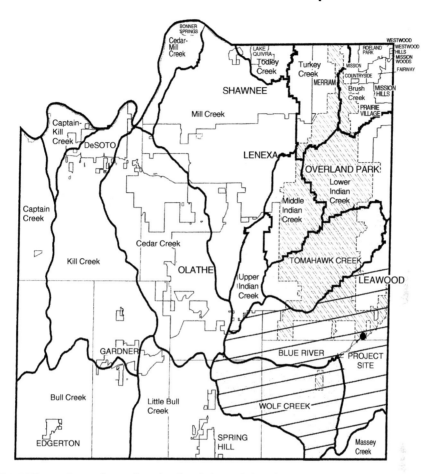

Figure II.8.4 The project is located in the floodplain of the Blue River, Johnson County, KS.

of the project was expanded to include the creation of a wetland and riparian ecosystem as a wetlands mitigation bank for 404 wetland permits, which would also be used as:

- An amenity for the Wilderness Valley residential development
- A regional park for public recreation
- A regional science center for public education
- A demonstration of using wetland systems for improvement of water quality from urban runoff

Project formulation

The wetlands mitigation bank concept was formulated to provide a source of revenue through the sale of credits to fund the other uses. According to URAMP (KS Water Office, 1997), most wetlands in Johnson County are "man-induced," or created from the activities of farming and ranching. Wetlands were typically associated with the construction of ponds and erosional deposition areas along streams. Therefore, they were of relatively low value and typically in the way of development. Although low in value, many are

Figure II.8.5 A typical gabion basket storm drainage project located in Lenexa, KS.

Figure II.8.6 A typical concrete-lined channel storm drainage project located in Overland Park.

classified as jurisdictional wetlands and require permitting by the U.S. Army Corps of Engineers (USACE), under Section 404 of the Clean Water Act.

Often, a problem exists with on-site mitigation of these wetlands because of the lack of long-term maintenance (USACE, 1992). Therefore, USACE, which was a participant in the URAMP meetings, had expressed an interest in using a wetlands mitigation bank to mitigate this problem. It was believed that a mitigation bank would provide a higher value for replacement wetlands in the region. The Johnson County Wetlands Mitigation Bank (Bank) was specifically designed to provide (1) an area to replace wetland disturbances and (2) high-quality wetlands and riparian ecosystems. It was designed to sell credits to other developers who needed a place to mitigate wetland disturbances off-site.

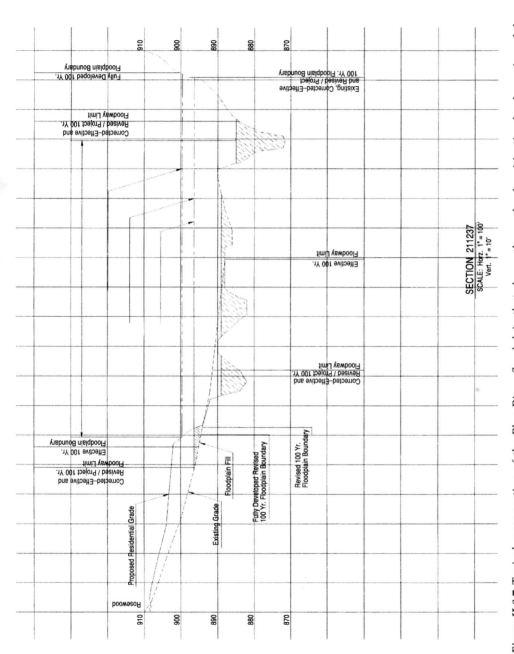

Figure II.8.7 Typical cross-section of the Blue River floodplain that shows the wetlands mitigation bank portion of the development provided to elevate the adjoining residential final grades above the elevation of the fully developed 100-year flood event.

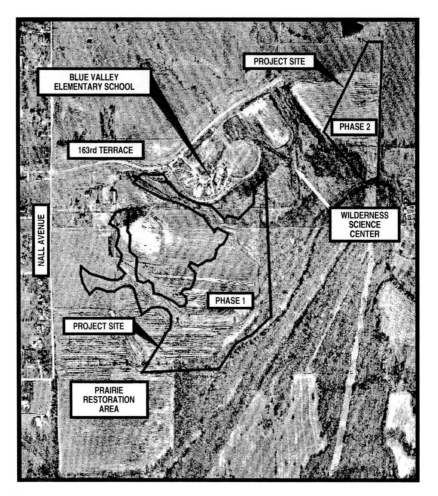

Figure II.8.8 Project development includes two phases: phase 1 construction to be completed by 2002 with approximately 53 acres of wetlands; and phase 2 construction to be completed by 2005 with approximately 10 acres of wetlands.

Project design

The Bank is approximately 63 acres consisting of two phases Figure II.8.8. Phase 1 is approximately 53 acres, and phase 2 is 10 acres. Phase 1 is separated into two construction periods, with approximately 30 acres completed in 2000 and the remaining 23 acres to be completed in 2001. Phase 2 will be completed in approximately five years following the completion of phase 1. The residential community will be completed in conjunction with the construction of each Bank phase.

The proposed wetland and riparian habitat types are shown in Figure II.8.9. Habitat development includes various wetland types, including wooded, shrub, and herbaceous, as well as pools of open water (Table II.8.1). Phase 1 also includes a preserved riparian habitat located along the Blue River. Habitat units are randomly distributed in swales, causeways, peninsulas, islands, and shallow zones around the edge of open water. Open-water areas will be approximately 8 or 10 ft deep and are expected to support fish and other aquatic life.

A watershed analysis and water budget were performed to establish optimum hydro-periods. These data are presented in Table II.8.2. One flashboard outlet structure and an

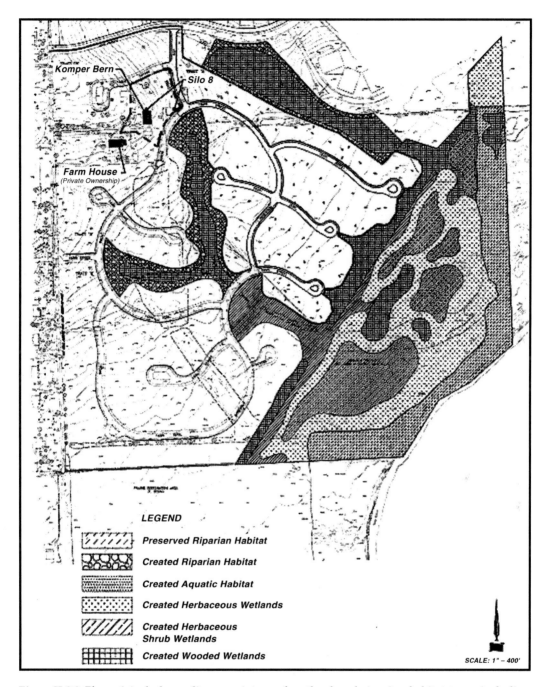

Figure II.8.9 Phase 1 includes a diverse mixture of wetland and riparian habitat types including preserved and created woodland, shrub, and emergent and open-water aquatic areas.

emergency spillway will control water levels. Equalization pipes connect several of the wetland cells that are separated by earthen causeways.

Wetland plantings include mixtures of native herbaceous and woody species, as well as recruitment of native seeds and rootstock. Native plant materials will be captured in topsoil that is salvaged and redistributed during construction of wetland cells. Swales are designed to function as natural stormwater channels and include rock weirs for erosion

Table II.8.1 Wetland and Riparian Habitat Acres

Habitat Types	Number of Acres
Riparian (preserved)	8.3
Riparian (created)	5.0
Aquatic (created)	4.6
Herbaceous wetlands (created)	18.0
Herbaceous shrub wetlands (created)	4.9
Created wooded wetlands (created)	12.7
Total acres	53.5

Note: Phase 1 of the project included a total of 53.5 acres of a diverse mixture of preserved and created wetland, riparian, and open-water aquatic habitats.

control. The residential-lot fills located at the fringe of the floodplain adjacent to the wetlands contain a buried rock berm to protect against slope failure and erosion during flood events. A series of trails and boardwalks is to be installed for maintenance and to provide access for the general public and students. The wetland and riparian zones located adjacent to residential lots will be separated by a wooden rail fence and 10-ft-wide native grass buffer.

Bank instrument review and approval

A bank instrument was prepared in accordance with USACE guidelines (1995) and submitted in October 1998. A mitigation bank review team (MBRT) was established to review the instrument and banking proposal. The MBRT included representatives from the U.S. EPA, U.S. Fish and Wildlife Service, U.S. Natural Resources Conservation Service, KS Dept. of Health and Environment, and KS Dept. of Wildlife and Parks. In addition, several local interests, including the Blue Valley School District (BVSD), the Johnson County Park and Recreation District (JCPRD), and the city of Overland Park, were involved in the review of the instrument.

The MBRT review required preparation of two revised documents that addressed concerns and interests of the various agencies. The final bank instrument was approved in April 2001, with a 401 Water Quality Certification received in February 2001. The final approval and certification required preparation of a water quality protection and monitoring plan. The wetland development must protect three water quality standards — wildlife, secondary contact recreation, and groundwater protection. The water quality protection and monitoring plan includes water quality monitoring stations at stormwater inlets and the principal spillway flashboard structure outlet.

The Bank service area includes all of Johnson County and portions of the Blue River watershed located outside the county. Approximately 53 wetland acre credits are available with various thresholds of project completion. Thresholds include permanent easements, grading as-built plans, documentation of hydrology, initial planting, replacement plantings, and final vegetation success. Vegetation success is measured by permanent transects that have been established within each wetland habitat.

The final bank instrument also includes intent-to-donate agreements with the BVSD and the JCPRD. The BVSD will receive the portion of the Bank that adjoins the existing Wilderness Science Center. The JCPRD will receive the remaining portion of the Bank, which is the first major land acquisition for the Blue River Streamway, a planned county regional park. The JCPRD is responsible for installing trails, boardwalks, and observation areas.

Table II.8.2 Monthly Water Budget Calculations — Wetland Cell Elevation 888.00

Pond Data				
Design surface area	= 6.74 acres	Wetland area	= 12.45 acres	Wooded area = 3.5 acres
D.W.S. elevation	= 888 ft			Herbaceous = 18.8 acres
Seepage	= 0.0002 in./hr	Baseflow (gal/day)	= —	Water = 6.74 acres
Drainage area	= 242 acres			Total wetland area = 15.56 acres
Conservation number	= 86	Starting elevation	= 888 ft	80% of wetland = 12.45 acres
Slope of channel	= 1.63			

	Inflows		Outflows				Results	
Event Date	Precipitation (in.)	Actual Runoff (in.)	Estimated Runoff (gal)	Estimated Evaporation (gal)	Estimated Transpiration (gal)	Estimated Seepage (gal)	Water Balance (gal)	Pond Elevation
January	1.14	0.27	1,784,495	339,671	293,393	2,312	1,149,119	888.00
February	1.43	0.72	2,925,837	251,291	306,708	2,312	2,365,525	888.00
March	2.41	1.89	7,677,887	321,878	564,738	2,312	6,788,959	888.00
April	3.91	4.35	16,193,322	741,164	1,342,326	2,312	14,107,520	888.00
May	5.87	8.64	28,170,118	1,099,746	1,624,855	2,312	25,443,206	888.00
June	5.73	12.79	27,309,488	1,736,806	1,911,996	2,312	23,658,373	888.00
July	4.12	15.45	17,450,827	2,673,032	2,143,936	2,312	12,631,546	888.00
August	3.9	17.90	16,130,550	3,005,927	1,854,568	2,312	11,267,744	888.00
September	4.56	20.96	20,076,061	2,718,650	1,453,683	2,312	15,901,415	888.00
October	3.49	23.05	13,713,684	2,064,263	1,063,272	2,312	10,583,837	888.00
November	3.03	24.73	11,089,899	1,061,577	596,015	2,312	9,429,994	888.00
December	1.66	25.33	3,933,619	528,693	361,989	2,312	3,040,625	888.00

Note: Data from Hillsdale Lake Reservoir (elevation = 918), water year 1997. A watershed and water budget analysis was performed for each wetland cell. The analysis of rainfall/runoff, evaporation, transpiration, and seepage values established the water-surface elevation for the optimum wetland hydro period.

Overview of Approval Process

- Zoning obstacles
  - Wetlands
  - Platting required
  - Rezoning required
  - Subdivision design
  - Single-family lots not allowed in the floodplain
  - Open drainage channels violate city's stormwater ordinances
  - Protests from neighbors

Figure II.8.10 Many city zoning and planning standards had to be addressed by detailed biological and engineering studies as well as variance requests that were reviewed and approved by the Overland Park Planning and Engineering staff, Planning and Zoning Committee, and City Council.

The final bank instrument also includes a trust fund that will provide long-term operation money to be transferred to the BVSD and JCPRD at the time of the property transfers. Money for the fund is obtained through a percentage of each Bank credit sale.

City and state review and approval

The design of the project included many unusual components that did not comply with city ordinances related to subdivision design and stormwater management. Most notably, the ordinances do not address or allow for the construction of wetlands systems and open, natural stormwater drainages (Figure II.8.10). Numerous revisions were made to the initial project to accommodate the zoning and planning restrictions. Nevertheless, the following key approvals were still required for the project to be successful:

- The wetland and mitigation bank area were rezoned Agricultural, which eliminated paying a significant excise tax.
- A stormwater variance was obtained for construction and maintenance of open drainage channels.
- The city's official floodplain maps were revised in order to allow fill in the fringe of the floodplain.
- The subdivision design included stringent compaction and erosion controls.
- The homeowner's association restrictions and covenants included measures to protect the adjacent wetlands from nutrients, trash, etc., from homes.
- Sales agreements included cautions about wetland pests such as mosquitoes, snakes, rodents, and unpleasant odors.
- Exemptions were required from the city's weed ordinance.

After extensive engineering and biological studies, including hydrology and hydraulic modeling, watershed and wetland function and values analysis, geotechnical study, and performance and construction specifications were submitted in support of the project, the city planning and engineering staff supported the project and the recommended approval of the required variances and applications.

Numerous state and county approvals were also required, including:

- Water rights permit
- Floodplain certificate
- Stream disturbance permit
- Sediment and erosion control plan and NPDES permit
- Cultural resource survey and clearance
- Threatened and endangered species clearance

Many state and local agencies considered this project an excellent opportunity to demonstrate the use of innovative wetland technologies to control water pollution from urban runoff. Therefore, many of the approvals and variances granted were considered a first for KS.

Summary and conclusion

The joint Wilderness Valley and Johnson County Wetlands Mitigation Bank Project provides many benefits. It is being constructed by a private partnership to create a public trust. In addition to preserving and enhancing a major reach of the Blue River, the wetlands serve as an excellent amenity for the adjoining housing development as well as an important regional county park and school science center. The county park will provide numerous recreation opportunities including birding, hiking, fishing, and nature study. The Wilderness Science Center will be expanded to include open-water studies and several new wetland types. The Bank also serves a critical need for other developers who need a quality wetland mitigation site.

When completed, the project is expected to provide excellent treatment of urban stormwater runoff, which will help to establish local guidelines for the use of similar treatment technologies to meet the new phase II Water Pollution Control Standards. The project has received much attention over the past two years. It received the 2000 Innovative Technology Award presented by the Kansas City Chapter of the American Public Works Association.

Acknowledgments

Project sponsors include Clay Blair Services Corporation, Developer; George Butler Associates, Engineer; Blue Valley School District, Donatee; Johnson County Park and Recreation District, Donatee; city of Overland Park, Project Reviewer; and Kansas Water Office, Project Reviewer.

Literature cited

Kansas Water Office and George Butler Associates, Inc., The urban resource assessment and management project (URAMP), Topeka, KS, 1997.

Final Banking Instrument, Johnson County Wetlands Mitigation Bank, U.S. Army Corps of Engineers, Kansas City District, MO, 1999.

Kansas Wetlands and Riparian Areas Alliance, Brochure regarding the background of wetlands in Kansas, Manhattan, KS, undated.

U.S. Army Corps of Engineers (USACE), Wetland mitigation banking concepts, prepared by Richard Reppert, Inst. for Water Resour., Alexandria, VA, 1992.

USACE, Federal guidance for the establishment, use, and operation of mitigation banks, Fed. Register, Nov. 28, 1995.

Response

Managing suburban watersheds for multiple objectives

The suburban edge is where development pressures are often the heaviest. As such, it is also the location where elements of water sensitive planning and design can be most beneficial in mitigating these pressures. Other chapters is this book highlight the merits of utilizing constructed wetlands for the treatment and management of stormwater runoff (Robert France and Philip Craul in Chapter I.7, Diana Balmori in Chapter I.8, Catherine Berris in Chapter I.9, and Glenn Allen in Chapter I.13). This chapter by Dennis Haag, Stephen Hurst, and Bryan Bear is significant in that it demonstrates how site-specific concepts of wetland design are effected through comprehensive watershed planning.

As Haag and his co-authors state, it is imperative to integrate different perspectives at the very start of water sensitive planning projects. The views and aspirations of such players as private developers, civil engineers, and state, federal, and local city water resource planners all need to be recognized, understood, and then balanced with the local community's wishes and desires. Watershed management planning, especially that which pushes the local wisdom base in its recommendations for innovative programs and policies — as was accomplished at this location in Kansas — can often be just as much about sociology as about ecology. This was also the case found in those studies described by Thomas Benjamin in Chapter II.4, Gail Krantzberg and Judi Barnes in Chapter II.6, David Blau in Chapter II.7, Kelly Cave in Chapter II.9, and Daniel Williams in Chapter II.11.

Another instructive element in this chapter is the presentation of a series of frameworks outlining straightforward, logical procedures to arrive at the goal of water sensitive planning. For example, it makes good heuristic sense to first assess the effects of urban development, then review existing urban development policies, and then provide alternative methods of resource management. Also, taking the phased approach of biological assessment, followed by watershed assessment, and then followed by public participatory workshops and discussions is also similar to the strategy that Benjamin adopts in Chapter II.4.

This chapter demonstrates the strength and importance of developing high-visibility demonstration projects as sentinel education tools for water sensitive planning. Finally, as Haag and associates imply, such projects will meet with the greatest success if they are truly motivated by encouraging and enthusiastically supporting the need for multiple objectives in addition to simply supporting stormwater treatment. In this case, the "ancillary" objectives were wetland mitigation banking, science education, public recreation, and a residential housing amenity creation.

chapter II.9

Urban watershed management (Detroit, Michigan)

Kelly A. Cave

Abstract

The Rouge River National Wet Weather Demonstration Project (Rouge Project) is a watershed-based effort, sponsored by the U.S. Environmental Protection Agency (U.S. EPA), to manage wet weather pollution to the Rouge River, a tributary to the Detroit River in southeast Michigan, which is designated as a significant source of pollution to the Great Lakes system by the International Joint Commission (IJC). The Rouge River watershed is largely urbanized, spans approximately 438 square miles, and is home to over 1.5 million people in 48 communities and 3 counties. Sources of pollution to the river include industrial and municipal point sources, combined sewer overflows (CSOs), stormwater runoff, interflow from abandoned dumps, discharges from illicit connections, discharges from failed on-site septic systems, and resuspension of contaminated sediment. The Rouge Project has expanded from a program to build and evaluate alternative approaches to control CSOs to a comprehensive watershed-based pollution abatement initiative.

The Rouge Project is demonstrating that a watershed-based pollution management program, which provides flexibility and real delegation of authority to local stakeholder agencies to decide *how* to achieve water quality goals, is achieving faster and more cost-effective restoration and protection of water resources. In addition, local involvement in addressing water quality problems is resulting in alternative designs for pollution controls, which incorporate multipurpose and aesthetic features that facilitate their acceptance by the general public. The Rouge Project is designing, constructing, and evaluating over 200 full-scale pilot pollution control and watershed restoration projects, including CSO control basins, stormwater best management practices (BMPs), wetlands, abandoned dump cleanups, habitat protection and restoration, and management of on-site sewage disposal systems. This chapter provides a summary of these pilot projects.

Addressing all of the sources of impairment in the Rouge River, however, necessitated strong consensus building among the 48 community governments, 3 county governments, state and federal governments, industries, environmental groups, and private citizens to show that they had a stake in restoring the Rouge River and that their participation was vital. This chapter also describes the Rouge Project efforts to build institutional and regulatory frameworks necessary to accommodate a watershed approach to wet weather pollution management. Consensus-building strategies, critical to the success of this effort, are also described here and were used to engage numerous stakeholders, gain their

Figure II.9.1 Location of Rouge River Watershed in southeast MI. (Courtesy of Rouge River National Wet Weather Demonstration Project, Detroit.)

support, provide them opportunities to influence decisions, and participate in actions to restore and sustain the Rouge River as a valuable community asset.

Introduction

In order to achieve water quality standards and associated designated uses within surface waters, it has become abundantly clear that pollution management must be addressed through a watershed approach. The watershed approach is a holistic approach that considers the impacts from *all sources of pollution and use impairment* in a receiving water. The historic implementation of water quality management programs in the United States at the federal and state levels has been to focus on point sources, which are the most obvious sources of pollution to water bodies. This program has worked well to control pollution from (large) point sources but has also left a patchwork of regulated and unregulated discharges of stormwater and nonpoint source pollution to surface waters. This patchwork is especially true in most urbanized areas where multiple local jurisdictions are located in the same watershed. More subtle sources of pollution, such as stormwater, are now emerging as the next priority for attention. The challenge is to develop innovative solutions to achieve water quality objectives that may be (1) more cost-effective, (2) implemented in a more timely fashion, and (3) better able to meet local needs. It has also become clear that water resources management must have the support of the general public in order to be effective and to become self-sustaining. A locally driven watershed approach to pollution management as a means to achieve management goals is an exciting concept that many have discussed but for which there is limited practical experience. This is particularly true in urban situations where there are multiple sources of impairment to a water body and stiff competition for limited local resources to address the pollution sources. The Rouge River National Wet Weather Demonstration Project (Rouge Project) in southeastern MI (see Figure II.9.1) has provided a unique opportunity for a watershed-wide approach to restoring and protecting an urban river system by using a cooperative, locally based approach to pollution control.

The Rouge Project has learned a great deal on what it takes to restore an urban waterway to its beneficial uses. The purpose of this document is to present some of the "lessons learned" to date. Each of these "lessons learned" was developed based on the extensive experience of the Rouge Project. By availing themselves of the information available from the Rouge Project, others will be able to save considerable time and money in the implementation of their own pollution control programs. Comprehensive information

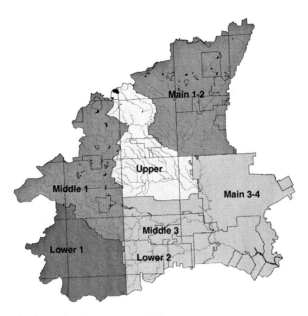

Figure II.9.2 Subwatersheds and communities of the Rouge River Watershed. (Courtesy of Rouge River National Wet Weather Demonstration Project, Detroit.)

on the Rouge Project, including technical reports and other materials, is available from the Web site http:\\www.rougeriver.com.

Background on the Rouge Project

The Rouge Project was initiated in 1992 by the Wayne County, MI, Dept. of the Environment. The project is a U.S. EPA grant-funded comprehensive program to manage wet weather pollution to restore the water quality of the Rouge River, a tributary of the Detroit River in southeast MI. The Rouge River has been designated as a significant source of pollution to the Great Lakes system. The Rouge River watershed is largely urbanized, spans approximately 438 square miles, and is home to over 1.5 million people in 48 communities and 3 counties. The communities and subwatersheds comprising the Rouge River watershed are shown in Figure II.9.2.

The eastern side of the watershed consists of much of the older industrial areas in southeast MI. The western side of the watershed consists of newer suburban development and areas under heavy development pressure. The diverse nature of the Rouge River watershed can be seen in Figure II.9.3. The Rouge River consists of 4 main branches totaling approximately 130 miles. All sanitary and combined sewers in the watershed are connected to the Detroit Wastewater Treatment Plant, which discharges outside the watershed into the Detroit River. Combined sewers serve 20% of the watershed. Separate sanitary and storm sewers serve most of the remaining areas of the watershed, with the exception of isolated pockets and rural areas in the headwaters that still have on-site septic systems. Historically, the major sources of pollution to the river were industrial and municipal point sources, wet weather sanitary sewer bypasses, and CSOs. The point sources have been successfully controlled by an aggressive National Pollutant Discharge Elimination System (NPDES) permitting process administered by the state regulatory agency (MI Dept. of Environmental Quality, MDEQ); however, the river still failed to meet water quality standards due to a wide range of sources such as CSOs, sanitary sewer overflows (SSOs), stormwater

Figure II.9.3 Diverse characteristics of Rouge River Watershed. (Courtesy of Rouge River National Wet Weather Demonstration Project, Detroit.)

runoff, illicit connections, failing septic systems, interflow from abandoned dumps, and resuspension of contaminated river bottom sediment.

The Rouge River had been identified as one of 43 tributary areas of concern (AOC) in the Great Lakes system by the IJC Water Quality Board in 1985. A remedial action plan (RAP) documenting water pollution problems and proposing corrective actions was prepared for the Rouge River in 1989 and updated in 1994. The Rouge River RAP cited the remediation of the CSO discharges in the combined sewer area of the lower watershed as a priority but also recognized the importance of controlling sources of pollution emanating from nonpoint and stormwater discharges in the upper watershed served by separate storm and sanitary sewers, and on-site septic systems (Bean et al., 1994). The Rouge Project was born out of a desire and critical need to manage the multiple sources of pollution in this large, urban watershed in a prioritized, comprehensive manner. The U.S. federal government began sponsoring this effort in 1992 to develop and demonstrate technical, institutional, and regulatory options and processes to protect and restore a large, multi-jurisdictional, urban watershed.

Early on, the Rouge Project focused on the control of CSOs in the older urban core portion of the downstream areas of the Rouge watershed. The Rouge Project initiated the watershed-wide management approach in southeast MI by facilitating CSO control and permitting based on common requirements throughout the watershed. Rouge communities served by combined sewers have entered into permits with the MDEQ and the EPA requiring a base level of abatement construction throughout the watershed followed by assessment of water quality impacts and future construction phases to meet public health and water quality standards. This approach is a significant departure from previous point source permits, which were typically issued on an individual basis and often not coordinated with other pollution control activities affecting the same water body.

CSO control is being implemented in phases, with phase 1 recently completed. Eight communities constructed 10 retention treatment basins to serve 35.1 square miles of combined sewer systems. Each of these basins is sized for different design storms and many also employ innovative technology. In addition, many of the CSO control basins are multipurpose facilities, such as the basin shown in Figure II.9.4, which provides a community park

Figure II.9.4 Aerial view of Inkster combined sewer overflow basin in Rouge River Watershed, Inkster, MI. (Courtesy of Rouge River National Wet Weather Demonstration Project, Detroit.)

and playground on the top of the basin. One retention/treatment tunnel is under construction in one community containing 3.2 square miles of combined sewers. Several communities separated their sewers; the drainage area of these projects totaled 3.4 square miles. A two-year evaluation study of the CSO control program was recently completed (Hufnagel et al., 2000; Kluitenberg et al., 2000). The design, operation, and cost information gained from the evaluation of phase 1 control facilities, coupled with efforts to control stormwater and other pollution sources in the watershed, will provide the basis for the phase 2 CSO control efforts.

Concurrent with the initial development of the CSO control strategy, the Rouge Project initiated a comprehensive data collection effort, which included gathering information about watershed features and characterizing existing water quality and ecosystem health. The first water quality sampling under the project began in 1993, and by the end of 1994 a supporting geographic information system and watershed modeling effort were in full operation with baseline, automated water quality monitoring sites located throughout the watershed (Mullett et al., 1994). Water quality and ecosystem health monitoring has involved an extensive effort in the collection, management, and analysis of data on rainfall, stream flow, in-stream water quality, CSO and stormwater quality, biological communities and habitat, in-stream bottom sediment, air deposition, and aesthetic conditions. In addition, the monitoring program includes measurement of the performance of various structural pollution controls, wetlands, and pollution prevention activities. The initial sampling — later confirmed during subsequent sampling in 1995 through the present — documented significant pollution problems in the Rouge River watershed upstream from the CSO discharges. State water quality limits for bacteria and dissolved oxygen were regularly exceeded even in dry weather periods in the upper watershed, and highly variable flows caused flooding, exacerbated bank erosion, and increased sedimentation that affected the lower river. This information, shown by the example in Figure II.9.5, confirmed the suspicions of many that the discharges from separated storm systems in heavily urbanized areas can be a significant sources of pollution, including coliform bacteria. The Rouge watershed assessment tools have proved to be critical in garnering public support for the river restoration efforts and have provided the general public, local

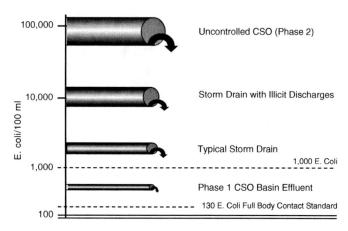

Figure II.9.5 Comparison of sources of *E. coli* bacteria. (Courtesy of Rouge River National Wet Weather Demonstration Project, Detroit.)

decision makers, and regulatory agencies with information to prioritize and tailor pollution control solutions to specific river reaches while coordinating efforts throughout the watershed.

Based on what was learned, the focus of the Rouge Project became more holistic to consider the impacts from all sources of pollution and use impairments in a receiving water. The Rouge Project began to identify the most efficient and cost-effective controls of wet weather pollution, while assuring maximum use of the resource. For example, over 60 pilot innovative stormwater control technologies are also being evaluated under the Rouge Project by 25 different communities and agencies. Categories of pilot stormwater management projects currently underway include wetlands creation and restoration, structural practices such as grassed swales and detention ponds, erosion controls, stream bank stabilization, and habitat restoration, to name a few. Figure II.9.6 shows a linear sand filter, one of the pilot stormwater management projects constructed and evaluated under the Rouge Project.

The Rouge Project has learned that illicit connections and failing septic tanks are major sources of pollution problems in the Detroit urban area (Johnson and Tuomari, 1998). Innovative ways to deal with these sources of pollution have been initiated.

A suite of computer models has been developed by the Rouge Project to simulate the water quality and quantity response of the Rouge River during wet weather events for existing and future conditions under various CSO and stormwater runoff management alternatives. This effort has led to a very useful public communication tool on water quality indices tied to actions needed to restore the Rouge River. A comprehensive geographic information system (GIS) and relational databases were designed and implemented to manage the wealth of data collected under the project. In addition, a special data exploration tool, DataView, was developed to support routine analyses of large time-series data sets (Rood et al., 1995). DataView is user-friendly and readily transferable to other locations. Related to DataView is the Rouge Information Manager, also a user-friendly, readily transferable tool (an "electronic file cabinet") for accessing multimedia information about the Rouge River restoration effort.

These tools have been vital in the success of the strong consensus-building activities necessary to show the 48 community governments, 3 county governments, state and federal governments, industries, environmental and community groups, and private citizens that they had a stake in restoring the Rouge River and that their participation was vital to the success of the comprehensive watershed management program. These consensus-building strategies were used to engage numerous stakeholders, gain their support,

Figure II.9.6 Linear sand filter retrofit to urban area for stormwater management, Wayne, MI. (Courtesy of Rouge River National Wet Weather Demonstration Project, Detroit.)

provide them opportunities to influence decisions, and participate in actions to restore and sustain the Rouge River as a valuable community asset. The Rouge Project's public education/public information program has been demonstrated to be a very effective component of the consensus-building process (Powell and Bails, 2000). Although many elements are included in the public education program, as illustrated in Figure II.9.7, one very important element is the involvement of youth in this effort. As one example, in the annual Rouge Water Festival held each May, 52 fifth-grade classes from over 27 schools (nearly 1,500 students) participate in this hands-on festival where they learn about the importance of water in all aspects of their lives. Over 100 schools are currently involved in education efforts of the project.

The Rouge Project has spent considerable effort to build institutional and regulatory frameworks necessary to accommodate a watershed approach to wet weather pollution management. Part of this framework is a watershed-based general permit for municipal stormwater discharges issued under the National Pollutant Discharge Elimination System (NPDES) program. This stormwater permit program was developed jointly by the Rouge communities and the MDEQ and is based on the concept of cooperative, locally based watershed management (Cave and Bails, 1998). Communities and agencies in over 95% of the watershed have applied for coverage under this innovative, watershed-based permit program. The MDEQ permit requires permittees to participate in watershed management planning for a self-determined subwatershed unit. The subwatershed management plans form the basis for implementing watershed goals and objectives that will result in

Figure II.9.7 Public education and involvement materials developed by the Rouge River National Wet Weather Demonstration Project. (Courtesy of Rouge River National Wet Weather Demonstration Project, Detroit.)

improved water quality and pollution control. The Rouge communities will also use these watershed management plans to achieve other program objectives, such as those under the federal TMDL program and the state Clean Michigan Initiative (Cave et al., 2000).

The local communities, agencies, industries, and citizens have been working together in seven subwatershed advisory groups to develop and implement management strategies for various segments of the river. The Rouge Project has recently been the catalyst for the Rouge Gateway Partnership, a collaborative effort among county government, corporations, local communities, and academic and cultural institutions, which is guiding redevelopment to restore the 7-mile section of the lower Rouge River. This section of the river includes a 4-mile concrete channel and a 3-mile section of navigable dredged waterway downstream of the channelized section. As shown in Figures II.9.8 and II.9.9, the Rouge Gateway Project will create a 7-mile greenway link for this area, providing the public access and linking the park system along Hines Drive to the Detroit River waterfront (Rouge Project, 2000). This effort will demonstrate to other communities how to reclaim waterways that have been essentially removed from public access and use.

Lessons learned to date

As stated earlier, the project participants have learned a great deal on what it takes to restore an urban waterway to its beneficial uses. The purpose of this chapter is to present some of the "lessons learned" so far. The approach used in determining the lessons learned was built on the idea of asking, "What do we know now that would have saved us time and money had we known at the start of the project?" Each of these 13 "Lessons Learned"

Chapter II.9: Urban watershed management (Detroit, Michigan)

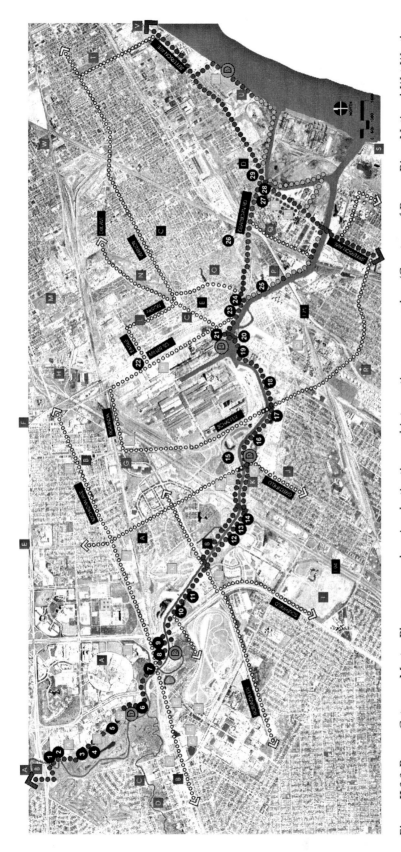

Figure II.9.8 Rouge Gateway Master Plan, a comprehensive destination and interpretive program plan. (Courtesy of Rouge River National Wet Weather Demonstration Project, Detroit.) (Color version available at www.gsd.harvard.edu/watercolors. Password: lentic-lotic.)

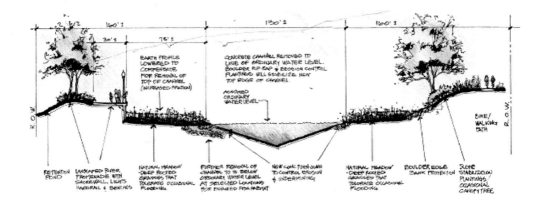

Figure II.9.9 Conceptual greenway along lower Rouge River, part of Rouge Gateway Master Plan. (Courtesy of Rouge River National Wet Weather Demonstration Project, Detroit.) (Color version available at www.gsd.harvard.edu/watercolors. Password: lentic-lotic.)

was developed from the vast experience of the Rouge Project. The background for each of these lessons is explained, citing specific experiences related to the lesson learned.

The watershed approach

Lesson learned

We have known for years that the only way to effectively achieve water quality and ecosystem protection in an urban river system is to look at all problems and their solutions from a holistic watershed perspective. The Rouge River Project experience gives us tangible measures of the benefits of this approach — hundreds of millions of dollars being saved.

The Rouge River Project has undertaken a number of actions to demonstrate how to restore an urban river. Those actions include, but are not limited to CSO controls, stormwater management (including flow reductions), stream restoration, and wetland creation and enhancement. The largest traditional point sources of pollution, CSOs, have been controlled or are programmed for control.

When the Rouge Project began, its main focus was on the control of CSOs because the perception was if CSOs — the largest point sources of pollution — were controlled, most of the river would meet water quality standards. The monitoring and modeling program quickly demonstrated that even if all of the CSOs were totally eliminated from discharging to the Rouge River, the designated uses of the river would still not be met due to other sources of pollution such as stormwater runoff and to lack of environmental habitat. In order to achieve water quality standards and associated designated uses within the Rouge River, it has become very clear that pollution management must be addressed in a holistic fashion that considers the interrelationship between the impacts from all sources of pollution and use impairments in a receiving water. Water quality protection at the watershed level is neither a single capital project, nor a series of projects. In essence, it is a "way of life," and the problems must be approached that way. The EPA's *Watershed Approach Framework* also discusses this philosophy (U.S. EPA, 1996).

The project has enough preliminary data to make a rough cost comparison between utilizing a watershed approach to achieve desired water quality objectives versus using the historical approach of addressing the causes of water quality degradation individually. These preliminary data indicate that cost savings for the Rouge River watershed citizens could easily approach several hundred million dollars.

Lesson learned

In a large watershed, it is most effective to restore or protect water quality and ecosystem health by looking at subwatersheds within the overall watershed. In the Rouge River Project, the focus on 20- to 50-square mile subwatersheds has allowed the project to move from a purely regulatory-driven CSO program to voluntary community-based efforts for stormwater management.

Environmental protection relies on a mix of federal and state resources as well as increasing levels of local resources. Use of local resources requires a search for and identification of common environmental interests. Most people have a strong environmental interest in their community; fewer have a strong interest in an environment with which they cannot identify. Therefore, subwatersheds give a means for focusing the local resources to address local problems due to the interest people have in their immediate surroundings. The EPA's *Top Ten Watershed Lessons Learned* document also discusses the need for identifiable goals to focus local resources (U.S. EPA, 1997).

Focusing on subwatersheds has many advantages. First, smaller areas are more manageable in terms of addressing water quality problems. Second, people identify more with a subwatershed than with a larger watershed. Local ownership of pollution problems and their solution is critical. In his article, *Crafting better urban watershed protection plans*, Schueler (1996) also suggests that subwatersheds having a drainage area of 518 to 3,885 hectares (2 to 15 square miles) in size provide the best scale from both a technical and a political basis on which to base management plans. Third, it is easier to analyze the various sources of water quality problems in the subwatershed and decide how to get a handle on the priority of dealing with those problems. It is critical to establish a hierarchy of pollution sources in subwatershed (point sources and nonpoint sources) based on the adverse water quality impacts of those sources. It is very important to keep reinforcing, at a subwatershed level, the concept of not randomly addressing pollution sources but to prioritize the control of those sources to get desired environmental protection. It may take a long time to correct some of these pollution sources, so it is important to prioritize the control programs. Fourth, it is critical to assess the cumulative watershed impacts to quantitatively assess the physical and biological processes and then fashion the subwatershed solutions. Fifth, before river restoration can be initiated, it is critical to understand the cause of stream disturbance and disequilibrium conditions. Without this understanding, the restoration often treats the symptoms instead of effecting a cure. Sixth, it is easier to manage a process that has a small set of stakeholders and competing interests. Seventh, it is easier to convince people that water quality improvements will require many little and possibly inexpensive actions instead of massive capital programs exclusively. Eighth, and finally, the tools needed to solve subwatershed water quality problems must be geared to that subwatershed. The management plan developed must be tailored to address subwatershed-specific problems.

Each of these advantages has associated cost implications. It is very important to use innovation in the watershed approach. By a consensus-based approach, focusing the local resources to address the local water quality problems will solve them faster and more inexpensively.

The control of combined sewer overflows (CSOs)

Lesson learned

The control program for CSOs must be flexible and tailored to the site-specific situation in order to achieve the most timely and cost-effective solution(s).

Michigan has had design standard requirements for the control of CSOs for a number of years. On the basis of those design standards, the MDEQ issued NPDES permits to the appropriate communities. Certain of the issued permits were challenged. There followed a period of negotiation between the permittees and MDEQ. The permittees, supported by the Rouge River Project, entered into negotiations with the state, which resulted in alternative design standards for certain CSO control facilities. These alternative design standards reflect site-specific considerations and will result in the same level of water quality protection. The modified design standards resulted in a savings of over $300 million to the CSO communities in the initial phases of the project.

CSOs are brief, intermittent pollution sources with long-term water quality consequences. The Rouge River Project has demonstrated that after sewer systems are optimized, CSO outfalls generally discharge for less than 100 hr/year. Most other communities in the United States will have similar CSO discharge duration. Cost-effective CSO control for the Rouge River has reduced the duration of discharge to 20 to 40 hr/year. This discharge has received first-flush capture, and the more dilute flow that follows the first flush is screened and disinfected, and has some degree of solids and biochemical oxygen demand (BOD) removal. These same CSO controls have provided an extra degree of insurance for water quality protection by virtually eliminating the chance for any dry-weather discharge from the combined sewer system.

One can look at the cost of CSO control in different ways. Using the Rouge River Project phase 1 experience, the capital cost is $15,000/acre, while the annualized cost amounts to $10,000/day of water quality protection. Either way, these represent extremely high fixed costs to deal with an intermittent source. The high fixed cost requires flexibility in the development of CSO controls because of the wide array of local conditions and priorities within a community. As stated previously, with design standards that reflect site-specific considerations, the potential for cost savings is enormous. Therefore, by developing a site-specific CSO control program that is integrated into an overall watershed management plan that prioritizes the pollution problems to be addressed, considerable cost savings can be realized.

Lesson learned

For large and costly CSO control programs, implementing the program in phases can save time and money and foster critical support from local communities.

The installation of CSO controls is very expensive. It is therefore critical that decisions on those controls be based on the best available information. Phasing the installation of controls in complex CSO situations allows for reasonable steady progress to achieve water quality goals in the shortest possible time in a cost-effective manner. Phasing allows time to understand and develop control programs on related sources of pollution that are adversely affecting water quality; to develop and implement sewer system optimization efforts; to determine what is necessary for public health protection; and to determine what is necessary for the achievement of water quality standards, considering all sources of impairment to the river system, in the most cost- and time-effective manner by allowing time to develop better baseline information for designing subsequent phases. The phased approach has some CSO controls installed on part of the overall system, a period of evaluating the effectiveness of the control technology used to meet the design standards, including evaluating the impact on the resource, and then making the decisions on the design standards and treatment technology to be used on the rest of the CSO system.

A fundamental premise of the Rouge River Project was to demonstrate various design standards and control technologies for CSO controls. This demonstration would occur in two phases. Phase 1 specified 17 communities to install CSO controls. Under phase 1, 6 communities have separated their sewers and 8 communities have constructed 10 retention treatment basins utilizing various treatment technologies and design standards. A two-year evaluation study of the completed CSO control basins was recently completed to assess compliance with the design standard used for each basin and to assess the positive water quality impacts on the receiving stream of the controlled CSO discharge. The results of the evaluation study, coupled with efforts to control stormwater and other pollution sources in the watershed, will provide the basis for the phase 2 CSO control program on the remaining CSO sources in the watershed. Phase 2 will require all of the CSO communities to install CSO controls necessary to protect public health by a certain date. Phase 2 controls will be based in part upon the information learned in phase 1. Phasing of the program also creates opportunities for communities to work together to develop joint projects that are mutually beneficial and more cost-effective in correcting pollution problems.

Preliminary results from the CSO basin evaluation study completed to date are providing extremely useful information. The information gained from the evaluation of design storms and control technologies will be extremely useful nationwide in determining cost-effective CSO controls to meet water quality standards. Information on the size of the CSO basins and their operation has the potential for savings hundreds of millions of dollars in the Rouge watershed. The nationwide savings will even be more substantial.

As stated earlier, it is critical to have strong local support for CSO control programs if they are to succeed. The best way to generate and sustain that support is to clearly demonstrate the benefits of the CSO control program. Those benefits are already becoming evident in the Rouge watershed. For example, over 100 miles of river are now CSO-free. This translates into the water quality downstream of controlled CSOs being as good as the water quality upstream. The result is that the natural beauty of the river, along with its ecological health, is being restored. In addition, the economic benefits are also becoming more evident because of the increased recreational usage of the river.

The control of stormwater

Lesson learned

In urban areas, stormwater discharges have just as large adverse water quality impacts as CSO discharges, but the control of stormwater sources is more politically and technically complex. The Rouge River Project recognizes these differences and is demonstrating the necessity of using a different approach to managing stormwater.

Stormwater runoff occurs during and after each rainfall event, whereas CSOs may occur only during the heavier storms. As has been discussed extensively in the literature (e.g., CWP, 2000), the sources of stormwater in urban areas are many and varied, as are the resulting adverse water quality impacts. The Rouge Project has documented that water quality standards are regularly exceeded because of discharges from stormwater in the watershed. Unlike CSOs, where the ownership of the outfall pipe is known, often the "ownership" of the stormwater source is not clear. The watershed approach has proven to be the most effective way to address stormwater issues. The watershed approach is locally driven, encourages voluntary participation by communities and other public agencies, requires the development of a comprehensive plan to address the problems, and defines actions and iterative steps leading to comprehensive, watershed-wide stormwater management. In the Rouge watershed, the local communities, working through the Rouge Project, assisted the MDEQ in the development of a watershed-based general stormwater NPDES permit (Cave and Bails, 1998). This regulatory framework encourages communities to cooperate in a watershed approach to address pollution problems. As discussed in the EPA's *Top Ten Watershed Lessons Learned*, "Watershed work is about partnerships . . . because partnerships equal power" (U.S. EPA, 1997).

Overall stormwater management includes structural, vegetative, or management practices to treat, prevent, or reduce stormwater runoff. Therefore, the solutions to stormwater problems are difficult and time-consuming to develop and implement.

The cost implications of this lesson are many. The associated potential for cost savings is therefore also evident. For example, use of the watershed approach to prioritize the water quality problems and then their solution has the potential to save a great deal of money. As will be discussed in the next lesson, tying stormwater control to other public works projects is a sure winner and money saver.

Lesson learned

Integrating stormwater control projects into ongoing public works plans and actions of local governments results in more timely management of stormwater.

Local governments often have a number of projects or actions planned or underway in the community such as street sweeping, drainage control programs, and related efforts. By showing the local officials the benefits of integrating the stormwater control program into other planned projects or actions, implementation of the stormwater program will proceed more quickly. This fosters local control over the outcome of the overall project or action, which also is a key to overall acceptance and then success.

This concept of integrating efforts is closely tied to the lesson learned on combining stormwater work with the phasing of work. For example, the Rouge River Project has demonstrated this by creating and then seeing the success of the three-county Roads and Drainage Roundtable, which has resulted in the road commissioners' integrating into their day-to-day thinking the needs of watershed protection as impacted by road and drainage work. In summary, local units of government can substantially improve in-stream water quality by reviewing and adjusting how they perform a number of their daily public works activities.

This lesson learned has obvious implications for substantial cost savings. Communities need to think of and create opportunities to marry public works and stormwater work, which results in significant cost savings and more timely environmental protection. Such a marriage involves other agencies whose prime mission is not water quality protection. This combining helps everyone involved.

General public support

Lesson learned

Achieving pollution abatement in a more timely, cost-effective fashion must have general public support, which can be assisted by combining the watershed approach with the concept of phasing needed pollution controls.

The public needs time to understand the complexities of restoring a degraded river and to then respond with the needed support, including the commitment will and financial support. The Rouge River Project has learned that the general public may not fully understand or appreciate a goal of "meeting water quality standards." What they do understand and appreciate is whether a waterway is fishable and swimmable. They accept that, if all of the sewage is not removed from the water, it is not swimmable; if toxics preclude fish consumption, the water is not fishable; and if habitat is destroyed, there will be no fish, and therefore, the water is not fishable. In summary, if the conditions of the river discourage fishing, swimming, and other recreation, attention must be directed at correcting the problems.

Generally, local commitments to address pollution problems in a watershed will come in small increments with the demand to demonstrate the value of those increments if support is to be sustained. Therefore, a phased program and the pursuit of a multitude of pollution sources/problems are essential. The Rouge River Project has taken this lesson to heart and has undertaken a number of locally driven watershed-based projects in order to demonstrate how to restore an urban river. Those projects include CSO controls, stormwater management, stream restoration, addressing failed septic tank problems, correcting illicit connections, wetlands creation and enhancement, and stormwater flow reductions. The project also has implemented a phased approach to the installation of controls.

The Rouge Project has demonstrated that it is cost-effective to spend adequate funds on developing base public support. Without that support, it will not be possible to achieve the needed water pollution control program to restore beneficial uses. This public support also will save money in the long run.

Public education and involvement

Lesson learned

Broad-based public education and involvement programs are critical to the overall success of watershed projects, particularly in urban areas.

One of the major goals of the Rouge River Project is public involvement and education. Each person who lives in the watershed needs to be educated as to how his or her individual daily actions affect the conditions of the river. Public support depends on public awareness and education. Education and involvement drive action. Therefore, public education efforts should be viewed as a campaign with intensity and high stakes. The approach employed should leverage the tremendous environmental education and media movement already in place.

The Rouge River Project solicits community input through a number of directed activities. Some successful programs are (1) getting education programs under way in the area schools, (2) getting people out to the stream to look at and experience it, (3) showing people some obvious things that have been done to get successes in improved water quality, and then widely publicizing the results so people see that things are being implemented instead of just planning and talking taking place, (4) making people aware of the river, which renews their interest in it, and (5) placing signs at each river crossing to increase awareness. A technical project such as this always faces the challenge of conveying data in a way so that the layperson understands the issues. The project has stressed the focus of restoring uses versus raising water quality. People understand restoring uses to the river versus the abstract concept of improving water quality. The project developed a graphical presentation on water quality indices, which has fostered a good focused debate on some key issues of what is the end objective to be achieved (Smith et al., 1997). This communication tool became the heart of the development of watershed management plans for subwatersheds.

Another important lesson has been on the need to measure, communicate, and account for progress. The EPA's *Top Ten Watershed Lessons Learned* states, "Having good data systems in place to measure and communicate progress is a critical part of watershed work" (U.S. EPA, 1997). They not only keep watershed issues on people's radar screens but assist in sharing successes and facing new challenges to the watershed.

Information systems

Lesson learned

A data management and information system that can effectively communicate to the broad public is critical to achieving success in watershed/water quality restoration projects.

In order to make informed decisions on water quality improvements in a watershed system, it is necessary to have access to, and be able to process, large amounts of data. It is important to consider *very early* in the process how the data will be compiled and analyzed. In other words, consider what type of data management and information system is needed to accomplish the desired objective of analysis and communication of results. What the data system demonstrates needs to come through as a clear message. The involvement of local officials in the process is critical to the long-term success of the information system. Most local governments have, or soon will have, geographic information systems (GIS) to aid them in decision making. Every effort should be made to bring together the local units of government in the watershed to establish ways to share information among data systems and to have commonality in information spreading into the various data/information systems. This cooperation fosters watershed-based decisions versus individual community-based decisions.

The Rouge River Project has developed two effective data management and information systems that are easily transferable to others for use. DataView is a user-friendly data exploration tool developed to support routine analysis of large time-series data sets. The system allows for tabular data viewing, data plotting, generating summary statistics, spatial display, and data export. The Rouge Information Manager is a user-friendly tool for accessing and displaying information. It serves as a portable electronic filing cabinet that anyone can use to access information on monitoring data, GIS maps, modeling alternatives, public involvement, wetlands, stream restoration, illicit connections, and other technical data. It contains maps, technical reports, photos, and videos, which can be displayed.

Information systems can be very costly. By shopping around it should be possible to adapt an already existing system to meet your needs. This will result in considerable cost savings. As stated earlier, the Rouge Project information system is easily transferable to others.

Also, cost savings will result by having a well-informed public that supports the water quality goals trying to be achieved.

Monitoring versus modeling

Lesson learned

Achieving a balance between monitoring and modeling can save money and time in implementing a wet weather water quality management program.

When addressing wet weather impacts on water quality in waterways, the natural tendency is to want to establish an extensive monitoring program to fully document the cause and effect of the water quality problems. Monitoring programs are time-consuming and expensive. Related to monitoring is the use of models to define present conditions and to project future options. The objective should be to achieve a balance between monitoring and modeling. The Rouge River Project has demonstrated several key aspects related to that balance.

First, think through how you will use *any* data collected from a monitoring program to make a subsequent decision. The natural tendency is to collect much too much data on a large suite of chemicals. After a series of hard questions, the amount of monitoring data collected can probably be reduced significantly, with the concomitant cost savings. Some of those hard questions are:

> How will I *specifically* use the data on this specific chemical to reach a decision?
> How much data do I need on this specific chemical to reach a decision?
> Do I need screening level data or extensive data to reach a decision?
> Are models available to help predict the future water quality conditions in this water body based on changed waste inputs?
> If so, what are the *minimum* data needed to support the model so that the prediction is within reasonable bounds, i.e., within a reasonable percentage of accuracy?
> How can I use the model to the maximum extent possible to save costs and only use limited monitoring to fill in gaps as needed?

By the same token, a project can be consumed by modeling and the need to develop the perfect model. Be careful not to continue to gather data to continually feed a model mindlessly. By asking a series of hard questions on modeling, you can have significant cost savings. Some of those hard questions are:

> Are models currently available to help predict the future water quality conditions in this water body based on changed waste inputs so that I do not have to develop a new model from scratch?
> If so, what are the *minimum* data needed to support the model, and what is the minimum modeling effort needed so that the prediction is within reasonable bounds, i.e., within a reasonable percentage of accuracy?
> How can I use the model to the maximum extent possible to save costs, remembering the concept of getting answers "within reasonable bounds?"

Remember to keep asking yourself, "What do I want to model?" If you do not keep asking that, your model could grow topsy. You must always remember that monitoring and modeling are tools to get to the end, but these tools should *never* be confused as *being* the end.

The idea is to use simple models at first, drawing on available water quality sampling information. Let the modeling show where additional data are needed to define location and parameters. Then continue a combination of modeling and monitoring to achieve the desired objectives.

As an example to demonstrate the economics of monitoring and modeling, the amount of money the Rouge River Project has spent to characterize the resource has been approximately $5,000 to $7,500 per mile/year (calculations based on total monitoring costs over the life of the project and the number of main stream miles). A long-term monitoring program is estimated to be approximately $2,500 per mile/year. Compare that to modeling costs, which can be calculated to be $2,000 to $3,000 per mile/year. The long-term modeling costs are $750 per mile/year. The ratio of monitoring costs to modeling costs increases from 2.5:1 for the short-term work to 3.3:1 for the long-term work.

In summary, local monitoring programs are needed. They can be minimized and still provide a good basis for planning and decision making. Some key steps to remember are:

1. Collect and analyze existing data.
2. Compare existing data to findings elsewhere such as the Rouge River Project and determine what conclusions can be drawn. Use simple models at this point to help make sense of the data.
3. Plan new data collection and monitoring to fill gaps in information and to further investigate known pollution problems in the watershed.

An overarching lesson associated with monitoring and modeling is the need to avoid "paralysis by analysis." Monitoring and modeling can do much to prioritize restoration activities, but certain common-sense activities should be implemented given only minimal data. Existing national data are often sufficient to flag those items. Some examples include septic field inspection and correction of problems, illicit connection removal, and downspout disconnection.

As stated in the previous example, the potential for cost savings through a mix of monitoring and modeling is enormous. In addition, a sound mix of modeling and monitoring will save time, which also translates into money saved.

Institutional and financial arrangements

Lesson learned

The toughest problems to be addressed and solved in wet weather and watershed protection programs are developing and implementing the institutional and financial arrangements needed to sustain the program. The technical issues are easy by comparison. Early and continued efforts should be directed toward developing workable institutional and financial arrangements.

The sources of stormwater runoff to surface waterways are many and varied, with "ownership" often unclear. CSO discharges are identifiable but can be costly to control. Other sources of waste that cause water quality degradation such as dry-weather pollutant sources (sediments, septic tank leakage, illicit connections) are a considerable problem in urban areas. Other considerations such as quantity of flow in an urban stream and the "flashiness" of that flow can adversely impact water quality and the program to restore beneficial uses. Habitat issues are an important consideration related to beneficial uses of the waterways. Urban growth and related land-use patterns can have major impacts on water quality restoration programs. All these challenges are related to the institutional and financial arrangements needed to accomplish the desired end objective of a restored waterway.

The Rouge River Project has learned some important lessons in this area. For example, it must be demonstrated to the residents of a watershed/urban area that all are paying their fair shares of the waterway restoration costs. It is generally accepted that every city and county must do a set of minimum activities to address water quality problems and that certain cities and counties will have to do more because of their specific set of water quality problems/issues. Residents need to be shown that the benefits of water quality improvement are worth the costs. Interjurisdictional agreements get local communities committed to the project's objectives and garners the day-to-day support needed to accomplish the necessary objectives.

In order for a watershed project to be successful, an "institution" to oversee the progress is not necessarily needed. What is critical is the need for effective institutional arrangements. These can be as simple as utilizing forms of interjurisdictional cooperation or remaking or combining existing institutional arrangements. These solutions will mirror the complexity of the problem to be solved.

A key element of the institutional and financial concerns deals with the federal and state regulatory agencies. Obtaining regulatory agency consensus on all aspects of a watershed project is critical to its overall success. Regulatory inertia is very strong toward continuing down the historical paths that have been followed in the past. Therefore, trying new methods of institutional and financial arrangements can be a tough sell. It is critical that the federal and state regulatory agencies have staff and managers fully involved in all aspects of a watershed project in order to witness the advantages of trying new approaches to address problems.

Substantial cost implications are associated with institutional and financial arrangements. The project has demonstrated that it is well worth the time spent upfront looking at these topics. Waiting until the end of a project and then using a piecemeal approach will be a waste of time and money. It takes time to solve these issues. Many meetings are necessary to address this topic, but they are worth the cost in the long run.

Flow as a water quality issue

Lesson learned

In urban areas, the quantity of stream flow is very important in assessing water quality and ecosystem health and the subsequent correction of problems. By solving problems associated with high flow, many water quality and ecosystem health problems will also be solved.

In urban areas, streams often experience high flows during wet weather events. These high flows cause or aggravate ecosystem health problems for several reasons, including scouring of banks and the resulting turbidity problems; subsequent sediment deposition and its related adverse impacts on habitat and beneficial uses; adverse impacts on wetland habitat; the physical destruction of the river banks, which adversely impacts land values; and the wide variation in flows, which adversely impacts swimming, wading, boating, and fishing. Wetlands in urban areas are very beneficial because they provide good storage for water, which tempers flow variability, improves water quality, and provides good habitat for aquatic life as well as other wildlife. Yet, highly variable flows adversely impact these needed wetlands.

Often, the damage caused by flooding and stream bank erosion on private property, the loss of recreational opportunities (i.e., flooded golf courses in flood plains), and traffic disruptions caused by flooding will quickly gain the public's attention and will be a stronger reason for citizen's commitment to invest in pollution control. By building detention ponds or other improvements to solve flooding problems, the pollution problems may also be largely solved. The Rouge River Project is showing the possibility of solving a number of its wet weather pollution problems in the context of regional detention and floodway improvement projects.

As stated in an earlier lesson, water quantity issues can often be integrated with other public works plans and actions so that the solution positively impacts water quality as well as water quantity. For example, cities need to control stream bank erosion in order to protect property. Integrating water quality considerations into how the erosion control is accomplished and maintained has a double benefit. Therefore, it is important to consider water quantity issues in the development of the programs to address urban quantity of flow issues.

Cost savings opportunities can be realized while addressing flow issues.

General

Lesson learned

Do not be afraid of taking advantage of dumb luck and legislative/political will.

This speaks for itself. The corollary is, bad luck also happens. When that occurs, learn from the "mistake" or event and move on in a positive, smarter fashion.

Conclusion

The Rouge Project in southeastern MI is a working demonstration of a watershed-wide approach to restoring and protecting an urban river system by using a cooperative, locally based approach to pollution control. Since 1992, the Rouge Project has implemented over $400 million of water pollution control, environmental restoration, and recreation projects. Total expenditures under the project are expected to exceed $530 million by the year 2003. (Note: these totals do not include related Rouge work by the city of Detroit for CSO controls outside of the Rouge Watershed, which is estimated to be in excess of $2 billion.)

A number of successes have been achieved to date, but a number of challenges remain. The year 1998 was a point when noticeable measurements of environmental progress along the Rouge River could first be made. Dissolved oxygen was higher at several sites along the river in the last two years, compared with earlier findings in 1994 and 1995. This success can be attributed to CSO controls, illicit connection elimination, stormwater management, and better public, industry, and community awareness of pollution prevention. Also, larger and more diverse species of fish are now sighted more often. The CSO basins have collectively captured and sent to treatment an estimated 4 billion gal of overflow per year since the first basins went online in 1997. The illicit discharge teams have eliminated 12.5 million gal/day of dry-weather discharges, and the Newburgh Lake project has eliminated the PCB fish advisory and restored a major recreational resource in the Rouge. The school-based education program operated by the Friends of the Rouge is in nearly 100 schools, and the annual Rouge Water Festival draws over 1,400 students/year.

The CSO data suggest that large storms may have a different water quality impact than small storms. This finding could result in significantly reduced costs when communities move from implementation of nine minimum controls to long-term control plans. Recreational use in the watershed has increased when a stretch of the river was opened to canoeing for the first time in 30 years. This action created new hopes and expectations for further increased use of the river. The individual stormwater communities have begun to understand how they are part of the water quality problems. Although they are traditionally very independent, they are finding ways to work together voluntarily to address issues such as illicit connections, public education, and implementation of BMPs. The Rouge Project is moving toward integration of all pollution sources and use attainment into a unified, consistent watershed management approach, but this has not yet been completely achieved.

The Rouge Watershed presents a unique management challenge because there are no significant point sources that can be controlled by the action of a single agency, or from which to readily establish an effluent trading scheme. Quite the opposite, the Rouge is dozens of communities, hundreds of major commercial, industrial, and institutional properties, as well as hundreds of thousands of residential homeowners. The environmental management goal of the project is the control of flow and wet weather pollution to achieve flow and quality to meet water quality standards. The institutional management goal is to find ways to effectively work with the parties within the watershed boundary to meet the environmental goal.

The Rouge Project is an overwhelming success so far. Water quality is improving, and the demonstration techniques have resulted not only in concrete and steel structures, but in real institutional changes that integrate the work of stormwater and watershed improvement into the basic institutions of government. The Rouge Project approach demonstrates that a watershed can be "managed." Most important, the Rouge River is being restored. It is hoped that by utilizing the information learned from the Rouge Project, others can save considerable time and money as they implement a watershed management program.

Acknowledgments

This chapter is a summary of select elements from the ongoing efforts of many individuals and organizations involved in the restoration of the Rouge River. The work of Dale S. Bryson to assist with development of the "lessons learned" is especially noted. The author is honored to represent the numerous contributors of this successful partnership to restore and protect a large urban waterway.

The Rouge River National Wet Weather Demonstration Project is funded, in part, by the U.S. EPA Grant #XP995743-01, -02, -03, -04, -05, -06 and #C995743-01. The views expressed by individual authors are their own and do not necessarily reflect those of the EPA. Mention of tradenames, products, or services does not convey, and should not be interpreted as conveying, official EPA approval, endorsement, or recommendation.

Literature cited

Bean, C., Schrameck, R., and Davidson, C., 1994 Rouge River remedial action plan update, MI Dept. Natur. Resour. and Southeast MI Coun. Govts., Detroit, 1994.

Cave, K.A. and Bails, J.D., Implementing a model watershed approach through a state general stormwater NPDES permit, *Proc. WEFTEC 98 Conf.*, Orlando, FL, 1998.

Cave, K.A., Bryson, D.S., and Ridgway, J.W., Achieving multiple objectives through a single watershed plan, *Proc. Watershed 2000 Conf.*, Vancouver, BC, 2000.

Center for Watershed Protection (CWP), *The Practice of Watershed Protection: Techniques for Protecting and Restoring Urban Watersheds*, 2000.

Hufnagel, C.L. et al., What performance monitoring tells us about how to improve the design of CSO storage/treatment basins, *Proc. Watershed 2000 Conf.*, Vancouver, 2000.

Johnson, B.A. and Tuomari, D., Did you know ... the impact of onsite sewage systems and illicit connections on the Rouge River, *Proc. Natl. Conf. Retrofit Opportunities for Water Resour. Protection in Urban Environ.*, Chicago, 1998.

Kluitenberg, E.H., Kaunelis, V.P., and Johnson, C.R., Evaluation of in-stream impacts of CSO control facilities, *Proc. Watershed 2000 Conf.*, Vancouver, 2000.

Mullett, Jr., N., Bristol, C.R., and Koleda, K.P., Project technical support: GIS/sampling/modeling, presented at 1994 Water Environ. Fed. Ann. Conf., Chicago, 1994.

Powell, J.A. and Bails, J.D., 2000. Measuring the soft stuff — evaluation public involvement in urban watershed restoration, *Proc. Watershed 2000 Conf.*, Vancouver, 2000.

Rood, S., Koleda, K., and Bristol, C.R., Data access and Analysis using RPO DataView, *Proc. WEFTEC 95*, Miami Beach, 1995.

Rouge River National Wet Weather Demonstration Project, Wayne County, MI, http:\\www.rougeriver.com, 1992–2001.

Rouge River National Wet Weather Demonstration Project, Rouge River Gateway Master Plan, Wayne County, MI, 2001.

Schueler, T., Crafting better urban watershed protection plans, CWP, *Watershed Protection Tech.*, 329, 1996.

Smith, E., Hughes, C., and Snyder, K., Communicating river water quality information to the public using a graphical indicator approach, *Proc. WEFTEC 97*, Chicago, 1997.

U.S. EPA, Watershed approach framework, EPA840-S-96-001, Washington, DC, 1996.

U.S. EPA, Top 10 Watershed Lessons Learned, EPA840-F-97-001, Washington, DC, 1997.

Response

Looking beyond the end of the pipe

Perhaps the clearest message in this chapter by Kelly Cave is the concept that the effective management of watersheds may often have just as much to do about managing people as about managing pollution. As Cave describes, it is essential to gather the wide consortium of interested players, including governments, industries, environmental groups, and private citizens, and include their viewpoints in consensus-building strategies. Community participation in identifying problems and suggesting solutions is often the key attribute separating those watershed management projects achieving success — such as the Rouge River in Detroit — and those that flounder. This is a recurrent theme discussed elsewhere in this book: notably by Gail Krantzberg and Judi Barnes in Chapter II.6 and by David Blau in Chapter II.7. Further, as Cave also points out, support from the general public is essential for water sensitive planning to be self-sustaining through time, a topic that Daniel Williams considers in Chapter II.11.

If such a process is to work, it is critical that the public become educated about the nuances of watershed management. As Cave identifies, there is a need for effective transfer of technology to the public in terms they can easily comprehend (e.g., converting phrases such as "meets water quality standards" to something like "fishable" or "swimmable").

The other major lesson from this chapter is the requirement to move beyond the belief in end-of-pipe, cure-all solutions to water quality management. Instead, much more comprehensive planning can be achieved by focusing on subwatersheds, by recognizing that advances are made incrementally in small steps, and by integrating stormwater control projects into ongoing public works plans.

chapter II.10

Modeling a soil moisture index using geographic information systems in a developing country context (Thailand)

John S. Felkner and Michael W. Binford

Abstract

Land-cover change has important influences and impacts on biodiversity, climate, ecosystem structure, as well as the land's capacity for sustained use. This is a particular problem in the developing world, where population, food, and land-use pressures are greatest.

The ability to characterize the spatial variability of soil water content in a simple yet physically realistic way is highly valuable. Soil moisture is one component of soil quality and agricultural suitability, and its measurement is important for land use, drought, agriculture, irrigation, and construction planning, as well as for evaluation of ecological system processes and health. Consequently, a need exists for techniques and models that can rapidly estimate soil moisture across a landscape, particularly in a context where access to rich data sources can be limited, and where inputs to spatially varying models of general land-use change are needed.

This chapter describes a model for continuous soil moisture variation estimation in a developing-country context, based on topographic, precipitation, land-cover, and soil quality data inputs. Building on previous terrain-based indices, the model represents static, or steady-state, conditions of relative soil water content and is thus not a water budget estimation but provides an output that is an index correlating with soil moisture availability. The model calculates average, not actual, runoff and water availability and provides a basic measure of general spatially heterogeneous soil moisture conditions in a developing-country context where data sources can be limited. The model inputs and outputs are transformed into geographic information system (GIS) raster grids, facilitating future incorporation into spatially explicit statistical or predictive land-use models.

The analysis was performed in two provinces of Thailand, which is an ideal country for such a study because it has experienced rapid rates of land-use change as well as strong population and economic growth in recent decades. Specifically, the model was performed in the Thai provinces of Sisaket and Chachoengsao, selected because of their high relative environmental and economic contrasts (Binford et al., 2002).

The model inputs are land cover, derived from classified satellite imagery, spatially varying precipitation data, spatially varying soil quality and permeability conditions interpolated from point sources, and topography in the form of a digital elevation model (DEM).

The land cover was derived using an ISODATA unsupervised clustering algorithm, with clusters identified by comparison with ground-truth data as well as with other land-cover images. The precipitation data were interpolated from a series of rainfall gauge stations in Thailand and were averaged to obtain representative conditions. Soil sample data points measuring soil field capacity were spatially interpolated and then converted to the U.S. Natural Resources Conservation Service (NRCS) Hydrologic Soil Group (HSG) categories, based on a observed relationship between field capacity and HSG categories. The HSG output was combined with land cover to obtain NRCS runoff code number (RCN) values. With these inputs, the GIS was used to measure the likely runoff and infiltration at each pixel, and the surface runoff water flow and accumulation were modeled, using established mathematical relationships derived by the NRCS. Then a soil moisture index was calculated as the tangent of the slope.

The results are consistent with expected soil moisture variation, considering land cover (forest and agriculture), topography, and low-lying wetland areas. Finally, the output was incorporated into a spatial statistical predictive model of land-use change and found to act as a significant variable "explaining" change over time. In particular, areas of low soil moisture were found to correlate with areas of deforestation due to agriculture, while areas of high soil moisture values correlated with no land-use change.

Introduction

In the last 20 years, land-cover change — whether it is deforestation in the tropics, urbanization, intensification of agriculture, or land degradation — has accelerated as a result of population pressure and economic development. Two hundred million hectares of land, most classified as agriculturally marginal, were brought under cultivation for the first time during the period from 1980 to 2000 (Dudal, 1980; Sanchez, 1976). Conversion of land on this scale, unmatched in human history, is large enough to contribute significantly to changes in global climate and global biogeochemical cycles (Turner et al., 1995; Woodward et al., 1996). The effects of this global-scale change are significant, powerfully impacting biodiversity, climate, ecosystem structure and the land's capacity for sustained use and its ability to regain its original cover (F.A.O., 1981; Stern et al., 1992; WCED, 1987).

Major land changes were centered primarily in the mid-latitudes of the Northern Hemisphere prior to the middle of the 20th century (Turner et al., 1990). In the last 50 years, however, the major land-cover changes have occurred in the tropics, where cropland and grassland/pasture expansion, deforestation, and urbanization, among other changes, are increasing rapidly (Woodward et al., 1996).

Land-cover change has an important influence on biogeochemical cycling of gaseous compounds of carbon, nitrogen, and other elements (Penner, 1994), and on water and energy balances, all at regional to global scales. Changes in the distribution of land cover alter regional — and possibly global — surface roughness, albedo, latent and sensible heat flux, and actual evapotranspiration, all of which are important parameters for general circulation models (Turner et al., 1995). Plants extract water from soils, returning it to the atmosphere through transpiration. In the Amazon rainforest, for example, water recycling by vegetation has had a direct impact on local and regional precipitation patterns (Victoria et al., 1991). Thus, changes in land cover at regional scales can have direct impacts on precipitation and local climate effects, and certainly land-use/cover change can affect atmospheric conditions (Foley et al., 1994; Henderson-Sellers, 1990).

This relationship is a problem in the developing, tropical world, where the land-use pressures of population increase are greatest. Consequently, a great need exists for the monitoring, quantitative assessment, and predictive modeling of land-use change (Behera et al., 1996; Rao, 1999; Robinson et al., 1994; Turner et al., 1995; Wald, 1999). Ideally, these models would be able to predict land-use change patterns and processes at a regional or local scale as well as continental or global, to quantify the changes and types of changes, and to provide insight into the drivers of these changes (Bockstael and Irwin, 1999). The derived information from such models can be used for more effective land-use planning, preservation of ecologically fragile areas, and possibly aid in poverty alleviation.

Soil moisture estimation is important

The importance of the variation in surface soil moisture, including the location of areas of very high water saturation, is well appreciated in disciplines including engineering, soil conservation, forestry, agriculture, and hydrology because of the effect on soil strength, erosion, soil aeration, plant ecology, secondary salinization, and storm runoff (O'Loughlin, 1986). The ability to characterize the spatial variability of soil water content in a simple yet physically realistic way is of major importance (Barling et al., 1994). There is strong support in the land-use change modeling literature for correlations between soil quality and the probability of land-use change. Superior soil quality tends to promote agriculture, which initially causes deforestation. Over time, however, rich soil can act as a brake on the loss of remaining forests because high agricultural productivity can reduce the economic need for deforestation (Crist, 1986; Katawatin et al., 1994, 1996a, 1996b; Rindfuss, 1998). Soil moisture is one component of soil quality and agricultural suitability. Thus, the estimation of soil moisture in a landscape is an important goal for effective land-use change modeling and land planning.

This goal is especially applicable in the developing world, which has the highest rates of land-use change and land degradation, and more than 70% of the world's total population (Skole, 1994). Soil moisture estimation in the developing world is important for drought, agriculture, irrigation, and construction planning (Binnie, 1997; Chiew et al., 1995; Damota et al., 1992; Fu and Gulinck, 1994; Kimmage and Adams, 1992; Kobayashi, 1996; Winkler and Ulehla, 1992). For example, land conversion to agricultural production in Thailand during the past two or three decades has resulted in severe soil erosion, land degradation, and decrease of the soil moisture content due to the lack of implementation of appropriate soil and water conservation measures (Kobayashi, 1996). Finally, soil moisture estimation can also be particularly useful in the delineation of wetland areas (Costanza and Sklar, 1984), which are in turn integral to the health of hydrologic and biotic ecosystems (Schulze and Mooney, 1993).

Goals of this modeling effort

Given this context, there is a need for techniques and models that can rapidly estimate soil moisture across a landscape, particularly in developing countries where both time and data are likely to be scarce (Binnie, 1997; Florinsky, 1998; Rango and Shalaby, 1998; Turner et al., 1995). The use of GIS and remote sensing tools offers a possible solution to meet this need, as lower costs and increasing power have developed dramatically in recent years (Goodchild, 1994; Goodchild et al., 2000; Robbins and Phipps, 1996). This includes the development of raster-format hydrologic models and watershed databases (Julien et al., 1995; Schloss, this volume; Mueller et al., this volume). In recent years, there has been considerable effort devoted to utilizing GIS to provide inputs (soils, land use, and topography) for comprehensive simulation models and to display model outputs spatially

(Arnold et al., 1998; Bonham-Carter, 1994; Cantwell, this volume; Julien et al., 1995; Moore and Hutchinson, 1991; Rewerts and Engel, 1991; Robbins and Phipps, 1996; Srinivasan and Engel, 1991; Srinivasan and Arnold, 1993; Tarboton et al., 1991).

The goal of this chapter is to describe the development and implementation of a model that will estimate relative soil moisture availability across a large landscape in a developing-country context where data are scarce. The soil moisture estimation output in spatially continuous raster GIS format would be useful as input to spatially explicit, statistical, predictive land-use models that predict future land use at a provincial scale (Agarwal et al., 2000; Chomitz and Gray, 1996; Dimyati et al., 1996; Felkner, 2000; Kaimowitz and Angelsen, 1998). Specifically, some land-use prediction models require a relative soil moisture estimate with values at all points on the landscape.

The soil moisture model described here relies primarily on topographic information, which, in the form of contour data or a DEM may be the only type of data available in a developing-country context (O'Loughlin, 1986). In the same vein, innovative and alternative methods for deriving spatial variation in precipitation patterns and in soil characteristics may have to be devised, as point data may be the only available data sources for these inputs. For this model, spatial variation in precipitation patterns and in soil qualities were derived from point data sources.

Our model uses the following data inputs: land cover derived from satellite remote sensing data; average annual precipitation data from meteorological stations; field samples of soil conditions; and a DEM interpolated from contour elevation information. The model is designed to produce raster GIS output. The approach is applicable at regional or provincial scales, rather than solely country or continental ones. The approach thus fulfills a perceived need in the literature of land-use change monitoring and modeling for regional, rather than country or larger-scale, efforts (Bockstael and Irwin, 1999; Kaimowitz and Angelsen, 1998; Lambin et al., 1999).

Previous research on soil moisture indices

The approach taken here relies heavily on topographic data to create a continuous grid of relative soil moisture. The importance of topography has been recognized as the primary determinant in regulating variation in surface soil moisture in several experimental studies (Anderson and Burt, 1977; Anderson and Burt, 1978; Beven, 1978; Dunne and Black, 1970; O'Loughlin, 1986). Topography plays an important role in the hydrologic response of a catchment to rainfall and can affect, for example, the location of zones of surface saturation and the distribution of soil water. The automation of terrain analysis and the use of DEMs have made it possible to quantify easily and accurately the topographic attributes of a landscape.

A number of terrain-based indices have been derived and relationships have been sought between these indices and a range of hydrologic processes (Barling et al., 1994; Beven and Kirkby, 1979; Burt and Butcher, 1986; Moore et al., 1988, 1991, 1993). The soil moisture index approach has typically been based on the use of simplified representations of the underlying physical processes (soil qualities, precipitation amounts, landscape cover, etc.) while at the same time striving to include most of the primary factors that regulate the system's behavior. This approach sacrifices physical sophistication and resolution but enables relatively easily computed estimates of the spatial patterns in a landscape (Barling et al., 1994; Moore et al., 1991, 1993; Moore and Hutchinson, 1991).

In some cases, particularly in developing-country contexts, a topographic map or DEM may be the only relevant information available for the area in question. Consequently, the terrain-based index approach may be best suited, and consistent, with the level of available data and the precision with which many of the management and planning questions both need, and can, be answered (Barling et al., 1994). Thus, this approach is logical for a developing-country context when a rapid assessment is needed as a component or input into larger land-use systems analysis.

This model represents a static, or steady-state, condition of relative soil water content and the location and size of zones of surface saturation (Beven and Kirkby, 1977, 1979; O'Loughlin, 1986). Thus, the model is not a water budget estimation but provides an output that is an index correlated with soil moisture availability. High values mean the soil is moist or even saturated; low values mean the soil is dry. The model calculates average, not actual, runoff and water availability and provides a basic measure of general spatially heterogeneous soil moisture conditions considering topography, generalized spatially heterogeneous average precipitation amounts, generalized spatially heterogeneous soil types, and specific land cover. The approach taken here modifies previous approaches to adjust to limited availability of input data, specifically a lack of appropriate and comprehensive soil data, and a lack of detailed, spatially heterogeneous precipitation data. The output is a GIS raster grid that provides a generalized reflection of the ability of the soil to hold water.

Description of study area and data sources

Thailand is an ideal country for such a study, because it has experienced extensive deforestation, land-use change, and urbanization within the last 40 years (Lombardini, 1994). Originally rich in tropical rainforests and high plant diversity, Thailand has become heavily agricultural since the 1960s. Also, Thailand has a wide variation in regional economic conditions, including income and education levels. Thus, analysis of different regions of Thailand provides a cross-section of tropical developing economic conditions.

The study area consists of two provinces of Thailand: Chachoengsao (5,351 km^2) and Sisaket (8,840 km^2). The larger land-use predictive model was applied to the two provinces in order to provide contrasting case studies, both for analysis and verification purposes. These two provinces were specifically selected to present a maximal contrast both economically and environmentally, based on a consideration of income levels and income growth, agricultural output, precipitation, and soil fertility (Binford et al., 2002).

The province of Chachoengsao is located immediately east of the Bangkok metropolis, while Sisaket is located in the poor northeast region, on the border with Cambodia (see Figure II.10.1). Whereas Chachoengsao is highly developed, with extensive infrastructure and industry, Sisaket is one of the poorest provinces in the country, with very limited infrastructure and urbanization. Chachoengsao boasts a broad industrial base, with intensive and diverse agricultural practices and excellent proximity both to ocean ports and to Bangkok. By 1990, more than 90% of the population of Sisaket was engaged in agriculture, compared with approximately 66% of Chachoengsao (Thai Census Dept., 1990). Educational attainment and literacy rates are considerably lower, and child mortality rates are higher, in Sisaket. Physically, the two provinces also contrast: Chachoengsao is on the flat coastal plain and has much richer soils than Sisaket, which is inland and has significant relief especially in the south (Daniere, 1999; Hirsch, 1993; Binford et al., 2002). Both provinces have been extensively deforested in the last 40 years (Lombardini, 1994), mostly for agriculture except in areas of topographic relief (Figures II.10.2 and II.10.3). Despite their relatively equal amounts of deforestation, the two provinces provide excellent case studies because of their simultaneous economic and environmental differences.

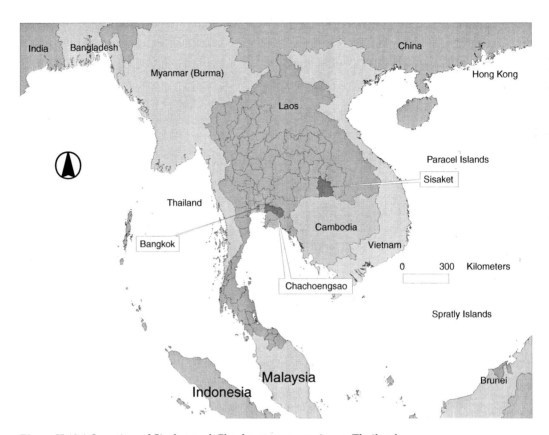

Figure II.10.1 Location of Sisaket and Chachoengsao provinces, Thailand.

Much of the data for this study were gathered by a research group headed by Dr. Robert Townsend of the University of Chicago, Dept. of Economics, with funding from the National Science Foundation and the National Institutes of Health (Binford et al., 1998; Townsend and Lim, 1998) in cooperation with the Thai government and the Thai Bank of Agriculture and Agricultural Collectives (BAAC). Precipitation data were interpolated from measurements taken by the Thai government at a series of rainfall gauges distributed throughout the country (Figure II.10.4) that included monthly precipitation amounts from the 1950s to 1986. Soil quality and conditions were also interpolated from samples taken in each of the provinces in a number of villages as part of an extensive socioeconomic and environmental survey conducted in 1997 (Binford et al., 1998) (Figure II.10.5). The soil samples were analyzed for field capacity (the volume of water remaining in a soil after gravity drainage of saturated soils, representing the maximum amount of water available for plants) among other soil qualities (see Figures II.10.6 and II.10.7).

Land-cover classes were derived from interpreted Landsat Multi-Spectral Scanner (MSS) and thematic mapper (TM) data (Figures II.10.2 and II.10.3). The DEM was interpolated from vectorized contour lines at 10-m intervals on 1:50,000 scale topographic quadrangles produced by the Thailand Dept. of Mapping (maps covering the Thailand–Cambodia border in southern Sisaket were unavailable).

The soil moisture procedure estimates *relative* soil moisture content conditions *on average* during the 1980s. This was done for purposes of a larger predictive land-cover modeling work, which predicted 1999 land-cover conditions based on historical trends in

Chapter II.10: *Modeling a soil moisture index using geographic information systems* 519

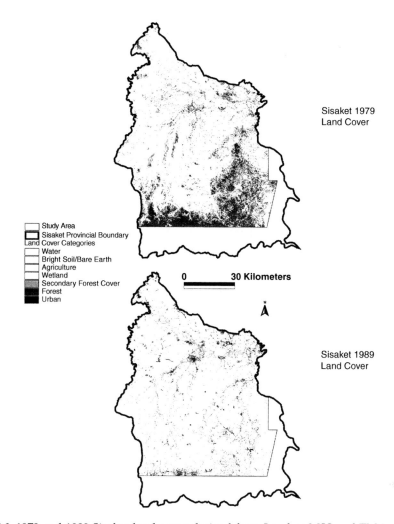

Figure II.10.2 1979 and 1989 Sisaket land cover derived from Landsat MSS and TM imagery.

land-use change from 1979 to 1989 (Felkner, 2000). Therefore, all inputs to the larger model were designed to represent approximate average conditions in the 1980s. Consequently, precipitation data for the soil moisture model was based on monthly averages from the 1980s. Also, land-cover classifications were derived from Landsat MSS images from 1979 and 1982, to approximate conditions in the early 1980s. Although the soil samples were obtained in 1997, we assume that the general spatial arrangement of intraprovincial soil conditions would not have changed appreciably since the early 1980s.

Methods

The model algorithm is shown in two flow charts in Figures II.10.8 and II.10.9. Figure II.10.8 depicts the process of preparing the four primary inputs to the soil moisture index calculation: a runoff curve number (RCN) grid reclassified from the HSG grid, in turn derived from a soil field capacity grid interpolated from the original soil samples; a precipitation grid interpolated from Thai rain-gauge measurements; and two input grids derived from

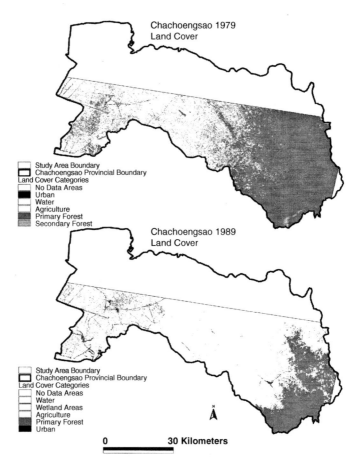

Figure II.10.3 1979 and 1989 Chachoengsao land cover derived from Landsat MSS and TM imagery.

the DEM — a slope grid (with every cell having the value of its slope in degrees) and a flow-direction grid (every cell contains a numerical representation of the direction vector of water flow across the surface) (Jenson and Domingue, 1988). Figure II.10.9 depicts the soil moisture modeling process once the input grids (which are rectified and have equal numbers and sizes of grid cells) are created. All grids used cell sizes of 60 by 60 m.

Creation of the DEM

Topography is the primary determinant of relative soil moisture content in the model. Specifically, the model requires the use of a DEM in order to measure slope, flow direction, and flow accumulation (the uphill area — in number of grid cells — from which water would flow into any particular cell) (Jenson and Domingue, 1988; Tarboton et al., 1991). The DEM was interpolated from vectorized contour lines, as described in the following paragraphs. Because of the need for the measurement of flow direction and flow accumulation, a "depressionless" DEM was generated.

First, contour lines were digitally extracted and vectorized (represented as vector GIS objects) from the Thai government's 50,000-scale topographic map quadrangles, using automated vector extraction software operating on scanned versions of the maps. Once the contour lines were vectorized, they were assigned appropriate elevation values. This process of extraction was manually supervised during the automated extraction routines,

Chapter II.10:   Modeling a soil moisture index using geographic information systems   521

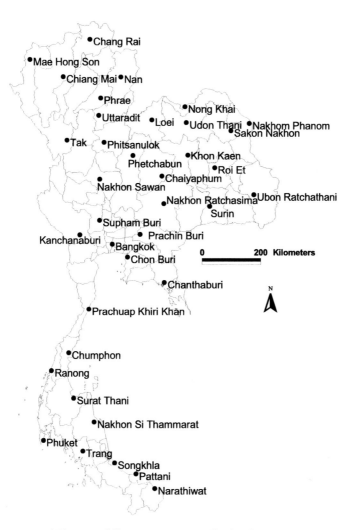

Figure II.10.4 Locations of Thai rainfall-gauge stations, Thailand.

and then the output was error-checked and "cleaned" by hand. The interpolation of a continuous raster surface was performed with the "topogrid" command in ArcInfo GIS software that is based on the ANUDEM program developed by Michael Hutchinson (Hutchinson, 1988, 1989, 1993, 1996; Hutchinson and Dowling, 1991). The method uses an iterative finite-difference interpolation technique that is essentially a discretized thin-plate spline approach (Wahba, 1990), where the roughness penalty has been modified to allow the fitted DEM to follow abrupt changes in terrain, such as streams and ridges. The procedure is specifically designed to consider hydrologic drainage features, which are the primary erosive and shaping force on a landscape.

The DEMs were used in the soil moisture index model for hydrologic analysis, so it was necessary to process the DEM surfaces iteratively to remove sinks. For many hydrologic analyses run on a GIS grid, the flow of water across the surface is often simulated as a function of gravity (a function of the elevation differences between grid cells). A sink is a location that will trap water running into it — essentially a well, depression, or pit on the surface of the DEM. Simulated water movement across the DEM surface is prevented from running uphill and will thus get "stuck" in the sink. Often, the sinks are

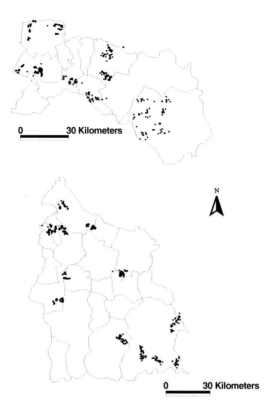

Figure II.10.5 1997 soil sampling sites, Sisaket and Chachoengsao provinces, Thailand.

artifacts of data processing errors and can be quite small (Mark, 1988). The process used for creating the depressionless grid is shown in Figure II.10.12.

The final output of the DEM creation process was smoothed to remove aberrant artifacts using cubic convolution smoothing (Hunter and Goodchild, 1997; Jenson and Domingue, 1988; Kyriakidis et al., 1999). Once depressionless DEMs were created, surface-analysis algorithms in ArcInfo GIS software were used to calculate slope and flow direction grids (Greenlee, 1987; Jensen and Domingue, 1988).

Land-cover classification

Land cover for the early 1980s for both provinces was classified from Landsat MSS data using the ISODATA unsupervised clustering approach implemented in the software program ERDAS Imagine, in combination with ground-truth data we collected in Thailand in 1998 (Sabins, 1987; Tou and Gonzalez, 1977). Two sequential Landsat MSS 1979 images, acquired on October 6, 1979, were a mosaic map that provided almost complete coverage of Sisaket. For Chachoengsao, a single Landsat MSS scene covered the southern two-thirds of the province, acquired on October 10, 1979. The satellite data were rectified to the UTM coordinate system, using the Indian 1975 datum and the Everest 1830 spheroid.

In August 1998, 80 and 40 widely dispersed ground-control and land-cover data points were collected in Sisaket and Chachoengsao, respectively, using a Global Positioning System (GPS) receiver, to allow geo-rectification and classification. The geographic locations of these points were tied in the field to their visible locations on the images, and then they were used as tie-point inputs into a bilinear interpolation for rectification (Jensen, 1996), with approximately a one-pixel root mean squared error (RMS).

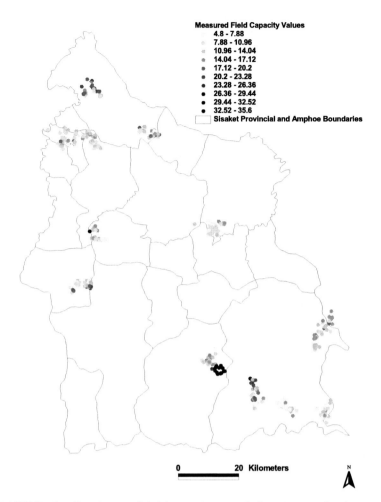

Figure II.10.6 1997 Sisaket Province soil field capacity sample locations and values.

Although the 1979 Sisaket MSS scenes were acquired on the same day, a multiple-date empirical radiometric normalization was performed using a regression approach with "pseudoinvariant" ground targets (water bodies; dry, barren soil; etc.) to minimize differences in atmospheric conditions, satellite acquisition angle, lighting conditions, and detector calibration errors (Jensen, 1996). The images were then were made into a mosaic map with a feathering interpolation system along the overlapping edges.

To aid in the process of assigning land covers to each cluster (agriculture, forest, built, or other), we used "natural-color" composite images in conjunction with the ground-control and ground-truth data. Normalized difference vegetation indices (NDVI) images helped distinguish forested areas (Rouse et al., 1973). "Tasseled cap," or Kauth–Thomas transforms (Crist, 1986; Crist and Kauth, 1986), were calculated for the raw imagery to identify the soil brightness index (SBI), the green vegetation index (GVI), the yellow stuff index (YVI), and the nonesuch index (NSI). In addition, "de-haze" images were generated as a product of a transform similar to Kauth–Thomas (Richards, 1993).

All these derived data layers were consulted closely in assigning labels to each cluster. Initially, 100 clusters were created for each image, and each was assigned a land-cover category. Extensive cluster busting (Jensen, 1996) separated the clusters that could not be clearly identified (e.g., extending both into built areas and into agriculture). The results are displayed in Figures II.10.2 and II.10.3.

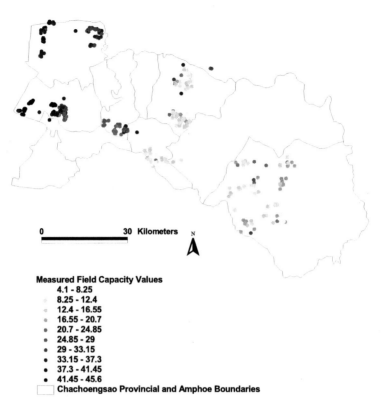

Figure II.10.7 1997 Chachoengsao Province soil field capacity sample locations and values.

Creation of the precipitation grid

Our model works most effectively with accurate spatially heterogeneous precipitation data, but these data were not available at the subprovincial level in Thailand. Rainfall data were obtained from a network of rain-gauge stations distributed throughout Thailand (Figure II.10.4). Monthly precipitation values during the 1980s were averaged for each precipitation station. Then, a simple inverse distance weighted (IDW) interpolation was performed from all the national rain-gauge stations to create a continuous grid of spatially varying precipitation values across each province. This method was the best option given readily available data inputs, although it yielded a poor degree of spatial heterogeneity for each study province.

Estimation of hydrologic soil groups for Thailand using field capacity measurements

Previous static soil moisture approaches have taken advantage of an empirical method and several derived mathematical relationships developed by the U.S. Natural Resources Conservation Service (NRCS, formerly Soil Conservation Service, or SCS). This approach uses HSG and empirically derived RCNs to estimate the ratio of runoff to infiltration for a given type of land-cover/soil-type combination (U.S. Dept. of Agriculture, 1972). The ratio of infiltration to runoff depends on both the degree of imperviousness of the land cover and the permeability of the soil type. Pavement or building roofs are practically impervious, while certain kinds of forest, prairie, or agricultural lands are highly permeable.

The HSG soil classification system was developed as a way of summarizing a soil series' hydrologic effects, and NRCS categorized every soil series in the United States into

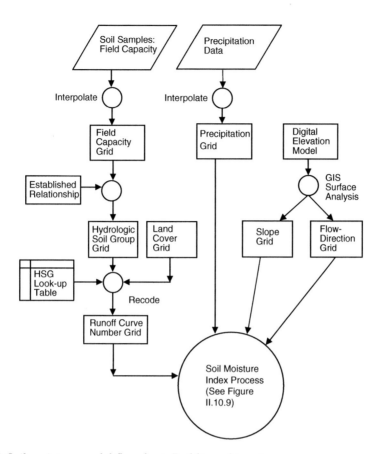

Figure II.10.8 Soil moisture model flowchart: Building of inputs.

four groups, lettered A through D (see Figure II.10.13). Group A has the highest infiltration capacity and is the least likely to generate runoff, while group D has the lowest infiltration capacity and highest runoff. Given a soil series' hydrologic soil group, the RCN is derived from a combination of HSG and land cover, according to published NRCS lookup tables (Ferguson, 1998). Thus, the NRCS curve number summarizes the combined effects of soil and land cover. RCN was derived from experimental studies establishing specific relationships between rainfall and runoff for different types of soil and land cover and can vary, theoretically, from 0 to 100 as soil and land-cover conditions generate increasing amounts of runoff. In practice, however, RCN is seldom less than 30, and the highest RCN, for impervious pavement and roofs, is 98.

HSG soil classifications do not exist for Thailand. Furthermore, accurate soil maps for each province were not obtainable; although, soil samples taken in 1997 in both provinces (Figure II.10.5) included a measurement of soil field capacity (among other soil variables), shown for each province in Figures II.10.6 and II.10.7. A method of estimating HSG for the entire area of each province was devised using field capacity measurements as a proxy for hydrologic group, based on an observed relationship between field capacity and HSG observed in soils from Puerto Rico (Soil Conservation Service, 1975), as shown in Figure II.10.14. Soils in Puerto Rico and Thailand, both tropical climates, share many similar properties, and the USDA conducts soil surveys in Puerto Rico, which include measurements of field capacity and simultaneous designation of hydrologic groups. HSG is primarily a classification of the soil's tendency to allow water permeability, and field capacity is related to soil texture and infiltration capacity (Dunne and Leopold, 1978).

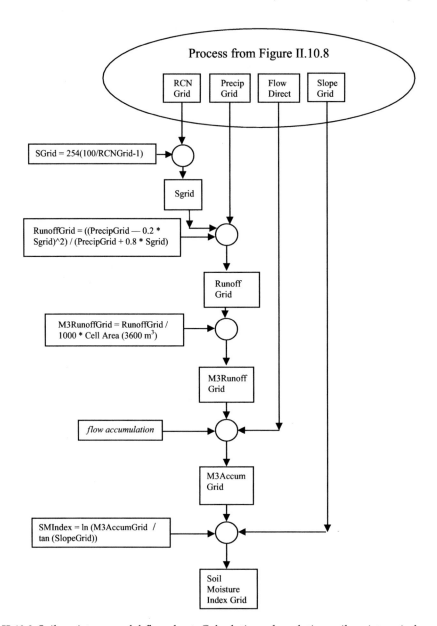

Figure II.10.9 Soil moisture model flowchart: Calculation of a relative soil moisture index.

Because the modeling process required a continuous and complete raster grid of the entire process, however, the first step in the estimation of HSG was the interpolation of the field capacity measurements to derive a continuous field for the entire provincial areas. This step was performed with a full recognition that soil spatial variation is by no means always continuous, yet without such an interpolation the model would have no soil spatial variation, whereas such an interpolation gives at least a rough approximation of provincial soil variation. Based on previous studies into the effectiveness of different point interpolation methods for soil samples (Laslett et al., 1987; McBratney and Webster, 1986), a kriging approach was used (Bonham-Carter, 1994; Myer, 1991). Once the point samples were effectively kriged to obtain complete raster surfaces, these were converted to HSG grids using a lookup table derived from the observed Puerto Rican relationship. The final

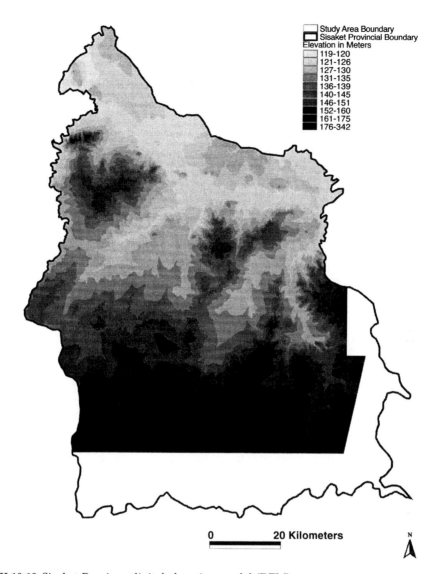

Figure II.10.10 Sisaket Province digital elevation model (DEM).

step in the process was to convert the HSG grids into RCN grids with the addition of the 1979 land-cover grids (see Figures II.10.2 and II.10.3): each pixel was assigned an RCN value as a function of both its HSG designation and its corresponding land cover according to the lookup tables (Ferguson, 1998). The resulting RCN grid was input into the soil moisture index calculation process, along with precipitation, slope, and flow direction grids. The final RCN grids are shown in Figure II.10.15.

Soil moisture index estimation process

The process of calculating the soil moisture index from the input grids is depicted in Figure II.10.9. The process was automated into Arc Macro Language (AML) and run in ArcInfo software. The approach combines the NRCS methodology for estimating runoff based on RCN and the calculation of the accumulation of uphill water flow into each "downhill" grid cell on a depressionless DEM in order to obtain the amount of runoff

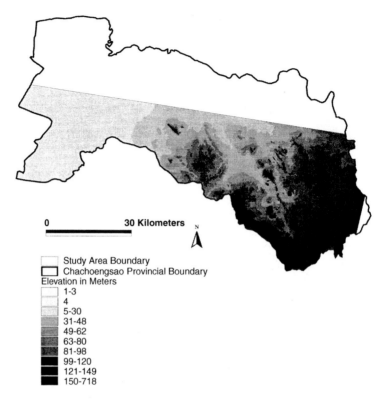

Figure II.10.11 Chachoengsao Province digital elevation model (DEM).

flowing through each grid cell. This "flow accumulation" grid (Jenson and Domingue, 1988) is divided by the tangent of the slope, and the natural log is taken of that quotient to produce the index. The NRCS methodology specifically calculates a retention factor, s, defined as (for precipitation in mm)

$$s = 254\left(\frac{100}{CN} - 1\right)$$

where CN is the runoff curve number. The retention Sgrid is next combined with the precipitation grid in the SCS curve number, equation which predicts surface runoff:

$$\text{Runoffgrid} = \frac{(\text{precipgrid} - 0.2 \cdot \text{sgrid})^2}{(\text{precipgrid} + 0.8 \cdot \text{sgrid})}$$

resulting in the creation of the runoffgrid. Finally, the soil moisture index is calculated as a function of the tangent of the slope:

$$\text{soil moisture index} = \ln\left(\text{m3accumgrid} \div \tan(\text{slopegrid})\right)$$

The final soil moisture index grids for Sisaket and Chachoengsao are shown in Figure II.10.16.

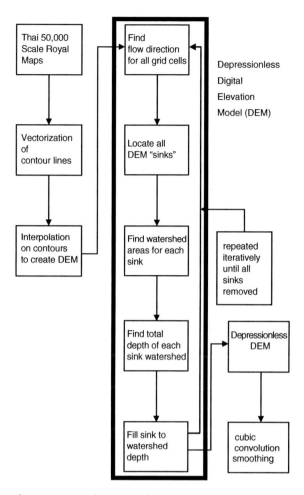

Figure II.10.12 Process for creating a depressionless DEM.

Results

Land-cover classifications for each province for 1979 and 1989 are shown in Figures II.10.2 and II.10.3. For the most part, the major land-cover classes were easily differentiated, especially forest and agriculture classes. Agriculture is the primary land use in both provinces, and extensive deforestation between 1979 and 1989 occurred as forest was converted to agricultural uses, as can be seen in the figures. Remaining forests tend to be clustered in the high topographic relief areas, in the south of Sisaket and the east of Chachoengsao. The densest urban areas tend to be clustered near the major rivers and, in the case of Chachoengsao, along the ocean shorelines in the southwest of the province. Despite the forest clustering in the higher relief areas, in general both provinces are relatively flat topographically and conducive to extensive agriculture.

Classification problems certainly were encountered, however, in the delineation of *built* land cover (urban, residential, dense settlement areas). It quickly became apparent that these were by far the most problematic to identify and label, as clusters that certainly appeared in areas that were clearly built (major urban areas) also often appeared in areas that were clearly agriculture, or even forest. Cluster-busting help to split these clusters along land-cover lines; although, certain clusters remained that were extremely difficult

Hydrologic Soil Group definitions, as used in the SCS method:

Group A:
Group A soils have low runoff potential. They have high infiltration rates even when thoroughly wetted. They consist chiefly of deep, well to excessively drained sands or gravels. This group also includes sand, loamy sand, and sandy-loam that have experienced urbanization but have not been significantly compacted.

Group B:
Group B soils have moderate infiltration rates when thoroughly wetted. They consist chiefly of moderately deep to deep, moderately well- to well-drained soils with moderately fine to moderately coarse textures. This group also includes silt loam and loam that have experienced urbanization but have not been significantly compacted.

Group C:
Group C soils have low infiltration rates when thoroughly wetted. They consist chiefly of soils with a layer that impedes downward movement of water and soils with moderately fine to fine textures. This group also includes sandy clay loam that has experienced urbanization but has not been significantly compacted.

Group D:
Group D soils have high runoff potential. They have very low infiltration rates when thoroughly wetted. They consist chiefly of clay soils with high swelling potential, soils with permanent high water tables, soils with clay pans or clay layers at or near the surface, and shallow soils over nearly impervious material. This group also includes clay loam, silty clay loam, sandy clay, silty clay, and clay that have experienced urbanization but have not been significantly compacted.

(Source: U.S. Conservation Service, *Urban Hydrology for Small Watersheds*, 2nd ed., Technical Release 55, Washington, 1986.)

Figure II.10.13 Hydrologic Soil Group (HSG) definitions.

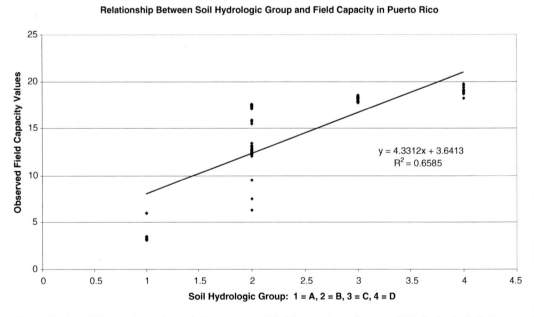

Figure II.10.14 Observed relationship between soil field capacity values and Hydrologic Soil Group (HSG) categories in Puerto Rican soils.

Chapter II.10: Modeling a soil moisture index using geographic information systems 531

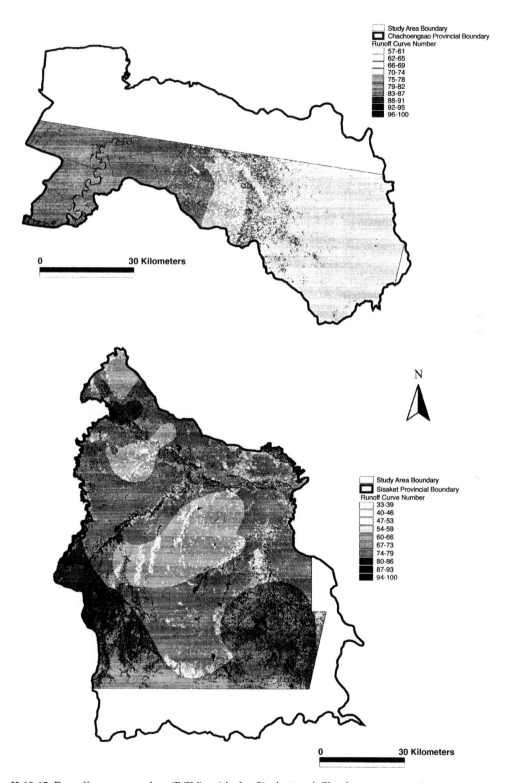

Figure II.10.15 Runoff curve number (RCN) grids for Sisaket and Chachoengsao provinces.

to classify. Thus, the classification of built land use in all of these classifications is certainly the class with the lowest accuracy confidence. The identification of forest and agriculture classes is much stronger.

The Sisaket and Chachoengsao RCN grids are depicted in Figure II.10.15. The influence of the major inputs that have delineated runoff code number can be seen in the outputs, namely land-cover and soil characteristic patterns. The figures reveal individual runoff/infiltration permeability patterns for each province. Sisaket shows higher degrees of runoff in the low-lying river/wetland areas in the north and near reservoirs and lakes in the south, as well as higher amounts of permeability in the southern forested areas, and lower amounts of permeability with higher runoff in the central agricultural regions. These trends are, however, mitigated in a number of cases by underlying soil quality patterns, based on interpolations from soil sample measurements, with soils in the south having less permeability than those in the central areas. This appears logical because the central provincial areas have been extensively used for agriculture in preference to the southern higher relief areas. The RCN models do not consider topography, however, and thus this may indicate that the central-area soils are superior for agriculture due to higher inherent permeability qualities, instead of only because they are topographically flat.

Chachoengsao shows generally higher runoff numbers in the western half of the province, which corresponds to the half that is much more highly developed, urbanized, and topographically flatter. Highest permeability appears as a central band running north–south in the center of the province and corresponds with areas used extensively for agriculture with little remaining forest cover even as of 1979.

The final soil moisture model output grids are depicted in Figure II.10.16. The results are consistent with a generalized representation of relative subprovincial soil moisture levels. In Sisaket, very high soil moisture values clearly appear in the major river valley in the northern portion of the province. At the same time, the higher-relief forested areas in the south have considerably lower soil moisture values, as would be expected due to less permeable soils and much higher slopes to encourage runoff. In general, the large agricultural areas in the central and west-central portions of the province have higher soil moisture values, corresponding to the general topographic pattern of the province as well as the pattern of primary agricultural uses established by 1979. On a smaller scale, the model has accurately increased soil moisture values in small valleys between ridges throughout the province.

Chachoengsao's results are somewhat distorted by the extremely flat portions of the DEM in the western part of the province, but generally the soil moisture output reveals that the eastern half of the province is drier. A comparison of the soil moisture output with the DEM for Chachoengsao reveals that the model has also accurately increased soil moisture values in lower elevation, with ridgetops appearing drier.

These results suggest the potential utility of a spatially continuous soil moisture output. Such information can be compared with data on land-use change, to look for correlations: if deforestation or land-use change does correlate with higher soil moisture values, then predictions can be made for future likely areas of deforestation. At the same time, inferences can be made regarding the need that agriculture has for high soil moisture values. In addition, if correlations between other types of land-use change — such as conversion of agriculture or forest to urban — and high or low soil moisture values can be established, such linkages could be useful for planning and management purposes.

However, a more rigorous method of assessment is needed beyond simply a visual assessment. Such comparisons are possible using GIS outputs, such as this soil moisture model grid output. The soil moisture models displayed in Figure II.10.16 were inputted into a spatial statistical model of land-use change for both Sisaket and Chachoengsao (Felkner, 2000). The approach carefully measured actual land-use change from 1979 to

Chapter II.10: Modeling a soil moisture index using geographic information systems 533

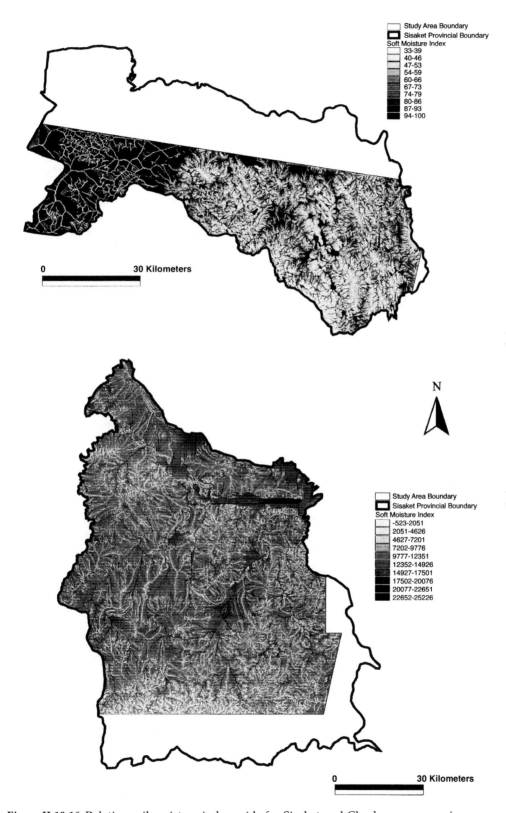

Figure II.10.16 Relative soil moisture index grids for Sisaket and Chachoengsao provinces.

1989 in both provinces along four basic change categories: conversion of forest to agriculture; conversion of forest to built uses; conversion of agriculture to built uses; and no change. Using this record of actual change as the dependent variable, a selected set of economic and environmental inputs were used as independent predictor variables using a classification-tree statistical technique (Venables and Ripley, 1994; Clark and Pregibon, 1992) to establish correlations between specific types of change and specific input variables. The soil moisture model constituted one of the "environmental" variables, along with topography, a proximity to forest edge model, and a proximity to water model. All inputs were weighted equally, and the statistical algorithm chose variables to partition the data into homogenous partitions by calculating splits along specific variables that resulted in the greatest reduction in variance between split data partitions.

The results revealed that the soil moisture index variable served as an important "explanatory" variable in deriving a model for land-use change, based on the inputs. While soil moisture played less of a role than did elevation or proximity to forest edges in explaining deforestation, it did factor significantly ahead of proximity to water bodies. The classification-tree algorithm chose soil moisture as a splitting variable by the fourth level of splits in the land-use change model "tree." Also, areas of lower-than-average soil moisture values were specifically highly correlated with conversion of forest to agriculture change, while areas of high soil moisture correlated highly with areas of no change. These results held for both provinces, but the derived tree models revealed that soil moisture played a more important role in "explaining" land-use change in Sisaket than in Chachoengsao and was specifically more significant (especially with high correlations with deforestation to agriculture) in southern Sisaket. The derived tree statistical model, using the soil moisture output displayed in Figure II.10.16, was then used to predict future land-use change with over 80% accuracy (Felkner, 2000).

These results suggest that, during the 1980s, deforestation to agriculture tended to occur in low soil moisture areas possibly because the relatively flat, low-lying high soil moisture areas (most conducive to agriculture) had already been cleared for crops in previous decades, and subsequent deforestation tended to occur in the drier, higher-relief areas with relatively impermeable soils. This could have been more significant in Sisaket than in Chachoengsao because, during the 1980s, Sisaket was relatively less deforested than Chachoengsao. Also, the correlation between high soil moisture with no land-use change could simply be a reflection of the clearing of high soil moisture areas for agriculture in previous decades: since these areas continue to be highly productive, they are not converted to other uses. Thus, the results should not necessarily be interpreted as indicating that low soil moisture inherently correlates with deforestation, but instead particular temporal and geographic economic conditions should be considered for context before conclusions are formulated.

Further analysis to consider more directly specific correlations between soil moisture variation and land use, or land-use change, could be achieved by running the classification-tree statistical algorithm using land use or land-use change as dependent variables and the soil moisture grids as independent predictors. The results would give very specific correlations between particular soil moisture index ranges and land use or land-use change types.

Thus, the approach taken here allows for the relatively rapid estimation of relative soil moisture amounts in a developing country context given limited data inputs, primary reliance on topographic map or DEM elevation data, and reliance on remote sensing for rapid and accurate land-cover derivation. Future research will focus on accuracy assessment and calibration of the model through actual on-site measurement and comparison with the model output.

Literature cited

Agarwal, C. et al., A review and assessment of land-use change models: Dynamics of space, time and human choice (publication draft), U.S. Forest Service Southern and Northern Change Program, 2000.

Anderson, M.G. and Burt, T.P., Automatic monitoring of soil moisture conditions in a hillslope spur and hollow, *J. Hydrol.*, 33, 27, 1977.

Anderson, M.G. and Burt, T.P., Toward more detailed field monitoring of variable source areas, *Water Resour. Res.*, 14, 1123, 1978.

Arnold, J.G. et al., Large area hydrologic modeling and assessment: Part I. Model development, *J. Am. Water Res. Assoc.*, 34, 73, 1998.

Barling, R.D., Moore, I.D., and Grayson, R.B., A quasi-dynamic wetness index for characterizing the spatial distribution of zones of surface saturation and soil water content, *Water Resour. Res.*, 30, 1029, 1994.

Behera, G. et al., Interfacing remote sensing and GIS methods for a sustainable rural development, *Intl. J. Remote Sensing*, 17, 3055, 1996.

Beven, K.J., The hydrological response of headwater and sideslope areas, *Hydrol. Sci. Bull.*, 23, 419, 1978.

Beven, K.J. and Kirkby, M.J., Considerations in the development and validation of a simple physically based, variable contributing area model of basin hydrology, *3rd Intl. Symp. Theor. Appl. Hydrol.*, Fort Collins, CO, 1977.

Beven, K.J. and Kirkby, M.J., A physically based variable contributing area model of basin hydrology, *Hydrol. Sci. Bull.*, 24, 43, 1979.

Binford, M. and Ervin, S., A static soil moisture index approach, in C. Steinitz, Ed., *Biodiversity and Landscape Planning: Alternative Futures for the Region of Camp Pendleton, California*, Harvard Univ. Press, Cambridge, MA, 1996.

Binford, M.W., Lee, T.J., and Townsend, R.M., In prep. Sampling design for an integrated socio-economic and ecological survey using satellite remote sensing and ordination. *J. Am. Stat. Assoc.*

Binford, M.W. et al., *Collaborative Research: Agroecological and Socioeconomic Systems Interactions in Thailand*, unpublished manuscript, 1998.

Binnie, C.J.A., Climate change and its potential effect on water resources in Asia, *J. Water Suppl. Res. Tech. — Aqua*, 46, 274, 1997.

Bockstael, N.E. and Irwin, E.G., Economics and the land use-environment link, in Folmer and Tietenberg, Eds., *Yearbook of Environmental Economics*, 1999.

Bonham-Carter, G.F., *Geographic Information Systems for Geoscientists: Modeling with GIS*, Pergamon/Elsevier Science, New York, 1994.

Burt, T.P. and Butcher, D.P., Development of topographic indices for use in semi-distributed hillslope runoff models, In O. Slaymaker and D. Balteanu, Eds., *Geomorphology and Land Management*, Gebruder Borntraeger, Berlin, 1986.

Chiew, F.H.S. et al., Simulation of the impacts of climate-change on runoff and soil moisture in Australian catchments, *J. Hydrol.*, 167, 121, 1995.

Chomitz, K.M. and Gray, D.A., Roads, land use, and deforestation: A spatial model applied to Belize, *The World Bank Econ. Rev.*, 10, 487, 1996.

Clark, L.A. and Pregibon, D., Tree-based models, In J.M. Chambers and T.J. Hastie, Eds., *Statistical Models in S*, Chapman and Hall, London, 1992.

Costanza, R. and Sklar, F.H., Articulation, accuracy and effectiveness of mathematical models: A review of freshwater wetland applications, *Ecol. Modeling*, 27, 45, 1984.

Crist, E.P., *Vegetation and soils information contained in transformed thematic mapper data*, IGARSS' 86 Symp., 1986.

Crist, E.P. and Kauth, R.J., The tasseled cap de-mystified, *Photogrammetric Eng. Remote Sensing*, 52, 81, 1986.

Damota, F.S. et al., Drought risks for the soybean crop in Rio-Grande-Do-Sul-State, Brazil, *Pesquisa Agropecuaria Brasileira*, 27, 709, 1992.

Daniere, A.G., Environmental behavior in Bangkok, Thailand: A portrait of attitudes, values, and behavior, *Econ. Devel. & Cult. Change*, 47, 525, 1999.

Dimyati, M. et al., An analysis of land use/cover using the combinations of Landsat and land use map — a case study in Yogyakarta, Indonesia, *Intl. J. of Remote Sensing*, 17, 931, 1996.

Dudal, R., *Soil Related Constraints to Agricultural Development in the Tropics*, IRRI, Los Banos, The Philippines, 1980.

Dunne, T. and Black, R.G., An experimental investigation of runoff production in permeable soils, *Water Resour. Res.*, 6, 478, 1970.

Dunne, T. and Leopold, L.B., *Water in Environmental Planning*, W.H. Freeman, San Francisco, 1978.

F.A.O., *Agriculture Towards 2000*, F.A.O., Rome, 1981.

Felkner, J.S., A digital spatial predictive model of land-use change using economic and environmental inputs and a statistical tree classification approach: Thailand, 1970s–1990s, unpublished doctoral thesis, Harvard Univ., Cambridge, MA, 2000.

Ferguson, B.K., *Introduction to Stormwater*, John Wiley & Sons, New York, 1998.

Florinsky, I.V., Combined analysis of digital terrain models and remotely sensed data in landscape investigations, *Prog. in Phys. Geog.*, 22, 33, 1998.

Foley, J.A., Kutzbach, J.E., and Coe, M.T., Feedbacks between climate and boreal forests during the Holocene Epoch, *Nature*, 371, 52, 1994.

Fu, B. and Gulinck, H., Land evaluation in an area of severe erosion — the Loess Plateau of China, *Land Degradation and Rehabilitation*, 5, 33, 1994.

Goodchild, M.F., Integrating GIS and remote-sensing for vegetation and analysis and modeling — methodological issues, *J. Vegetation Sci.*, 5, 615, 1994.

Goodchild, M.F. et al., Toward spatially integrated social science, *Intl. Reg. Sci. Rev.*, 23, 139, 2000.

Greenlee D.D., Raster and vector processing for scanned linework, *Photogrammetric Eng. and Remote Sensing*, 53, 1383, October 1987.

Henderson-Sellers, A., Modeling and monitoring 'greenhouse' warming, *Trends in Ecol. and Evol.*, 5, 270, 1990.

Hirsch, P., *Political Economy of Environment in Thailand*, Journal of Contemporary Asia Publishers, Manila, 1993.

Hunter, G.J. and Goodchild, M.F., Modeling the uncertainty of slope and aspect estimates derived from spatial databases, *Geograph. Anal.*, 29, 35, 1997.

Hutchinson, M.F., Calculation of hydrologically sound digital elevation models, *3rd Intl. Symp. on Spatial Data Handling*, Columbus, OH, 1988.

Hutchinson, M.F., A new procedure for gridding elevation and stream line data with automatic removal of spurious pits, *J. Hydrol.*, 106, 211, 1989.

Hutchinson, M.F., Development of a continent-wide DEM with applications to terrain and climate analysis, *Environmental Modeling with GIS*, Goodchild, M. et al. (Eds.). Oxford Univ. Press, New York, 1993.

Hutchinson, M.F., A locally adaptive approach to the interpolation of digital elevation models. *3rd Intl. Conf. on Integrating GIS and Environ. Modeling*, Santa Fe, NM, 1996.

Hutchinson, M.F. and Dowling, T.I., A continental hydrological assessment of a new grid-based digital elevation model of Australia, *Hydrol. Processes*, 5, 45, 1991.

Jensen, J.R., *Introductory Digital Image Processing*, 2nd ed., Prentice-Hall, Upper Saddle River, NJ, 1996.

Jenson, S.K. and Domingue, J.O., Extracting topographic structure from digital elevation data for geographic information system analysis, *Photogrammetric Eng. Remote Sensing*, 54, 1593, 1988.

Julien, P.Y., Saghafian, B., and Ogden, F.L., Raster-based hydrologic modeling of spatially-varied surface runoff, *Water Resour. Bull.*, 31, 523, 1995.

Kaimowitz, D. and Angelsen, A., *Economic Models of Tropical Deforestation — A Review*, Center for International Forestry Research, Jakarta, 1998.

Katawatin, R., Crown, P.H., and Grant, R.F., Simulation modeling of land suitability evaluation for dry season peanut cropping based on water availability in northeast Thailand, *Soil Use and Manage.*, 12, 25, 1996a.

Katawatin, R., Crown, P.H., and Klita, D.L., Mapping dry season crops in Thailand using Landsat-5 TM data, *Canad. J. Remote Sensing*, 22, 450, 1996b.

Katawatin, R. et al., Regional cropping potential in NE Thailand, *Canad. J. of Soil Sci.*, 74, 358, 1994.

Kimmage, K. and Adams, W.M., Wetland agricultural production and river basin development in the Hadejia-Jama Are Valley, Nigeria, *Geograph. J.*, 158, 1, 1992.

Kobayashi, H., Current approach to soil and water conservation for upland agriculture in Thailand, *Jarq-Japan Agricul. Res. Quart.*, 30, 43, 1996.

Kyriakidis, P.C., Shortridge, A.M., and Goodchild, M.F., Geostatistics for conflation and accuracy assessment of digital elevation models, *Intl. J. of Geograph. Inform. Sci.*, 13, 677, 1999.

Lambin, E.F. et al., *Land-Use and Land-Cover Change (LUCC) Implementation Strategy* (48), International Geosphere-Biosphere Programme (IGBP), Stockholm, 1999.

Laslett, G.M. et al., Comparison of several spatial prediction methods for soil pH, *J. of Soil Sci.*, 38, 325, 1987.

Lombardini, C., Deforestation in Thailand, in K.B.A.D. Pearce, Ed., *The Causes of Tropical Deforestation, the Economic and Statistical Analysis of Factors Giving Rise to the Loss of Tropical Forests*, Univ. College London Press, London, 1994.

Mark, D.M., Network models in geomorphology, *Modeling in Geomorphological Syst.*, John Wiley & Sons, New York, 1988.

McBratney, A.B. and Webster, R., Choosing functions for semi-variograms of soil properties and fitting them to sampling estimates, *J. Soil Sci.*, 37, 617, 1986.

Moore, I.D., Burch, G.J., and Mackenzie, D.H., Topographic effects on the distribution of surface soil water and the location of ephemeral gullies, *Trans. Amer. Soc. Agricul. Engineers*, 31, 1098, 1988.

Moore, I.D., Grayson, R.B., and Ladson, A.R., Digital terrain modeling: A review of hydrological, geomorphological and biological applications, *Hydrol. Processes*, 5, 3, 1991.

Moore, I.D. and Hutchinson, M.F., *Spatial extension of hydrologic process modeling, Intl. Hydrol. and Water Resour. Symp.*, Austria, 1991.

Moore, I.D., Norton, T.W., and Williams, J.E., Modeling environmental heterogeneity in forested landscapes, *J. Hydrol.*, 150, 717, 1993.

Myer, D.E., 1991, Interpolation and estimation with spatially located data, *Chemometrics and Intelligent Lab. Syst.*, 11, 209, 1991.

O'Loughlin, E.M., Prediction of surface saturation zones in natural catchments by topographical analysis, *Water Resour. Res.*, 22, 794, 1986.

Penner, J., Atmospheric chemistry and air quality, in Meyer, W.B. and Turner, B.L., Eds., *Changes in Land Use and Landcover: A Global Perspective*, Cambridge Univ. Press, Cambridge, 1994.

Rango, A. and Shalaby, A.I., Operational applications of remote sensing in hydrology: Success, prospects and problems, *Hydrol. Sci. J. (J. des Sci. Hydrologiques)*, 43, 947, 1998.

Rao, D.P., Sustainable development and remote sensing, *Intl. Arch. of Photogrammetry and Remote Sensing, Special Unispace Vol.*, 32, 1999.

Rewerts, C.C. and Engel, B.A., ANSWERS on GRASS: Integrating a watershed simulation with a GIS, ASAE paper No. 91-2621, Amer. Soc. Agricul. Eng., St. Joseph, MI, 1991.

Richards, J.A., *Remote Sensing Digital Image Analysis*, 2nd ed., Springer-Verlag, Berlin, 1993.

Rindfuss, R., *Soils, water, people and pixels: A study of Nang Rong*, Univ. NC — Chapel Hill, National Institutes of Health (NIH), http://www.cpc.unc.edu/projects/nangrong/nangrong_home.html, 1998.

Robbins, C. and Phipps, S.P., GIS/water resources tools for performing floodplain management modeling analysis, *AWRA Symp. on GIS and Water Resources*, Ft. Lauderdale, FL, 1996.

Robinson, J. et al., The need for projections of land-use and land-cover change, in W.B. Meyer and B.L.Turner, Eds., *Changes in Land Use and Landcover: A Global Perspective*, Vol. 4, Cambridge Univ. Press, Cambridge, 1994.

Rouse, J.W. et al., Monitoring vegetation systems in the Great Plains with ERTS, *3rd ERTS Symp.*, 1973.

Sabins, M.J., Convergence and consistency of fuzzy c-Means/ISODATA algorithms, *IEEE Trans. Pattern Anal. Machine Intelligence*, 9, 661, 1987.

Sanchez, P.A., *Properties and Management of Soils in the Tropics*, John Wiley & Sons, New York, 1976.

Schulze, E.D. and Mooney, H.A., *Biodiversity and Ecosystem Function*, Springer-Verlag, New York, 1993.

Skole, D.L., Data on global land-cover change: Acquisition, assessment, and analysis, in W.B. Meyer and B.L. Turner, Eds., *Changes in Land Use and Landcover: A Global Perspective*, Cambridge Univ. Press, Cambridge, 1994.

Soil Conservation Service, U.S.D.A., *Soil Survey of Arecibo Area, Northern Puerto Rico* (Soil Survey Report), in cooperation with the College of Agricultural Sciences, Univ. PR, San Juan, 1975.

Srinivasan, R. and Arnold, J.G., Integration of a basin scale water quality model with GIS, *Water Resour. Bull.* (accepted for publication), 1993.

Srinivasan, R. and Engel, B.A., A knowledge based approach to extract input data from GIS, ASAE paper no. 91-7045, Amer. Soc. Agricul. Eng., St. Joseph, MI, 1991.

Stern, P.C., Young, O.R., and Druckman, D., Eds., *Global Environmental Change: Understanding the Human Dimensions*, National Academy Press, Washington, DC, 1992.

Tarboton, D.G., Bras, R.L., and Rodriguez-Iturbe, I., On the extraction of channel networks from digital elevation data, *Hydrol. Processes*, 5, 81, 1991.

Thai Census Dept., *The Thai National Census*, Bangkok, 1990.

Tou, J.T. and Gonzalez, R.C., *Pattern Recognition Principles*, Addison-Wesley, Reading, MA, 1977.

Townsend, R. and Lim, Y., General equilibrium models of financial systems: Theory and measurement in village economies, *Rev. Econ. Dynamics*, 1, 1998.

Turner, II, B.L. et al., Two types of global environmental change: Definitional and spatial-scale issues in their human dimensions, *Global Environ. Change*, 1, 14, 1990.

Turner, B.L., Skole, D., Sanderson, S., Fischer, G., Fresco, L., and Leemans, R., *Land-Use and Land-Cover Change (LUCC): Science/Research Plan* (HDP Report 7), IGBP, Stockholm, 1995.

U.S. Dept. of Agricul., S.C.S., *National Engineering Handbook* (SCS/ENG/NEH-4), Soil Conservation Service, Washington, DC, 1972.

U.S. Soil Conservation Service, *Urban Hydrology for Small Watersheds*, Technical Release 55, 2nd ed., Washington, DC, 1986.

Venables, W.N. and Ripley, B.D., *Modern Applied Statistics with S-PLUS*, 1st ed., Springer-Verlag, New York, 1994.

Victoria, R.L. et al., Mechanisms of water recycling in the Amazon Basin: Isotropic insights, *Ambio*, 20, 384, 1991.

Wahba, G., *Spline Models for Observational Data*, Society of Independent Applied Mathematics, Philadelphia, 1990.

Wald, L., Data fusion for a better exploitation of data in environment and earth observation sciences, in ISPRS, Ed., *Unispace III — ISPRS Workshop on "Resource Mapping from Space,"* Vol. 32, ISPRS, Vienna, 1999.

WCED, *Our Common Future*, World Comm. Environ. and Develop., 1987.

Winkler, L. and Ulehla, J., A comparison of the methods for computing soil-moisture under field crops with actual soil-moisture data, *Rostlinna Vyroba*, 38, 885, 1992.

Woodward, F.I. et al., *Natural Disturbances and Human Land Use in Dynamic Global Vegetation Models*, IGBP, Stockholm, 1996.

Response

Incorporating scientific information into land-use planning

This chapter by John Felkner and Michael Binford, like that of Amir Mueller, Robert France, and Carl Steinitz (Chapter II.15), is important in reminding us that water sensitive planning can involve concerns related to subsurface- as well as surface-water resources. Further, as Felkner and Binford state, given the rapid alterations in land-use patterns and processes occurring in developing nations, an important need exists to create rapid appraisal methods to be able to identify and measure repercussions while they are being wrought instead of waiting to assess damage done afterward. Such concerns are particularly apt with respect to soil resources and agricultural planning.

This chapter provides an important lesson for water sensitive planners in clearly instructing that comprehensive and realistic GIS analysis modeling should not be approached as if it were a simple, completely honed tool to be blindly applied to any situation without alteration. Instead, this example shows that in order to be effective, GIS techniques should be tailored to the specifics of the study problem and region in question. Felkner and Binford's soil moisture index was created in a challenging geographic context through the innovative application of firm scientific knowledge and incorporation of local specifics based on the best available data. This perhaps is the most useful take-away message from this chapter — namely, that the use of computers for developing spatially explicit models should not occur at the expense of the inclusion of a healthy amount of imagination on the part of the applicants involved in applying and testing those models. The simple and unimaginative use of GIS analysis will generate nothing more than simple and unimaginative results whose real-world significance may be marginal at best. We, as water sensitive planners, should expect more, though, as this chapter demonstrates, the way to that end will almost certainly be neither simple nor simplistic.

chapter II.11

The design of regions: a watershed planning approach to sustainability

Daniel Williams

Abstract

The knowledge and understanding gained from the region's natural system and applied to watershed planning creates an excellent model leading to water sustainability and urban smart growth. The battle lines between the developers and the preservationists have been drawn for years. The same old and tiresome conflicts will come up again if a consensus cannot be reached between the two. The decisions clearly need to be made on an interactive systems basis and an agreement of interdependency be declared. The water supplies, demands, and reuses within a watershed can be accomplished with planning and design criteria that are integrated in building and zoning codes while creating desirable regional and neighborhood form.

The regional design projects described here are compelling examples. They connect many disciplines working together to find a common vocabulary, one that is understood by architects, landscape architects, planners, ecologists, economists, developers, clergy, and neighbors. Regional watershed design will be of great interest to citizens in that it can empower them to act, to plan for the next 100 years, and to create alternatives to the present trends and growth patterns that have raped their land, segmented their neighborhoods, and stressed their economy.

For the last 50 years, lineal, nonsystems thinking has driven growth and development. During this time, development has been about housing, not community and not about the natural resources and processes replaced by the development. In these developments, which represent an overwhelming majority of the housing stock in the country, the site preparation permanently changed the vegetation, soil, geo-hydrological, and social patterns. Water has been rerouted from river to reservoir, mountain to desert, and region to region for urban and agricultural consumption — resulting in the drying up of groundwater aquifers, some of which have not been recharged in thousands of years. This water is now transported for use to regions less abundant in precipitation and permanently lost to systems such as estuaries and riparian zones — instead irrigating deserts for golfing, urban sprawl, and agricultural crops.

No cities or towns within any watershed supply their own potable water without pumping, diverting, or damming the regional watershed. It is not just necessary to plan at the larger scales but essential for the preservation, protection, and increased quality of life.

Introduction

> All things arise from the principle of water.
>
> — Vitruvius

The U.S. Environmental Protection Agency has identified the need for a comprehensive watershed approach to ecosystem management and protection (Light and Dineen, 1994). In cities from Seattle to Miami, the impact of unplanned growth on the regions' water regime and habitat is severe, unnecessary, and expensive (Figure II.11.1). In the past, there has been the perception of the availability of unlimited energy, water, land, and other natural resources. This apparent abundance combined with the seemingly limitless powers of technology was seen as sufficient to balance the recognized consequences of boundless growth and development. When an environmental problem arose, a technological solution was sought. The profession of planning, following this lead, became little more than developing a patchwork of land uses, separated and segregated as color patches. Lacking both vision and "sense of place," this planning helped create the opportunity for sprawl and placelessness.

The grand scale of degradation in recent years (Figure II.11.2) emphasizes the need to recognize the limits of natural resources. Technological solutions often cause greater harm, requiring even more technology to resolve them (Carter, 1974). The effects of unbridled, poorly designed growth are unavoidable when the approach is piecemeal and nonsystem-defined. These effects have been revealed in relatively modest ways, such as water shortages, oil shortages, power outages, and higher taxes. It is, however, more profoundly illustrated in the breakdown of whole biological systems, which threaten the extinction of species and destruction of natural resources linked to short-term economic gain. These are clear signs that the human process has significantly infringed on the processes of nature. The natural production, cycling, and recycling of materials and energy is being short-circuited. More is being taken from the environment than is being returned; the natural capital — that which sustains all life — is declining (Hawken, 1995). Now, faced with limited resources, a dwindling tax base, and the increasing costs of government applying technology indiscriminately, we will only increase these negative attributes.

Designing regions afford the opportunity to refocus, to discover how to do more with less, and how to successfully design within the natural system sciences. The new challenge is to integrate urban design with natural sciences taking full advantage of the free work of nature. Water is the common resource in this approach.

Connecting land use and water resources

> ...watershed design requires a change in the perception of what we have thought land to be to what are the common functional, ecological and regional connections that must be reconnected? (Williams, 1998a)

Water has the remarkable ability to be deaf to political boundaries (Figure II.11.3). This uniqueness makes it the most effective common denominator of the elements that sustain

Chapter II.11: *The design of regions: a watershed planning approach to sustainability*

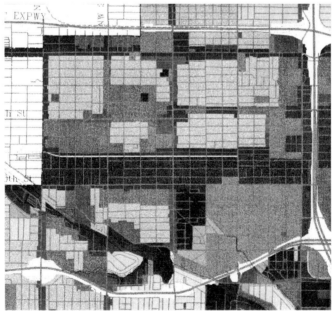

Figure II.11.1 One hundred years of draining wetlands has created cheap land for sprawl and permanently reduced the water storages.

life. Where water managers and planners typically get into trouble is when the lines of water supply and demand do not reflect the boundaries of the users. Virtually all municipalities get their potable water supply from a watershed outside their own development boundaries. New York City, as an example, has been purchasing land in the Catskills watershed for decades. This action that protected its potable watershed has saved in the order of $5 to $8 billion for additional treatment for their potable water supply (Robert, Yaro, personal communication, 1996).

Figure II.11.2 The completed development of the area from the Everglades to the Atlantic will be impervious surface by 2025, virtually eliminating recharge in the entire region.

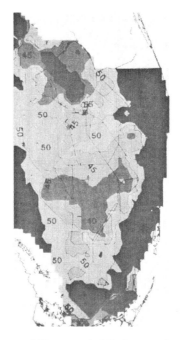

Figure II.11.3 A hydrographic topo of 29-year rainfall data — showing amorphic rainfall patterns. Note the greater averages over the urban impervious area along the coastal zone.

Hawken states that "... in any endeavor, good design resides in two principles. First, it changes the least number of elements to achieve the greatest result. Second it removes stress from a system rather than adding it" (Hawken, 1995). In that urban settlements have overwhelmingly disregarded these most basic concepts, the initial task is to reconnect the natural systems principles to the natural system's regional and local patterns.

Regional watershed planning takes the approach of reconnecting elements lost in previous plans and healing the system while creating more livable communities. Designing local and regional watershed connections that reinforce the interaction of land use and water is a first step in the design of regions. The objectives are:

- To provide a comprehensive understanding of the relationships between land use and regional water management objectives while building consensus among local, state, and federal interests as to the actions necessary to achieve regional water management objectives
- Develop a regional design vision that illustrates the reconnection to natural resources and general systems ecology (Williams, 1992)

In the relationship of water quantity and quality, the relationship of water to land use is the single relationship over which humans have some control. Because it is the land that receives, stores, and distributes water, it is the land planning, particularly the master planning, that decides the fate of the water crop. Only land, the watershed, can make this claim.

If the water quantity is insufficient but a renewable supply exists, then the receiving land area must be increased. If the quality is insufficient, then the land must store and clean the water through biological uptake. If there is insufficient water for use and the supply is insufficient, the water must be recycled to make up for the lack of supply. If sustainability is desired — all of the above must be accomplished. In other words, water sustainability requires regional self-sufficiency — the water that lands on the area is the working budget for all uses — additional consumption of water over that allotted amount is deficit spending. This deficit spending reduces the net total capacity of the system and, in that it is accommodating additional growth, the deficit can only get worse as the demand grows while the recharge area is reduced. In the budgeting cycle, first the natural system and hydrological integrity should be satisfied, then the combined needs of the agricultural and urban systems. Without regional systems solutions, the loss to all systems will far exceed the gain to one.

Case study: Florida

> Water, not growth management controls, will be the ultimate factor in determining how big Florida can grow. "Water management is part of the overall effort in Florida to get a handle on where this state is, and where we're headed to, and to get out ahead of our tumultuous growth, once and for all," Lt. Gov. Buddy MacKay said. "It gives you an idea that whatever it is that's going to limit growth in Florida, it's not going to be growth management . . . the availability of that resource is probably going to be the ultimate limiting factor." (*The Ledger*, 1994)

In Florida, there is a large section of uniquely beautiful yet dying piece of land known as the Everglades (Figure II.11.4). The Native Americans called this area "a river of grass," which many years ago became the title of Marjory Stoneman Douglas's book — a work that, until now, saved it from development.

The issues of the Florida Everglades and the drainage associated with the urban and agriculture use in its surrounding watershed have been well studied (Carter, 1974). This type of drainage has been done for the entire county and most of the state.

Figure II.11.4 The Everglades, a 5000-year gentle flow of water falling 20 ft in 120 miles.

Figure II.11.5 The Miami River, once flowing from the Everglades basin, now a drainage ditch delivering stormwater from thousands of miles of drainage canals.

Figure II.11.5 is of the Miami River in South Florida in 1996; remarkably not a single drawing was done to guide this city's future. More time is spent deciding on the quantity of parking spaces than considering the future of the city for future generations. The issue then is not if a place will change, but when it changes, can it be in a way that the citizens desire. Planning, in a sense, is taking the necessary steps to assure one gets what is desired, rather than what happens by default.

Water is a vital element of Florida's natural and human-made environments. Fresh water underlies the state and periodically inundates much of its interior with an expansive freshwater marsh, the Everglades, which covers much of the state's southern interior. The surface and underground freshwater reservoirs of the region are essential to the integrity of near-shore ecosystems (Figure II.11.6). These waters flow into coastal estuaries, mixing

Figure II.11.6 Illustration showing relationship between rainfall and surface- and groundwater recharge. Remove the recharge and saltwater intrusion occurs.

with the vast saltwater seas that surround the state's 1,200 miles of coastline. At the interface between salt and fresh water, sensitive and finely balanced estuaries provide nurseries for abundant marine life and, consequently, commercial fisheries. As the foundation of Florida's natural environment and cornerstone of the state's recreation and tourism industries, water is essential to the state's sustained health and prosperity. Due to its seeming abundance, water, as Plato said, "... gets the least amount of care."

Historically, wetlands were considered not only worthless, but also a threat to human survival (Carter, 1974). Estuaries (Figure II.11.7) were mosquito-infested mangrove swamps and in dire need of improvement. Thus, the last 100 years of Florida's history have been characterized by attempts to tame nature, marked by endeavors to adapt the natural landscape solely to human uses. A result of this is a drainage and flood control system in South Florida that is inadequate for flood control, is inadequate for water supply, and provides insufficient water quality protection. This is occurring at a time when demand for flood protection, water supply, and water quality standards is increasing exponentially while the water losses exponentially decline due to drainage. This demand goes hand in hand with the pressures exerted upon the present system by the rapid and unchecked outward expansion of urban settlements into the water recharge areas.

Opportunity

Following Hurricane Andrew on August 24, 1992, the opportunity arose to apply a watershed approach as a critical element of the rebuilding effort in South Dade County. This "adversity to opportunity" approach seeks to successfully integrate land-use development and natural resource protection by preserving, protecting, and, when necessary, reestablishing the connections on a regional scale. Thus, watershed planning would become the conceptual foundation for the local, state, and federal planning initiatives on smart growth, all connected to the need for potable water (Williams, 1993).

A watershed plan represents an initial step in developing a comprehensive development plan for the restoration and protection of Everglades National Park and Biscayne Bay while future growth occurs. It will also begin the educational process, which is necessary to build consensus within the local communities. This consensus provides the power to the local agenda for developing and implementing watershed management strategies.

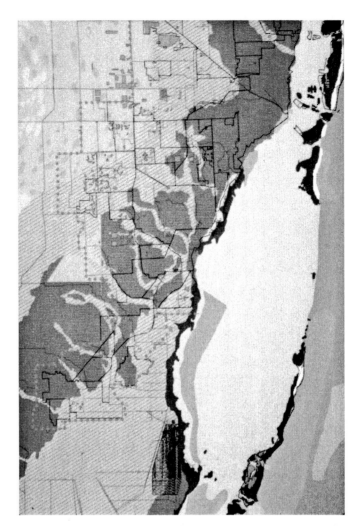

Figure II.11.7 The South Dade regional system illustrating coastal and estuary systems connecting to make a vital ecosystem.

The objectives of the watershed plan are:

1. To provide a comprehensive understanding of the relationships between land use and regional water management objectives
2. To build consensus among local, state, and federal interests as to the actions necessary to achieve regional water management objectives in South Dade County
3. To select a hydrologic basin within the South Dade County watershed for future study and the development of future demonstration projects (Williams, 1994).

Watershed planning: A process

A first step is to find out where the community wants to be in 50 to 100 years. In Lewis Carroll's *Alice in Wonderland*, Alice asked the Cheshire Cat, "Would you please tell me which way I ought to go from here?" The cat replies, "That depends a good deal on where you want to get to." The second step is to find out how the system functions and what elements are in the regional resources inventory.

Figure II.11.8a Aerial photo showing development around low-lying wetlands and flood control canals.

An additional million people will pave over an additional 21 square miles of the remaining pervious area. This is 21 square miles (Figure II.11.4) of present recharge area lost that once received, stored, and distributed water to sustain the regional water budget. Use without recharge is not sustainable. More important, the water budget is truly a function of a continuous supply. Water budgets that include the huge storages of the aquifer are not budgets that can be sustained. In an economic analogy, this is similar to living off the principal rather than the interest. With this agricultural recharge comes the associated runoff and pollutants, but the urban system may be many times worse. Agriculture, while a major consumer of water, is also an important water recharge area. These open agricultural land areas, which are pervious surfaces, allow percolation of resident rainfall to add to and maintain the regional storage. It is estimated that up to 7 times more runoff is associated with urban land use. Degradation of the water in the runoff is equivalent to secondarily treated effluent (Light and Dineen, 1994).

The plan

One of the questions citizens asked over and over is, "How will this affect my daily experience? What will sustainability look like?" These greenway areas (Figure II.11.8a and b, showing before and after) allow for the storing and release of water to the Biscayne Bay, mimicking the historic natural system. These areas will also increase the watersheds' ability to store water, while providing additional flood protection. Because this water is being stored, it can be cleaned up through biological uptake prior to its conveyance to the canals, to recharge areas, and then on to estuaries.

Shown within this vision (Figure II.11.9) are areas composed of three land-use types: urban, agricultural, and reclaimed wetland storages. In this concept, all of the land uses are networked through the storage, distribution, and clean-up of water. This creates a networking of water use and reuse. The wetlands represent local treatment plants while the canal littoral areas and larger surface storage areas would be sized as localized storage for clean-up as the pollutant loading would dictate.

Designs of the greenways can take many forms and be many sizes. The size would depend on the pollutant loading of the land use within the watershed that drains to the

Figure II.11.8b The wetlands have been reclaimed, improving flood protection, adding open space, and increasing property value.

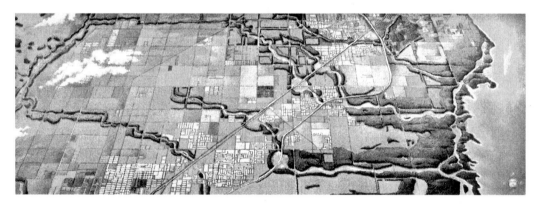

Figure II.11.9 The Watershed Interactive Network Plan — the WIN plan, illustrating systems planning and design that reconnects all components of land use into a working community of urban, agricultural, and natural system functions.

canal. These greenways also act as habitat for various species, including humans. Creating pathways rich in diversity and function — including lineal parks, greenways, bike paths and local flood protection (Figure II.11.10) are funded by stormwater management funding.

At the coastal zone, canals would no longer flow directly into the Biscayne Bay but through an archipelago of sorts. By diverting the flow and creating storages upstream, the water quantity and quality will be enhanced. Simultaneously, considerably more lineal area is created for the food-chain nursery ground.

This new edge condition in the urban segment where the canal and littoral zone meet would create identity and image for the community while increasing the flood protection and neighborhood open-space amenities. The combined water storage areas and open spaces not only help define better neighborhoods but increase property values at the same time (Figure II.11.11a, b, and c).

Chapter II.11: The design of regions: a watershed planning approach to sustainability 551

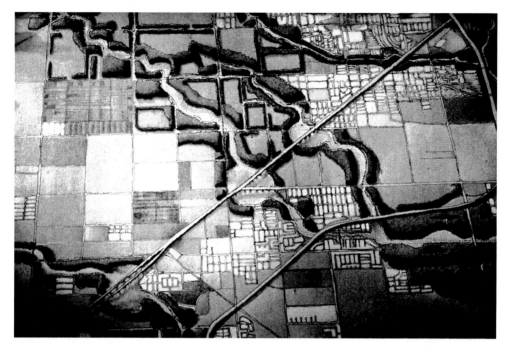

Figure II.11.10 A detail of the WIN plan illustrating linkages between urban design, hydric open space, gray and black water reclamation, and transportation.

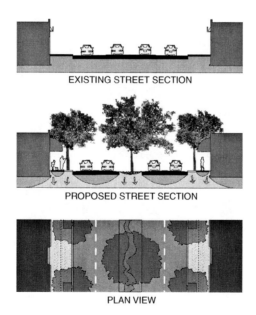

Figure II.11.11a Illustration of conversion of a typical road system to a WIN plan street — recharges the aquifer, reduces heat island impacts, creates desired microclimate, and increases property values.

The agricultural areas would again be defined by the natural water regime in the area. Agriculture would employ reverse-flooding irrigation, which will simultaneously waste less water and control nematodes and other pests while rebuilding soil. The flows of water to Biscayne National Park would filter through these reclaimed wetlands and be better

Figure II.11.11b A hydric boulevard — stormwater is gravity-fed to the center green area for storage, clean-up, and aquifer recharge. Microclimate, quality of life, and property value all increased at a cost lower than the typical street. (See Figure II.11.11c.)

Figure II.11.11c The typical, undesirable street — same right-of-way as Figure II.11.11b but environmentally costly.

timed to the bay. This regional watershed plan illustrates all the interactive, networked, and sustainable ideas incorporated into a single vision. It is a vision that the citizens, scientists, and planners through consensus consider affordable, practical, and desirable.

Conclusion

The future . . . it's not working! Even developing new and wonderful walkable communities will not have any significant change in the disastrous path being taken, nor will energy-saving homes, or commercial conservation, or drip irrigation, or hydrogen fuel cells, or paying-our-way in fossil fuel costs — none of it is enough. It all comes from the same trend, the same flawed conceptual base — individuals rule — the land has no voice. Typically, development is thought of as roads and houses neatly or not so neatly placed upon the land. In most cases this development obscures the land's beauty and its deeper value. The powerful ethic of house and home, that basic need to place a roof over one's head and provide places to work and have community, has disregarded the value of the regional systems connectivity, its *place*, and it is costing huge amounts.

The relationship between land use and water is the essential tool needed to design and plan for the symbiotic relationships presented here. The missing links are the next steps, which require the planning profession's proactiveness on the regional scale. The design of regions creates the drawings and makes the design decisions necessary to protect the region so that it can work as an organism and applies the understanding of these relationships to short- and long-term planning.

The watershed design approach develops many successful connections:

- Protection of the natural systems functioning within regional and urban greenways increases the value of the adjacent properties while providing for regional recharge.
- Storing water within these hydric greenways will increase water supplies for future use in an economic way while increasing the potable water supply and receiving the bonus of community open space.
- The preservation and protection of parks and conservation areas integrated within the urban and regional pattern provide critical livability, health, and economy to the entire region.
- Planning and zoning must integrate water management with land-use decisions and, while doing so, create strong growth management boundaries informed by conservation science and design.
- Ecosystem management and watershed planning simultaneously address the complex issues of urban quality of life and preservation of natural systems.

Developing a common vision, illustrating it, and establishing the incremental steps for its implementation are essential to urban and regional sustainability. Watershed planning provides a process that informs the community-based vision while providing for an economic and environmentally sustainable future.

Live locally, think globally, act regionally!

Literature cited

Blake, N.M., *Land into Water — Water into Land: A History of Water Management in Florida*, Univ. Presses of Florida, Tallahassee, 1980.
Carter, L.J., *The Florida Experience: Land and Water Policy in a Growth State*, John Hopkins Univ. Press, Baltimore, 1974.
Douglas, M.S., *The Everglades: River of Grass*, 1996.
Hawken, P., *The Ecology of Commerce: A Declaration of Sustainability*, Island Press, Washington, DC, 1995.
The Ledger, Water limits may halt growth, Lakeland, FL, Sept. 26, 1994.
Light, S.S. and Dineen, J.W., Water control in the Everglades: A historical perspective, in Davis, S.M., and Ogden, J.C., Eds., *Everglades: The System and Its Restoration*, St. Lucie Press, Delray Beach, FL, 1994.
Miami Herald, Andrew: Recovery by the numbers, August 24, 1994.
Williams, D., *The Watershed Interactive Network: Towards a Sustainable Watershed Design Methodology*, Univ. Miami, Center for Urban and Community Design, Miami, 1992.

Williams, D., *Regional Studies — Introduction, The New South Dade Planning Charrette: From Adversity to Opportunity,* Univ. Miami Press, Miami, 1993.
Williams, D., South Dade Watershed Project Workshop report, Center for Urban and Community Design, Miami, 1994.
Williams, D., South Dade Watershed Project Workshop report, Center for Urban and Community Design, Miami, 1995.
Williams, D., Sustainability panel, *APA: FL Conf.,* Tampa, 1998a.
Williams, D., The Eastward Ho vision, FL Dept. of Comm. Affairs, proposal, Univ. FL, Miami Education and Research Center, Miami, 1998b.

Response

Expanding planning vision in space and time

Sometimes even watersheds are not large enough for truly effective water sensitive planning. Due to the unnatural rerouting of water across watershed boundaries, from locations where there is less infiltration to locations where the water is used in toilets and then discharged into pipes and carried to a third location, Daniel Williams believes that planning at larger spatial scales is essential for human and environmental sustainability. Also important is the need to broaden the time frame of planning given that aquifer recharge may take place on the scale of centuries, not years or even decades. With these two expanded visions, Williams asks the provocative question as to whether we can ever know for sure if any of our development projects are truly sustainable.

Past planning has been predicated, Williams instructs, on the misconception of unlimited supply and an unhealthy faith in techno-fix solutions as the means to solve all problems; that is, almost all large communities now draw water from outside their watersheds, and if the amount decreases, well, one can always build a longer intake pipe, can't one? Such a planning approach has helped to create a lack of "sense of place" and has promoted sprawl.

Modern water sensitive planning needs a much greater comprehensive understanding of the relationships (both physical and social) between landscapes and water assets and uses. Such understanding, Williams believes, can be found only at the regional scale.

Implicit in this chapter is a lesson of great importance that underlies this entire book; namely, that water is not a renewable resource, and until it is stopped being treated as such, we will continue to suffer from various, so-called water crises, both in North America and abroad. Toward this end, Williams makes a plea for water sensitive planning to shift its focus from housing and community development to natural resources and processes. For, in the end, it is the water/landscape that should be regarded as the ultimate client, not the humans who wish to reside there.

chapter II.12

GIS watershed mapping: Developing and implementing a watershed natural resources inventory (New Hampshire)

Jeffrey A. Schloss

Abstract

Watershed planning, management, and protection efforts have generally been driven by more "reactive" approaches in which actions and concerns stem from perceived or actual impairment and existing threats. Decisions are often made using a very limited set of data that may or may not be directly related to the specific watershed of concern. Typically, these decisions are directed with a bias toward land capability analysis. A more proactive approach requires a sufficient watershed natural resources inventory that would allow for the documentation and analysis of landscape activities and features of importance along with information on the extent and condition of in-water and riparian resources and their associated habitats. This then would allow for informed preemptive planning and protection. It is also the only way to approach community stewardship and planning for the more pristine watersheds where limited degradation has occurred and the emphasis is to maintain it that way.

The advent and refinement of geographic information systems (GIS) and related spatial technologies, such as remote sensing and global positioning systems (GPS), have brought forth the capability to conduct watershed inventorying and analyses that can allow watershed stakeholders to become better informed on the current state of the watershed and predict the future consequences of change.

This chapter provides a basic overview of these technologies and introduces the concept of a Watershed Natural Resources Inventory (WNRI) as a useful watershed assessment tool for proactive community planning, minimum impact site design, and the overall management and stewardship of watershed resources. This procedure can also be utilized for meeting local decision-making needs, for developing protection strategies for watershed lands and source waters, and for the design and implementation of monitoring programs. Case studies incorporating WNRIs for a large multijurisdictional watershed, for a small lake watershed involving three communities, and for a series of small wetland watersheds in a single community setting are discussed. The chapter concludes with an outline of the steps involved in the implementation of a critical resources analysis, which represents the optimum combination of "top-down" (capability/limitations) and "bottom-up" (resource co-occurrence) GIS-aided watershed assessment.

Introduction

The surface waters of New Hampshire represent a valuable resource contributing to the state's economic base through recreation, tourism, and real estate revenues. Some lakes and rivers serve as current or potential water supplies. For most residents (as indicated by boating and fishing registrations and surveys) our waters help to ensure a high quality of life. New Hampshire currently leads all of the New England states in the rate of new development and redevelopment (according to 2000 U.S. Census data recently posted on the Web). The long-term consequences of the resulting pressure and demands on the state's precious water resources remain unknown. Of particular concern is the response of our waters to increasing nonpoint source pollutant loadings due to watershed development and land-use activities.

In New Hampshire, as in most New England states, we rely on local decision makers to act as stewards for our community and watershed resources. These local decision makers are primarily elected or appointed, or they may be volunteers. They often do not have, or lack access to, the proper resource information and education on which to base their decisions. Yet, these local officials are responsible for evaluating development, subdivision, residential, industrial, commercial, and recreational projects, all of which can have significant impact on a community's natural resources. They are also responsible for the implementation of land-use and zoning regulations and the development of the community's master plan that affects the future land use of their community. Thus, while local decision making is the key to watershed-based community resource protection, the information and education required to make informed decisions have often been lacking or severely limited. This lack of information has resulted in a more reactive approach to watershed resources protection where problems are addressed when they occur or are perceived to occur or when resources are threatened. To correct this situation, it is imperative that these decision makers are provided with the specific information of the location and status of critical watershed resources that can then allow for the optimum proactive protection through comprehensive planning and management.

The natural resources inventory approach was originally developed as a process for communities to document and locate important resources for developing resource protection plans and for updating town master plans (Auger and McIntyre, 1992; updated recently incorporating basic GIS analysis and mapping approaches by Stone, 2001; see also the following section). Our work, initiated concurrently, has specifically focused on using a water resources sensitive approach that expanded upon this original resources inventory concept and focused on the GIS analyses that would be useful to use at the watershed and subwatershed levels (Schloss and Rubin, 1992; Schloss and Mitchell, 1996).

We have developed the Watershed Natural Resources Inventory (WNRI) as a useful watershed/waterside assessment tool for proactive community planning, minimum impact site design, and the overall management and stewardship of watershed resources. This procedure can also be utilized for meeting local decision-making needs, developing protection strategies for watershed lands and source waters, and the design and implementation of monitoring programs.

Recent developments in computer speed and capacity, access to the Internet, and the development of more "user-friendly" technical software have made mapping and spatial analysis technologies more available. These technologies can facilitate the WNRI process and allow watershed stakeholders to become better informed on the current state of their watersheds and predict the future consequences of change.

Chapter II.12: GIS watershed mapping

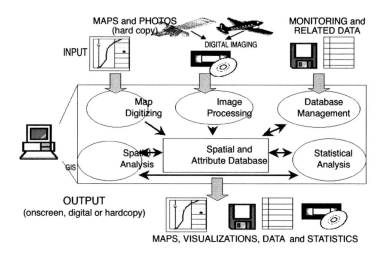

Figure II.12.1 Schematic diagram of a geographic information system.

An introduction to GIS and related spatial technologies

GIS

A *geographic information system* (*GIS*) can be defined as the hardware, software, and data that assimilate descriptive information with location referencing to allow for spatial analysis and visualization. Figure II.12.1 displays a schematic diagram of a GIS system. More simply, GIS can be thought of as a system that allows for digital mapping along with the capability to perform analyses on the elements of the map as well as related descriptive information. These graphic map elements include polygons (watershed extent, lakes, wetlands, land parcels, political boundaries, etc.), lines (roads, trails, pipelines, etc.), and points (wells, outfalls, sampling stations, etc.). The related data can be further divided into two categories: descriptive data that contain attribute information about the element (i.e., lake name, road name, station number) and spatially linked reference data. The reference data contain characteristics of the map elements (town population, wetland vegetation, soil characteristic, well yield, station water quality results, etc.). It is the ability to manipulate both the map elements and reference data that allows for some powerful analysis to take place (see examples in Chapters II.10, II.14, and II.15).

In addition to the manipulation of map elements and their associated data, a GIS also allows for spatial manipulation and analysis. These are derived through topology analysis and include the ability to derive, detect, or perform the following spatial queries:

Connectivity. The intersection or overlap of selected elements allows for the location of points such as where a road intersects a stream. This can be used to find culvert locations throughout the watershed or to help select water sampling stations.

Proximity. The closeness of one object to another allows for selecting elements by their locations relative to other elements. Example uses of this analysis might allow for the detection of salt storage areas within 500 ft of a water body or hazardous waste storage within a quarter mile from a school building.

Adjacency. The bordering of one element to another allows for the detection of edges and transitions. This operation can be used to find land parcels that border a wetland or to link subwatersheds with the receiving water into which they empty.

Table II.12.1 GIS Data Commonly Available

- Base Map (Background) Data
  — USGS topographic maps and components (i.e., 1:24K scale)
  — Orthophoto (corrected and digitized aerial photography)
- Infrastructure
  — Roads/rail/trails
  — Water/sewer/gas/oil
  — Transmission lines
- Geology
  — Surficial
  — Bedrock
- Hydrography
  — Lakes, ponds, rivers
  — Streams
- Soils (w/attributes)
  — County surveys
  — Regional surveys
- Wetlands
  — National Wetlands Inventory (U.S. Fish and Wildlife)
  — USGS Digital Line Graph data (DLG) (from "topo" maps)
  — Derived from soils
  — Derived from remote sensing
- Groundwater resources
  — Aquifers, wells
  — Transmissivity/yield/quality
- Demographic (Census)
- Land cover/land use
  — Satellite/aerial photo derived land cover
  — Regional/local planning agency land-use/zoning
- Digital elevation models
- Watershed delineations
  — USGS Hydrologic Unit (HU) delineations

Source: USDA Natural Resources Conservation Service.

Buffering allows for the derivation of a boundary of a set width around a point, line, or polygon element. This operation can provide additional map elements representing setbacks, zones, and regions.

Neighborhood and network analysis allows for the derivation of a linear path through an analysis that relates a specific area, or point, to those around it. Slope and aspect are common examples of neighborhood operations, whereas flow and transport corridor determinations are examples of network analysis that expands upon those operations.

It is this range of analyses described above as well as the ability to work with and process a series of map layers (overlay analysis) that separate GIS from other spatial databases such as computer-aided design (CAD) and computer-aided mapping (CAM) systems. No other system allows us to combine relational and spatial queries that can create a new layer (map) from two or more existing layers (Congalton and Green, 1992).

Table II.12.1 lists the typical GIS data that are more readily available. Most GIS maps utilize existing base map components derived from U.S. Geological Survey (USGS) topographic map quadrangles. Besides the availability of these digitally scanned maps (commonly referred to as digital line graphs, or DLGs) certain elements of these maps (roads, water bodies, wetlands, etc.) may also be available in separate GIS overlays (often referred to as themes or coverages). Soil maps can be quite useful at the resolution they are generally mapped at (usually at a 1:24,000 scale with 2- to 5-acre soil units delineated) by the USDA Natural Resources Conservation Service (NRCS; formerly Soil Conservation Service). Information contained in their associated soil characteristics data can include how wet the soils are, slope, depth to groundwater, erodibility, and suitability for septic installation or homesite development. The U.S. Fish and Wildlife Service has cataloged and digitized much of the country's wetlands at a similar scale from aerial photographic interpretation. Other data may have been generated on a statewide basis from state agencies or cooperative groups.

Although it would appear that the USGS, NRCS, the Census Bureau, U.S. Fish and Wildlife, or the U.S. Environmental Protection Agency would be your first stops when gathering available GIS data, it is more important to find out where (or if) you have a regional or statewide GIS data repository. This can be a critical time and resource saver, for often these centers offer data at low or no cost. In addition, they may already have processed the data from sources listed above to be compatible with the map projection and coordinate system commonly employed for locally produced GIS maps and data. This is most important, as you need every data layer to register correctly to perform any accurate mapping or analysis. State GIS repositories may be located at a university, state library, or agency (state or regional planning, resource management, resource protection, or transportation agencies are some of the more common locations). In some cases the data may be distributed among a group of agencies with no centralized clearinghouse.

In addition to the government agencies listed previously, private companies also provide GIS data for a range of fees. Due to the rapid growth in both the power and storage capacities of desktop computers and with multiple paths available for the conversion of CAD/CAM data to GIS, it is not uncommon to find larger towns and local municipalities with their own GIS systems generating their own data on infrastructure and local resources. In most cases, state or regional planning agencies act to support these efforts and/or act as GIS service providers for their member towns.

When working with GIS data it is important to understand the limitations of those data. Coverages are derived from data with varying resolution due to the original spatial scale employed. A typical USGS 7.5-min topographic quadrangle (1:24,000 scale) is usually spatially accurate to about 50 ft or so. Temporally, the data may reflect aerial photo interpretation that can be over a decade or more old. This may mean that the roads or the forested areas shown are no longer located there. The least accurate layer used limits the accuracy of any GIS analysis. Typical data sets provide the resolution that allows for watershed and subwatershed analysis, region or town-wide analysis, and possibly large "zones" within these areas. They do not, however, allow for anything close to site-specific determinations. This is a very important and sometimes abused limitation as GIS systems allow for production of products that exceed the scale of resolution. Although no direct conclusions should be based on GIS data for a single site or house lot, if GIS data suggest a certain resource or condition exists at a site, you may want to be sure to visit that site to check out the possibility of the occurrence. This may be especially critical if you only have the resources or time to visit a limited number of sites. Thus, the GIS data may be employed as a screening tool for situations beyond the data's spatial resolution.

Remote sensing

A second spatial technology, *remote sensing*, is the use of airborne instrumentation (airplane or satellite) to collect information by photography or digital sensors that can later be interpreted to create (after processing, correction, and geo-referencing) additional spatial

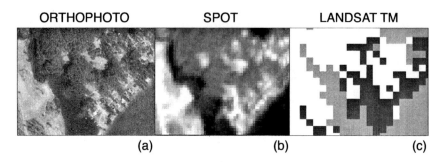

Figure II.12.2 Comparisons of resolution from (a) digital orthophotoquads from aerial photography (1-m pixel size), (b) SPOT satellite panchromatic imagery (10-m pixels), and (c) LANDSAT Thematic Mapper satellite imagery (30-m pixels).

data layers. Land-cover GIS layers are often derived from satellite remote sensing. These coverages estimate the generalized land cover (forest, wetlands, water, crops, urban, etc.) over an area. Older classifications relied on LANDSAT Thematic Mapper (TM) imagery, which offered 30-m resolution (each picture element or "pixel" approximately represented a 30 × 30-m area on the ground). More recent, SPOT satellite imagery, which offers 20-m color and 10-m panchromatic (black and white) resolution, has been made available, and currently, the newest satellites, such as IKONOS, now offer 1- to 5-m resolution, which in many cases is the equivalent of standard aerial photography. Figure II.12.2 allows for the comparison of data from this range of resolutions. In all cases, care must be taken when working with these data. Atmospheric conditions and clouds can limit areas interpretable in the coverage. Forest canopy cover can hide low-density development, wetlands, and streams. Steep-facing slopes are sometimes misinterpreted as water or wetland. Most notably, these scenes represent a moment in time; the date of the available imagery may not reflect current conditions, and different classification procedures may have been used on different imagery sets. Thus, care must be taken when comparing the available data.

This is not to imply that the data are not useful. Figure II.12.3 displays the use of Landsat TM derived land cover for the relatively small Bow Lake, NH, watershed. A GIS query was run using the satellite-derived, generalized land-cover data for those features known to be associated with nonpoint source pollution (active agriculture and highly developed areas) that were 50 ft or closer to a shoreline area, wetland, or lake tributary. This then allows for the informed selection of water quality sampling sites that can bracket these presumed "hot spots."

For watershed-specific work, other, less commonly available, spatial data coverages may be important. Aquatic and riparian habitat areas, sampling stations, and their related water quality data are some examples of information that may be available, digitally or not, that can assist in inventory and analysis. Resource agencies, universities, recreational, wildlife, gardening and sporting clubs, angling and hunting guides, and other members of the watershed community may be important resources that can assist in collecting and documenting habitat and species occurrence data. Protection agencies, universities, volunteer monitoring groups, and watershed or landowner associations may be able to assist in providing water quality and related data. In urban and suburban areas, storm drain system maps can be very important in relating land-use activities to water quality impacts.

GPS

To create additional spatial data or to "ground-truth" remotely sensed and GIS data, *global positioning system* (GPS) technology is used. GPS is a highly accurate locationing method

Chapter II.12: GIS watershed mapping   563

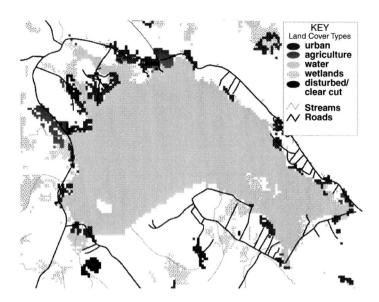

Figure II.12.3 The use of Landsat Thematic Mapper satellite-derived land cover for the Bow Lake, NH, watershed. Features known to be associated with nonpoint source pollution (active agriculture, cleared land, and highly developed areas) 50 ft or closer to a shoreline area, wetland, or lake tributary are identified by shading. (Color version available at www.gsd.harvard.edu/watercolors. Password: lentic-lotic.)

that utilizes a special computerized radio device. This device receives satellite data from the U.S. Department of Defense's constellation of 24 very high altitude satellites orbiting the earth, which continuously transmit their positions and associated information. At any given time and place, if at least four satellites are in "view," a GPS receiver can pick up the signals and use them to determine a position on the earth's surface. Initially, the Department of Defense degraded the public-use GPS satellite signals, on occasion, out of security concerns. This "selective availability" occurrence has recently been halted indefinitely. As a result, even recreational GPS units can now provide a relatively accurate location fix.

As with remote sensing above, atmospheric conditions can greatly affect the accuracy of GPS. Also, the arrangement of the satellites being accessed in relation to your position on the ground will affect results (some GPS receivers come with software that will allow you to plan your time of readings when satellite positions are optimal). With a range of features available to account for potential errors, the accuracy of GPS receivers can vary widely. Recreational units are generally accurate to 10 to 30 m. Mapping quality GPS units commonly employ differential correction in which, after collection, data are compared to data collected at a nearby base station location at the exact same time to allow for precise adjustments and higher (sub-meter) accuracy. Additional signal processing options including carrier phase analysis can further improve accuracy. Some mapping GPS receivers and marine GPS navigational units are designed to receive and process in real time the signal from a nearby base station, Coast Guard beacon, or commercial satellites. The highest resolution can be found on survey quality differential GPS receivers and base station combinations that can provide centimeter or greater accuracies in real time.

The remainder of this discussion highlights examples of how these spatial technologies have been employed to conduct and implement watershed natural resource inventories. They have been selected because they represent a variety of project types, occur through a continuum of small to large watershed projects, and cover a range of community concerns and resulting initiatives.

GIS watershed inventory and analysis of the Squam Lakes watershed

Background

In this first example, we explore a large multijurisdictional watershed where we had sufficient resources to obtain and/or digitize a wide variety of GIS data layers to explore what data were most critical and how GIS analysis could provide the needed information for watershed protection and planning. As part of a model watershed study under the direction of the NH Office of State Planning, a multiagency task force worked to create a GIS based resource inventory of the Squam Lakes watershed (Scott et al., 1991). The state's GIS repository, GRANIT (for Geographically Referenced Analysis and Information Transfer), is housed at the University of NH (UNH) but linked to state agencies and regional planning commissions. Data "layers" used in this GIS study included bedrock geology, hydrology (streams, wetlands, lakes, ponds, and aquifers), soils, elevation, land-use zoning, land cover (from aerial photographs and satellite images), and wildlife habitats. This was in addition to a base map of roads and political boundaries. Also included was over 10 years of water quality data, collected weekly during the ice-free season throughout the lake, by volunteer monitors of the Squam Lakes Association under the direction of the NH Lakes Lay Monitoring Program coordinated out of UNH.

The Squam Lakes watershed comprises the land surrounding Squam Lake, Little Squam Lake, and three smaller tributary lakes located in what is known as the "Lakes region" in central New Hampshire. The watershed is truly multijurisdictional because it lies in three counties with six towns bordering the lake and an additional two towns located in upland areas of the watershed. The watershed covers 42,418 acres, of which 7847 compass water area.

Conventional land capability analysis

In the initial study, a conventional "top-down" GIS analysis of land capability was undertaken to display all of the developable area remaining in the watershed (Figure II.12.4). The process involved "subtracting out" all areas in the watershed that were limited to further development. This included areas of water, wetlands, and the required buffers around them as per town and state regulations (easily facilitated through GIS proximity, buffering, and neighborhood analysis), areas restricted due to excessive slopes or poor soils, protected lands, and already developed areas. The GIS was also used to analyze information on zoning specific to each town (i.e., land area required for each house lot) and provide a full "buildout scenario" that could estimate the maximum number of new houses allowed by town and by subwatershed, and it even estimated the resulting increase in population under this development scenario.

The build-out analysis made quite an impressive series of maps. However, it only took a simple pie chart to translate the bottom line: for the Squam Lakes watershed, about 12% of the watershed was currently developed or protected; about 52% was constrained or restricted to development; and almost 37% of the watershed was still left to be developed (Figure II.12.5). Although, as a whole, the lake displayed excellent water quality and was relatively pristine in nature, areas with less desirable water quality conditions existed within the lake. Thus, the problem was defined: areas of the lake were already showing signs of water quality degradation, yet current laws and regulations would allow development within the watershed to expand over three times the area of what was already developed. What was still needed was a method to locate critical lake areas and produce additional GIS products to educate and support decision makers and their communities.

Chapter II.12: GIS watershed mapping

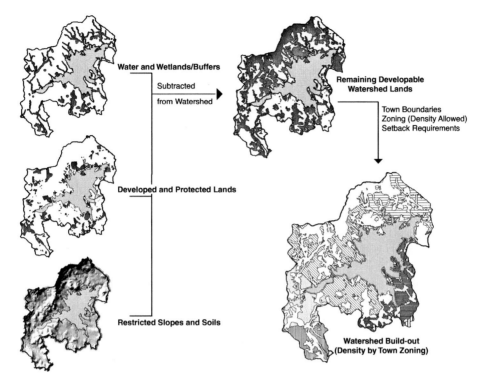

Figure II.12.4 Land capability analysis of the Squam Lake watershed facilitated through GIS analysis. (Color version available at www.gsd.harvard.edu/watercolors. Password: lentic-lotic.)

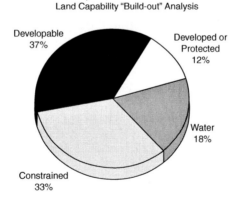

Figure II.12.5 Pie diagram of results from the Squam Lakes watershed land capability analysis. (Color version available at www.gsd.harvard.edu/watercolors. Password: lentic-lotic.)

Geographic display of data and GIS visualization

The project recommendations included suggestions on taking into account land-based resources such as productive forest and agricultural soils, and aquifer recharge areas. Still needed was a method to locate critical lake areas and produce additional GIS products to assist decision makers concerned with water quality and watershed resource protection. With that in mind, it was time to go beyond the traditional GIS approaches and "push the envelope" by exploring GIS data display and visualization. Displaying the water

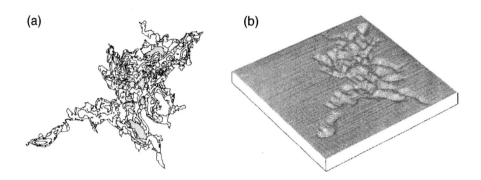

Figure II.12.6 (a) Bathymetric map (depth contour plot) of Squam Lakes, NH, and (b) GIS-derived 3D model and visualization of the lake bottom structure derived from the bathymetric data.

quality data spatially, it became apparent that many of the small coves and embayments were areas of more degraded water quality. The data suggested that the lake did not react uniformly to watershed inputs; that it was not just one big reaction vessel or "bathtub" as is commonly assumed for many large lake systems. This concept was further enhanced by taking the bathymetric map (depth contour plot) of the lake and using the GIS to create a 3-D model of the lake bottom (Figure II.12.6). No experience in topographic readings was necessary to be able to see how the lake was really made up of multiple basins connected together and that each of these basins had high sills around it. Many of these sills reached up to shallow water depth, well above the basin thermocline, effectively separating the bottom waters of the lake basins for most of the year. Thus, the GIS allowed for the definition of functionally separated lake basins within the lake system. Of the 18 lake basins defined, 17 had been monitored through the volunteer program, so further GIS spatial analysis of water quality conditions could be done.

Geographic referencing and spatial analysis

With the basins defined, they could be associated through the GIS with the abutting subwatersheds. This would allow for analyses of what characteristics of the land around the basins had an influence on the basin's water quality. Although our study team had the luxury of an extensive GIS database of land cover (right down to the type of tree stand from aerial photography), we started with some basic GIS analysis using information that would be more readily available to localities across the state. With some relatively simple data analyses, areas of the lake that react more critically to nutrient loading were defined (Schloss and Rubin, 1992). We found good relationships between water clarity for each basin and the mean depth (derived from the GIS by dividing the surface area of the basin by the volume of the basin) and between algae levels and a combination of mean depth and the ratio of the basin surface area to the area of land that drained into it (abutting subwatersheds). This explained much of the data variation in all basins.

A generalized land-cover analysis was also undertaken to determine the land use water quality relationships that occurred throughout the watershed. It was found that the shoreland zone land cover (an arbitrary 250-ft area from the lakeshore created by the GIS) explained less water quality variation than the total subwatershed land cover. This implies that although shoreline regulations are important for the Squam Lakes (and most likely our other pristine lakes) activities *throughout* the watershed also have major impacts.

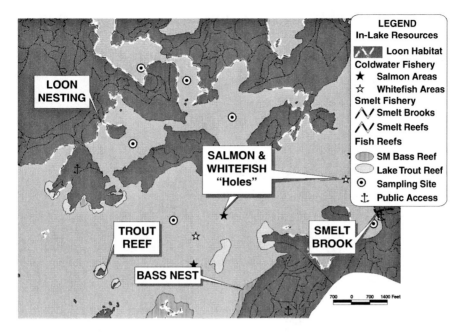

Figure II.12.7 Section of GIS map displaying the elements collected for the in-lake wildlife resources inventory of the Squam Lakes, NH. (Color version available at www.gsd.harvard.edu/watercolors. Password: lentic-lotic.)

In-lake resource inventories and integration of the data layers

Through community advisory groups we learned that other aspects of the watershed besides water quality held equal if not greater importance. To that end, a GIS layer of loon habitat (provided by volunteers of the NH Loon Preservation Society), bass nesting areas, cold-water fish reefs and deep holes, and smelt brooks (from NH Fish and Game and volunteer surveying) was created (Figure II.12.7). The GIS could then reference the various in-lake and shoreline wildlife extent contained in each of the basins. Now the GIS Watershed Natural Resources Inventory was complete, with information of in-lake water quality conditions, susceptibility to additional nutrient loading, and significant wildlife resources. From this information, the GIS was used to perform a "bottom-up" analysis to locate the lake's most critical areas. For each basin and adjoining subwatersheds the GIS simply averaged together all of the criteria scores. Depending on the priorities of any specific town or stakeholder group, however, a similar process may be employed that uses a weighting multiplier to favor higher-priority concerns. The resulting integration was best visualized for stakeholders by draping a color (light or "cold" for less critical, reddish or "hot" for most critical areas) over the 3D plot of the lake basins. Figure II.12.8 is a grayscale representation of this type of plot.

Useful watershed natural resource inventory GIS products

Whereas we now use many of the GIS visualizations in snazzy animated computer presentations for our statewide and regional education programs, putting the information to use in implementing local decisions and management plans was a primary goal. The materials produced for the community associations and decision makers in the watershed had to be more functional. Many towns and most citizens still do not have easy access to GIS systems, so a more "low-tech" set of products was developed. For the town decision

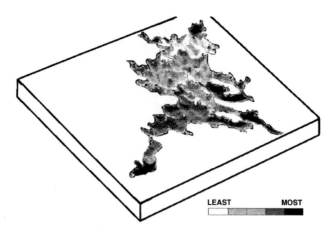

Figure II.12.8 GIS visualization of critical lake basins from integrated analysis of the Squam Lakes Natural Resources Inventory data. Darker shading represents the more critical waters. (Color version available at www.gsd.harvard.edu/watercolors. Password: lentic-lotic.)

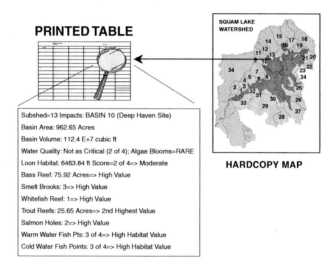

Figure II.12.9 Diagram of "low-tech" GIS products created for utilization of Squam Lakes Natural Resources Inventory data by local decision makers and stakeholders. (Color version available at www.gsd.harvard.edu/watercolors. Password: lentic-lotic.)

makers, a map of the watershed area was provided, delineating the various subwatersheds and subbasins of the lake labeled by a unique number. These numbers were then referenced to a printed table that contained the water quality and resource information of both the basins and the abutting subwatersheds (Figure II.12.9). Tabled information could also be captured to a spreadsheet or a database system, and digital maps could also be provided to those with GIS display systems. Now, instead of having to decide on the approval of a project based solely on information provided by the applicant, the decision maker can look up the subwatershed where the project is being proposed, check on the important lake resources that may be impacted, weigh the benefits and concerns, and have the applicant address specifically how he or she will minimize loss or impacts to that resource. Additionally, with critical areas of the lake and associated subwatersheds catalogued, a targeted land protection effort using voluntary and (if possible) regulatory approaches can be implemented.

Figure II.12.10 Photograph of Lake Chocorua, NH, with Mount Chocorua in the background. (Color version available at www.gsd.harvard.edu/watercolors. Password: lentic-lotic.)

Results and implementations

Information gained from this study has benefited the lake association, the surrounding towns, state agencies, and researchers. The lake monitoring group has utilized the GIS study to reevaluate its monitoring strategy by conducting more frequent sampling at the more susceptible "hot-spot" sites. The lake association has also used the information to call for better enforcement of boat speed rules for the more critical areas defined by the GIS. One of the towns is reevaluating its zoning and planning laws given the results of this study. Another town has used the watershed natural resources inventory data in planning board meetings. Information from the GIS study was used by our state environmental enforcement agency to evaluate its permitting of a large development project in the watershed. Also, one basin that did not follow the expected outcome of the analysis. We later learned from investigation that a significant amount of fill used to cover over a wetlands area at the end of this bay impeded the flow of a tributary that acted to flush the bay out during the high-flow events that follow the spring snow melt. This was the only embayed basin with a restricted tributary. It serves as an example of the damage that can result when wetland areas are filled in.

The Lake Chocorua watershed project

The Squam project afforded us the knowledge of which GIS data layers and analyses proved most useful to determine various watershed characteristics and related water quality conditions. For the Chocorua project, where we had limited resources, our first step was to define the issues that we needed to address and then determine what GIS analysis approach would be necessary.

Background

The watershed for Chocorua Lake is 13.2 square miles and is well protected except in a few vulnerable areas. The still-pristine quality of Chocorua Lake and its view of Mount Chocorua (Figure II.12.10) are widely known as an icon of New Hampshire. The area attracts many tourists, especially in the autumn when the lake reflects the chiseled mount surrounded by the multicolored leaves of autumn. Compared with most NH lakes, relatively few houses are located in the watershed. Those near the lake have conservation easements and covenants, which require extensive setbacks for housing and septic systems and limit shoreland clearing. The lake provides local residents and visitors an area for swimming, fishing, and low-impact boating.

Well over half of the watershed is uncut forest, the highest-quality watershed. About 95% of this land is protected by conservation easements, which have preserved upland wooded buffers all around the lake, except on the eastern shore where State Route 16 borders the lake. Several large wetland complexes protect the lake by providing natural filtration and assimilation. Although the lake is surrounded by protective forest cover in most of its watershed, areas in the upper watershed are prime for development. Chocorua Lake is very shallow, with an average depth of 12 ft. As a result, sunlight reaches through most of the water column. Thus, even low concentrations of nutrients are readily available to algae and other plant life that occur throughout the lake. Visual surveys of the entire shoreline suggested that Route 16 was one of the main sources of erosion and nutrients into the lake; however, before the NH Department of Transportation (NH DOT) would agree to mitigate the runoff and erosion situations, they wanted to be sure that this was a major problem for the lake compared to runoff and tributary inputs from the rest of the watershed.

Lake water quality and diagnostic study

Long-term sampling through participation in the NH Lakes Lay Monitoring program (mentioned in the Squam example) has shown a continued decrease in lake water quality with symptoms such as increasing algae levels and aquatic weed growth, and reduced water transparency. In addition, synoptic sampling near the lakeshore indicated the highest concentrations of total phosphorous were occurring along the lake's eastern shore, which borders the highway. In order to be complete and to provide additional watershed information to allow the Chocorua Lake Conservation Foundation (CLCF) to develop a land protection strategy, a water and nutrient budget diagnostic study was initiated.

The GIS Natural Resources Inventory employed for this project was directed toward compiling the information from which GIS analysis could provide the necessary data to feed into the water and nutrient budget calculations and the predictive runoff and loading diagnostic modeling. Of particular importance were the conservation lands, generalized land cover (derived from remote sensing that was combined with an "on the ground" watershed survey to determine development densities), and the extent of wetland complexes. In addition, the road and highway culvert locations were located and digitized using GPS to facilitate the delineation of subwatersheds that contributed to the road runoff.

The results of the lake watershed diagnostic study confirmed that for its areal extent, as well as for its relative water contribution, the Route 16 runoff measured through the drainage culverts contributed the greatest amount of nutrients into the lake (Figure II.12.11); however, it also disclosed that the northern watershed area contributed the majority of the nutrients overall. Additionally, it documented the importance of the wetlands in those subwatersheds acting to both reduce the nutrient loading as well as shunt the nutrients until the end of the growing season. All the pollutants were not necessarily generated and coming directly off the highway, because there was evidence of sediment deposition occurring from up-watershed sites bordering the highway, but the culverts were acting to channel everything directly into the lake. This information provided the NH DOT with the justification it needed to assist in the mitigation of the highway sites. It also directed the CLCF to concentrate its land protection efforts on the northern watershed lands, especially those

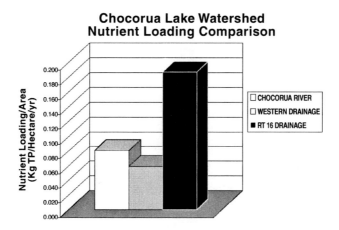

Figure II.12.11 Bar graph depicting relative areal total phosphorus loadings from various subwatersheds of the Lake Chocorua watershed. (Color version available at www.gsd.harvard.edu/watercolors. Password: lentic-lotic.)

that contained the large wetland complexes. Strategies to be used for this effort include outright purchases, covenants and easements, and the reevaluation of existing zoning regulations.

GIS analysis of local wetland buffer options for Deerfield, NH

The examples discussed previously involve a single major watershed focus of concern and analysis of the component subwatersheds. The following example is concerned with protecting specific water resources: wetlands, within a single jurisdiction, the town of Deerfield, NH. It was chosen to highlight additional analysis approaches used to help implement resources protection (see also Chapter II.14 for an additional GIS study of protective buffer strips).

Background

A collaborative effort between the Audubon Society of NH, the NH Office of State Planning, UNH Cooperative Extension, and the USDA Natural Resources Conservation Service produced a guidance document on riparian buffer function that included recommendations for regulatory and nonregulatory buffer widths (Chase et al., 1995). It provides municipalities with both a scientific rationale and practical actions for protecting and preserving naturally vegetated upland areas that border surface waters and wetlands. Through a review of the current scientific literature and recommendations of other states, and with priority focused on water quality protection, a "reasonable" minimum buffer width of 100 ft is recommended in the buffers guide. A larger buffer is recommended for sensitive wetlands (bogs, fens, white cedar swamps), prime wetlands, endangered or threatened species protection, or to support wildlife habitat more thoroughly. Ultimately, local decision makers will need to determine the most appropriate buffers to suit their needs and the means for establishing them (see also Chapter II.1 for an additional discussion of protective shoreline buffers). In an effort to demonstrate how GIS might be used to assist in the decision-making process, a pilot project was undertaken and the results presented to a statewide audience at a GIS workshop for decision makers sponsored by the NH Office of State Planning and the University of New Hampshire.

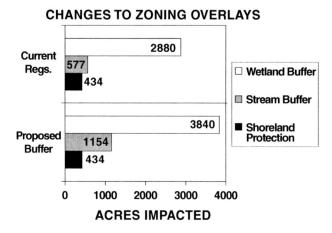

Figure II.12.12 Graph showing pre- and post-implementation effects of incorporating 100-ft buffer overlay for wetlands and low-order streams in the town of Deerfield, NH. Each bar in set represents land acreage impacted by a specific regulation.

The town of Deerfield, NH, completed a comprehensive inventory of its natural resources in the spring of 1991 extensively using GIS (see Appendix D in Auger and McIntyre, 1992). For investigating the various buffer scenarios it was first necessary to take an inventory of the water resources of concern. From the GIS base map, surface waters are already delineated. The GIS soils coverages were used to delineate wetlands areas (from hydric soils classifications). The inventory of Deerfield disclosed that wetlands comprise 86% of the town's water resources acreage and many are connected and lie within stream corridors that run throughout the town.

Existing lands under regulatory restriction

The existing regulatory buffers and setbacks in the town were analyzed using GIS. Two sets of state laws and regulations are already concerned with maintaining a vegetated buffer at the shoreline of lakes and streams. The Comprehensive Shoreline Protection Act requires that a minimum tree basal area must be maintained at greater than 50% within 150 ft from the shore of lakes greater than 10 acres and fourth-order or greater streams (except those in the NH Rivers Program). State forestry regulations also maintain this requirement for land within 50 ft from a perennial stream or brook. A setback of 75 ft for buildings and septic systems bordering wetland areas is also required under town regulations. The GIS display of these overlay zones indicates the existing acreage of these areas as 434 acres under the Shoreland Protection Act, 577 acres under the forestry regulations, and 2,880 acres bordering wetlands with town-mandated setback restrictions.

Adopting the 100-ft buffer recommendation and exploring other scenarios

Through the use of GIS, maps were produced that visualized the extent of lands that would be impacted by the new recommendations. Imposing the 100-ft buffer overlay for wetlands and streams about doubles the protective acreage around streams and adds another 1,000 acres that border wetlands (Figure II.12.12). This represents a 40% increase in the protected areas. This initially might appear to have a large impact and involve many landowners, but once already protected lands are subtracted out and large land tracts are considered, the number of landowners impacted is greatly reduced. Thus, using an overlay of the town tax map, the decision makers are now able investigate the degree to which different lands might be affected by various regulatory approaches.

The buffers document suggests a buffer larger than 100 ft for more critically important or prime wetlands and for wildlife corridors. Our study explored the use of a 200-ft buffer in our analysis. An overlay of this buffer was created to visualize the impact and to discern whether this size chosen was adequate to serve both water quality and wildlife habitat concerns. The resulting analysis indicated that with the 200-ft buffer some wetlands in the sample area would be connected to each other, but others would not. If habitat considerations are a goal, the GIS analysis indicated that other, perhaps nonregulatory, methods would be needed to establish habitat connections among all of the critically important wetlands in the town.

Optimizing nonregulatory protection approaches

Nonregulatory approaches to buffer protection were also explored in detail with GIS analyses. For purposes of wildlife habitat and travel corridor protection and to maximize the benefits of conservation lands, acquisition of larger buffer areas may be required. To achieve this level of protection a town may have to rely on land acquisition and/or conservation easements. Use of the GIS information regarding the wetland and stream locations, existing and proposed buffer overlays and habitat land-cover information, along with property or tax map overlays and existing conservation lands, can help decision makers choose the most cost-effective way of achieving their goals. Creating GIS visualizations of the extent of impacted areas and the functional connectivity of the lands can provide stakeholders with the perspective needed to gain community acceptance of the proposals.

Critical watershed land and resources analysis

The ultimate synthesis of a GIS Watershed Natural Resources Inventory analysis is to target those most critical lands and water resources that are at the greatest risk. The risk may be the total loss or partial degradation that can occur due to development or change in nearby land-use activities. The identification process utilizes a combination of the "top-down" and "bottom-up" approaches already discussed in the examples here but proceeds through a series of processing steps that integrate the analysis outcomes to yield the results. We have been piloting a version of this approach in our coastal watershed communities, where it is predicted that the greatest near-future expansion of our state population is most likely to occur.

Step 1: Land capability analysis

Initially a conventional "top-down" land capability analysis is undertaken as previously described in the Squam Lakes Watershed Study. Lands already developed, permanently protected lands, and land constrained from development due to local or state regulations as well as water and wetlands and their associated buffers are located. The land that remains is cataloged as developable.

Step 2: Prioritize land and water areas of special concern

From the watershed natural resources inventory select and, if warranted, weight the list of important areas of concern. These may include (but are not restricted to):

- Wetlands (designated as prime, highly functional, large, or part of a cluster or corridor)
- Wildlife habitats (in-water, riparian, and upland)

- Important lands (unfragmented lands, forest, agricultural lands, significant lands)
- High-quality waters
- Sensitive waters (low assimilation capacity, vulnerable to development)
- Endangered or unique plant or animal communities
- Important recreation areas
- Commercially important plants and animals (aquatic and terrestrial)

Once the selection and weighting is decided a GIS analysis is employed to produce a co-occurrence data layer. This coverage indicates lands that contain (or in terms of the water resources, abut) the most types of, or highest-priority (if weighted) resources. Typically, this is displayed on a GIS map as a continuum of shading or coloring. Again, this is similar to the critical basin process employed for the Squam Lake analysis. This result is a "bottom-up" resource analysis product.

Step 3: Determine areas most likely to become developed

By incorporating data on existing infrastructure such as roads, power lines, and water and sewer service, those areas most likely to become developed next can be predicted. This is based on the assumption that it is cheaper to build new development where you are close to existing infrastructure than to cover the expense of expanding that infrastructure out to lands farther away. This also may be considered a "bottom-up" approach only because it concerns "on the ground" infrastructure resources.

Step 4: Process all of the elements together

Combining the results of steps 1 through 3 together creates an analysis of those developable lands (step 1) that have the highest resource value (step 2) and are the most likely to be developed in the near future (step 3). This may be presented to stakeholders on a map using various shading or colors for co-occurrences, by employing a series of mylar overlays, or by viewing manipulated GIS layers directly from the computer.

The final step would be to develop a protection strategy (or strategies) to target these critical lands. Further analyses using land parcel data or different analysis "runs" may have to be employed to target the appropriate critical lands for the specific strategy or approach undertaken.

Lessons learned and future predictions

GIS and related spatial technologies have proven themselves as valuable tools for analysis, assessment, and communicating to the public. Although "high-tech" analyses are critical to successful watershed natural resources inventory projects, the "low-tech" information transfer products have a much wider utility in regards to facilitating implementation of management plans and protective measures. The GIS-based watershed natural resources inventory data analysis consistently resulted in a proactive approach instead of a reactive response. It allowed for an informed watershed community to become an empowered community with fact-based concerns that could more easily result in consensus and support. An additional important lesson was that all the possible data in the world is not needed for a useful outcome. The most critical element for success is to understand the problem to be addressed and then to determine what data will be required to facilitate informed decision making. As demonstrated in the examples here, as well as in the symposium roundtable discussions that follow, a key to optimizing the power of GIS spatial analysis is to combine the "bottom-up," resource-based data analyses and the "top-down" land capability analysis to define those most critical lands and watershed resources to direct management, regulatory, and nonregulatory attention and efforts.

A more complete understanding of landscape-scale interactions within watersheds as well as their relation to water quality and in-water resources can be provided through GIS-based spatial analysis. Currently, our resolution is limited for most elements such that site-specific scaled analyses are not yet within our reach. However, given the rate at which the technology is improving (more detailed and near real-time remote sensing; more powerful "user friendlier" GIS software and related spatial analysis tools), the necessary resolution to achieve that level of accuracy is just on the horizon.

Literature cited

Auger, P. and McIntyre, J., *Natural Resources: An Inventory Guide for New Hampshire Communities*, Upper Valley Land Trust and Univ. of NH Coop. Ext., Durham, NH, 1992.

Chase, V., Deming, L.S., and Latawiec, F., *Buffers for Wetlands and Surface Waters: A Guidebook for New Hampshire Municipalities*, Audubon Soc. of NH, Concord, NH, 1995.

Congalton, R.G. and Green, K., The ABCs of GIS: an introduction to geographic information systems, *J. Forestry*, 90, 13, 1992.

Schloss, J.A. and Mitchell, F., Promoting watershed based land use decisions in New Hampshire communities: Geographic information system aided education and analysis, in *Proc. Watershed '96*, Baltimore, 1996, 830–833.

Schloss, J.A. and Rubin, F.A., A "bottom-up" approach to GIS watershed analysis, in *Proc. 1992 GIS/LIS Conf.*, San Jose, CA, *Amer. Soc. Photogrammetry and Remote Sensing* (and others), Vol. 2, 672–680, 1992.

Scott, D. et al., Squam Lakes watershed plan, NH Office of State Plan., Concord, NH, 1991.

Stone A., *Natural Resource Inventories: A Guide for New Hampshire Communities and Conservation Groups*, Upper Valley Land Trust, NH Div. Forest and Lands, USDA Forest Serv. and Univ. of NH Coop. Ext., Durham, NH, 2001. (A revision and update of Auger and McIntyre, 1992.)

Response

Janus planning:
Using computer tools to look backward and forward simultaneously

Geographic information systems, or GIS, has emerged as an extremely powerful tool in water sensitive planning. The ability to map watershed attributes as varied as soil moisture (Chapter II.10 by John Felkner and Michael Binford), riparian buffers (Chapter II.14 by Robert France, John Felkner, Michael Flaxman, and Robert Rempel), or surface permeability and aquifer recharge potential (Chapter II.15 by Amir Mueller, Robert France, and Carl Steinitz) is an important step in identifying resources for protection from future development. This is the "bottom-up" approach that Jeffrey Schloss promotes in this chapter. Unfortunately, GIS has also been used as an excuse to identify areas that are imagined to be capable of absorbing more development — what Schloss criticizes as the "top-down" approach.

As Schloss identifies, resource or bottom-up GIS analysis leads to preemptive planning and protection, which in turn helps toward fostering watershed stewardship. Stewardship comes about through education, and it is here where GIS serves its most important role in water sensitive planning. Through providing convenient visualizations of complex and occult attributes, GIS allows stakeholders to become informed not just about the current status of their watershed, but also about the status of past conditions as well as the predicted status of future conditions given different development scenarios (see Chapter II.15 by Mueller, France, and Steinitz).

The case studies described in this chapter demonstrate a range of GIS applications such as build-out analyses, lake bathymetry and subwatershed mapping, sensitive species locations, road effects, wetlands, and riparian buffer options. Schloss presents a logical framework for undertaking watershed planning through GIS analysis: critical lands analysis → land capability analysis → prioritization of areas of special concern → determination of areas most likely to become developed → combination of all elements into a final planning summary.

Schloss also briefly touches on a serious weakness in GIS technology that offers a pertinent, cautionary lesson for water sensitive planners. For it is the very elements that make GIS so effective as a communication and education tool that also seduce many into using the techniques in a superficial and cavalier fashion. There is a cardinal need to move beyond the common use of GIS analysis to produce snazzy computer-based presentations, and instead to push the tool toward new and carefully thought-out directions in watershed management (e.g., see Chapter II.10 by John Felkner and Michael Binford, and Chapter II.13 by Margot Cantwell). In the end, it is important to remember that GIS is a means toward another end, never the end itself.

chapter II.13

The effect of spatial location in land–water interactions: A comparison of two modeling approaches to support watershed planning (Newfoundland, Canada)

Margot Young Cantwell

Abstract

Two landscape models of land–water interactions were compared for their usefulness in assisting planners in zoning watersheds and evaluating future development proposals. A number of lake and water quality models have widespread acceptance by watershed managers (e.g., Dillon et al., 1986; U.S. EPA, 1988; Riley, 1988). These models are fit to a particular lake and watershed using landscape, climatic, and basin characteristics as well as point and nonpoint source pollutant inputs. In many jurisdictions, these models have been used to predict lake carrying capacity and establish maximum target levels of pollutant input to maintain a desired water quality level. This maximum target level is then described in planning documents in terms of the maximum amount of development (or land use) to be permitted in the watershed (e.g., Kings County, Nova Scotia, and Muskoka Lakes Region, Ontario). One question facing watershed planners implementing the target level is the effect of spatial location on land-use or development impact — does it matter where in the landscape the development/land use goes? Two approaches were compared in this study. The first assumed that control of land-use type and quantity was sufficient; the second approach considered the implications of land-use location. The second approach was determined to be a useful tool to zone the landscape and evaluate development proposals. The approach builds on the standard lake water quality modeling approach, allowing managers to predict the amount of pollutant input attributable to a particular quantity of land use in a specific location within a lake catchment. The model approaches were evaluated for the Government of Newfoundland and Labrador together with Environment Canada, using data for Gander Lake and its catchment.

Introduction

In 1994, the Government of Newfoundland and Labrador faced opposition from a logging company operating within the Gander Lake catchment. The company challenged the provinces' forestry setback of 150 m (DELA, 1989, 1991) from Gander Lake, stating that the standard setback was not scientifically defensible. The challenge raised the question of the appropriate spatial configuration of land-use restrictions within a catchment, and whether or not water bodies were adequately protected (and scientifically defensible) as defined by the policy set in the Provincial Environment and Lands Acts. A number of studies were undertaken to explore this question, including this study, which was to identify and apply a landscape model of land–water interactions that enabled the spatial location of development restrictions to be considered.

Gander Lake, an approximately 50,000-m-long × 2,000-m-wide × 290-m-deep lake is the drinking water supply for the town of Gander and several smaller towns and villages in central Newfoundland (48.5° latitude; 55.5° longitude). The lake and Gander River, which bisects it, are a world-class Atlantic salmon fishing resource. These two resource functions allowed Gander Lake and River to be studied over the past 20 years by Fisheries and Oceans Canada, Memorial University, and the Newfoundland and Labrador Department of Environment. The town of Gander has a Canadian Air Force presence and so Transport Canada records of treatment plant performance and airport and town stormwater runoff quality were documented and available. The catchment represented an excellent opportunity to consider the spatial effects of land use.

The project involved a number of tasks, published in the Watershed Management Plan for Gander Lake and its catchment (EDM, 1996). Water quality analysis was undertaken by Jacques Whitford Environment Limited of St. John's, Newfoundland, and three water quality models (Dillon, MinLake, and WASP4) were fit to the lake by R. H. Loucks Oceanology of Halifax, Nova Scotia. This work indicated that the lake was oligotrophic, dimictic, and considered resilient to the pressures of land development. The lake currently meets or exceeds Canadian drinking water quality guidelines. Phosphorous was identified as the limiting nutrient and as the focus for water quality management.

Circulation characteristics and water quality results indicated a total annual phosphorous loading for the lake of approximately 11,730 kg/year. Of this total, phosphorous input from precipitation was estimated at 2800 kg/year. The only point source inputs are from the sewage treatment plants (none of which discharge directly to the lake). Using treatment plant records together with input stream water quality results and assuming all cottages (of which there are only a few) have failing on-site sewage systems, the maximum possible loading to the lake attributable to sewage was estimated at 130 kg/year. This left approximately 8600 kg/year of phosphorous in the lake attributable to runoff from catchment land uses. Further modeling indicated that the lake could sustain an approximate 50% increase in phosphorous input before its trophic state moved from oligotrophic to mesotrophic.

With a model fit to the lake, the question for the watershed managers was how to assign land uses within the catchment to sustain the target water quality level. In planning documents, land uses are assigned by zoning maps that have the following attributes: land-use type (what use and at what concentration/density?), area (what extent?), and location (where specifically can development occur?). In making decisions about assigning zones, managers must consider both how much development (or preservation) should be allowed and specifically where in the landscape that development/preservation should be located (see Chapters II.7 and II.12).

The following questions were considered:

1. What is the source of the phosphorous in the catchment runoff and how might it be distributed spatially?
2. Are standard land-use export values used to calibrate lake water quality models adequate for policy development, including zoning the landscape?
3. Is a method that includes a spatial component a better approach for predicting the contribution of phosphorous from the different land uses?
4. What is the likely impact of proposed future land uses on the phosphorous balance of the lake?
5. What is the most scientifically defensible approach to zoning for future land uses within the catchment?

Method

Two approaches were considered for managing the Gander Lake catchment:

1. The fact that phosphorous input is primarily described by land-use type and location is a minor consideration. Watershed policy created using this assumption would typically set a maximum development level in the policy document (e.g., maximum number of acres of development, or a total number of cottages). This is a common approach to watershed modeling and policy development. For example, each of the three water quality models fit to Gander Lake used this approach to establish the land-use input and the carrying capacity of the lake. This approach was referred to as the land-use model.
2. The fact that phosphorous input is described as much by land-use location in the catchment as by the actual land-use type is a consideration. Watershed policy created under this assumption would prescribe both how much development or land use and in which specific areas of the catchment. Policy might also allow for increases or decreases in land-use type or extent depending on specific locational criteria. This approach was referred to as the spatially explicit model.

Phosphorous is commonly the limiting nutrient in watersheds managed for water clarity (see Chapter II.17). Similar to the Gander Lake catchment, the various land uses in the watershed are frequently a large contributing source. Of the land-use fraction (after direct precipitation and point sources are subtracted) soil is typically the primary source, as phosphorous is a natural soil nutrient and phosphorous tends to attach itself to soil particles (Dillon et al., 1986; Hickman, 1987; Leopold, 1994; Loucks, 1993). Soil reaches the water column though erosion processes that occur naturally at relatively low rates and often are accelerated when soils are exposed, such as during construction for development, field tilling in agriculture, and silvaculture or road-building practices in forestry (see Chapter II.16). In addition, phosphorous released from organic matter (e.g., garbage on streets, tree leaves on yards and in forests, or pet feces), as well as phosphorous added as a soil fertilizer (e.g., to lawns, gardens, and agriculture fields), also tends to attach itself to soil particles. Thus, eroding soil itself, as well as dirt washed from streets and other land surfaces, tends to carry the significant phosphorous load associated with land use. Therefore, although phosphorous is derived from many sources, its attachment to a soil particle and its delivery via erosive processes make it possible to model it as part of a soil erosion process. This is particularly the case in nonurbanized catchments where phosphorous associated with soil erosion is normally the primary source of phosphorous in lake waters (Leopold, 1994; Dunne and Leopold, 1978).

To compare the different approaches, two different landscape models describing land–water interactions were fit to the Gander Lake catchment. The first model was a standard land-use export approach, referred to as the land-use model. Land-use type is considered by many researchers to be the most sensitive variable in determining how much pollutant a given land area will contribute to a lake and considered sufficient to establish lake water quality models (e.g., Dillon et al., 1986; U.S. EPA, 1988). Reference values for phosphorous and suspended solids contributed from different land uses may be found in the literature. To evaluate the first approach, land-use export values were developed for the different land uses found in the catchment. The export values were derived from similar landscapes in New England and Nova Scotia and calibrated to the Gander Lake catchment using water quality data and land-use areas from a local sub-catchment. The land-use model is defined by Eq. II.13.1.

$$\mathrm{TSS} = \sum_{1}^{n}(E \times A) \qquad (\mathrm{II}.13.1)$$

where
TSS = total annual suspended solids input to the lake in kg/yr
n = number of land uses
E = export value of suspended solids for each land use (kg/ha/yr)
A = area of each land use (ha)

To explore the second approach, a landscape model for land–water interactions with a spatially explicit component was used. The modeling approach taken was similar to that used by Binford (1989). The approach is a variation on the Universal Soil Loss Equation (USLE), with a modification to consider sediment delivery to a lake. The USLE is a tool developed for agriculture to estimate the potential of the landscape to erode. The USLE, by far the most widely used method for predicting soil loss, was developed by Wischmeier and Smith (1978). The USLE has been used primarily in agriculture watersheds to predict soil loss in fields (e.g., Mellerowicz et al., 1994). In recent years, it has been applied to other land uses, including forestry and urban and suburban development. To date, only limited application of the USLE to regional-scale planning has been used, although the approach has merit (e.g., Binford, 1989; Rees, 1996).

The USLE has four locational factors: an R-index describing how erosive the local rainfall is, (essentially how hard the rainfalls, and thus its likelihood to dislodge soil particles); a K-index describing the physical nature of the soil; and two factors, L and S, describing the shape of the terrain (both the slope and the effect of gravity on slope effect, referred to as the hill slope-length and hill slope-gradient factors). These four factors represent a sort of "thumb print" of the catchment. Nothing can be done about these four factors, and thus they represent an inherent sensitivity or a landscape unit sensitivity.

Two factors are used to consider land-use and property management. The C, or cropping-management factor, is a factor that considers the effect of the particular land-use type. The P, or erosion-control practice factor, allows for a modification of C when certain management practices are used. The P factor is well developed for agriculture, where different cropping techniques (e.g., contour farming) have a demonstrated effect on the soil loss. P-factors are being developed for forestry land uses reflecting different silvaculture practices (e.g., shelter wood and no duff disturbance practices as opposed to clearcuts and forest tilling). *P*-factors for development landscape types are less well developed.

The USLE is

$$A = (RKLS)(CP)(2471) \qquad \text{(II.13.2)}$$

(Inherent sensitivity) (Land use and Management)

where
A = soil loss (kg/ha)
R = rainfall erosive index
K = soil erodibility index
L = hill slope-length factor
S = hill slope-gradient factor
C = cropping-management factor
P = erosion-control practice factor

For the Gander Lake catchment, the USLE was solved in the GIS for each 1-ha grid cell across the landscape. Obviously, all of the eroded soil does not end up in the water column of Gander Lake as suspended solids. Total soil loss was translated to a total annual suspended solids (TSS) loading by multiplying the total soil loss by a delivery ratio (the fraction of the eroded soil likely to reach the lake) and then multiplied by the percentage likely to remain in suspension (i.e., the settlable solids fraction was removed). This result was then summed across the catchment for a total annual TSS contribution (as per Eq. II.13.3). The map algebra used in the spatially explicit model is illustrated in Figure II.13.1.

$$\text{TSS} = \sum_{1}^{n}(A \times D \times S) \qquad \text{(II.13.3)}$$

where
TSS = total annual suspended solids input to the lake in kg/year
n = number of 1-ha grid cells
A = soil loss (kg/ha/year)
D = delivery ratio
S = % of soil in suspension (kg/ha/year)

Equations II.13.1 and II.13.3 can also be used to estimate the total annual phosphorous loading to a lake by using published export values for phosphorous in Eq. II.13.1 (as opposed to suspended solids), and by multiplying A by the fraction of the soil that is phosphorous in Eq. II.13.3. Total suspended solids (TSS) was chosen for use in the Gander Lake catchment because there were better local export values (E, in Eq. II.13.1) for TSS than for phosphorous, and the primary objective was to compare two landscape models of land–water interaction for their usefulness in preparing plan area policy.

Both modeling approaches were calibrated by fixing the annual soil loss in the catchment and then calibrating for the land-use variables. The annual soil loss in the catchment was established as follows. First, using the average measured TSS concentration in several subcatchment input streams together with the total estimated catchment flow, a value of 7,077,000 kg/year of TSS was derived (5.8 mg/l TSS × 1.212 E^{12} l/year = 7.077 E^6 kg/year).

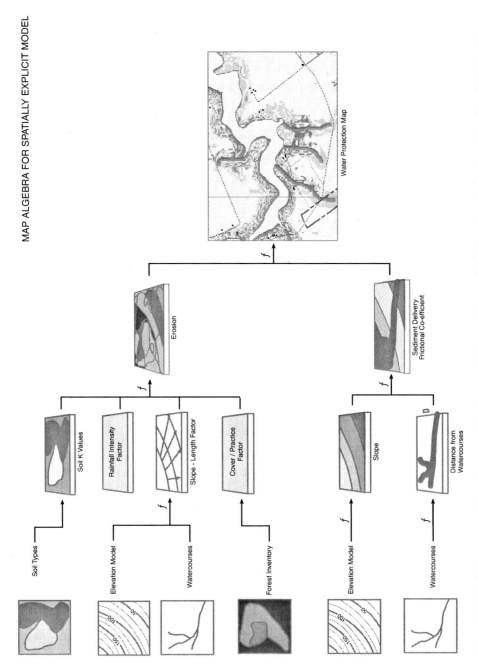

Figure II.13.1 GIS map combinations used in the spatially explicit model of land–water interaction (Eq. II.13.3). (From EDM, Environ. Design and Manage. Ltd., Land-usep plan for the Big Cove First Nation, 2000.) (Color version available at www.gsd.harvard.edu/watercolors. Password: lentic-lotic.)

Table II.13.1 Expected Annual Soil Loss from the Gander Lake Catchment Based on Literature Values from Catchments of a Similar Size

Watershed Management Plan for Gander Lake and Its Catchment	
Gander Lake catchment area	77,250 ha
	298 sq. mi.

U.S. Water Resources Council, Estimated Soil Loss from Drainage Areas in the North Atlantic Region

	Soil Loss (nonurbanized catchments)	
	tons/sq. mi./year	kg/year
High value	1210	327,321,253
Average value	250	67,628,358
Low	30	8,115,403

Estimated Soil Loss from Glacial Till Lithologic Type

	Soil Loss (nonurbanized catchments)	
	tons/sq. mi./year	kg/year
High	300	81,154,030
Medium	200	54,102,686
Low	180	48,692,418

Sources: Dunne and Leopold (1978), and Leopold (1994).

Then, from particle size analysis of soil type data [15], the approximate percentage of settlable solids (the sand and most of the silt, given the residence time in the lake) was estimated to be about 87%. The TSS of 7,077,000 kg/year was thus assumed to be derived from a total annual soil loss of approximately 54,500,000 kg/yr (7.077 E^6 kg/year × 7.69 = 5.44 E^7). This annual soil loss value (54,500,000 kg/year) was then compared with literature values of expected soil loss from similar sized catchments (see Table II.13.1). The resulting correlation was very good, and the TSS and total soil loss in the catchment values were adopted and fixed for the catchment.

Both land–water interaction models were set to achieve this resultant level of soil loss or lake water quality impact. Both were then calibrated by adjusting land-use variables. Characteristic values for E-factors in Eq. II.13.1 (land-use model), taken from New England (Borman and Likens, 1979) and Nova Scotia (Vaughan Engineering, 1993), were modified to reflect measured water quality results in a local subcatchment. C-factors and D-ratios in Eq. II.13.3 (spatially explicit model) were also modified. No published C-factors were available for Newfoundland land uses, and so more general literature values (Dissmeyer and Foster, 1981; Dunne and Leopold, 1978) and values from Acadia National Park (Binford, 1989) were used as a starting place. P-factors were not used in the model (assumed to be 1, or no modifying practice factor) due to a lack of published values.

Calibration for delivery ratios (D in Equation II.13.3) proved to be a difficult exercise, moving the project more into the realm of planning than science (i.e., Are we more right than wrong, as opposed to, can we demonstrate this for certain?). Similar to C-factors, literature on sediment delivery ratios is very sparse, although there is some ongoing work to better define them (e.g., Arbour, 1996; Binford, 1995). In addition, models for routing sediment (and other materials) are being modified for application to urban and developed land uses. For this study, delivery ratios were developed following the method of Snell (1985). Working in an Ontario agricultural landscape, Snell assigned a high delivery ratio to areas that were within 100 m of a stream or water body; a medium delivery ratio to areas of steep slope (greater than 5%) that were set back from a water body by at least a 100-m forested buffer; and, a low delivery ratio to areas that were tributary to a depression, or had shallow slopes (less than 5%) and were set back from a stream or water body by at least a 100-m forested buffer. Snell further refined delivery ratios to reflect hydrologic soil properties in the absence of a buffer zone. This condition (no buffer) existed so rarely in the Gander Lake catchment that this refinement was not included.

Resultant GIS landscape maps and values describing the current land–water interactions were compared. Then both modeling approaches were used to evaluate proposed future developments in the catchment and consider the potential effect of the spatial component.

Results

Figure II.13.2 illustrates the GIS output map that results when the land-use model (Eq. II.13.1) was applied across the catchment, using land-use areas and land-use export values for suspended solids delivery. Final export values that achieved calibration are shown in Table II.13.2, together with the literature values and values from a local sub-catchment referred to as the airport ditch from which the calibration values were derived.

Figure II.13.2 reflects the areas of concern if the premise is that pollutant delivery to the lake is primarily correlated to land-use type. The figure illustrates the expected export of suspended solids based on the range of land uses currently existing in the catchment. The map result for phosphorous export produces a similar pattern. Values range from a high export value associated with developed and cleared landscapes to low export values associated with forest and bog landscape types. It is interesting to note that the range of potential contribution is relatively narrow, with a high of 1200 kg/ha/year and a low of 0–35 kg/ha/year.

Figure II.13.3 illustrates the map that results when the spatially explicit model (Eq. II.13.3) is applied across the catchment. Values used to calibrate, as well as literature values from which the calibration values were derived, are shown in Table II.13.3. The R-index used was 870 as published for Gander (Wall et al., 1983). The K-indices used for each soil type were derived from local soil surveys (refer to Table II.13.3). The L- and S-factors were considered together. A slope map for the catchment was generated from the digital terrain model (DTM) provided by the Newfoundland and Labrador Department of Environment, Water Resources Division. L was defined as the width of the grid cell in the GIS (in this case 100 m). Using tables in Dunne and Leopold (1978) and Wischmeier and Smith (1978), the LS (length-slope index) was derived. (A refinement to this approach where the computer calculates the hill slope-length, as opposed to using the grid cell width, has since been developed by the author (EDM, 1999, 2000). Final C-values derived during calibration, as well as literature values used as a starting point, are also shown in Table II.13.3. The final values used align well with the literature values, with the following exceptions: (1) the C-value for development was lowered due to the suburban nature of development in the catchment; and (2) the value for burned areas was lowered because most of these areas are very old burns with little soil left to erode.

Delivery ratios for Snell's high, medium, and low categories were estimated based on literature values reported by Lowrance et al. (1988) and Cooper et al. (1987), and with advice from Arbour (1996). A setback value of 200 m was used (as opposed to Snell's 100 m), given the more recent literature on the effectiveness of various buffer widths (e.g., Lowrance et al., 1988; Cooper et al., 1987; Toth, 1990) and considering specific site conditions in the Gander catchment (a more complete discussion of this is provided in the

Chapter II.13: The effect of spatial location in land–water interactions

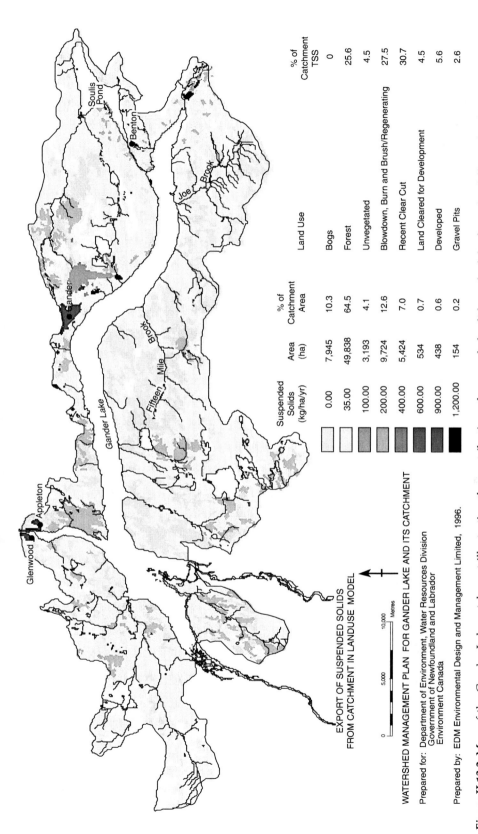

Figure II.13.2 Map of the Gander Lake catchment illustrating the contribution of suspended solids to the lake from the different land uses, as predicted by the land-use model (Eq. II.13.1). Note that the town of Gander and the adjacent Gander airport are both predicted to have a relatively high impact. (From EDM, Environ. Design and Manage. Ltd., R.H. Loucks Oceanol. Ltd., and Jacques Whitford Environ. Ltd., Watershed management plan for Gander Lake and its catchment, Govt. of Newfoundland and Labrador together with Environ. Canada, St. Johns, Newfoundland, 1996.) (Color version available at www.gsd.harvard.edu/watercolors. Password: lentic-lotic.)

Table II.13.2 Export Values (E in Eq. II.13.1) Used in the Land-Use Model and Published Export Values from Which the Calibrated Values Were Derived

Watershed Management Plan for Gander Lake and Its Catchment

Export (E) Values Calibrated for the Gander Lake Catchment Land Uses

Area (ha)	Land Use	Calibrated TSS Export Values (kg/ha/year)	Total for Land Use (kg)
154	Gravel pits	1200	184,800
49,838	Forest	35	1,744,330
3693	Blowdown	200	738,600
7945	Bogs	0	0
1510	Burn	200	302,000
3193	Unvegetated/bedrock	100	319,300
534	Land cleared for development	600	320,400
5424	Recent clearcut	400	2,169,600
438	Development	900	394,200
4521	Brush/regenerating	200	904,200
	In the water (approx. 13% of total)		7,077,430
	Remaining 87% assumed lost to sediment		47,364,339
	Total soil loss (kg/year)		54,441,769

Export (E) Values from the Literature

	Suspended Solids (kg/ha/year)		
Land Use	New England	Nova Scotia	Airport Ditch
Forested	33	10–100 (low)	—
15% cleared	—	low–medium	—
Agriculture/golf	—	medium–high	—
Residential unserviced	—	500	—
Development	—	500–1000	—
Land cleared for development	—	low–high	—
Gravel pits	—	low–high	—
Recent clearcut	380	—	—
Brush/regenerating	190	—	—
40% development/60% forested	—	—	326

Sources: New England, Borman and Likens (1979); Nova Scotia, Centre for Water Resources (CWRS), as cited by Vaughan Eng. (1993); Airport ditch, Gander International Airport, Safety and Technical Services, 1995–1996 monthly stormwater runoff data.

published document). In addition, low-value areas were split into two categories, Low Value 1 for areas tributary to a bog, and Low Value 2 for other low-value areas. This was done because very large bogs cover extensive areas of the catchment and export from lands draining to these bogs will be less than export from other low-value areas. The final delivery ratios uses are as shown in Table II.13.3. In Eq. II.13.3, S represents the portion of the total soil loss that remains in suspension in the water column. The settlable solids fraction of the total soil loss was estimated at 87%; thus 13% was assumed to remain in suspension.

Figures II.13.2 and II.13.3 present a very different picture of where in the landscape the suspended solids (and thus phosphorous) found in the lake are coming from. A number of comparisons can be made:

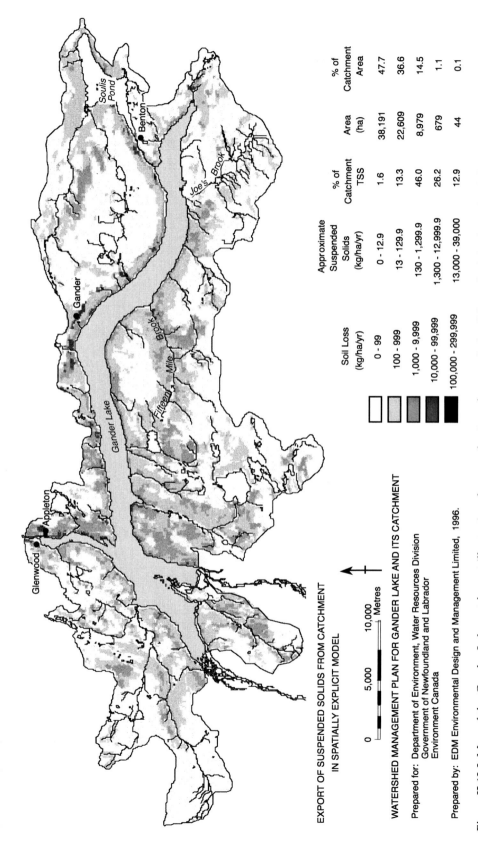

Figure II.13.3 Map of the Gander Lake catchment illustrating the contribution of suspended solids to the lake from the different land uses, as predicted by the spatially explicit model (Eq. II.13.3). Note that the town of Gander and the adjacent Gander airport are both predicted to have a relatively low impact. (From EDM, Environ. Design and Manage. Ltd., R.H. Loucks Oceanol. Ltd., and Jacques Whitford Environ. Ltd., Watershed management plan for Gander Lake and its catchment, Govt. of Newfoundland and Labrador together with Environ. Canada, St. Johns, Newfoundland, 1996.) (Color version available at www.gsd.harvard.edu/watercolors. Password: lentic-lotic.)

Table II.13.3 K-indices, C-factors, and D-ratios (for Eq. II.13.3) Used in the Spatially Explicit Model as Well as Published Values from Which the Calibrated Values Were Derived

Watershed Management Plan for Gander Lake and Its Catchment

K-Index Approximation

Soil Types	K-Index Approximation*	K-Index Assumed**
Bn, Bu, Wg	0.34	—
Ga, Gw	0.37	—
Ep	0.065	—
Fb, Gb, Sp, Sw	0.27	—
Py, Tn	0.12	—
Su, Bo, Ho	0.44	—
Peat, muck	—	0.35

* Soil descriptions and cabability classes described in the 1972 Soil Survey were applied to tables for estimating the K-index in Dunne and Leopold (for areas on map sheets 2D15 and 2D16). For the 2D14 map sheet (not covered in the 1972 Soil Survey), more general descriptions and values in the 1993 soil survey were used.

** Descriptions and values for peat and muck are not available; value was assumed from extrapolation of tables.

Sources: Dunne and Leopold (1978); Wells and Heringa (1972).

C-Factor Estimate

	Acadia National Park*	Other Values	Calibrated for Gander
Land cleared for development	0.1	0.038–0.055***	0.05
Development	0.5	—	0.2
Gravel pits	—	—	0.2
Recent clearcut	—	0.004–0.115**	0.02
Brush/regenerating	0.01	0.01–0.04***	0.01
Burn	0.01	0.003–0.011***	0.007
Blowdown	0.01	—	0.01
Forested	0.001	0.001***	0.001
Unvegetated/exposed bedrock	0.002	—	0.002
Bog	N/A	N/A	N/A
Ponds	N/A	N/A	N/A

* Values used by Binford (1989).

** Dissmeyer and Foster (1981).

*** Dunne and Leopold (1978).

D-Ratio Estimate

	Ontario*	Calibrated for Gander
High value	80%	80%
Medium value	50%	50%
Low value 1	20%	20%
Low value 2 (bogs)	—	10%

*Values used by Snell (1985).

1. The range of values in each figure is significantly different. In Figure II.13.2, TSS values range from 0 to 1,200 kg/ha/year. In Figure II.13.3, TSS values range from 0 to 39,000 kg/ha/year. The land-use model approach presents a much more even distribution of impact than the spatially explicit model approach.
2. The models describe a significantly different distribution of impact. When spatial implications are considered, the lowest impact areas are reduced in terms of the total impact they contribute, and the highest impact areas are increased. For example, both models have only a small percentage of the catchment producing high impact (generating high TSS). In the land-use model (Figure II.13.2), only 0.8% of the landscape is in the top two categories; corresponding with 1.2% in the top two categories in the spatially explicit model (Figure II.13.3). Both models also have a relatively high percentage of the landscape producing low impacts on the lake. In the land-use model, 78.9% of the landscape produces 100 kg/ha/year or less; similarly in the spatially explicit model, 84.3% of the catchment produces 100 kg/ha/year or less. The different modeling approaches vary significantly, however, in terms of the overall impact of these high and low areas. In the land-use model, the 0.8% of the landscape in the top categories is responsible for only 8.2% of the total TSS; while in the spatially explicit model, the 1.2% of the landscape in the top categories produces 39.1% of the total TSS. Similarly, for the low-impact areas, in the land-use model, the bottom 78.9% produces 30.1% of the impact, where in the spatially explicit model, the bottom 84.3% is responsible for only 14.9% of the impact.
3. The spatial location of the land use has a significant modifying effect on the land use as described in the spatially explicit model. For example, in the land-use model (Figure II.13.2) the town of Gander and the adjacent Gander airport are predicted to deliver significant suspended solids to the lake (between 900 and 1,200 kg/ha/year). In the spatially explicit model (Figure II.13.3), the hill slope-length and hill slope-gradient effect (the town and airport are on very flat ground) and the delivery ratio (which directs the majority of the runoff through bogs, wetlands, and streams, to Soulis Pond before it eventually reaches the lake at the very eastern end) reduce the impact of these land uses to less than 13 kg/ha/year.

To determine the importance of landscape features in the spatially explicit model, the sensitivity of the various factors in the USLE ($A = RKLSCP$) was considered. In the model, R and L are constant throughout the catchment. Variation in soil type (K), while important, can only explain a change in soil loss of six times, given the range of soil types in the catchment. P-factors (used to modify C) were not used. This left the LS (variables describing slope or landscape location) and C (cover or land-use type) factors to consider.

Slope sensitivity was investigated by dividing the total number of slope values (n) into five equal categories according to area ($n1, n2, \ldots n5$), where each represents approximate 200 km^2 area. Slope values in the catchment range from 0 to 60%: $n1$ = 0 to 2%; $n2$ = 3 to 5%; $n3$ = 6 to 8%; $n4$ = 9 to 14%; and $n5$ = 15 to 60%. Using these slope values, together with the K-value for Gander Series soil (the most common soil type in the catchment), and the actual range of C-factors (land-use/cover types) in the catchment, Table II.13.4, catchment sensitivity to slope was generated. In the table, the USLE for each slope category was solved. Table II.13.4 and the accompanying chart illustrate how sensitive soil loss in the catchment is to slope. The table and chart are interesting in that they explain the enormous variation seen in the spatially explicit model. When soils are completely exposed, the C-factor (no cover) can describe a 200-times increase in erosion over a forested

Table II.13.4 Sensitivity of the Spatially Explicit Model to the Variables Describing Slope (Location) as Compared with Those Describing Cover (Land Use)

Watershed Management Plan for Gander Lake and Its Catchment

Soil Loss (kg/ha)

USLE Factors				Cover Values[†]				
(n1...n5) Slopes	R	K*	LS	Forested 0.001	Burn 0.007	Regenerating 0.01	Clearcut 0.02	Gravel Pit 0.2
1	870	0.34	0.18	20	137	195	391	3,910
4	870	0.34	0.61	66	464	662	1,325	13,249
7	870	0.34	1.50	163	1,140	1,629	3,258	32,580
11	870	0.34	2.10	228	1,596	2,281	4,561	45,612
21	870	0.34	2.70	293	2,053	2,932	5,864	58,644

* Gander soil type (Orthic Humo-Ferric Podzol); 67% of the catchment is in this soil type.

† Actual range of cover values used in the spatially explicit model.

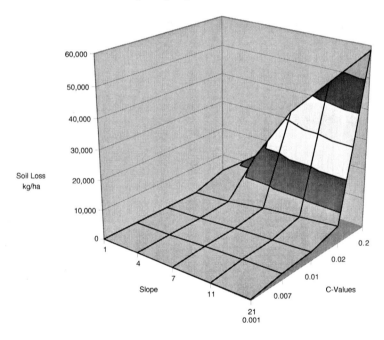

landscape. For the majority of the landscape types in the catchment, however, the different cover types explain a range of soil loss from 1 to 20 times. The slope factor consistently is responsible for a range of soil loss of 3 to 15 times. Thus, in this catchment, slope (location) is as important to consider as land use. When steep slopes are combined with very sensitive land uses, the combined effect is significant, as demonstrated by the peak in the chart of Table II.13.4.

One concern with the spatially explicit model is the lack to data to calibrate both the C-factor and, in particular, the D-ratio. In order to further evaluate the approach, a map result considering only the "inherent sensitivity" component of Eq. II.13.2 was prepared (see Figure II.13.4). The inherent sensitivity is a measure of landscape sensitivity irrespective of land use. It asks, which areas are more sensitive to any land-use change, by virtue

Chapter II.13: The effect of spatial location in land–water interactions

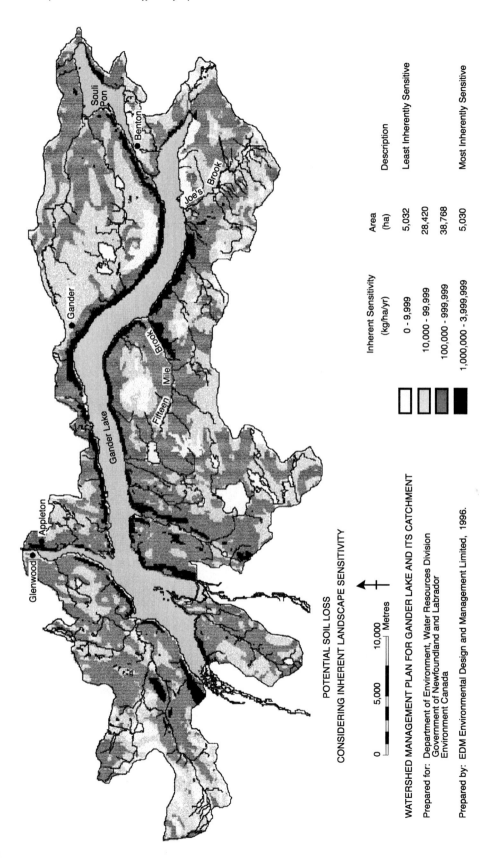

Figure II.13.4 Map of the Gander Lake catchment illustrating the inherent sensitivity of the catchment to soil loss. (From EDM, Environ. Design and Manage. Ltd., R.H. Loucks Oceanol. Ltd., and Jacques Whitford Environ. Ltd., Watershed management plan for Gander Lake and its catchment, Govt. of Newfoundland and Labrador together with Environ. Canada, St. Johns, Newfoundland, 1996.) (Color version available at www.gsd.harvard.edu/water-colors. Password: lentic-lotic.)

of their physical attributes? All the factors in the equation $A = RKLS$ are well established in the literature. The map result (Figure II.13.4) illustrates the outcome if the soil is left exposed across the catchment. The method is consistent with that used by Coote et al. (1992) in preparing the water erosion risk map of the Maritime Provinces of Canada. Figure II.13.4 is thus an excellent "thumb print" of the catchment, illustrating those areas where soil loss will be minimal and those areas where it may be extreme if subjected to similar land-use and management practices.

The usefulness of each approach was further considered by evaluating the effect of proposed future land-use changes in the Gander catchment. The proposed changes are as shown in Table II.13.5. Of note is a large forest clearcut planned on the southern shore of the lake, recreation development on the steep slopes between the town of Gander and Gander Lake, and cottage development on the lakeshore in cottage management areas. In a land-use only approach, the impact of these changes is identical, irrespective of location. Table II.13.5 illustrates the potential impact of these land uses using the export values as calibrated in the first model run.

If, however, the land uses are mapped and considered in the second modeling approach, a different picture arises. Figure II.13.5 is the mapped result of the spatially explicit model. C-values for both cottage development and the recreation area were assigned as "development." Former clearcut areas were recategorized as "regenerating." The spatially explicit model predicts the planned large clear to have less impact on the lake than that predicted by the land-use model, even though the clearcut is immediately adjacent to the lake. The clearcut is proposed for one of the more resilient areas of the landscape based on the inherent sensitivity map (Figure II.13.4). On the other hand, the land clearing for recreation on the sensitive steep slopes between the town and the lake, as well as the very sensitive lands proposed for cottage development may result in a 46% increase in sediment loading, even though the area affected is less than 100 ha.

Discussion

Both landscape models of land–water interaction were calibrated to the Gander Lake catchment. Both approaches describe very different landscape areas as the source of suspended solids and phosphorous to the lake.

A landscape modeling approach where land–water interactions are described by assigning export values to land uses (the land-use approach) has been adequate to calibrate hydrodynamic lake models (e.g., Dillon et al., 1986); however, the same landscape modeling approach may be inadequate for watershed planning policy. This is largely because a catchment total is all that is required for water quality model calibration. The precise location of pollutant input is not critical, and assuming a relatively uniform terrain across the catchment, averaging will tend to even out the site-specific landscape differences.

Planning policy, on the other hand, is a very site-specific task. Applying zoning or other mechanisms to remove land from development or restrict land use has important economic consequences, both for individual landowners and for the regional economy. In addition, policy that is too restrictive can lead to low-development densities, which can increase other environmental concerns such as landscape fragmentation and development sprawl. For these reasons, planners must be able to clearly defend the specific restrictions they apply to watersheds. Evaluation of proposed future land-use changes and development proposals is also a very site-specific task. The question that must be answered by the planner is whether to permit a specific land use in a specific location (for examples, see Chapters II.10, II.12, II.14, and II.15). Planners are not able to rely upon averaging

Table II.13.5 Impact of Proposed Future Land-Use Changes as Predicted by the Land-Use Model Watershed Management Plan for Gander Lake and Its Catchment

Land Use	E-Values kg/ha	Area 1995 (ha)	TSS Total, 1995, kg/year each land use	Area 2005 (ha)	TSS Total, 2005, kg/year each land use	Change 2005–1995 kg/year	% Change Total 1995–2005
Gravel pits	1,200	154	184,800	154	184,800	0	0.00
Forest	35	49,838	1,744,330	45,785	1,602,475	−141,855	−8.13
Blowdown	200	3,693	738,600	3,693	738,600	0	0.00
Bogs	0	7,945	0	7,945	0	0	0.00
Burn	200	1,510	302,000	1,510	302,000	0	0.00
Unvegetated/bedrock	100	3,193	319,300	3,193	319,300	0	0.00
Land cleared for development	600	534	320,400	534	320,400	0	0.00
Recent clearcut	400	5,424	2,169,600	3,944	1,577,600	−592,000	−27.29
Development	900	438	394,200	438	394,200	0	0.00
Brush/regenerating	200	4,521	904,200	9,945	1,989,000	1,084,800	119.97
Recreation and cottage clearing	900	N/A		109	98,100	98,100	9810000.00
Total TSS for catchment 1995, kg/year			7,077,430				
Total TSS for catchment 2005, kg/year					7,526,475		
Total increase for catchment, kg/year						449,045	
Total % increase							6.34%

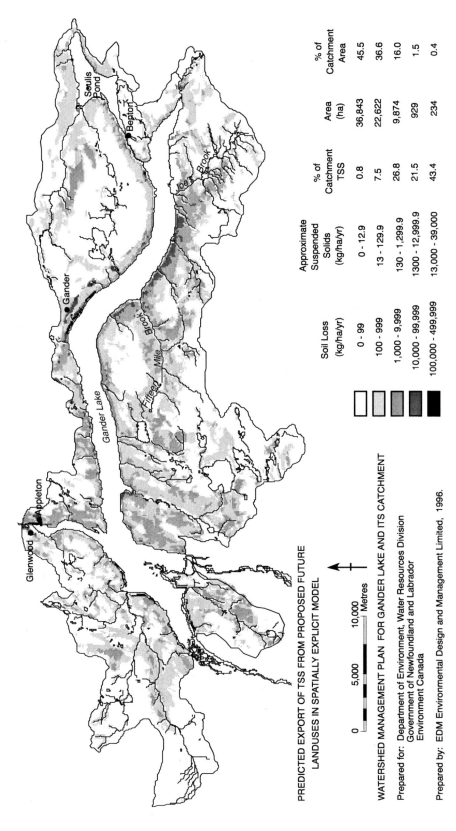

Figure II.13.5 Map of the Gander Lake catchment illustrating the contribution of suspended solids to the lake from proposed future land uses, as predicted by the spatially explicit model (Eq. II.13.3). (From EDM, Environ. Design and Manage. Ltd., R.H. Loucks Oceanol. Ltd., and Jacques Whitford Environ. Ltd., Watershed management plan for Gander Lake and its catchment, Govt. of Newfoundland and Labrador together with Environ. Canada, St. Johns, Newfoundland, 1996.) (Color version available at www.gsd.harvard.edu/watercolors. Password: lentic-lotic.)

across the landscape; they must defend site-specific decisions. Planners are also under enormous pressure to ensure maximum use of the watershed, both for economic and environmental reasons.

Insufficient data on C-factors and D-ratios are available in the Gander Lake catchment to fully embrace the spatially explicit model. Of particular concern is the lack of C-factors for urban landscape types. The model does offer enormous promise, especially for largely undeveloped catchments. The model results, which suggest highly variable soil loss across the landscape supports anecdotal evidence in the construction industry. Construction managers of large road projects frequently claim that there are vast areas of the landscape where the erosion and sediment control measures are simply not necessary — nothing is eroding anyway, and then there are other areas where no matter what they do, the earth just keeps falling away and they exceed their construction permits (e.g., Rushton, 1996). The results of the spatially explicit model are also supported by data reported by Snell (1985) and discussion in a report by Wall et al. (1978), both indicating that a very small portion of the fields in an Ontario agricultural landscape were responsible for the majority of the stream sediment.

Whereas uncertainty surrounds the C-factors and D-ratios in the spatially explicit model, much less uncertainty surrounds the RKL- and S-factors. The inherent sensitivity of the landscape ($A = RKLS$) may therefore be considered a valid description of the potential of the landscape to erode. For the Gander Lake catchment, the inherent sensitivity was reclassified (see Figure II.13.6) and adopted as the zoning plan. The zoning plan preserves 7.6% of the catchment (red areas) and places severe development restrictions on an additional 9.1% (blue areas) that are very sensitive to any land-use change. Note that many of the restricted areas are at a distance from the lake edge — a very different spatial configuration than the buffer setback restriction in previous policy.

If soil is the primary source of phosphorous from the catchment, then the spatially explicit model of land–water interaction must be considered very seriously as an appropriate model of the spatial distribution of the land-use sources of phosphorous and suspended solids in lake catchments. This modeling approach in support of planning has been adopted for Newfoundland and Labrador and to date has been defensible to land-use proponents.

Acknowledgments

This study was funded under the Canada–Newfoundland Agreement Respecting Water Resource Management. The entire project, including water quality sampling, lake modeling, land–water interaction modeling, and recommendations for management, titled Watershed Management Plan for Gander Lake and Its Catchment, was published as a government document by the Government of Newfoundland and Labrador together with Environment Canada in 1996. Haseen Khan, in the Water Resources Division of the Government of Newfoundland and Labrador, and Joe Arbour at Environment Canada were both active and supportive throughout the project. The author acknowledges Dr. Ron Loucks for his careful lake modeling and critical input to the land–water interaction approach. A special thanks to all of the team at EDM: Jeff Pinhey for the water balance and input, advice, and support; Nancy Griffiths for her careful research; and Paula Dykstra, who ran the models in the geographic information system (GIS). A very special thanks to Professor Michael Binford, now at the University of Florida, who first applied the spatially explicit approach in Acadia National Park, and who continues to find time and energy to support former students as they apply his teachings.

596 Handbook of water sensitive planning and design

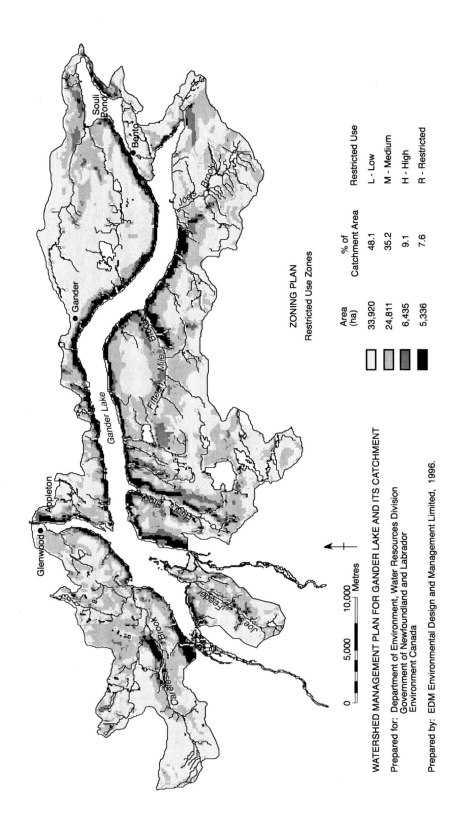

Figure II.13.6 Zoning plan for the Gander Lake catchment illustrating areas restricted to development. Note the difference between this map and a standard lakeshore buffer approach. (From EDM, Environ. Design and Manage. Ltd., R.H. Loucks Oceanol. Ltd., and Jacques Whitford Environ. Ltd., Watershed management plan for Gander Lake and its catchment, Govt. of Newfoundland and Labrador together with Environ. Canada, St. Johns, Newfoundland, 1996.) (Color version available at www.gsd.harvard.edu/watercolors. Password: lentic-lotic.)

Literature cited

Arbour, J., Personal communication, Water Resour. Div., Conservation Branch, Environ. Canada, 1996.
Binford, M., Spatially explicit models of land–water interactions for evaluating development proposals, unpublished paper, 1989.
Binford, M., Personal communication, Dept. of Landscape Architect., Grad. Sch. of Design, Harvard Univ., 1995.
Borman, H.F. and Likens, G.E., *Pattern and Process in a Forested Ecosystem*, Springer-Verlag, New York, 1979.
Cooper, J.R. et al., Riparian areas as filters for agricultural sediment, *Soil Sci. Soc. of Amer. J.*, 51, 416, 1987.
Coote, D.R. et al., *Water Erosion Risk: Maritime Provinces*, Agriculture Canada, Ottawa, ON, 1992.
The Dept. of Environ. and Lands Act (DELA), Newfoundland and Labrador, 1989.
DELA, Newfoundland and Labrador, 1991.
Dillon, P.J. et al., Lakeshore capacity study, trophic status, Res. and Special Proj. Branch, ON Min. of Environ., Ontario, 1986.
Dissmeyer, G.E. and Foster, G.R., Estimating the cover management factor (C) in the universal soil loss equation for forest conditions, *J. Soil and Water Conserv.*, 36, 235, 1981.
Dunne, T. and Leopold, L., *Water in Environmental Planning*, W.H. Freeman and Co., New York, 1978.
EDM, Environ. Design and Manage. Ltd., Land use plan for the Big Cove First Nation, 2000.
EDM, Environ. Design and Manage. Ltd., The Western Common land use plan, 1999.
EDM, Environ. Design and Manage. Ltd., R.H. Loucks Oceanol. Ltd., and Jacques Whitford Environ. Ltd., Watershed management plan for Gander Lake and its catchment, Govt. of Newfoundland and Labrador together with Environ. Canada, St. Johns, Newfoundland, 1996.
Hickman, R.E., *Loads of Suspended Sediment and Nutrients for Local Nonpoint Sources to the Tidal Potomac River and Estuary, Maryland and Virginia, 1979–81 Water Years*, U.S. Geol. Surv., 1987.
Loucks, R., The First Lake total watershed management project, phase I: neighborhood stewardship, 1993.
Leopold, L., *A View of the River*, Harvard Univ. Press, Cambridge, MA, 1994.
Lowrance, R., McIntyre, S., and Lance, C., Erosion and deposition in a field/forest system estimated using Cesium-137 activity, *J. Soil and Water Conserv.*, 43, 195, 1988.
Mellerowicz, K.T. et al., Soil conservation planning at the watershed level using the universal soil loss equation with GIS and microcomputer technologies: A case study, *J. Soil and Water Conserva.*, 49, March–Apr., 194, 1994.
Rees, H.W., Personal communication, Land Resource Unit Head, Centre for Land and Biol. Resour. Res., Res. Branch, Agricul. Canada, 1996.
Riley, M.J., MinLake: External memorandum no. 13, Dynamic Lake Water Quality Simulation Program, St. Anthony Falls Hydraulic Lab., Univ. of MN, Minneapolis, 1988.
Rushton, D., Personal communication, Nova Scotia Dept. of Transport. and Commun., 1996.
Snell, E.A., Regional targeting of potential soil erosion and nonpoint-source sediment loading, *J. Soil and Water Conserv.*, 40, Nov.–Dec., 520, 1985.
Toth, R.E., Hydrologic and riparian systems: The foundation network for landscape planning, Intl. Conf. on Landscape Planning, Univ. of Hannover, Germany, 1990.
Transport Canada Gander Intl. Airport, Stormwater quality monitoring results, Transport Canada, Airports Group, 1995.
U.S. EPA, *A Hydrodynamic and Water Quality Model*, WASP4 (water analysis simulation program), Environ. Res. Lab., Athens, GA, 1988.
Vaughan Engineering, Ltd., Shubenacadie Lakes planning/pollution control study, Halifax, NS, 1993.
Wall, G.J., Dickinson, W.T., and Greuel, J., Rainfall erosion indices for Canada east of the Rocky Mountains, *Canad. J. Sci.* 63, May, 271, 1983.
Wall, G.J., van Vliet, L.J.P., and Dickinson, W.T., Contribution of sediments to the Great Lakes from agricultural activities in Ontario, Intl. Joint Comm., Windsor, ON, 1978.

Wells, R.E. and Heringa, P.K., Soil survey of the Gander-Gambo Area, Newfoundland, Report no. 1: Newfoundland soil survey, Res. Branch, Canada Dept. of Agricul. and Agricultural and Rural Develop. Act, 1972.

Wischmeier, W.H. and Smith, D.D., *Predicting Rainfall Erosion Losses: A Guide to Conservation Planning*, Agriculture Handbook no. 537, U.S. Dept. Agricul., Washington, DC, 1978.

Response

Linking land use to landscapes for water quality protection

In this chapter, Margot Cantwell illuminates what has been a somewhat odd dichotomy in water quality land-use planning. On the one hand, many models have relied on averaged estimates from landscape and land-use classifications in their predictions of development threats to lakes; on the other hand, planning policy is rooted in a strong site-specific tradition. Cantwell's comparative analysis provides a much needed assessment of the strengths and shortcomings of adopting these differing approaches for water sensitive planning. An important take-away message from this chapter is demonstration of the utility of incorporating science into the process of land-use decision making for lake protection. In this respect and similar to John Felkner and Michael Binford's work in Chapter II.10, and Robert France and Philip Craul's work in Chapter I.7, Cantwell's study demonstrates the need to pay careful attention to soils when engaging in water sensitive planning and design; in other words, literally "ground-truthing" projects.

The GIS-based process Cantwell uses is very much in the spirit of the "bottom-up" method championed by Jeffrey Schloss in Chapter II.12. Cantwell ably demonstrates the power of GIS mapping when combined with landscape rate equation models. All too frequently, GIS analysis is used to produce nothing more than attractive, colorful maps that do no more than inform about the blatantly obvious. What this particular chapter demonstrates is the evolution of functional GIS map-modeling. No longer is GIS mapping regarded as the end in itself; instead is it used as the tool it was always intended for, as a means to spatially link landscape processes to environmental predictions. This is a very important lesson and one that Cantwell uses to support the contention raised by Neil Hutchinson in Chapter II.17. That is the truly important questions involved in water sensitive planning should be more concerned with where development resides on the land than perhaps simply how much of it may be there in the first place. In a sense, this is the parallel argument to that put forward by Larry Coffman in Chapter I.5 in terms of considering the benefits of using spatially explicit low-impact development designs over averaged percentage imperviousness to effectively manage stormwater runoff.

chapter II.14

Spatial investigation of applying Ontario's timber management guidelines: GIS analysis for riparian areas of concern

Robert France, John S. Felkner, Michael Flaxman, and Robert Rempel

Abstract

Riparian zones are not only ecotones situated between adjacent ecological systems, they are also the areas on the landscape where conflicting interests about natural resource management abut. For many regions of the boreal forest in Ontario, the timber industry is the major employer and generator of revenue, whereas sport fishing is the major commercial recreational pursuit. In order to ensure protection of the latter, the former is regulated by precluding logging from occurring within near-shore riparian areas. Despite site-specific studies on the extent of ecotonal coupling between boreal forests and water bodies as affected by clearcutting, there is little understanding about what the effects of establishing such riparian "areas of concern" might be for the timber industry on a landscape scale. Despite forestry officials publicly stating fears that as much as 30% of the terrestrial land area would be included in such buffer strips, our GIS analysis for a region of northwestern Ontario showed this amount to be closer to 10% of the surface area. This in turn was estimated by our analysis to correspond to a reduction of about 8% in the amount of merchantable timber that could be harvested.

> In many parts of Ontario merchantable timber occurs in areas adjacent to aquatic environments (shoreline areas). Many of these aquatic environments, including wetlands, lakes and streams, provide or can potentially provide valuable habitat for fish.
>
> — Timber management guidelines,
> Ontario Ministry of Natural Resources (1988)

Introduction

Natural resource management may often involve tradeoffs made between the economic viability of resource extraction industries and the ecological integrity of the environment from which the resources are obtained. Perhaps nowhere is this dichotomy felt more strongly than in the regulation of timber harvesting near water bodies. This is natural because there is widespread recognition that the land–water interface is of critical importance for environmentally sensitive planning (Chapters II.1 to II.4). Specifically, after reviewing the copious scientific information demonstrating the potential for deleterious effects of clearcutting upon aquatic biota, land-use planners have agreed that there is a need to establish riparian buffer strips around rivers and lakes in order to protect them from forestry operations (Brewin and Monita, 1998). This is typically the extent of the agreement between ecologists, planners, and foresters. Contention arises in trying to determine the precise size, or width, that is deemed necessary to enable those buffer strips to perform their job effectively.

Environmentalists argue that when it comes to protective buffer strips, bigger is better, especially in the absence of absolutely unequivocal scientific data about the possible long-term effects of near-shore logging. Forestry companies counter by questioning the implicit and often unsupported assumptions that may characterize the environmentalists' platform, at the same time as decrying the potential economic losses due to timber reserves being "locked up" in such buffer strips. Arguments in both cases are substantially weakened by lack of actual landscape data upon which to engage in an intelligent dialectic. The purpose of the present study was to provide a preliminary spatial analysis of forest patterns in northwestern Ontario as an aid toward facilitating more meaningful discussions for future land-use planning decisions in this region of the boreal forest.

At present, an alarming disjunction exists between the scale at which scientific investigations of the effects of riparian forestry have been conducted and the scale at which industrial forestry operations are being implemented. For example, although studies have explored the site-specific consequences of riparian deforestation on litter production, erosion, wind, solar energy, as well as nutrient and food resources (Chapter II.16; France, 1997a; France and Peters, 1995; France et al., 1998; Steedman et al., 1998; Steedman and France, 2000), the wider consequences of what such effects may be at the landscape scale are not well known. Forestry officials, bemoaning the, in their minds, overly stringent guidelines for riparian areas of concern, have provided unsupported statistics to the press (e.g., "The Communications, Energy and Paperworks Union argue that more than 30% of the land in Northern Ontario is protected when buffer zones around lakes, rivers and streams are included in the calculation [of protected land]" (Mackie, 1998)).

The present research was directed toward one simple goal — to examine the validity of the claim made by the forestry industry that 30% of the land area in northern Ontario is contained within the riparian areas of concern under the present timber management guidelines (OMNR, 1988).

Ontario forestry

Ontario contains 1.5% of the world's forests, a major portion of which is classified as being boreal and is, therefore, representative of the largest forest type on the planet. The forestry industry in Ontario contributes a considerable amount of revenue to the economy and employs thousands. Simultaneously, in certain regions of the boreal forest, sport fishing represents the second-largest resource industry. Because of a recent history of contentious (and even litigious) debate between environmental groups and the timber industry and government resource managers, issues of land-use planning and forest and fisheries management have come into the spotlight. In particular, the response of forests to natural and anthropogenic processes and their sustainable management have both received much needed attention (Perera et al., 2000; OMNR, 2000).

Special focus has been directed toward understanding the possible interactions between riparian timber removal and the health of aquatic systems (e.g., France, 1997b; Steedman, 2000; Steedman et al., 2001). Ontario's recommended guidelines (OMNR, 1988) for riparian buffer strip widths (termed "areas of concern") to protect lakes containing commercially important lake trout are based on the upper values suggested in relation to surface slope and inferred transport distance of eroded soil (discussed in Chapter II.16). These widths are 30 m for slopes of 1 to 15%, 50 m for slopes of 16 to 30%, 70 m for slopes of 31 to 45%, and 90 m for slopes of 46 to 60%.

Study area

The study area is located in northwestern Ontario, 150 km east of the Experimental Lakes Area and near the OMNR's Coldwater Lakes Field Station (Steedman et al., 2001; Figure II.14.1). The area is in the Lake of the Woods Plains Ecoregion and is transitional between the northern boreal and Great Lakes/St. Lawrence forests. Fire disturbance plays an important role in structuring vegetation communities here. The region is typified by small headwater lakes situated in catchments characterized by shallow organic and coarse-textured mineral soils with abundant bedrock outcrops and topographic relief generally not more than about 60 m. Riparian forests are predominantly composed of black spruce and jack pine, with trembling aspen and paper birch prevalent in backshore areas (France et al., 1998). The present study area is within a region previously investigated through GIS analysis for understanding the response of forest vertebrates to landscape-level patterns in forest disturbance (Rempel et al., 1997; Voigt et al., 2000).

Methods

Given existing riparian buffer guidelines, as well as forestry industry claims regarding the proportion of the total land area (or the total forested area) existing under current buffer guidelines, the methodology described here had as its goal the physical measurement of the approximate proportion of total land area protected by current buffer guidelines, and the obtaining of the approximate proportion of that area actually wooded, using current land-cover data in a 315-km^2 study area in northern Ontario (see Figure II.14.1). To achieve this goal, land-cover, topographic, and hydrologic data sets were geo-rectified and overlaid in a geographic information system (GIS) and a modeling approach was devised.

The basic structure of the approach was initially conceptualized as follows:

1. Measure the slope of all land bordering water bodies (streams, lakes, and wetlands).
2. Use a GIS buffering function to accurately buffer all water bodies according to existing riparian timber buffering guidelines to the distance specified by those guidelines as a function of slope, thereby obtaining the total protected buffered area for the study site.
3. Examine both the composition (hardwood/softwood) and the merchantability status (regular production, reserve production, age status, etc.) of timber stands on those riparian buffer areas to estimate the percentage of merchantable timber in the total protected buffer areas.

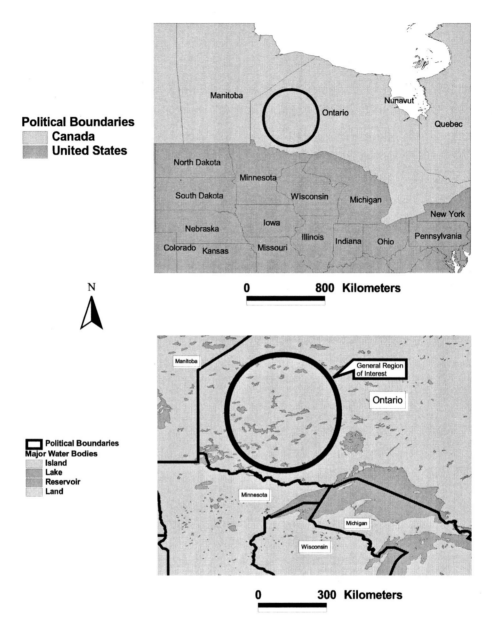

Figure II.14.1 Context map showing location of northwestern Ontario and region of study focus.

Once these goals were achieved, further analysis could be performed to derive the proportions and percentages of softwoods, hardwoods, and merchantable stand types in relation to various slope and water body edge categories. The final modeling approach used in this analysis is depicted in Figure II.14.2. Figure II.14.3 displays a larger (595.6-km^2) context region, showing lakes, streams, and wetlands. The smaller (315.0-km^2) focus study area is displayed in Figure II.14.4, and appears to be similar in terms of shoreline density and land-cover composition to surrounding areas

For the analysis, topographic data, and land features (lakes, streams, wetlands, and their boundaries) were obtained from the Ontario Digital Topographic Database (DTDB) as part of the Ontario Basic Mapping (OBM) Program for the Province of Ontario, constructed by

Chapter II.14: Spatial investigation of applying Ontario's timber management guidelines

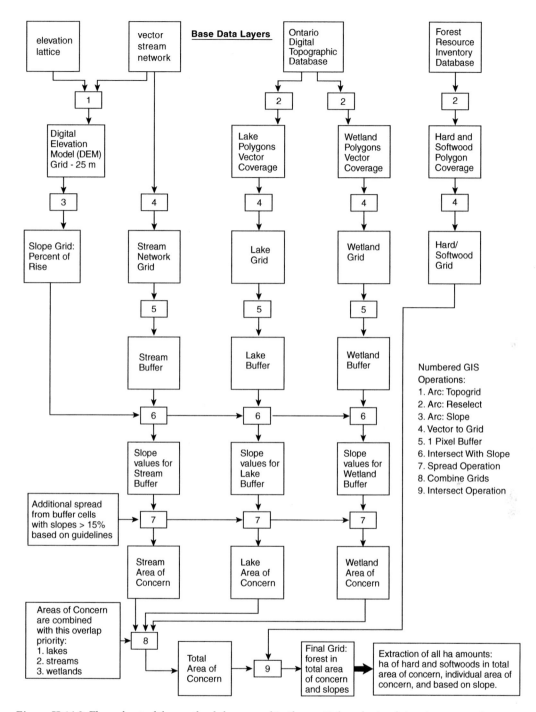

Figure II.14.2 Flow chart of the methodology used in the spatial analysis of riparian areas of concern.

the Provincial Mapping Office of the Natural Resources Information Management Branch, under the Ontario Ministry of Natural Resources (OBM, 1996). Geographic data on forest cover, including the timber merchantability condition and the hardwood/softwood species composition of individual tree stands, were obtained from the Forest Resource Inventory (FRI) (R. Rempel, unpubl. data).

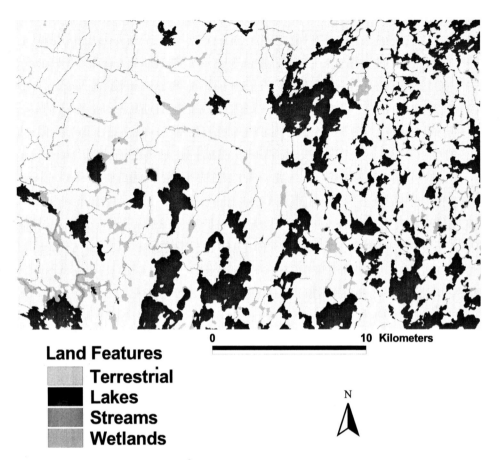

Figure II.14.3 Map of larger (595.6 km^2) study area showing landform and water body designations.

Digital data from both sources were available georectified to a Universal Transverse Mercator (UTM) projection system, and all data were geo-registered in a GIS. This data allowed the calculation of basic surface-area proportions and percentages including those of water to land; lakes versus wetlands versus streams versus total area; total land–water interface amounts; amounts of the total buffer areas versus total study area; and breakdowns for percentages of varying slope amounts on land–water interface areas on the 315-km^2 study area, as discussed next.

Both the OMB and the FRI data sources provided topographic, waterbody and timber stand location and composition data in GIS vector format. The OMB topographic data were provided in an irregular lattice of mass points format, with point elevation values obtained at approximately 100-m intervals, and then additional elevation points on the crowns of hills and bottoms of depressions, as well as along the edges of water bodies.

Because the derivation of topographic slope values was crucial for the model, the first step in the analysis was the creation of an accurate digital elevation model (DEM). The DEM was created using the "topogrid" function in ArcInfo software that is based on the ANUDEM program developed by Michael Hutchinson (Hutchinson, 1988, 1989, 1993, 1996; Hutchinson and Dowling, 1991). The method uses an iterative finite-difference interpolation technique which is essentially a discretized, thin-plate spline approach (Wahba, 1990), where the roughness penalty has been modified to allow the fitted DEM to follow abrupt changes in terrain, such as streams and ridges. The procedure is specifically designed to consider hydrologic drainage features, which are the primary erosive

Chapter II.14: Spatial investigation of applying Ontario's timber management guidelines 607

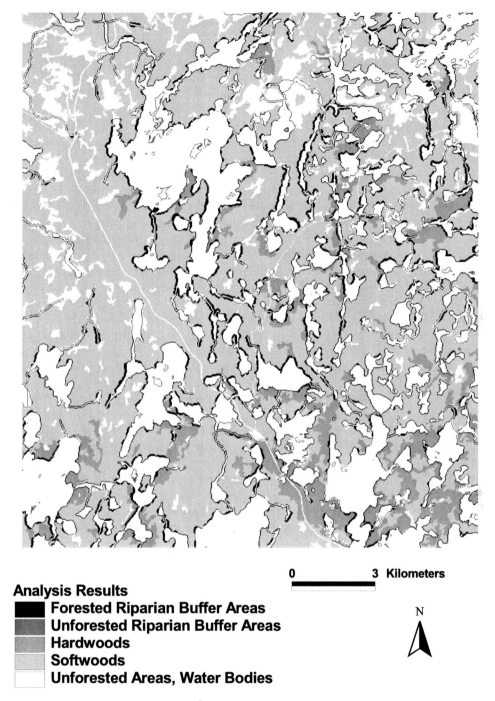

Figure II.14.4 Map of smaller (315.0 km²) study area showing landform and water body designations, forest types, and riparian areas of concern based on slope.

and shaping force on a landscape. Consequently, vector stream network data from OMB were used as an input into the DEM creation program. Furthermore, the program assumes that all unidentified sinks are errors, because sinks are generally rare in natural landscapes (Goodchild and Mark, 1987). The DEMs were generated with raster grid cells of 25 m, because this was deemed an appropriate level of spatial precision given the lattice elevation point spacing.

Because the analysis depended on the summing of areas of varying slope values all water body and forest stand vector data were also converted to 25-m raster grid layers to facilitate comparison with the grid of slope values derived from the DEM. Stream networks, lakes, and wetland polygons were extracted from the OMB data and converted to separate raster grids with 25-m cell sizes.

Tree stand polygons of standing merchantable production timber were selected from the FRI data and their respective hardwood/softwood compositions were considered for categorization as either predominantly hardwood or predominantly softwood. These stand polygons, once thus classified, were gridded at the 25-m cell resolution.

Next, it was necessary to determine the approximate slope at the edges of all riparian bodies, as the OMB buffer guidelines fluctuate depending on slope. To accomplish this, a one-pixel buffer was created around water bodies in the stream, lake, and wetland grids, respectively. These buffers were independently combined with the slope grid derived from the DEM to determine the slope values for each buffer pixel (effectively the slope at the water's edge). For every pixel with a derived slope value greater than 15%, a further spread function was performed from those high slope cells according to the specified guidelines (see above) for each of the three buffer grids, respectively. This final spread function was performed as a linear distance, according to the guidelines (30 m, 50 m, 70 m, or 90 m) *across* the DEM — that is, considering topography rather than "as the crow flies" distance (although with such small spread distances, no spread was greater than three or four pixels, and thus the topographic data were unlikely to greatly influence which cells were included in the spread). This spread provided the appropriate estimated buffer areas of concern for streams, lakes, and wetlands, respectively. To obtain the total buffer widths for the study area, the buffers for lakes, streams, and wetlands were derived in separate steps and then were combined into a single grid.

The final step was the intersecting of the hardwood/softwood grid for the standing merchantable production timber stands with the total areas of concern grid. The output from this intersection (Figures II.14.4 and II.14.5) allowed the calculation of the total area and percentage of merchantable hardwoods and softwoods in official riparian buffer areas, with specific breakdowns for softwoods and hardwoods.

The accuracy of this analysis depends directly, it should be noted, on the accuracy of the input data layers. All input data sources — the elevation lattice points and the land-cover and forest data — were obtained by Canadian government environmental management agencies in accordance with established data collection procedures. Nonetheless, potential error in the input data sources could affect the final analysis. This degree of error in the final analysis is dependent directly upon the following:

1. The degree of accuracy of the vectorized land-cover data — lake, stream, wetland, and forest stand polygons
2. The degree of accuracy of the elevation mass points
3. The degree of accuracy of the DEM interpolation method

Furthermore, accuracy of final measurements is certainly reduced by the process of converting the original vector data into 25-m raster cells, and then measuring riparian buffer widths as spreads in increments of 25-m raster cells, because of the generalization of surface area that occurs with pixelization. Future research will attempt to overcome this potential measurement error by deriving buffer widths through uniform vector spreads.

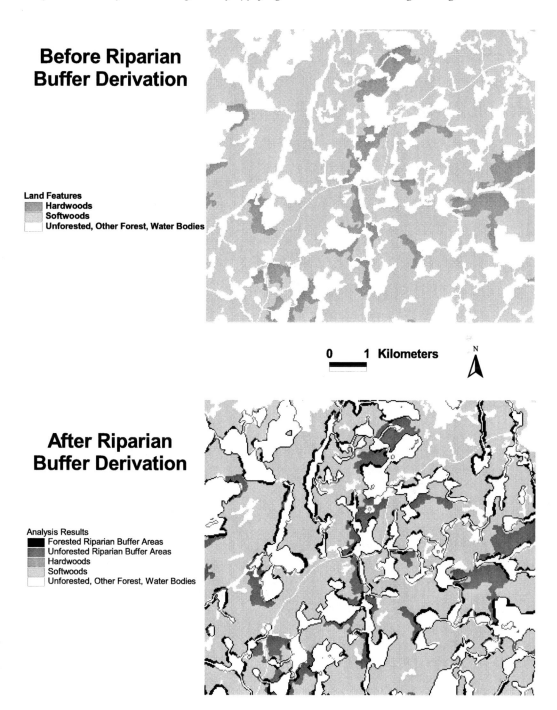

Figure II.14.5 Close-up of study area showing application of riparian areas of concern in relation to Ontario's Timber Management Guidelines.

Results and discussion

In the larger 595.6-km^2 context area (Figure II.14.3), 76.7% of the total surface area is land and 23.3% is collectively occupied by water bodies (Figure II.14.6). Lakes comprise 17.8% of the area, with streams and wetlands respectively subsuming 1.8 and 3.7% of the total

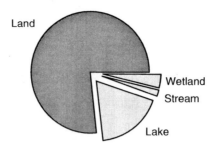

Figure II.14.6 Proportion of land and water in the study region of northwestern Ontario.

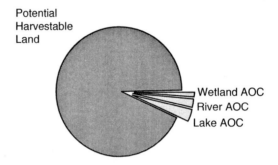

Figure II.14.7 Proportion of land occurring within the riparian areas of concern for the study region of northwestern Ontario. AOC = areas of concern.

area. Within the study region, there are a total of 32.3 km of land–water interface: 15.4 km for lake-riparian contact, 10.5 km for stream-riparian contact, and 6.4 km for wetland-riparian contact. Spatially, for the terrestrial surface area, only 6.7% of the total landscape is contained within the riparian areas of concern (Figure II.14.7), half this amount being associated with lake riparian zones. Reciprocally, the amount of land that is potentially available for timber extraction is 93.3% of the surface area.

Clearly, the forestry industry's contention that "30 per cent of the land in Northern Ontario is protected when buffer zones around lakes, rivers and streams are included in the calculation," is untenable for this characteristic subregion study area of northwestern Ontario. Such a large value occurs only if the proportion of the landscape covered by water bodies is added into the calculation. For the riparian areas of concern, from 97 to 100% (depending on the type of water body) of the riparian land has surface slopes of less than 15%, meaning that the minimum riparian buffer strip widths of 30 m were almost exclusively applied in this analysis.

It is, of course, not really the land that is of most interest to the forestry industry, but rather that which grows atop it. For the 315.0-km^2 area selected for detailed analysis, 26.2% of the total area is occupied by water bodies, and 11.9% of the total area is located within the riparian areas of concern. Of the remaining land surface area, 7.3% does not contain trees of sufficient merchantability, with the majority of this (4.6%) occurring within upland areas unassociated with riparian zones (Figure II.14.8). For the land that does contain merchantable trees (both upland and riparian), the majority of these (87%) are softwoods.

Our analysis for this particular region of northwestern Ontario found that only 8.3% of the total realized merchantable timber reserve was collectively situated within the designated riparian buffer areas of concern: 6.1% for lakes (comprised of 80% softwoods), 1.4% for streams (comprised of 86% softwoods), and 0.9% for wetlands (comprised of 88%

Chapter II.14: Spatial investigation of applying Ontario's timber management guidelines 611

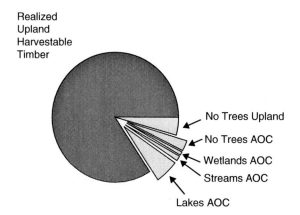

Figure II.14.8 Proportion of merchantable timber reserves occurring within the riparian areas of concern in the study region. AOC = areas of concern.

softwoods). Again, this is far less than the 30% reduction in timber harvest (inferred from an assumed 1:1 correspondence between land-surface area and amount of merchantable timber) supposed by some industry officials to prevail following continued adoption of the riparian buffer widths specified in the timber management guidelines. Consequently, it remains to be proven that "in some cases, posited by the OMNR" (1988).

The present analysis was designed to provide a first-stage assessment of potential economic tradeoffs between multiple-use resourcism that might exist in this region of Ontario. One obvious shortcoming to our present analysis pertains to appraisal of merchantable timber reserves based solely on GIS designations of forest typology classifications with no consideration given to estimations of actual timber yields. We plan next to undertake a more detailed GIS-economic analysis of how projected timber yields might be influenced through compliance to the timber management guidelines, as well as to how alternate scenarios of increased or reduced stringency in prescribing buffer strip widths might affect financial returns.

Acknowledgments

This study was supported by the Milton Fund and the GSD Dean's Fund, both of Harvard University, and by the Ontario Ministry of Natural Resources. The authors thank Paul Cote for his helpful suggestions throughout the study.

Literature cited

Brewin, M.K. and Monita, D.M.A., Eds., Forest-fish conference: Land management practices affecting aquatic ecosystems, Can. For. Serv. Info. Rep. NOR-X-356, 1998.

France, R.L., Land-water linkages: Influences of riparian deforestation on lake thermocline depth and possible consequences for cold stenotherms, *Can. J. Fish. Aquat. Sci.*, 54, 1299, 1997a.

France, R.L., Stable carbon and nitrogen isotopic evidence for ecotonal coupling between boreal forests and fishes, *Ecol. Freshw. Fish*, 6, 78, 1997b.

France, R.L. and Peters, R.H., Predictive model of the effects on lake metabolism of decreased airborne litterfall through riparian deforestation, *Conserv. Biol.*, 9, 1578, 1995.

France, R., Peters, R., and McCabe, L., Spatial relationships among boreal riparian trees, litterfall and soil erosion potential with reference to buffer strip management and coldwater fisheries, *Ann. Bot. Fenn.*, 35, 1, 1998.

Goodchild, M.F. and Mark, D.M., The fractal nature of geographic phenomena, *Annal. Assoc. Am. Geograph.*, 77, 265, 1987.

Hutchinson, M.F., Calculation of hydrologically sound digital elevation models, presented at the 3rd Intl. Symposium on Spatial Data Handling, Columbus, OH, 1988.

Hutchinson, M.F., A new procedure for gridding elevation and stream line data with automatic removal of spurious pits, *J. Hydrol.*, 106, 211, 1989.

Hutchinson, M.F., Development of a continent-wide DEM with applications to terrain and climate analysis, in *Environmental Modeling with GIS*, Oxford Univ. Press, New York, 1993.

Hutchinson, M.F., A locally adaptive approach to the interpolation of digital elevation models, presented at the 3rd Intl. Conf./Workshop on Integrating GIS and Environ. Modeling, Santa Fe, NM, 1996.

Hutchinson, M.F. and Dowling, T.I., A continental hydrological assessment of a new grid-based digital elevation model of Australia, *Hydrol. Proce.*, 5, 45, 1991.

Mackie, R., Loggers also care about the wilderness, says union leader, *The Globe and Mail National News*, p. A7B, Dec. 1, 1998.

OBM, *Ontario Digital Topographic Database: A Guide for Users*, ON Min. of Natur. Resourc., Toronto, 1996.

Ontario Min. of Natur. Resour. (OMNR), Timber management guidelines for the protection of fish habitat, Toronto, 1988.

OMNR, Forest sustainability beyond 2000 conf., www.forconfor2000.org, 2000.

Perera, A.H., Euler, D.L., and Thompson, I.D., Eds., *Ecology of a Managed Terrestrial Landscape. Patterns and Processes of Forest Landscapes in Ontario*, UBC Press, Vancouver, 2000.

Rempel, R.S. et al., Timber-management and natural-disturbance effects on moose habitat: Landscape evaluation, *J. Wild. Manage.*, 6, 517, 1997.

Steedman, R.J., Effects of experimental clearcut logging on water quality in three small boreal forest lake trout (*Salvelinus namaycush*) lakes, *Can. J. Fish. Aquat. Sci.*, 35, 190, 2000.

Steedman, R.J. and France, R.L., Origin and transport of aeolian sediment from new clearcuts into boreal lakes, northwestern Ontario, Canada, *Wat. Air Soil Pollut.*, 122, 139, 2000.

Steedman, R.J. et al., Effects of riparian deforestation on littoral water temperatures in small boreal forest lakes, *Boreal Environ. Res.*, 3, 161, 1998.

Steedman, R.J., Kushneriuk, R.S., and France, R.L., Initial response to catchment deforestation of small boreal forest lakes in northwestern Ontario, *Canada. Verh. Intl. Verein. Limnol.*, in press, 2001.

Voigt, D.R. et al., Forest vertebrate responses to landscape-level changes in Ontario, in Perera, A.H., Euler, D.L., and Thompson, I.D., Eds., *Ecology of a Managed Terrestrial Landscape. Patterns and Processes of Forest Landscapes in Ontario*, UBC Press, Vancouver, 2000.

Wahba, G., *Spline Models for Observational Data*, Soc. Independ. Appl. Math, Philadelphia, 1990.

Watson, D.F. and Philip, G.M., A refinement of inverse distance weighted interpolation, *Geo-Process*, 2, 315, 1985.

Response

Size matters

As Frank Mitchell outlined in Chapter II.1, shoreline buffers provide a host of beneficial values to lakes, wildlife, as well as humans, especially if planned on the scale of watersheds rather than on fragmented locals. The visualized display of spatial data through GIS analysis offers an extremely powerful tool to decision makers toward informed buffer strip management in a watershed context, as Jeffrey Schloss points out in Chapter II.12. Such approaches can be particularly useful when examining previously held beliefs founded solely on untested assumptions rather than on empirical observations.

The case study described by Robert France, John Felkner, Michael Flaxman, and Robert Rempel demonstrates the utility of GIS analysis in testing assumed beliefs involved in contentious environmental management. That the forest resource industry had voiced their beliefs so strongly and assuredly, and that these in turn had been picked up and circulated to a wide audience in the national press, provides an interesting lesson in the needed role of science in the rhetoric of environmental debates.

The scale of timber resource extraction in this region of the boreal forest is truly immense. Small differences in the size of protective riparian buffer strips could therefore have immense repercussions on the total amount of forest cleared and on the total amount of revenue gleamed. This chapter provides a first investigation of some very basic questions often required for effective land-use management and water sensitive planning: What is the area of water on the landscape? Are rivers, lakes, or wetlands predominant? What is the extent of land–water interface in the study region? The GIS approach used to answer these questions in this chapter is really pretty basic in scope. As the GIS analysis found, however, careful attention was needed toward examining the methods used in order to obtain as precise data as possible before these questions could be addressed. The lesson here for planning professionals is an important one — even the simplest questions may be difficult to answer in a defensible manner (especially if part of a contentious debate) given limitations in the available data.

For water sensitive planning, this study underscores the importance of scaling up site-specific results to a regional perspective larger than a single watershed. As France et al. mention, however, "it is of course, not really the land that is of most interest to the forestry industry, but rather that which grows atop it." In other words, what are the economic implications of these spatial results? Their findings, admittedly just a preliminary examination of the question, suggest that the losses of timber revenues may not be as significant as imagined by the logging industry. Further, much more precise, research is suggested by this analysis. As a first cut at the problem, however, this chapter demonstrates a horizon direction for land-use planners; namely, the intersection of scientific and economic data. The role of GIS in bridging between these two fields of resource management will become increasingly important as an avenue of study.

chapter II.15

Aquifer recharge management model: Evaluating the impacts of urban development on groundwater resources (Galilee, Israel)

Amir Mueller, Robert France, and Carl Steinitz

Abstract

Generating different alternatives for regional development plans is insufficient without a process for evaluating the cumulative impact of each alternative. Therefore, it is necessary to evaluate the impacts of each plan on the full spectrum of socioeconomic and environmental issues. In this chapter, we develop a model that focuses on groundwater as a keystone indicator for the cumulative impacts of urban development. The aquifer recharge management model (ARMM) uses a geographic information system (GIS) of overlays combined with surface features to evaluate the impacts of different land uses on the aquifer underlying a region of Galilee, Israel. The model uses imperviousness as a means for measuring the of impacts of urban development on aquifer recharge areas. The increase in the total area of impervious surfaces associated with urban development heightens the intensity of floods and increases pollutant concentration from point and nonpoint sources. Polluted urban runoff degrades the quality of surface water in streams, rivers, and lakes. Furthermore, it contributes to the pollution of groundwater when it is recharged into the aquifer.

Introduction

One of the challenges in regional and landscape planning is in adequately comparing the impacts of a wide range of feasible alternatives. Therefore, it is necessary to evaluate the impacts of each plan on the full spectrum of socioeconomic and environmental issues. The purpose of this study was to develop a tool that models the impacts of urbanization on aquifer recharge areas. The ARMM focuses on groundwater as a keystone indicator for the cumulative impacts of urban development.

The threat to groundwater resources is exacerbated by the increase in the amount of impervious surfaces associated with urban development (Chapter I.1). Increases in the intensity of floods and pollution episodes are directly related to increases in the total amount of impervious surfaces. Because water resources are easily degraded by pollution from anthropogenic sources, it is important to develop strategies and practices that integrate land-use decisions with a comprehensive water management plan (Chapters II.7, II.8, II.11, II.13, and II.17).

Degradation of groundwater quality

Degradation of groundwater occurs when pollutants are introduced into the aquifer from diffuse sources such as agricultural and urban runoff, leaching of pollutants from landfills and industry, and hydrological deficits that alter groundwater flow (Carman et al., 1997; Berg et al., 1999). In short, anthropogenic impacts on aquifers result in the overall degradation of groundwater quality and the capacity of the aquifer to function effectively (Goldenberg and Melloul, 1994). Inappropriate surface-water management may contribute to the accumulation of pollutants in the filtration and saturated zones of an aquifer (Goldenberg and Melloul, 1994; Kuylenstierna et al., 1997). Reducing or even eliminating water pollution as well as increasing long-term viable water resources requires altering current water-use management and land-use policies.

To understand how land-use policies may improve groundwater quality, and perhaps increase groundwater reserves, it is important to view groundwater as an integral part of the hydrologic cycle. When water precipitates over land, some of it is lost through evapotranspiration, some flows above ground as runoff in the form of rivers and streams, and the remainder infiltrates through the surface into the ground. When water reaches an impermeable layer such as clay or marl, it accumulates in the saturated zone. When the saturated layer is permeable enough to allow the storage and transport of significant amounts of water, an aquifer is formed (Gordon, 1984; The Conservation Foundation, 1987). Most aquifer recharge results from precipitation percolating through the ground surface, seepage from streams, and surface waters.

Existing land uses and land cover as well as the permeability of soils and overlying geology are important factors in identifying aquifer recharge areas. Recharge in undeveloped areas is critical since it tends to be relatively free of pollutants and can be easily monitored and controlled. It is important to note that not all land-use activities pose the same threat to groundwater resources. Threats depend on the inherent hazards associated with each activity, where the activity takes place, and the number of people it affects (Gordon, 1984; Jaffe and DiNovo, 1987; Carman et al., 1997; Berg et al., 1999).

Although groundwater is less frequently polluted, once it is polluted, it is much slower to recover from pollution episodes that would be considered mild for surface water in lakes and streams. Little dilution or attenuation of pollutants occur in groundwater due to the slow rate of lateral movement through the saturated zone. Furthermore, pollutants in the ground may persist for years, and even decades (Jaffe and DiNovo, 1987).

The duration, type, and intensity of pollution will determine the degree of risk to groundwater quality and quantity. Typically, groundwater pollutants are introduced from landfill, sewage treatment facilities, and agricultural and industrial runoff (Jaffe and DiNovo, 1987; Issar, 1993; Goldenberg and Melloul, 1994; Berg et al., 1999). Groundwater contamination also results from the recharge of untreated urban runoff into the aquifer (Carman et al., 1997). Potential contaminates come from vehicles, road oiling and salting, pet wastes, industrial activities, and erosion from construction sites. Pollutants include phosphates, sulfates, and nitrates from fossil fuels, fertilizers, and detergents. In addition, bacteria, pathogens, heavy metals, and organic and inorganic compounds are commonly found in urban runoff (Jones and Clark, 1987; The Conservation Foundation, 1987).

Imperviousness

Urbanization and its subsequent increase in impervious surfaces is affecting an increasing number of watersheds and aquifers throughout the world (Jones and Clark, 1987). The increase in impervious surfaces (roofs, streets, parking lots, etc.) has resulted in a significant increase in the amount of rainfall that is now being discharged into streams as surface runoff, therefore lost to aquifer recharge. Even in the absence of direct industrial and municipal runoff, water quality is adversely altered. Suspended sediments in an urban watershed tend to be an order of magnitude higher than in natural, vegetated watersheds. Increased sedimentation, especially of fine silts and clays, will increase the amount of impervious surfaces once it settles in catchments and debris basins, as well as stream flood plains (Schlosser, 1991; Jones and Clark, 1987).

The health of a watershed, and to a certain extent the aquifer recharge area, is directly related to the amount of urban development (Chapter I.3). Schueler and Claytor (1997) have developed a model by which the amount of impervious surfaces in a watershed may determine the quality of runoff and thus the quality of water recharged into the aquifer.

Imperviousness is a useful tool for measuring the impacts of land development on watersheds and aquifers. Furthermore, imperviousness is one of a few variables that can be easily quantified, managed, and even controlled throughout all stages of development (Chapters I.1, I.3, and I.5). Impervious surfaces collect and accumulate pollution, therefore, they become a source for pollutants during storms. A direct link exists between the amount of impervious surfaces in a watershed and levels of pollution (Watershed Protection Techniques, 1994).

In this model, imperviousness is used as an indicator for the impacts of urban development on aquifer recharge areas. It can be easily inferred from geology, mapped, and displayed in a series of colored maps that identify potential areas of aquifer recharge. Furthermore, imperviousness can be used as a measure of environmental health since it known to contribute to the overall degradation of urbanized watersheds and urban streams.

Model structure

Before any meaningful evaluation of future alternatives can take place, it is important to develop a framework that will allow for a comprehensive comparison of the impacts each alternative has on groundwater resources (Figure II.15.1). For the ARMM to be useful, it has to be able to evaluate the effectiveness of policy decisions in protecting aquifer recharge areas. The possible range of alternatives is classified into 10 strategies based on the "escape of tigers." Policy decisions are classified into the 5 levels of reform derived from the proceedings of a Symposium on the Integrity of Water held in 1975 by the U.S. EPA. The application of the two models to the ARMM is described next.

In "The escape of tigers: An ecological note" (Table II.15.1) Haddon (1970) describes 10 strategies that deal with different levels of intervention to prevent the harmful transfer of energy. These strategies describe different ways to modify the world in order to reduce or eliminate economic losses associated with the natural interaction of people and the environment. Strategies 1 to 3 focus on policy decisions; strategies 4 to 7 describe the range of design and technical solutions; and strategies 8 to 10 are a combination of technical solutions coupled with monitoring and warning mechanisms. In the ARMM, Haddon's tiger classification is applied to two forms of energy release: increased urban runoff and the pollution of groundwater.

The Federal Water Pollution Control Act of 1972 was designed to "restore, maintain and protect the integrity on the nation's water and water resources." As is the case with most forms of legislation, the meaning of the word "integrity" was not made clear in the language of the act. A symposium held by the EPA in 1975 was meant to clarify the issue. The symposium focused on two interrelated aspects of the concept of integrity: as a desirable characteristic of natural ecosystems and as a moral or cultural principle. Later, Regier and France (1988) classified the various perspectives of the symposium participants

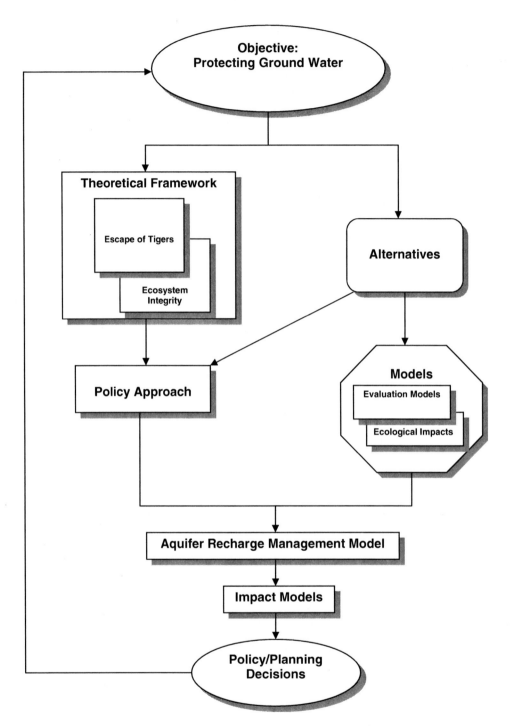

Figure II.15.1 Framework for examining alternative impacts on groundwater resources.

into five categories (Table II.15.2). Their classification is based on the degrees of reform that would achieve integrity and on the level of societal change or technology used as implied by the act's wording.

The 10 classes of the escape of tigers were coupled with the five categories of ecosystem integrity to yield the basis for reclassifying the alternatives and their impacts (Table II.15.3).

Table II.15.1 Escape from Tigers Management Options

1. Prevent the marshaling of the energy.
2. Reduce the amount of energy marshaled.
3. Prevent the release of the energy.
4. Modify the rate or spatial distribution of the release of energy.
5. Separate in space or time the energy being released.
6. Separation by interposition of a material barrier.
7. Modify appropriately the contact surface.
8. Strengthen the structure.
9. Detection and evaluation by generating a signal that a response is required.
10. Return to preevent conditions.

Source: Haddon (1970).

Table II.15.2 Strategies for Managing Ecosystem Integrity

1. Deep reform — protection of water as a moral obligation for its inherent worth
2. Partial reform — pragmatic, sectoral, step-wise societal change within a specified general direction
3. Incremental advances — recognition that integrity is a multidimensional concept that is not value-free, thus economics may sometimes dictate ecologics
4. Slowing the rate of retreat — undertaking inexpensive but visible initiatives to protect an image of concern

Source: Regier and France (1990).

Aquifer recharge management model

Many of the existing integrated GIS models require varying degrees of empiricism, functional relationships, and deterministic descriptions of hydraulic processes to accurately analyze the impacts of development. Furthermore, these models require lager amounts of data that are not always available without a major investment in site-specific, base-line studies (Table II.15.4). Simple models that are capable of simulating spatial processes and identifying the distribution of nonpoint sources of pollution at a watershed scale are scarce (but see Chapters II.10 and II.13).

In contrast to most GIS models, the ARMM, a raster GIS model, utilizes the grid functions in ArcView and Arc/Info and requires only three layer of input: a digital elevation model (DEM), geology, and high-altitude aerial photograph (Figure II.15.2). These data are reclassified to yield the basic building blocks of the model. Vector GIS coverage of topography and geology is converted into grids. This conversion allows for a more effective handling of the data by the ArcView grid module. The interactive GIS interface in ArcView allows for a systematic input of spatial data and for the graphic display of impacts.

Unlike traditional GIS models that require uniform data throughout the study area, the ARMM can operate on data with high levels of variability taking advantage of the reclassify option in ArcView. This technique overcomes data limitations, which are based on high-altitude aerial photography that tends to be high quality in the center and distorted along the edges. Furthermore, the model uses the cell as the smallest unit of surface attribute. This allows for greater flexibility and accuracy than the traditional point line and polygon approach of most traditional GIS models.

The DEM provides the base for stream networks and watersheds. Geology is reclassified into an infiltration rate based on the percolation characteristics of each rock type. The aerial photograph provides the base for a land use/land-cover plan of existing conditions. Each of the land-use/land-cover classifications is given a runoff coefficient based on empirical data of surface attributes. Empirical data shows that on average only 30% of precipitation is naturally recharged into the groundwater. In areas where chaparral is

Table II.15.3 Classification of Development Alternatives and Their Impacts

Pollution Region	Site	Escaped Stormwater Region	Site
1. Prevent the marshaling of the energy — deep reform			
I Remove polluting activities from aquifer recharge areas	I Collect/treat runoff II Don't build on aquifer recharge areas III No agriculture, cars, polluting activities	I Eliminate impervious surfaces II Re-vegetate with chaparral III No new settlement	I On-site small-scale active recharge II Injection wells
2. Reduce the amount of energy marshaled — deep reform			
I Reduce the number of polluting activities II Treat pollution	I Reduce the amount of polluting activities II Small-scale development in high density III Treat pollution	I Reduce the amount of impervious surfaces II Reduce number of settlements	I On-site retention and recharge II Minimize impervious surfaces III Collect surface runoff from site
3. Prevent the release of the energy — deep reform			
I Remove pollution activities on aquifer recharge areas	I On-site wastewater treatment and discharge II Allow only nonpolluting activities	I Regional, large retention basins with active recharge	I On-site retention and recharge II Surround site with French drains to collect runoff
4. Modify the rate of energy or spatial distribution of energy release — partial reform			
I Build off of aquifer recharge areas II Build in a dispersed pattern	I Small buildings II Fewer cows, goats, sheep III Less industry IV Large open spaces	I Allow runoff onto aquifer recharge areas II Dispersed pattern, large green spaces	I Single-family housing on large lots (>1 acre) II Retain runoff on-site and discharge slowly
5. Separate in space or time — partial reform			
I Phase construction along with mitigation measures II Alternate between grazing sites	I Locate polluting activities off recharge areas	I Dispersed pattern of settlement II Phase construction along with mitigation measures	I Small footprint high-rise II Build only on impervious geology
6. Separate by interposition of a material barrier — incremental advances			
I Fence off aquifer recharge areas	I Treat runoff before recharge	I Build drains to collect and divert runoff	I Cover development with a canopy and divert water to recharge areas
7. Modify appropriately the contact surface — incremental advances			
I Eliminate grazing areas II Eliminate roads (viaducts) III Pollution-neutralizing surface	I Cover site with impervious material to prevent infiltration of pollutants II Pollution-neutralizing surfaces (I.e. wetlands)	I Eliminate imperious surfaces II Maintain grazed areas III Eliminate frosts from aquifer recharge areas	I Build underground II Cover roofs/buildings with sod
8. Strengthen the structure — holding the line			
I Combined regional sewage stormwater treatment	I On-site sewage/stormwater treatment II Full containment of pollution III Closed systems	I Regional reservoirs recharge aquifer actively	I Connect storm drains to sewage runoff treatment and injection wells
9. Detection and evaluation by generating a signal that a response is required — slowing the rate of retreat			
I Monitor streams and wells II Stop/modify activity	I Monitor drains and wells II Add clean water from alternate source III Recharge with desalinized/reclaimed water	I Monitor peak floods and pump water to reservoirs II Increase aquifer recharge areas as needed by overpumping	I Monitor drains and sewers II Build ditches and systems as demand increases
10. Return to preevent conditions — slowing the rate of retreat			
I Develop alternative water source II Desalinization III Water reclamation IV Abandon aquifer V Adequate water pricing	I Pay for cleanup II Evaluate cost-benefit of aquifer protection	I Allow current development and recharge II When threshold reached, fix it	I Strengthen/fix storm drains and recharge systems II Abandon settlements if threshold is reached

Source: Adapted from Haddon (1970) and Regier and France (1990).

Table II.15.4 Data Requirements

Aquifer Recharge Management Model	Stormwater Management Model[1]
1. Digital terrain model	1. Weather data: precipitation, snow melt, wind speed, snowmelt coefficient, snow distribution, and other melt parameters
2. High-altitude aerial photo	2. Surface quantity: area, imperviousness, slope, Manning's roughness coefficients, Horton or Green–Ampt infiltration parameters
3. Geology layer	3. Subsurface quantity: porosity, field capacity, wilting point, hydraulic conductivity, watertable elevation, ET perimeters, coefficients for groundwater outflow
	4. Channel/pipe quantity: linkages, shape, slope, length, Manning's roughness, storage volumes at manholes and other structures.
	5. Storage sedimentation quantity: stage-area-volume-outflow relationships, hydraulic characteristics of outflows
	6. Surface quality: land use, total curb length, catchbasin volume and initial pollutant concentration, street sweeping interval, dry days prior to initial precipitation, friction perimeters for each land use
	7. Dry-weather flow constants variations
	8. Particle size distribution, Shields' parameter decay coefficients for channel/pipe quality
	9. Storage/treatment: parameters defining pollutant removal equation, individual treatment options.
	10. Storage/treatment cost: parameters for capital and operation and maintenance costs

[1] Adapted from: *EPA SWMM Windows Interface User's Manual.*

the main land cover, however, 80% of precipitation is recharged into the aquifer. By contrast, in areas with dense forest cover, nearly all precipitation is lost due to evapotranspiration. This provided the basic reclassification scheme for the impact of vegetation type on the aquifer recharge (different vegetation types reduced the infiltration factor of the underlying substrata by as much as one level; for example, forests on high infiltration substrates were classified as "medium"). The model looks at imperviousness as an indicator for loss of water to aquifer recharge, so an increase in the total area of impervious surfaces indicates a decrease in the total surface area available for water to infiltrate through the filtration zone.

The model treats all hard surfaces such as buildings and roads as being impervious unless a policy or technology is implemented that results in bottom-line change of the surface characteristics. For example, a policy may require all new development to retain and recharge all runoff on site.

For purposes of simplification, some general assumptions influence the structure of the model:

- All surface runoff flows from areas of no infiltration to areas of maximum infiltration, and all runoff reaches the water table in these areas.
- Groundwater protection provides a surrogate for protecting other resources such as wildlife habitat and stream corridors.

GIS Layers

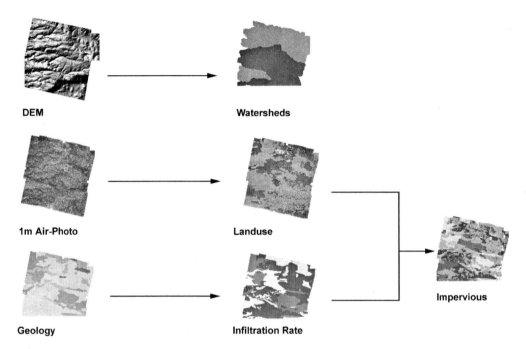

Figure II.15.2 GIS-layered approach used in this study.

Case study

As a case study, to test the impacts of conventional development, we reexamined four alternative land-use development futures that were produced in the Harvard Design School ISCAR Studio (Steinitz, 1997). The study area is located in the Western Galilee in northern Israel and is roughly 250 km^2, containing seven watersheds. It is home to nearly a hundred thousand people in several Arab Druze and Jewish settlements. Tefen Industrial Park is located in the center of the study area and is the main place of employment in the area. All alternatives assume tripling of the current population at plan buildup without the addition of new settlements. Furthermore, as population growth is measured as a factor of area, there is no change in current densities and development practices. Existing land-use and the resulting imperviousness provide the base against which all alternatives are compared (Figures II.15.3 and II.15.4).

ISCAR alternatives

1. Ring road (RR)

- An expansion of the roads that connect the villages surrounding the center of the study area in order to decentralize economic growth
- Some urban development is diverted to the north and south of the study area
- Impacts are concentrated along this road network where the majority of urban development occurs

Chapter II.15: *Aquifer recharge management model*

Figure II.15.3 Existing land use. (Color version available at www.gsd.harvard.edu/watercolors. Password: lentic-lotic.)

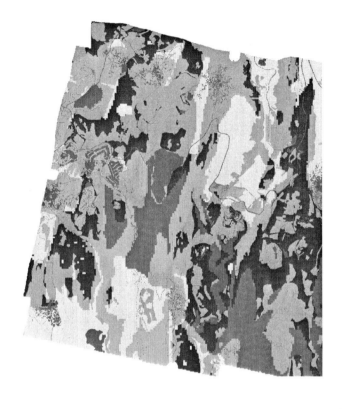

Figure II.15.4 Existing imperviousness. (Color version available at www.gsd.harvard.edu/watercolors. Password: lentic-lotic.)

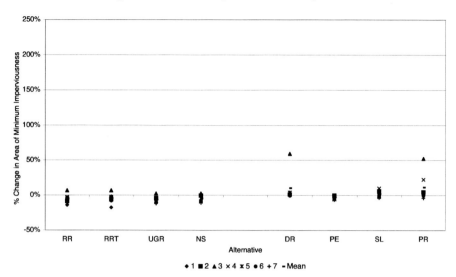

Figure II.15.5 Percentage change in maximum imperviousness under different development alternatives: RR = ring road; RRT = ring road through Tefen; UGR = upgrade existing roads; NS = north–south road; DR = deep reform; PE = return to preevent; SL = slowing the rate of retreat; and PR = partial reform.

Overall, this alternative increases the total amount of impervious surfaces by as much as 63% on average per watershed. The most impacted watersheds are 1 and 5, where the total area of impervious surfaces more than doubles (Figures II.15.5 and II.15.6).

2. Ring road through Tefen (RRT)

- A shorter ring road that cuts through Tefen is expanded to allow development of the periphery
- Most of the development is concentrated south of the center of the study area
- Impacts are concentrated along this road network where the majority of urban development occurs

This alternative has a slightly lower impact, increasing the amount of impervious surfaces by as much as 62% per watershed. In watershed 1, the total area of impervious surfaces has increased by as much as 150%; in watersheds 4 and 5 the total impervious area increases by 100% (Figures II.15.5 and II.15.6).

3. Upgrade existing roads (UGR)

- Requires no major infrastructure investment beyond the expansion of the existing road network to handle the increase in traffic associated with population growth
- Most of the impacts are concentrated in the center of the study area around Tefen and its surrounding settlements

Despite no investment in new infrastructure, this alternative increases the amount of impervious surfaces by an average of 66% per watershed. Watersheds 4 and 5 increase by as much as 100%, and watershed increases by more than 200% (Figures II.15.5 and II.15.6).

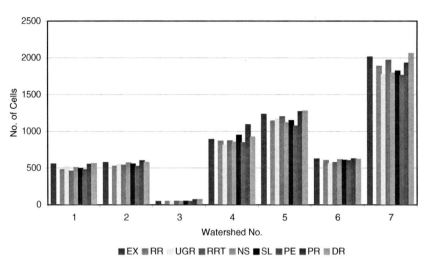

Figure II.15.6 Distribution of maximum runoff by watershed for different development alternatives (identified in Figure II.15.5). (Color version available at www.gsd.harvard.edu/watercolors. Password: lentic-lotic.)

4. North-south road — NS

- Requires a major investment in infrastructure by building a high-speed, limited-access road that connects the two urban areas in the north and south of the study area
- Impacts are concentrated mainly in the north and the south and the center remains relatively undeveloped

Of all the ISCAR alternatives, the north–south road (Figures II.15.7 and II.15.8) has the highest impact. Overall it increases the total amount of impervious surfaces by nearly 77% per watershed on average. In watershed 1 it increases by more than 200%, and in watersheds 4 and 5 it increases by more than 100% (Figures II.15.5 and II.15.6).

Policy alternatives

The next four alternatives were selected from the larger set of possibilities (Table II.15.3) to test the impacts of policy decisions that favor the protection of aquifer recharge areas. Natural resource management is not simply about protecting or exploiting resources, it is also about achieving some dynamic balance between people and natural processes. In this view, degradation of aquifer recharge areas can be controlled, reduced, and even eliminated by implementing land-use management programs. As the pumping of groundwater tends to be cheaper than the divergence of surface water, watershed management becomes a challenging task due to the biophysical, social, organizational, and economic phenomena associated with water and land uses. The difficulty in watershed management comes from the diffused use by different segments of a population of a common resource (all ethnic groups in the region have equal access to the aquifer). Successful implementation of watershed management depends on appropriate incentives that will resolve the conflicts between public and private interests that are always associated with water use (Kozub et al., 1987).

Chapter II.15: Aquifer recharge management model

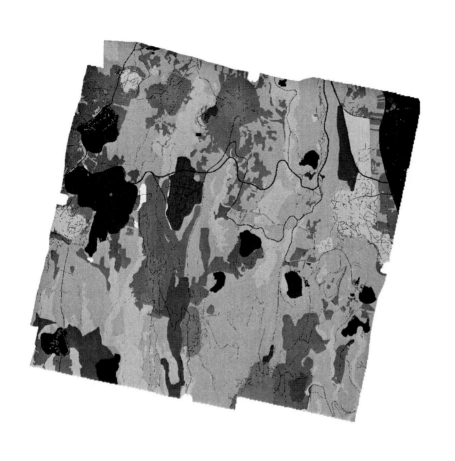

Figure II.15.7 Land-use plan for the north–south road development alternative. (Color version available at www.gsd.harvard.edu/watercolors. Password: lentic-lotic.)

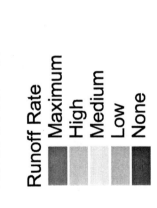

Figure II.15.8 Imperviousness resulting from the north–south road development alternative. (Color version available at www.gsd.harvard.edu/watercolors. Password: lentic-lotic.)

These alternatives test the ability of the model to deal with the implementation of different policies. The four policies were developed to mitigate the impacts of development on aquifer recharge through the implementation of practices that encourage retaining, treating, and artificially recharging urban runoff into the aquifer. The effectiveness of the policy is measured by level of reform and compliance.

In the following alternatives, the level of compliance with aquifer recharge management practices varies by alternative and ethnic group. The basic assumption is that it is easier to enforce compliance in the Jewish sector than in the Druze and Arab sector. This is due in part to politics and in part to economics. The level of income in the Jewish sector can absorb the increase in prices due to the use of expensive technology to treat urban runoff. Mandating the use of the same technology in the Druze and Arab sectors may put new development out of the reach of local residents. The alternatives also differ in the level of reform or technology implemented for mitigation purposes.

1. Deep reform (DR)

- The plan assumes no population growth since the protection of the aquifer is the overriding consideration
- All existing structures and settlements are retrofitted to treat polluted runoff and active aquifer recharge

The impact of this plan (Figures II.15.9 and II.15.10) is to improve the aquifer recharge over existing conditions. In all watersheds, the total amount of impervious surfaces decreases by 12% per watershed on average (Figures II.15.5 and II.15.6). In watershed 3, the total area of no impervious surfaces increases by more than 50%.

2. Return to pre-event (PE)

- All new development should comply with a policy that requires minimal treatment of polluted runoff and recharge. The driving force behind this alternative is its requirement for more stringent compliance to avoid threshold impacts. Half of all new development must comply with a law that changes the bottom-line characteristics of impervious surfaces to low imperviousness.
- No existing structures are retrofitted to treat runoff and provide recharge

The overall impact of this alternative is similar to existing conditions (Figures II.15.5 and II.15.6).

3. Slowing the rate of retreat (SL)

- Most new development is not required to comply with the policy
- In the Druze and Arab sector, only 20% of new development is required to install minimal recharge technologies. The Jewish sector has higher compliance requirements with more efficient technologies.

The overall impact is to increase the amount of impervious surfaces by 7% on average per watershed conditions (Figures II.15.5 and II.15.6).

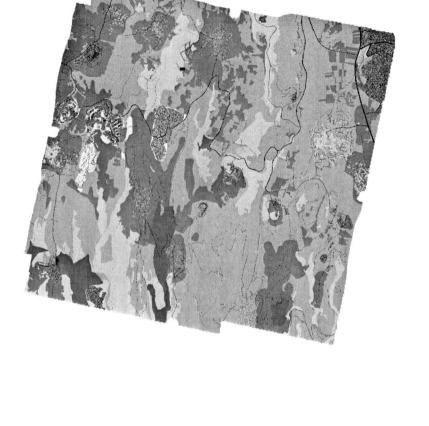

Figure II.15.9 Land-use plan for the deep reform development alternative. (Color version available at www.gsd.harvard.edu/watercolors. Password: lentic-lotic.)

Chapter II.15: *Aquifer recharge management model* 631

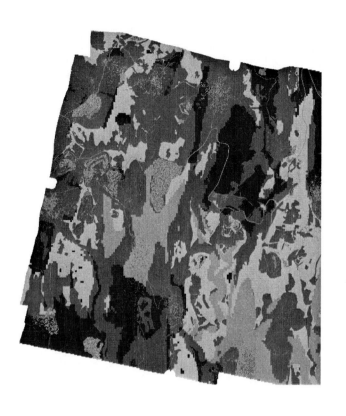

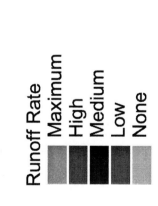

Figure II.15.10 Imperviousness resulting from the deep reform development alternative. (Color version available at www.gsd.harvard.edu/watercolors. Password: lentic-lotic.)

4. *Partial reform (PR)*

- Most new development is not required to comply with the policy.
- Those that do comply must achieve a high level of recovery of runoff and aquifer recharge. In the Arab and Druze sectors, 20% of new development is required to implement technology that would result in a high level of aquifer recharge and 10% is required to employ minimal mitigation. In the Jewish sector, 40% of new development is required to employ the highest available levels of mitigation and 10% is required to use the medium levels.

The overall impact of this alternative is to increase the amount of impervious surfaces by as much as 45% on average per watershed (Figures II.15.5 and II.15.6). In watershed 3, the total area of none impervious increases by 50%.

Conclusion

Under existing practices of development and even under some of the less stringent policy alternatives, the degradation of aquifer recharge areas due to urban development continues. This is evident in the change in the number of cells with maximum runoff (and therefore, the degree of consequent infiltration).

The ARMM provides planners and decision makers with a simple tool to identify critical areas of conflict between aquifer recharge and urban development. It also provides a means to identify the magnitude of the impacts and it informs decision makers of the most desirable alternative. As is the case with most models, the results are only as good as the alternatives tested. It is the responsibility of planners and decision makers to ensure that the widest range of alternatives is tested before meaningful decisions are made.

Literature cited

Berg, R.C., Curry, B.B., and Olshanksy, R., Tools for groundwater protection planning: An example from McHenry County, IL, *Environ. Manage.*, 23, 321, 1999.

Carman, N., Shamir, U., and Meiron-Pistiner, S., Water-sensitive urban planning: Protecting groundwater, *J. Environ. Plann. Manage.*, 40, 413, 1997.

The Conservation Foundation, *Groundwater Protection*, Washington, DC, 1987.

Goldenberg, L.C. and Melloul, A.J., Hydrological and chemical management in the rehabilitation of an aquifer, *J. Environ Manage.*, 42, 247, 1994.

Gordon, W., *A Citizen's Handbook on Groundwater Protection*, Natural Resource Defense Council, New York, 1984.

Haddon, W., On the escape of tigers: An ecologic note, *Technol. Rev.* 6, 45, 1970.

Issar, A.S., Recharging and salivation processes in the aquifers in Israel, *Environ. Geo.*, 21, 152, 1993.

Jaffe, M. and DiNovo, F., *Local Groundwater Protection*, American Planning Assoc. Washington, DC, 1987.

Jones, R.C. and Clark, C.C., Impacts of watershed urbanization on stream insect communities. *Water Resour. Bull.*, 23, 1047, 1987.

Kozub, J.J., Meyers, N., and D'Silva, E., *Land and Water Resource Management*, Econ. Develop. Inst. of the World Bank, Washington, DC, 1987.

Kuylenstierna, J.L., Bjorklund, G., and Najlis, P., Future sustainable water use: Challenges and constraints. *J. Soil Water Conserv.*, 52, 151, 1997.

Mostaghimi, T.U.S.S. and Shanholtz, V.O., Identification of critical nonpoint pollution source areas using geographic information systems and water quality modeling, *Water Res. Bull.*, 28, 877, 1992.

Regier, H.A. and France, R.L., Perspectives on the meaning of ecosystem integrity in 1975, in Edwards, C.J. and Regier, H.A., Eds., *An Ecosystem Approach to the Integrity of the Great Lakes in Turbulent Times*, Great Lakes Fishery Commission, Ann Arbor, MI, 1990.

Schlosser, I.J., Stream fish ecology: A landscape perspective, *BioScience*, 41, 704, 1991.

Schueler, T. and Claytor, R., Impervious cover as a urban stream indicator and watershed management tool, *Watershed Development Effects*, 1997.

Steinitz, C., Alternative futures in the Galilee, Israel, www.gsd.harvard.edu/faculty/steinitz/steinitz.html, 1997.

Watershed Protection Techniques, The importance of imperviousness, 3, 100, 1994.

Response

Planning by examining alternatives

In this chapter, Amir Mueller, Robert France, and Carl Steinitz present a heuristically appealing model that differs from many others currently in use due to its overt simplicity. Model development always involves tradeoffs between realism and generality. The aquifer recharge management model described here for a particular site in Galilee is easily adaptable to other locations due to its simple reliance on imperviousness as the metric of most importance. This is the same strategy employed in generation of the open-space and impervious surfaces model described by Jennifer Zielinski in Chapter I.3. Models such as these have great utility due to the absence of hidden assumptions and complex mathematics that are sometimes employed in other cases, thereby giving false impressions of much greater accuracy than really exists. Transparency in the derivation of water sensitive planning models is necessary if they are to be communicated to, and subsequently accepted by, the nonprofessional public. This is particularly true when undertaking land-use planning in rural communities where the education base may not be as high as it is in urban settings.

Another important message in this chapter concerns the temporality involved in water sensitive planning. The technique used in this study is referred to as "alternate futures scenario planning," a methodology that is becoming increasingly popular. In simplified terms, it is based on the assessment of current conditions and the use of that data in order to predict what the future conditions might become given certain assumptions. This approach is similar in a way to the resource-driven, "bottom-up" GIS analyses described by Jeffrey Schloss in Chapter II.12. These studies represent a maturity in the discipline of land-use planning in their shift in focus from being merely reactive to damage to instead offering proactive guidelines in order to preempt environmental disturbances before the latter become manifest. "Planning," in its most basic definition, is really all about concern for the future, and the model developed in this chapter goes far toward illustrating just how such an approach may be initiated.

chapter II.16

Factors influencing sediment transport from logging roads near boreal trout lakes (Ontario, Canada)

Robert France

Abstract

Ontario's guidelines for protection of aquatic resources from riparian timber harvesting are based on adoption of buffer strip widths in relation to surface slope from a study conducted over 40 years ago. That work measured the maximum distance traveled by 36 erosion plumes arising from a single logging road in New Hampshire. Because statistical reanalysis of that earlier study indicated that surface slope only explained 16% of the variation in sediment transport, a new study was undertaken in northwestern Ontario of the factors influencing erosion plumes from logging road washouts. Sixty-eight erosion plumes were measured along a 4.3-km stretch of logging road in the riparian zones of three boreal lakes. Road-generated sediment was rarely found to move more than 30 m, thereby suggesting that the current guidelines, which have a 30-m minimum "area of concern" buffer strip width, are for the most part adequate. Sediment transport distance was found to be unrelated to surface slope but was significantly influenced by both road gradient and downslope orientation, as well as by the presence of large surface obstructions blocking plume flow. A set of simple design criteria is suggested to minimize or perhaps even eliminate chances of sediment input to water bodies, a small but still real possibility under the present guidelines.

> "Our number one water quality problem in the national forest system is roads."
>
> — M. Dombeck, former U.S. Forest Service Chief
> (in Young-Petersen, 2000)

Introduction

Clearcutting has been a prevalent means of timber harvesting since the beginnings of forestry in Ontario (Burgar, 1983). Today, the magnitude of such logging operations in the Boreal Forest on the Canadian Shield is immense, exceeded by only a few locations elsewhere in the world. Because forestry and sport fishing are the two most important generators of economic revenue for much of the Canadian Shield, the Ontario Ministry of Natural Resources (OMNR) has developed series of *timber management guidelines* to ensure protection of aquatic habitats during logging of watersheds (OMNR, 1985, 1986, 1988a, 1991, 1995).

Work in western North America has demonstrated that one of the most overt consequences of clearcutting can be substantial increases in the production of sediment (Patric et al., 1984), much of which originates from the construction of logging roads (Beschta, 1978; Rothwell, 1983; Reid and Dunne, 1984; Bilby, 1985). Although research in northwestern Ontario has investigated the role of litterfall in mitigating interrill and aeolian erosion following riparian timber harvesting (France, 1997; France et al., 1998; Steedman and France, 2000), in addition to estimating whole-basin sedimentation rates after forest removal (Blais et al., 1998), no published studies are available for this region regarding the extent of sediment yields arising from logging roads on sloped ground.

Lakes, once envisioned as self-contained "microcosms" (Forbes, 1887), are now considered to be integrated components in a dynamic continuum of landscape processes (e.g., Likens, 1984). From this awareness comes recognition that "the maintenance of vegetation near waterbodies can mitigate many of the potential negative effects of [timber] harvesting . . . [such that] the presence of a vegetated area adjacent to waterbodies acts to buffer the waterbody from the effects of harvesting" (OMNR, 1988b). The present study investigates landscape influences on sediment transport arising from logging roads on sloped terrain in northwestern Ontario, and examines the effectiveness of existing shoreline (riparian) buffer strip reserves (OMNR, 1988a, 1991). Environmental guidelines for forest access roads crossing water bodies, not considered in this study, are described elsewhere (OMNR, 1995).

Concomitant with the increased scale of timber removal in Ontario has been a substantial decrease in the recommended buffer strip widths from near 200 m around all water bodies two decades ago, to as little as 30 m at the present time for "cold water" lakes, which are known to support trout (reviewed in France et al., 1998). Today, Ontario's recommended buffer strip widths for trout lakes are based on the upper values suggested by Trimble and Sartz (1957) in relation to surface slope: 30 m for slopes of 1 to 15%, 50 m for slopes of 16 to 30%, 70 m for slopes of 31 to 45%, and 90 m for slopes of 46 to 60%. Trimble and Sartz's recommendations arose from measurements made of the maximum distance traveled by 36 sediment plumes downslope from small culverts along a single logging road in New Hampshire. What was not calculated by Trimble and Sartz (1957), nor mentioned in OMNR's uncritical adoption of these guidelines, however, is that slope explained only 16% of the variation in sediment transport. Later research on logging road erosion (e.g., Haupt, 1959) has questioned Trimble and Sartz's conclusions regarding buffer strip effectiveness. Trimble and Sartz also recognized the limitations of their own data, acknowledging that "distances of sediment deposition may at times be greater than the maximum estimated from the observations." Therefore, OMNR's (1988a) assurance that adoption of such guidelines will "prevent any input of sediment" into water bodies could be regarded with skepticism.

Given the questionable scientific background for this particular aspect of the present *timber management guidelines*, OMNR have themselves admitted that "the effectiveness of the guidelines in mitigating potential negative effects of harvesting has not been determined" (OMNR, 1988b), such that "more information on the effectiveness of riparian reserves in northern Ontario is necessary in order to determine if the guidelines are being followed and are working" (Pike and Racy, 1989). This supports a survey of timber management operations in Canada, wherein Freedman (1982) concluded that "while the usefulness of buffer strips is generally acknowledged, the widths that are required for them to be effective are still controversial." The goal of the present study, therefore, is to obtain information useful for planning the effective widths needed for protective buffer strips in northwestern Ontario.

Methods

This study was conducted in northwestern Ontario, 45 km northwest of the town of Atikokan and 150 km southeast of the Experimental Lakes Area, in a region transitional between the Boreal and Great Lakes/St. Lawrence forest types (France, 1997; France et al., 1998; Blais et al., 1998; Steedman and France, 2000). The landscape is composed of gently rolling knolls of Precambrian bedrock interspersed with boggy plains and glacial deposits. The logging roads surveyed in this study ring three small headwater lakes whose watersheds had been experimentally clearcut three years previously as part of an OMNR research project (Steedman and France, 2000; Steedman et al., 2000). These lakes support 5 to 7 species of fish, including lake trout (*Salvelinus namaychush*). Road surfaces are composed of local sand and gravel and have received little or no motorized traffic since logging operations ceased in 1996.

During 1999, 4.3 km of logging roads were walked and all observed washout erosion plumes (predominantly composed of unconsolidated sand) were measured. In addition to maximum sediment transport distance, plume widths were assessed at several locations in order to estimate surface-area coverage. Detailed, plume-specific maps were drawn and are displayed in CD-ROM format (with accompanying photographs) in France (in prep). Factors that could potentially influence sediment erosion were measured in three basic locations: upslope of the road, the source road itself, and downslope from the road along the path of the erosion plume. In all cases, the factors were selected for ease of measurement, suitable for inclusion in a large synoptic regional survey.

Upslope factors included those that might influence minicatchment hydrology: the overall shape (concentrating, dissipating, flat), average slope determined along 30-m lengths at the center axis and on each side edge at 45°, and extent (low, medium, high) of small water collection reservoirs; and those with a potential influence on sediment generation and movement: vegetation density and height (low, medium, high) determined at six locations along the slope transect lines.

Source road factors included those that could potentially influence the generation, amount, type, and transportability of sediment: road width, gradient, and orientation to downslope plume (perpendicular, angled, parallel), whether the road was sunken or raised, upslope (i.e., cut) embankment height, distance to the nearest erosion washout, and the particle size (coarse, medium, fine) and surface compaction state (low, medium, high) of the road.

Downslope factors included those that might be expected to influence the ease of sediment movement: slope gradient, slope obstruction index (number of branches, boulders, etc. along the plume length, as in Haupt, 1959), estimation of the extent of surface depressions (difference between straight and surface contoured distances over complete plume length), forest floor roughness (low, medium, high), vegetation cover and height (low, medium, high) determined at three locations, shape of general area (concentrating, dissipating, flat), and plume shape (determined from four cross-width measurements), number of entry washouts and side rills.

Analyses consisted of examining correlation coefficients for variables treated independently and also in ANOVA's for variables treated interactively. Significance was considered at $p < 0.05$. Only those variables with the most demonstrable influence on sediment transport distance are discussed next. The purpose of analyses undertaken in this study was to highlight, in a nonpredictive way, the major factors influencing length of erosion

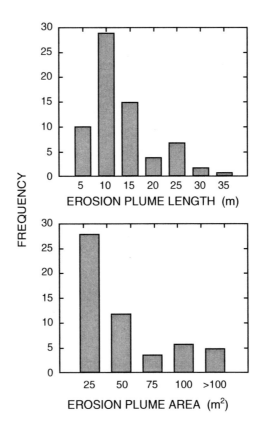

Figure II.16.1 Frequency distributions of lengths and surface areas of 68 erosion plumes originating from logging road washouts in northwestern Ontario.

plumes. Other studies concerned with developing predictive empirical models of sediment movement have used multiple regression techniques (e.g., Haupt, 1959; Packer, 1967; Elliot et al., 1999).

Results

Of 4.3 km of surveyed roads surrounding the study lakes, 104 individual road drainage exits were observed, comprising 68 specific erosion plumes. About half of these plumes were isolated (and therefore independent) from one another, being separated by distances of over 30 m; about a quarter, however, were situated within 10-m road distance from each other (and therefore influenced by the same drainage flow). The remaining quarter were in between these two categories. The average surface area of the plumes was not great, most covering less than 50 m^2 size (Figure II.16.1, bottom panel). Plumes varied considerably in shape in relation to both source and downslope surface characteristics (illustrated in France, in prep). In some cases, a single washout spread to multiple lobes, whereas in other cases, multiple washouts conjoined into a single plume. Other erosion plumes were intermediate between these extremes; for example, proximal parallel washouts with either considerable or little intermixing of their sediment loads.

Maximum length of erosion plumes varied between 3 and 32 m. Most important, only 2% of all measured erosion plume distances exceeded the smallest buffer strip width guideline of 30 m, with over half of all plumes being less than 10 m in length (Figure II.16.1, upper panel).

Chapter II.16: *Factors influencing sediment transport from logging roads* 639

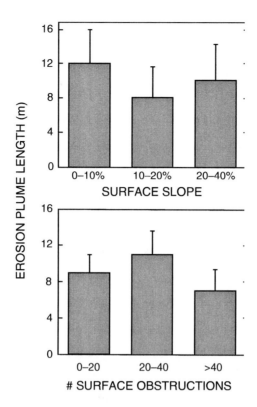

Figure II.16.2 Influence of surface slope and number of obstructions on the length of erosion plumes. Error bars are ± 2 SD. No significant differences (ANOVA tests; $p = 0.05$) exist among groupings.

Of all the factors examined for their influence on sediment transport distance, none displayed any statistical interaction, and only three were independently significant. Surprisingly, neither the landscape characteristics of downslope surface slope nor the number of surface obstructions played a demonstrable role in affecting plume length (Figure II.16.2).

In contrast, measurements of the sediment road source were found to strongly influence erosion plume length. The orientation of the road was important in that plumes originating from washouts perpendicular to the aspect of the road were only half the length of those originating from washouts situated either angled or parallel (illustrated in France, in prep) to road direction (Figure II.16.3, upper panel). Road gradient also effected sediment transport, with slopes between 10 to 15% producing plumes about twice the length of those coming off roads having slopes less than 5% (Figure II.16.3, lower panel). The worst situations in terms of sediment movement occurred when steep gradient roads abruptly turned to become contoured with the surface slope. In such cases, a series of washouts developed as the sediment from road surface did not make the bend and continued downslope after building up considerable energy (illustrated in France, in prep). Here, runoff rills were also frequently observed on the road surface, and sometimes erosion scouring could be seen downslope (photographs in France, in prep). Transported sediment plumes in these locations were often in the order of 15 to 35 cm deep.

Finally, although smaller surface obstructions were not found to significantly influence overall sediment transport distance, whether or not the plume abutted against a major terminal obstruction was important. Plumes that ended in natural or logging terminal obstructions averaged about 8 m in length compared to about 13 m for those plumes that did not end abruptly (Figure II.16.4). Natural termini included bedrock walls, boulder

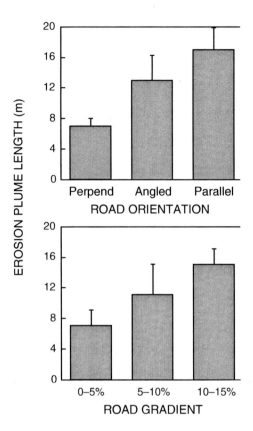

Figure II.16.3 Influence of road orientation (see text) and gradient on the length of erosion plumes. Error bars are ± 2 SD. Significant differences (ANOVA tests; $p = 0.05$) existed between perpendicular and both angled and parallel orientated roads (which did not differ from each other), and between 0 to 5% and 10 to 15% road gradients.

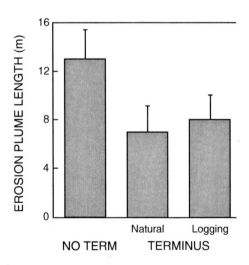

Figure II.16.4 Influence of terminal obstructions on length of erosion plumes. Error bars are ± 2 SD. Plumes ending in either natural or logging-produced barriers (see text) were significantly (ANOVA tests, $p = 0.05$) shorter than plumes with no such terminal obstructions.

fields, small water bodies, and mature forest trees, and termini created by logging included slash piles, skidder ruts, excavated pits, and plowed hills (illustrated in France, in prep).

Discussion

This survey found very few cases in which erosion plumes from logging roads exceeded 30 m in length, the minimum width recommended for protective buffer strips in Ontario. The straightforward conclusion is that adherence to Ontario's existing guidelines for buffer strip widths (OMNR, 1988a) should generally be adequate for prevention of most sediment transport to aquatic systems (at least for this particular study region of northwestern Ontario). Empirical studies such as the present example have a further use in being able to identify the most important influences of erosion and to provide a design framework for reducing sediment transport to levels well below the legal requirements, thereby satisfying the heightened safety concerns of environmental managers and public interest groups.

The role of downslope surface gradient in influencing sediment transport is not as straightforward as Trimble and Sartz's (1957) early work may have suggested. Whereas some work (e.g., Nieswand et al., 1990) is conceptually based on the assumption of a strong influence of slope being the primary determinant of erosion plume distance, other empirical work (e.g., Haupt, 1959) has, like the present study, found no existence of such a relationship between surface slope and transport distance. In Haupt's (1959) study, it was the downslope surface obstruction index that was the most significant regulator of sediment flow distance, a finding not duplicated, however, in the present study. Additionally, although other work has highlighted the importance of vegetation density in influencing sediment transport (e.g., Packer, 1967), this did not seem important in northwestern Ontario. The present work did find that whereas small surface obstructions or ground cover characteristics were unimportant in influencing sediment transport, large surface obstructions were significant in their role in ultimately blocking plume movement.

The demonstrable influence of the source road on sediment transport was a finding similar to both the present study and previous research. As was the case observed for northwestern Ontario, both Packer (1967) and Bilby et al. (1989) considered road gradient to be extremely important in determining the maximum distance of sediment movement. Such a result is largely a function of sediment-laden runoff moving rapidly down road surfaces at high velocity, resulting in gullies being formed, which in turn produces more sediment movement, which then adds to further road cutting and sediment transport, etc. For the present study site, when such steep slopes occurred in relation to roads that were oriented parallel to the surface gradient, the surface area of potentially erodible sediment was much greater than that for contoured roads situated perpendicular to the surface gradient; i.e., perpendicular road widths were generally only about 6 m compared with unobstructed runout distances of well over 20 m for downslope-oriented roads.

The present results can be important for developing management strategies for limiting sediment erosion from logging roads in northwestern Ontario, because all the factors found to be significant determinants of plume length can be easily designed to maximize their effectiveness in reducing the possibility of extended sediment transport. In other words, the following brief list of simple design criteria should be examined by professional foresters to mitigate erosion damage to aquatic systems:

1. Avoid, whenever possible, running roads downslope toward the water.
2. If this cannot be achieved, limit where downslope roads occur to only those locations having gentle surface gradients.

3. If downslope roads of steep gradient are absolutely unavoidable, then reduce the likelihood of extensive runoff plumes developing by:
   a. Building a gradual rather than sharp turn of the road where it meets the contoured grade so as to limit serious erosion washouts
   b. Grading the contoured road surface down on the upslope (cut) edge to form a raised downslope (fill) lip over which sediment will have more difficult time moving
   c. Bisecting the downslope road surface by numerous cross-ditches in order to prevent the buildup of high runoff energy by channeling water flow away
   d. Constructing buttresses of logging slash and other debris (e.g., boulders) or create deliberate excavation pits at the road bends to thwart the "straight-on" runout movement of sediment further downslope

Experimental testing and monitoring of these simple, common-sense measures for roads on slopes could provide information needed for improved protection for Ontario's aquatic resources under the existing guidelines for buffer strip widths (further best management practices applicable to roads crossing streams are detailed in OMNR, 1995). These measures are not new, having been previously suggested for mitigating road erosion in other regions: shallow grade (Packer, 1967) and insloped (Elliot and Tysdal, 1999) roads, cross-ditching (Elliot et al., 1999), and brush barriers (Swift, 1986). Casual surveys of logging operations by a variety of companies over a wide area of northwestern Ontario suggest, however, that more careful attention paid toward implementing these measures might be beneficial toward more effectively limiting sediment movement.

Sediment plume lengths measured in this study in northwestern Ontario are generally shorter than those for other areas characterized by steeper terrain and greater rainfall (tabulated in Elliot et al., 1999). Nevertheless, in the absence of more careful attention paid to logging road design and plume barrier construction, there still remains the possibility that some sediment might reach water bodies in this region. Therefore, another management option to guarantee no inputs of sediment to Ontario's water bodies is to simply increase the smallest recommended buffer strip width from 30 m to some, as yet to be determined, width. In such a situation, detailed attention to site-specific design of logging roads would not be of as much paramount importance as under the present guidelines of probable, but not definitive, effectiveness.

Given that surface slope was found to be relatively unimportant in influencing sediment transport distance, does it then make sense to maintain buffer strip guidelines that are largely based on this factor? Yes, because although small-scale surface undulations (and also obstructions) may not play a major role in washout plume length, the overall surface slope will affect road gradient and orientation, two factors of marked importance for determining the likelihood of sediment transport. Therefore, notwithstanding the implicit managerial/regulatory problems of dealing with a sliding scale of buffer strip width recommendations (rather than having a single width value irrespective of the local landscape), it makes sense to use slope as a reflection of sediment transportability due to its controlling influence on effective road design.

Finally, the issue of what elements should actually comprise an effective buffer strip has not been addressed in this study (but see France et al., 1998) (i.e., Do buffer strips really have to be forested to work?). The present results suggest that the role of buffer strips as sediment filters (only one of the many important protective roles played by riparian forests) may actually not need trees to effectively trap sediment. In such cases, buffer strips, or perhaps more appropriately — "areas of concern" (OMNR, 1988a), can perform this function without an intact forest being present; however, it is important to recognize that trees, although perhaps unimportant in terms of reducing road erosion,

Acknowledgments

This work was supported by grants from the Ontario Ministry of Natural Resources and the Milton Fund and Design School Dean's Fund of Harvard University. R. Steedman is thanked for providing research facilities at the Coldwater Lakes Camp and many discussions about forestry operations, as is F. de Kock for generous field assistance.

Literature cited

Beschta, R.L., Long-term patterns of sediment production following road construction and logging in the Oregon Coast Range, *Water Resour. Res.*, 14, 1011, 1978.

Bilby, R.E., Contributions of road surface sediment to a western Washington stream, *Forest Sci.*, 31, 827, 1985.

Bilby, R.E., Sullivan, K., and Duncan, S.H., The generation and fate of road-surface sediment in forested watersheds in southwestern Washington, *Forest Sci.*, 35, 453, 1989.

Blais, J.M. et al., Climatic changes in northwestern Ontario have had a greater effect on erosion and sediment accumulation than logging and fire: Evidence from 210Pb chronology in lake sediments, *Biogeochem.*, 43, 235, 1998.

Burgar, R.J., Forest land-use evolution in Ontario's Upper Great Lakes Basin, in Flader, S., Ed., *The Great Lakes Forest: An Environmental and Social History*, Univ. of MI Press, 1983.

Elliot, W.J. and Tysdal, L.M., Understanding and reducing erosion from insloping roads, *J. Forestry*, 97, 30, 1999.

Elliot, W.J., Hall, D.E., and Graves, S.R., Predicting sedimentation from forest roads, *J. Forestry*, 97, 23, 1999.

Forbes, S.A., The lake as a "microcosm," *Bull. IL Natur. Hist. Surv.*, 15, 537, 1887.

France, R.L., Potential for soil erosion from decreased litterfall due to riparian clearcutting: Implications for boreal forestry and warm- and cool-water fisheries, *J. Soil Water Conserv.*, 52, 452, 1997.

France, R.L., Peters, R., and McCabe, L., Spatial relationships among boreal riparian trees, litterfall and soil erosion potential with reference to buffer strip management and coldwater fisheries, *Annales Botanni Fennici*, 35, 1, 1998.

Freedman, B., *An Overview of the Environmental Impacts of Forestry, with Particular Reference to the Atlantic Provinces*, I.R.E.S. Research Dalhousie Univ., Halifax, NS, 1982.

Haupt, H.F., Road and slope characteristics affecting sediment movement from logging roads, *J. Forestry*, 57, 329, 1959.

Likens, G.E., Beyond the shoreline: A watershed-ecosystem approach, *Internationale Vereinigung fur Theoretisch and Angewandte Limnologie Verhandlungen*, 22, 1, 1984.

Nieswand, G.H. et al., Buffer strips to protect water supply reservoirs: A model and recommendations, *Water Resour. Bull.*, 26, 959, 1990.

Ontario Ministry of Natural Resources (OMNR), *Class Environmental Assessment for Timber Management on Crown Lands in Ontario*, Toronto, 1985.

OMNR, *Timber Management Planning Manual for Crown Lands in Ontario*, Toronto, 1986.

OMNR, *Timber Management Guidelines for the Protection of Fish Habitat*, Toronto, 1988a.

OMNR, Prevention, mitigation and remedy of potential negative effects. Statement of evidence panel X: Harvest. OMNR, Thunder Bay, ON, *Environ. Impact Assess. II*, 512, 1988b.

OMNR, *Code of Practice for Timber Management Operations in Riparian Areas*, Thunder Bay, ON, 1991.

OMNR, *Environmental Guidelines for Access Roads and Water Crossings*, Toronto, 1995.

Packer, P.E., Criteria for designing and locating logging roads to control sediment, *Forest Sci.*, 13, 2, 1967.

Patric, J.H., Evans, J.O., and Helvey, J.D., Summary of sediment yield data from forested land in the United States, *J. Forestry*, 76, 101, 1984.

Pike, M. and Racy, G.D., *Why Have Riparian Reserves in Timber Management? A Literature Review*, OMNR, Thunder Bay, ON, 1989.

Reid, L.M. and Dunne, T., Sediment production from forest road surfaces, *Water Resour. Res.*, 20, 1753, 1984.

Rothwell, R.L., Erosion and sediment control at road-stream crossings, *The Forestry Chron.*, 25, 62, 1983.

Steedman, R.J. and France, R.L., Aeolian transport of sediment from new clearcuts into boreal lakes, northwestern Ontario, Canada, *Water, Air and Soil Pollution*, 122, 139, 2000.

Steedman, R., Kushneriuk, R., and France, R., 2000, Initial response to catchment deforestation of small boreal forest lakes in northwestern Ontario, Canada, *Internationale Vereinigung fur Theoretisch and Angewandte Limnologie Verhandlungen*, in press.

Swift, L.W., Filter strip widths for forest roads in the southern Appalachia, *Southern J. Appl. Forestry*, 10, 27, 1986.

Trimble, G.R. and Sartz, R.S., How far from a stream should a logging road be located? *J. Forestry*, 53, 339, 1957.

Young-Petersen, T., Ripping roads for watershed health, *Orion Afield*, 4, 23, 2000.

Response

Empirically testing planning assumptions

Shoreline buffers are a recognized and established best management practice to protect water bodies from sediment generated as a result of riparian development. As Frank Mitchell outlines in Chapter II.1, such shoreline buffers can also play a variety of other important roles such as providing wildlife corridors and offering human amenities, the latter a topic elaborated on by Charles Flink in Chapter II.3. Issues of how wide buffer strips need to be in terms of providing their protective function is an issue touched upon by Jeffrey Schloss in Chapter II.12, and by Robert France, John Felkner, Michael Flaxman, and Robert Rempel in Chapter II.14.

Particularly, it is the erosion caused by the construction and continued operation of roads that is regarded as being the major source of sediment and concordant nutrient pollution to lakes and rivers from land use. In Chapter II.12, Jeffrey Schloss explores the implications of one such road on the water quality of a lake in New Hampshire through use of GIS analysis. In the present chapter, Robert France examines the implications of road construction and sediment transport at the much smaller scale of a few individual sites.

As France discusses, the width recommendations for Ontario's buffer strip guidelines, based as they are on a single, possibly flawed, study in New Hampshire over four decades ago, needed to be critically reexamined. The lesson here for water sensitive planners is an important one. Assumptions can, through time, become easily codified into planning recommendations that may be unrealistic. It is always important, therefore, to critically reexamine such entrenched assumptions with fresh, modern eyes and methods.

In this case, France determined that the established buffer strip width recommendations, notwithstanding their dubious origin, nevertheless *did* seem adequate in terms of preventing sediment transport from logging roads to receiving waters. The approach taken in this chapter, namely that of an empirical survey of numerous sites, is an investigative tool that is well adapted to objectively testing assumptions of water sensitive planning. It is easy to imagine such an approach being adapted to a wide range of investigations, for it transcends idiosyncrasies of a specific site and enables broader conclusions to be drawn about land management in general. Further, as France demonstrates in this chapter, it is possible, and indeed seems natural, to move from empirical scientific analyses to design recommendations, as, for example, with respect to the construction of logging roads.

One of the most interesting and provocative results from this study is the finding that the role of buffer strips as sediment filters may occur independent of whether or not standing trees are actually present. The finding that surface obstructions such as piles of logging slash can effectively trap sediment, and therefore reduce the potential for aquatic damage from road erosion, is of major interest to timber companies already fearing the economic implications of what they imagine to be overly protective buffer strips surrounding water bodies in this region (see Robert France, John Felkner, Michael Flaxman, and Robert Rempel in Chapter II.14).

chapter II.17

Limnology, plumbing and planning: Evaluation of nutrient-based limits to shoreline development in Precambrian Shield watersheds

Neil J. Hutchinson

Abstract

The concept of using water quality as a planning tool for recreational lakes has been in active practice in Ontario and parts of the United States for approximately 25 years. In practice, assumptions regarding anthropogenic loadings of phosphorus to a watershed (generally septic systems servicing shoreline development) are linked to estimates of natural phosphorus loading. The resultant model estimates total phosphorus concentration and the response of trophic status indicators such as water clarity and dissolved oxygen in specific lakes. Linking the model to a water quality objective allows planners to set capacities for anthropogenic phosphorus loads, and hence shoreline development such as cottages, resorts, or permanent homes. This chapter presents an example of how the concept can be applied in practice, based on the application of the author's experience to a test watershed in south-central Ontario. Practical examples are given to show the development and calibration of accurate trophic status models, the use of monitoring data to set ecologically valid water quality objectives and their translation into shoreline development capacities, and to show the strengths and weaknesses of the approach.

The availability of a scientifically based water quality model has overemphasized water quality as a planning tool and generated unrealistic expectations of a single-capacity determinant among the public. Recent advances in our understanding of the geochemistry of domestic septic systems indicates that less phosphorus may be mobile than was previously assumed. In addition, as alternative septic technologies for phosphorus abatement are developed a refocusing of capacity determinants will be required. A combination of land-use regulations and a scientifically based management program is recommended as an alternative to a single, phosphorus-focused approach. These could address stresses to the ecology of the riparian and littoral zones and acknowledge the importance of social determinants such as noise, crowding, powerboats, and the wilderness aesthetic. This would promote a diversity of planning approaches, shift the existing focus away from plumbing and septic systems, and provide a more holistic management program which protected more components of the lake system.

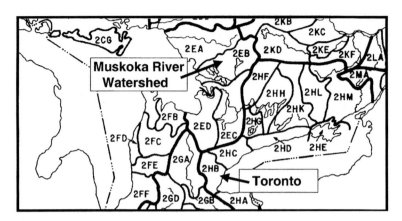

Figure II.17.1 Southern Ontario tertiary watershed divisions showing Muskoka River watershed draining west to Georgian Bay.

Introduction

Lake-based recreation and tourism is a fundamental cornerstone of the economy for many small rural communities situated on the Precambrian Shield in Ontario. The economic stimulus provided by tourism has diversified the economy of these communities, which were historically dependent on declining revenues from forestry or mining. Tourism on the southern portions of the Shield draws from the large and prosperous demographic of southern Ontario, in particular the "Golden Horseshoe" of 7 million people, which extends from east of Toronto to Niagara Falls, but also draws from the Great Lakes states of Michigan, Ohio, and New York.

The Precambrian (or "Canadian") Shield is made up of Precambrian gneiss and granite bedrock that lies near or on the surface throughout most of the northern areas of the province. Landscapes on the southern Shield are dominated by bedrock ridges and knolls, thin soil cover, and pine and hardwood forests. Lakes are also a dominant presence on the landscape. The Province of Ontario has over 260,000 lakes greater than 1 ha in size (Cox, 1978) and most of these are situated on the Shield. Thin soils and acidic bedrock mean that most of these lakes are nutrient-poor and possess high water clarity. Clear waters, rocky shorelines, and forest cover produce exceptional recreational attributes and, as a result, southern portions of the Shield have served as a focus for shoreline recreation in the form of cottages and resorts.

The Muskoka River Watershed was chosen as a test case to illustrate the development, validation, and application of a trophic status model to govern recreational shoreline development. The watershed covers approximately 5737 km^2 and is located on the southernmost fringes of the Precambrian Shield, approximately 180 km north of Toronto (Figure II.17.1). Four large (>5,000-ha) lakes — Lake Muskoka, Lake Rosseau, Lake Joseph, and Lake of Bays — form the recreational focus of the region. These were opened up to summer recreation with the arrival of railways in the 1880s and the formation of steamship service on larger lakes shortly thereafter. Approximately 150 smaller lakes also support shoreline recreation. Three major towns are located in the watershed, each with a population exceeding 8,000 permanent residents. Eight smaller communities of 500 to 2,000 people also exist. Some of these are serviced with sewage treatment plants discharging to the watershed, while others are serviced by septic systems.

Development on these lakes consists of approximately 18,000 shoreline residences, 4,300 resort units, and 1,400 trailer sites. Approximately 4,800 ha of urbanized areas are located adjacent to shorelines and 2,700 kg/year of phosphorus are added from point source STPs in urban centers. The Lake Rosseau watershed also includes approximately 4,800 ha of agricultural land use. There are approximately 4,400 vacant shoreline lots across the watershed, which represents a substantial resource base of future development potential. Approximately 1,400 back lots (i.e., set back from the shoreline) exist, and about one third of these are vacant.

The Province of Ontario and various municipal governments in recreational areas have maintained water quality programs since the late 1970s, to manage recreational growth in recognition of the important economic link of tourism to water quality. These programs generally consist of four elements:

1. Policies to maintain water quality through limits to shoreline development
2. Predictive models linking shoreline development to water quality
3. Lake-specific policies, including development objectives based on water quality and,
4. Monitoring programs to track changes in water quality in lakes

This chapter describes a process for developing a water quality model, validating its predictions of water quality, and setting development objectives on the basis of water quality. Issues of water sensitive land-use planning with respect to rural lakes are also covered in Chapters II.1, II.12, II.13, II.14, and II.16.

The author has made use of water quality and land-use data for a set of lakes situated within the Muskoka River watershed. The concepts and observations herein are those of the author alone and are intended to guide technical practitioners of water quality planning in recreational lakes in Ontario and elsewhere. They are not presented as specific recommendations for water quality planning for lakes in the Muskoka River Watershed — but the Muskoka watershed is used as an example of how these models can be applied throughout the Precambrian Shield in Ontario and elsewhere.

The history and origins of lakeshore capacity planning in Ontario

The first lakeshore capacity planning initiatives in Ontario grew out of the efforts to control eutrophication of Lake Erie in the 1970s. Lake Erie was a large and visible example of the threats posed by enrichment of surface waters with the algal nutrient phosphorus and of the success of remedial programs that were centered on managing the lake's phosphorus budget. In the same era, the eutrophication of inland lakes was also documented in response to inputs of partially treated sewage effluent. Among these was Gravenhurst Bay on Lake Muskoka, which suffered a history of algal blooms until tertiary sewage treatment was implemented in 1972 (Michalski et al., 1975) and enhanced treatment and relocation of the outfall were implemented in May 1994.

The primary water quality concern in Ontario's cottage country is also nutrient enrichment. Excessive phosphorus input promotes the growth of algae, causing a loss of water clarity. The lake user sees this as "greener" water of less aesthetic appeal or as surface blooms of nuisance algal growth. Algae settle to the bottom of the lake, where their decomposition consumes oxygen, reducing the amount of cold, oxygen-rich habitat available for sensitive aquatic life such as lake trout (*Salvelinus namaycush*) and triggering remineralization of sediment-bound phosphorus. Residential or cottage development on a shoreline may increase the input of phosphorus to a lake. Domestic septic systems may be a significant component of the loading, but clearing of the shoreline, fertilizer application, and increased erosion are also important.

The Lake Erie experience prompted an examination of phosphorus sources to inland lakes, in the understanding that unchecked recreational development and phosphorus loadings from septic systems might impair recreational water quality. The "Dillon–Rigler" model (Dillon and Rigler, 1975) was the first model to specifically address the relationship between potential eutrophication of Ontario's Precambrian Shield lakes and the density of residential development on their shorelines. Its rapid recognition and acceptance by the international scientific community led to the development of Ontario's "Lakeshore Capacity Study" (1976 to 1980) in the belief that substantial predictive relationships might be developed for other responses of lakes to shoreline development. The Lakeshore Capacity Study was coordinated by the Ontario Ministry of Municipal Affairs and Housing and published in 1986 (Downing, 1986). It produced predictive models for land use, fisheries exploitation, wildlife habitat, microbiology, and water quality and integration of all components into one capacity model. Although several of these models were very useful, the water quality model was the only one adopted for routine use by management agencies in Ontario. It was a minor variant of the Dillon — Rigler model (Dillon et al., 1986).

The Ontario Lakeshore Capacity Model was based on the nutrient or "trophic status" of a lake. It provided an accurate and quantitative linkage between the amount of shoreline development, phosphorus and chlorophyll "a" concentrations in a lake, and resultant water clarity. Both the Dillon–Rigler and Ontario models were developed and calibrated for small Precambrian Shield lakes in the Muskoka and Haliburton recreational areas of Ontario. Subsequent work refined the models and added significant new areas of understanding, most notably a component linking shoreline development and oxygenated lake trout habitat (Molot et al., 1992) and testing of model predictions on the original study lakes (Dillon et al., 1994) and an independent set of study lakes (Hutchinson et al., 1991).

Lakeshore capacity planning in Ontario

Lake-based recreation and tourism are fundamental to the economics of municipalities in the southern Precambrian Shield region of Ontario, and so water quality is also linked with tourism. Shoreline residents buy property with an expectation of stability in the water quality of their lake and of stability in their social environment. Under Ontario's Planning Act, municipalities have the legislative jurisdiction to monitor and regulate land use. Ontario's Provincial Policy Statement also requires that municipal planning policies recognize and protect water quality.

As part of this responsibility, some municipalities undertook programs to implement variants of the province's "Lakeshore Capacity Model" in the late 1970s and early 1980s. In areas of the province that were not organized into municipalities, the province's Ministry of the Environment implemented lakeshore capacity planning based on the province's model. The amounts of shoreline development allowed by these approaches are termed "lakeshore capacities" or "development capacity." Management of lakeshore capacity is intended to protect water quality directly. Shoreline development capacities, however, whether based on trophic status or other means, also help to maintain the social stability desired by shoreline residents by assigning a finite limit to the amount of shoreline development.

The provincial and municipal models have enjoyed widespread popular support since their implementation. Adaptations to personal computers were made in the late 1980s and early 1990s, but no thorough technical reviews or evaluations have been undertaken. Periodic reviews and reworking of these models are required, however, to incorporate recent advances in scientific understanding.

Trophic status models and shoreline development policy

One of the significant breakthroughs achieved by the Dillon–Rigler model and its variants was their perceived ease of use and the accuracy of their predictions of water quality. Although the model is conceptually simple, its application can be complex. The model is supported by the results of approximately 25 years of detailed measurements on calibrated watersheds in south-central Ontario, and these calibrations must be validated periodically. The model was originally developed and calibrated on headwater lakes but, in practice, it is used in a watershed context (Dillon and Rigler, 1975; Dillon et al., 1986). Watershed modeling is conceptually straightforward, and the calculations eased by the use of personal computers, but implementation of policy in a watershed context, often between municipal government boundaries, is very complex. The model must be supported by monitoring programs to ensure the validity of its predictions. The management endpoints, both for water quality and development capacity, must be substantiated and defensible. Finally, implementation of a water quality program must be done by policy that is clear, fair to all resource users, and defensible. Therefore, the conceptual simplicity of the model may not be carried through into its application. This is reflected in four requirements for managing shoreline development by trophic status:

1. An accurate and defensible model based on sound inputs and data sources and defensible assumptions (The model must be able to distinguish natural sources of phosphorus [which are not manageable] from anthropogenic sources of phosphorus [which are the intent of the management program])
2. A process of model calibration and operation to validate predictions against measured water quality
3. Water quality end points, expressed as both water quality and allowable limits of development (as either anthropogenic phosphorus load or number of lots or development units)
4. An implementation strategy, formal planning instruments, and a process to guide implementation and ensure fair allocation of capacity

Technical basis of the models

The water quality models used in Ontario are all mass balance models that predict the trophic status of lakes at steady state with their phosphorus and hydrologic loadings. In summary phosphorus loading from the atmosphere and from the watershed as a function of soils and geology is linked with the lake's water load. Phosphorus loss from the water column is modeled as a settling velocity, specific to whether the hypolimnion is oxic or anoxic. The result is a prediction of "natural" or "background" phosphorus concentration. Human influence is added in the form of phosphorus loading from septic systems servicing shoreline residences, as a function of a per-capita phosphorus contribution of 800 gm/year, a count of the number of shoreline residences and estimates of their usage (as capita years per year, Dillon et al., 1986). A schematic of model operation is given in Figure II.17.2.

Although the mass balance principles of phosphorus loading and expression are generic, the resultant predictions of phosphorus and chlorophyll "a" concentration, Secchi depth transparency and hypolimnetic oxygen in the Ontario models were all derived from regression relationships specific to the Muskoka–Haliburton region of central Ontario (Hutchinson et al., 1991; Dillon et al., 1994). Phosphorus export coefficients for the natural landscape will also change as a function of wetlands, soil types, geology, and land use

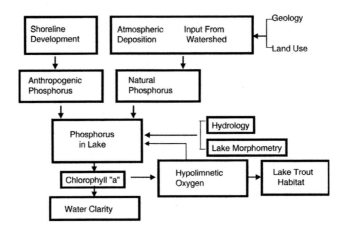

Figure II.17.2 Schematic of water quality models used in Ontario. Lake trout habitat is not modeled by municipalities, as fish habitat management is a federal government responsibility in Ontario.

(Dillon and Molot, 1997). Finally, the settling velocity for loss of phosphorus to lake sediments used in the models is an average of local estimates made from calibrated watersheds and lakes with oxic and anoxic hypolimnia (Dillon et al., 1986). In-lake phosphorus retention may vary regionally and from lake to lake. The use of these models must always, therefore, be preceded by calibration to specific local conditions.

Three major assumptions underlie the use of the Ontario Trophic Status model:

1. The first is that 100% of the phosphorus loaded to a shoreline septic system will ultimately be expressed as increased trophic status (Dillon et al., 1986, 1994). This assumption has only been tested indirectly as a function of the fit of predicted with measured phosphorus in study lakes (Hutchinson et al., 1991; Dillon et al., 1994). Recent investigations of septic system geochemistry (Robertson et al., 1998) and the mechanisms of phosphorus mineralization in soils (Isenbeck-Schroter et al., 1993; Jenkins et al., 1971) suggest that this assumption is debatable where soils are present between a septic system and a water body and that 100% phosphorus export is, in fact, unlikely.
2. The second assumption is that all anthropogenic phosphorus sources within 300 m of the lakeshore, or any inflowing tributary, must be included in the lake's phosphorus budget (Dillon and Rigler, 1975; Dillon et al., 1986). Although this assumption is based on the need to place boundaries on the inclusion of phosphorus sources, the distance of 300 m is arbitrary and has neither been substantiated nor tested.
3. Finally, the lakes used to calibrate the Ontario model are small, headwater lakes (Dillon et al., 1986). Although straightforward adaptation of the model to an entire watershed is recommended (Dillon and Rigler, 1975; Dillon et al., 1986), published validations have focussed only on headwater lakes (Dillon et al., 1994) or on simple watersheds with a limited number of lakes (Hutchinson et al., 1991). The additional complexity inherent in expanding the original concept and development allocation schemes to large watersheds must be addressed as part of any policy development process.

These assumptions form the basis of challenges to the concept of determining lakeshore capacity through trophic status (Michalski, 1994) and guided much of the technical recasting of trophic status models reported herein.

In summary, trophic status models have been developed that can accurately predict the responses of lakes to natural and anthropogenic phosphorus sources. The models are regionally specific and must be validated against local conditions before they can be used with confidence. In addition, assumptions guiding anthropogenic loading estimates and watershed implementation are open to some debate on both empirical and mechanistic grounds and should be addressed before beginning any management exercise.

Model validation

The accuracy of predictions made by a trophic status model must be confirmed against measurements of water quality in the subject lakes before the models can be used with confidence in policy setting. This requires:

1. Establishment of a water quality monitoring program to determine existing levels of phosphorus in lakes for comparison against present-day model predictions
2. Maintenance of the monitoring program for the long term to determine any trends in water quality
3. Calibration of natural phosphorus loadings and basic model operation in undeveloped lakes, with no human phosphorus sources
4. Calibration of the model on developed lakes to determine if assumptions on anthropogenic phosphorus sources are valid
5. A process to resolve inaccuracies in model predictions and to update the model on the basis of monitoring results

Model validation must also meet the requirements of the planning process it supports. The intent of a lake management program is to achieve stable and predictable water quality. This must occur in concert with the requirements for a stable and consistent planning and policy environment. Water quality programs should set stable targets for a minimum of 20 years — time to resolve new steady states in water quality in a monitoring program and to provide a stable economic and planning environment, without jeopardizing the resource through over-allocation of development.

Monitoring programs

The monitoring program supporting a lakeshore capacity policy must strike a balance between practical implementation, accuracy, and expense. Ontario's water quality models were developed on the basis of an intensive, long-term research program on a small number of study lakes which was undertaken by the provincial Ministry of the Environment in the Muskoka and Haliburton regions of Ontario. The program was based on dedicated laboratory procedures and analytical staff, long-term personnel, and routine scientific review. In contrast, a municipal program may have to be implemented with limited financial support, summer or term staff, a variety of commercial or government laboratory analyses, lack of in-house expertise, and staff who manage water quality only as one aspect of their career. None of these, on its own, jeopardizes the integrity of a water quality program, but all represent the potential for error and the need for stable policy support. The water quality program must, therefore, be supported by a cost-effective monitoring program that can be maintained for the long-term.

The water quality model is based on predictions of total phosphorus, and so the best comparisons of model accuracy are obtained by measuring total phosphorus directly. Published water quality relationships for Muskoka–Haliburton lakes (Clark and Hutchinson, 1992) suggest that long-term trends in water quality can be determined by making one phosphorus measurement each year at the time of spring overturn, when the lake is completely mixed from top to bottom, thus reducing program costs. Accordingly, an effective water quality monitoring program may consist of:

1. An annual spring overturn measurement of total phosphorus for comparison with water quality predictions made by the model
2. Biweekly measurements of Secchi depth during the summer to track long-term changes in water clarity, the recreational attribute that forms the basis of the water quality program
3. An annual measurement of the dissolved oxygen profiles made at the end of summer, when oxygen stress is most likely, to determine the oxygen status of lakes for model input, to establish the suitability of aquatic habitat, and to track long-term changes

The results of the monitoring program can be used to calibrate the water quality model and to identify those lakes for which the model does not produce accurate estimates of trophic status.

Model calibration

The primary requirement for a water quality model, once accuracy has been demonstrated, is that it be based on a solid mechanistic understanding of watershed and lake dynamics. A purely empirical approach, in which understanding and technical substantiation are ignored and the model "fit" is the only rationale for model acceptance, may not be technically defensible. The coefficients and assumptions that make up the predicted water quality must be clear and documented, as they will form the basis of challenges to the model. Clear technical rationale is also required so that all model users can understand the model and improve it in the future. Wherever possible, the water quality model should be substantiated by reference to the primary scientific literature.

The Ontario Lakeshore Capacity Study calibrated its trophic status model (Dillon et al., 1986) on research lakes that had no shoreline development. This was done in order to quantify basic lake processes in the absence of the additional uncertainty inherent in assumptions of human phosphorus loading from shoreline septic systems. While this is mechanistically sound and represents the ideal approach, it may not be practical for municipalities or other jurisdictions. Water quality programs are focussed on those lakes where shoreline development is present, because they are the lakes that require management activities such as development capacities. Uninhabited lakes are not generally monitored as part of a water quality program, posing difficulties in calibrating basic model elements such as settling velocity, natural watershed loads, or in-lake retention. These factors must be well quantified before considerations of the uncertainties in quantifying human phosphorus loadings.

Where the number of uninhabited lakes in the calibration set is not considered adequate for model development, then the calibration exercise should include lakes in which only limited shoreline development is present. A cutoff where less than 10% of the total potential load is added from shoreline development may suffice in these cases. Measurements of water quality may be available for larger numbers of these "sparsely inhabited" lakes to assist with model calibration.

Calibrating natural phosphorus sources. Consideration of two factors is recommended to improve model fit for natural phosphorus loading.

Incorporation of wetlands to describe natural phosphorus export. Recent research on south-central Ontario watersheds showed that phosphorus export from wetlands (plus atmospheric deposition) determined natural phosphorus loading to a lake. This was stated as the following estimate of phosphorus load (from Dillon and Molot, 1997):

Dissolved Organic Carbon Determines Total Phosphorus

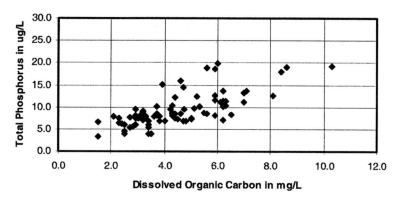

Figure II.17.3 Influence of dissolved organic carbon (DOC) on average long-term total phosphorus concentrations in Precambrian Shield lakes.

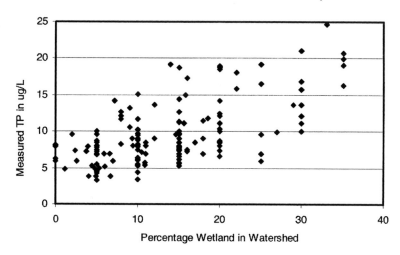

Figure II.17.4 Relationship of average long-term measured phosphorus in Precambrian Shield lakes with wetland area in catchment.

$$\text{kg TP/year} = \text{catchment area (km}^2) * (3.05 + (0.54 * \% \text{ wetland}))$$

This relationship is driven by the export of phosphorus with dissolved organic carbon from wetlands in the catchments of the lakes. This is shown for lakes of the Muskoka River watershed in Figure II.17.3. Total phosphorus concentrations were significantly related ($p < 0.000001$, $r^2 = 0.39$) to the amount of wetland in the catchments of lakes in the Muskoka River watershed (Figure II.17.4.). Natural phosphorus loading from all catchments containing wetland can therefore be estimated from wetland area.

Phosphorus retention in shallow lakes. Both the Dillon–Rigler and Lakeshore Capacity Study models were developed and calibrated for use in lakes that are deep enough to stratify. The models are not intended for use in shallow lakes, in which phosphorus does not settle to hypolimnetic waters. The major requirement for modeling a shallow lake is to devise an accurate estimate of in-lake retention of phosphorus. The models estimate retention as a function of the areal water load (m^3 of runoff/year per m^2 of lake surface) and the settling velocity of phosphorus, as modified by the hypolimnetic oxygen status (Dillon et al., 1986). For shallow lakes, this method overestimates retention as it does not allow for wind-driven resuspension of phosphorus.

Several methods of estimating retention in shallow lakes can be attempted, most of which involve modification of the apparent settling velocity. In the end, shallow lakes are best calibrated individually. In-lake phosphorus retention can be modified to achieve a best fit between measured and predicted estimates of total phosphorus concentration. Calibration of the many shallow lakes present in the watershed, most of which have no shoreline development, provides the final improvement to the model and, together with the other modifications, provides the confidence in model accuracy to allow its use for setting development capacities.

Atmospheric phosphorus loading. Trophic status models must include a phosphorus contribution from the atmosphere directly to the lake surface. Atmospheric loading includes contributions from precipitation and from dry loading (or dust). Values such as 20.7 $mg/m^2/year$ for the Precambrian Shield in south-central Ontario can be obtained from published monitoring studies, such as those of the Ontario Ministry of the Environment (Dillon et al., 1992).

Hydrologic loading. Trophic status models require input of hydrologic loading for each lake. The models are long-term steady-state models and so are cast to use average annual depth of runoff. These values can be obtained from published figures (Canada Dept. of Fisheries and Environ., 1978). They can then be coded into the model in a "lookup" table and determined for each lake on the basis of latitude and longitude of the lake, as entered by the user. They should be reviewed periodically, however, in light of the potential changes in runoff stimulated by climate change.

Calibrating anthropogenic phosphorus sources. Consideration of wetlands and shallow lakes improves the fit of the model so that measured and modeled phosphorus concentrations do not vary in undeveloped lakes. Review of model results for developed lakes showed that the measured phosphorus concentrations were lower than the concentrations predicted by the model but higher than those predicted for a case where no development was present. From this, it can be concluded that some portion of the potential anthropogenic phosphorus load is being expressed in the lake but that the model must be recalibrated to account for the portion of the development load that is not expressed. Several modifications are required to reduce the large positive bias (overprediction of measured phosphorus) observed in developed lakes.

Incorporation of phosphorus retention by soils. The water quality model for lakes in the Muskoka River watershed incorporates a substantial departure from other Ontario models in its geochemical assumptions regarding phosphorus movement from septic systems to lakes. Both the original Dillon and Rigler (1975) and Ontario models (Dillon et al., 1986) assumed that all septic system phosphorus generated within 300 m of the shoreline would ultimately migrate to the lake. This assumption may be considered reasonable as a conservative approach but has never been tested directly.

Since the publication of the original models, direct monitoring studies and mechanistic understanding of soil and phosphate interactions have provided evidence that conflicts with the original assumptions. Mechanistic evidence (Stumm and Morgan, 1970; Jenkins

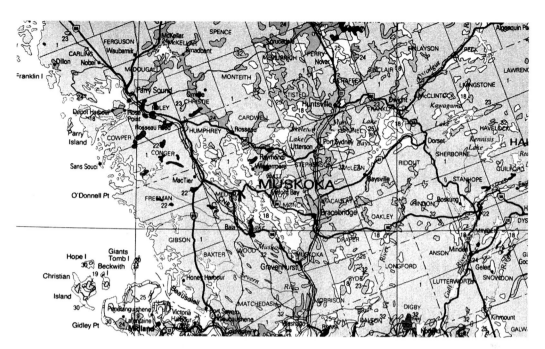

Figure II.17.5 Quaternary Geology of Muskoka, Ontario (from Barnett et al., 1991). Pink areas (1) denote igneous and metamorphic bedrock, which is exposed or covered with thin drift. Blue areas (24) are glaciolacustrine deposits of clay and silt. Remaining colors are glaciolacustrine deposits of outwash and till, sand and gravel (18, 23, 25) or organic peat and muck (32).

et al., 1971; Isenbeck-Schroter et al., 1993) and direct observations made in septic systems (Willman et al., 1981; Zanini et al., 1997; Robertson et al., 1998) all show strong adsorption of phosphate on charged soil surfaces and mineralization of phosphate with Fe and Al in soil. The mineralization reactions, in particular, appear to be favored in acidic and mineral-rich groundwater in Precambrian Shield settings (Robertson et al., 1998), such that over 90% of septic phosphorus may be immobilized. The mineralization reactions appear to be permanent (Isenbeck-Schroter et al., 1993), and direct observations suggest that most septic phosphorus may be stable within 0.5 m of the tile drains in a septic field on the Precambrian Shield (Robertson et al., 1998).

The mechanistic and geochemical evidence is supported, in part, by trophic status modeling. Dillon et al. (1994) reported that only 26% of the potential loading of phosphorus from septic systems around Harp Lake, Muskoka, could be accounted for as measured phosphorus in the lake. The authors attributed the variance between measured and modeled estimates of phosphorus to retention of septic phosphorus in thick tills in the catchment of Harp Lake. Although the Muskoka watershed is frequently characterized as an area of thin to no soils over bedrock, this description is in no way universal. The central corridor of the watershed (in which Harp Lake is located) occupies a glacial outwash plain of alluvial sands and gravels, and many catchments contain substantial soil deposits. Western and southwestern Muskoka represent the more typical topography of thin soils and granite ridges and outcrops (Figure II.17.5). Even in these areas, however, tile fields are often, by necessity, built on imported fill and so some attenuation is possible.

Revisions to trophic status models should use the findings of these recent studies to improve the positive bias (i.e., overprediction of measured phosphorus) in the model by accounting for a 74% retention of septic phosphorus for those lakes with suitable soils in their catchments (Dillon et al., 1994). The positive bias is apparent in model results for all developed lakes, but most pronounced in heavily developed lakes. All of the study lakes can be located upon a map of surficial geology (i.e., Southern Ontario Engineering Geology Terrain Study; Mollard, 1980). Those lakes situated within the same types of thick till as are found around Harp Lake, and those lakes situated within outwash or alluvial plains can also be assumed to retain 74% of their septic phosphorus in soils without expression in the lake. The model should therefore be coded to reduce the septic phosphorus contribution by 74% for those lakes. Soil classifications (from Mollard, 1980) for which 74% retention can be assumed are:

MG/R = ground moraine over bedrock
LD = glaciolacustrine delta
GO = outwash plain

Adoption of soil retention provides substantial improvement to the fit between measured and modeled estimates of phosphorus concentration in recreational lakes. The estimate of 74% may still be considered conservative, as Robertson et al. (1998) and Wood (1993) both reported retentions well in excess of 90% for septic systems located in the Precambrian Shield in general and at Harp Lake, specifically.

Distance of development from shorelines. Much of the recreational shoreline development in Ontario is located in one tier around a lakeshore, with a minimum setback of 15 to 30 m between the septic system and the water's edge. As development intensifies, however, and prime shoreline space is occupied, development may occur in a second or third tier back from the lakeshore, with water access from communal docking or bathing facilities. Although it is reasonable to assume that second- and third-tier development will have some influence on water quality, it is unlikely that it will have the same influence as development located on the shoreline. The water quality model, therefore, must find a means of accommodating distance of development from a lake while still including some contribution of phosphorus to the lake.

The Dillon–Rigler and Ontario models both adopted a convention of assuming that 100% of the septic system phosphorus generated within 300 m of a lake would ultimately migrate to the lake. Although it is a useful figure to define the limits of a modeling exercise, it is very difficult to defend technically, given the knowledge of phosphorus geochemistry described above. It also leads to counterintuitive interpretations, in which a septic system located 299 m from the shore has a 100% impact, while one located 301 m back has no impact. The absence of substantiation or a mechanistic basis for the assumption, and the complexity involved in site-specific derivation for a watershed modeling exercise, make it very difficult, however, to propose a valid alternative approach.

Setback of development can be accommodated by modifying the assumption of phosphorus contribution with distance; for example, by coding the model so that:

1. Development within 100 m of the shoreline provides 100% contribution of septic phosphorus (as modified by soil thickness and type, see above)
2. Development between 100 and 200 m has its phosphorus contribution reduced by one third
3. Development between 200 and 300 m has its phosphorus contribution reduced by two thirds
4. Development beyond 300 m has no phosphorus contribution

Although this approach still suffers from arbitrary distinctions it does accommodate the concept of distance of development from the shoreline and phosphorus attenuation by soils.

Table II.17.1 Usage of Shoreline Residences in the District Municipality of Muskoka

Zone 1: Outlying area	0.82
Zone 2: Close to major highway	1.23
Zone 3: Close to major urban center	2.09
Resort unit usage	1.23
Trailer and camp sites	0.41

All values are given in capita years per year for each residential type.

Revised per-capita phosphorus contribution. The original Dillon–Rigler and Ontario models used a figure of 800 gm/C/year as an estimate of per-capita contributions of phosphorus to septic systems from human waste and household cleaning products. This figure was originally derived, in part, from measurements of total phosphorus in septic tanks (Dillon et al., 1986) made between 1965 and 1980. Phosphorus concentrations ranged from 5 to 21.8 mg/L, with a mean of 13.2 mg/L. More recent research conducted by the Ontario Ministry of the Environment. (Gartner Lee Limited, 2002, in preparation) found a range of 4.3 to 13.3 mg/L of total phosphorus in septic tanks serving shoreline residences and an overall average of 8.2 mg/L. The more recent values reflect the limitation of phosphate in laundry detergent in the early 1970s and represent 62% of the phosphorus concentrations used in Ontario's Lakeshore Capacity Study. Strict application of the reduced concentration produces an estimate of 500 gm/C/year (i.e., 800 * 0.62). No figure will be completely accurate, however, and so 100 gm/year was added to the measured value in order to maintain a protective and conservative approach to estimating phosphorus loadings. A water quality model should therefore consider a per-capita phosphorus contribution of 600 gm/year to the septic system.

Shoreline development also adds phosphorus to a lake from the conversion of a forested landscape by clearing and lawn planting, and hardening of soils. Previous Ontario models did not include a contribution from these sources. A model should account for this clearing by including an estimate of 2000 m^2 for the average size of the developed portion of each shoreline lot and an increased export coefficient of 17 mg/m^2/year (= 34 gm/lot/year) from those areas.

Validation of cottage usage figures. The final requirement for estimating human phosphorus loadings from shoreline development is to obtain estimates of the number of days that residences are used in a year. Seasonally occupied residences will be occupied for fewer days per year than permanent homes, and different usage figures apply to resort units or trailer sites. In the 1970s, the Ontario Lakeshore Capacity Study estimated seasonal and permanent usage of shoreline residences figures as 0.89 and 2.55 capita years/year, respectively, on the basis of a cottage survey conducted in the Muskoka–Haliburton region of Ontario (Downing, 1986). Twenty years later, the District Municipality of Muskoka (1995) undertook a similar study and determined that the overall lot usage had not changed substantially (Table II.17.1). Usage figures did not support the commonly held perception that large numbers of cottages were converting from seasonal to permanent use to accommodate retirees or "telecommuters." Only two lakes had high proportions of permanent residents, and these were close to a major urban center (the town of Huntsville). Regionally specific lot usage surveys should accompany development of lake trophic status models to ensure accuracy.

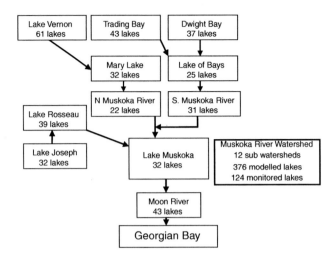

Figure II.17.6 Schematic of Muskoka River Watershed.

Other phosphorus sources. The model should also include anthropogenic sources of phosphorus from other activities in the watershed:

1. Loadings from sewage treatment plants which can be obtained from operating records.
2. Areas of agricultural land use can be determined from air photo interpretation. A coefficient of 45 mg/m^2/year can be used to describe export from agricultural land use (Winter and Duthie, 2000).
3. Urban runoff of 45 mg/m^2/year for those portions of urban areas within 300 m of a lake or a river. These loadings are not reduced with setback distance, as urban areas are hardened and offer little opportunity for attenuation.

Final model results

The Muskoka River drains a watershed with a total area of 5,737 km^2. The watershed model is built around 12 subwatersheds of the Muskoka River, as defined by the major river or lake at the downstream end of each (Figure II.17.6). The hydrologic and nutrient output from each subwatershed enters the next subwatershed downstream, from the western fringes of Algonquin Park at the furthest extent upstream, to Georgian Bay of Lake Huron. The total area of surface water included in the model is 56,000 ha, and the size of modeled lake segments ranges from 3 to 6,200 ha.

The model is both large and complex. It provides explicit characteristics and model output for 376 individual lakes, rivers, or portions of each. The model contains nearly 12,500 individual calculations or data points to calculate total phosphorus concentrations, and approximately 15,000 additional calculations to estimate and track development capacities.

Model accuracy

The final version of the Muskoka Watershed water quality model produced excellent estimates of total phosphorus concentrations in 123 lakes for which monitoring data were available. Mean negative bias in predictions (i.e., underestimate of measured water quality) occurred in 3 lakes and was within 1% of measured values (Table II.17.2). The median

Table II.17.2 Summary Statistics Showing the Percentage Agreement between Modeled and Measured Estimates of Water Quality in the Muskoka Watershed Model

	All Lakes		D.I. < 1.21		D.I. ≤ 1.11	
	+Error	−Error	+Error	−Error	+Error	−Error
Average error	14.74	−0.84	5.89	−0.28	2.83	−0.28
Median error	8.08	−0.51	2.81	−0.28	2.34	−0.28
No. lakes	120	3	47	1	32	1
No. > 40% error	9	0	1	0	0	0

overestimate was 8.1% in 120 lakes. The average overestimate of 14.7% included 9 lakes for which the bias exceeded 40%. The criterion of 40% was chosen as the mean coefficient of variation in measured phosphorus concentration for all lakes. Model errors in excess of 40% were considered unacceptable as they were outside the range of natural variance in water quality. These errors only occurred as a result of positive bias, however, as an indication that not all of the potential human phosphorus load modeled was actually expressed in the lake.

Model accuracy was further expressed by comparison of model error with the proportion of the total phosphorus load contributed by human sources. These sources were described using the "development index" (DI), which was the ratio of total phosphorus load to potential anthropogenic phosphorus load for each lake. A DI of 1.0 represents a lake with no anthropogenic loading, a DI of 1.5 denotes a 50% increase, 2.0 a doubling, and so on.

For 32 "undeveloped" lakes (DI < 1.11) the average overprediction was <3%. Positive bias increased with development intensity such that median and average error were 2.8 and 5.9%, respectively, for 47 moderately developed lakes (DI < 1.21, Table II.17.2). A scatterplot comparison of measured and modeled phosphorus concentrations (Figure II.17.7) shows a) the good correspondence between the two and (b) the tendency for the model to overpredict phosphorus concentration. The positive bias persisted, even after accounting for attenuation of septic phosphorus (see Model Validation section above).

The absolute difference between measured and modeled phosphorus was <1 μg/L for 75 lakes (Figure II.17.8), and error for 77% of the lakes was within 2 μg/L. All of these errors represent positive bias, however, such that the model has a strong tendency to overestimate phosphorus concentrations in developed lakes.

The positive model bias was related to the estimate of phosphorus loads from shoreline development. Median and average positive bias increased with the intensity of development (DI, Table II.17.2), but negative model bias did not change. Positive bias in excess of 10% was confined to developed lakes. The trend to overprediction appeared on undeveloped and sparsely developed lakes but was less than 5% and likely reflected model variance.

The positive bias was reduced by accounting for the retention of phosphorus in soils (see Model Validation). Average and median errors were 9.7 and 5.7%, respectively, for the 66 lakes in which 74% phosphorus retention was assumed on account of the soil characteristics. Error only exceeded 40% on two of these lakes. For lakes where retentive soil mantles were not assumed in the model, average and median error increased to 21.2 and 13.0%, respectively, and 8 lakes had positive bias above 40%. This suggests that some retention may be occurring around all lakes and not just those with thick soil mantles in their catchments.

In spite of the positive bias in model results, it is clear that shoreline development has influenced the phosphorus concentrations in lakes to some extent. Phosphorus concentrations have increased with development, but by far less than predicted on the basis of an

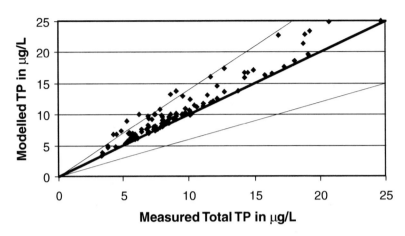

Figure II.17.7 Accuracy of water quality predictions made with final version of the Muskoka watershed water quality model. Thick line shows perfect fit (1:1), and narrow lines enclose accuracies of ±40%.

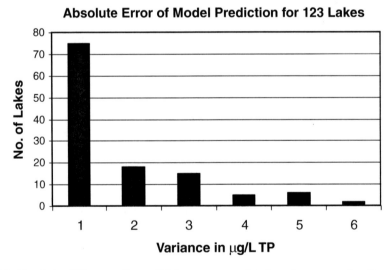

Figure II.17.8 Absolute difference (in µg/L) between modeled and measured estimates of total phosphorus concentrations in Precambrian Shield lakes.

assumption that 100% of all septic phosphorus is mobile. This observation was supported by the modeling exercise, by the detailed observations of phosphate mobility in septic systems summarized in Robertson et al. (1998) and by geochemical descriptions of phosphate behavior (Isenbeck-Schroter et al., 1993; Stumm and Morgan, 1970). The setting of water quality objectives and development limits must account for these understandings in phosphate behavior and be aware that significant overestimates of the impacts of shoreline development may hinder the implementation and defense of water quality-based development limits.

Setting water quality objectives and development limits

The intent of a water quality program is to use the monitoring and modeling exercises to support water quality-based shoreline development capacities. Some review of approaches to setting phosphorus objectives is therefore warranted.

Surface-water management in Ontario

The Ontario Ministry of the Environment (MOE) manages environmental quality primarily through two pieces of provincial legislation: The Environmental Protection Act and The Ontario Water Resources Act. Policies and procedures for management of surface water quality that arise from this legislation are elaborated in implementation documents such as *Water Management: Policies, Guidelines, Provincial Water Quality Objectives of the Ministry of Environment and Energy* (OMEE, 1994). The goal of surface water management in Ontario is *"to ensure that the surface waters of the province are of a quality which is satisfactory for aquatic life and recreation"* (OMEE, 1994).

Ontario established Provincial Water Quality Objectives (PWQOs) in the 1970s in order to meet this goal. The first objectives were mostly adopted from other agencies, such as The International Joint Commission, but were later developed in Ontario (OMEE, 1992).

> *"Provincial Water Quality Objectives (PWQOs) are numerical and narrative ambient surface water quality criteria. They are applicable to all waters of the province (e.g., lakes, rivers and streams) except in those areas influenced by MOEE approved point source discharges. In specific instances where groundwater is discharged to surface waters, PWQOs may also be applied to the groundwater. PWQOs represent a desirable level of water quality that the MOEE strives to maintain in the surface waters of the province. In accordance with the goals and policies in Water Management (OMEE, 1994), PWQOs are set at a level of water quality which is protective of all forms of aquatic life and all aspects of the aquatic life cycle during indefinite exposure to the water. The Objectives for protection of recreational water uses are based on public health and aesthetic considerations"* (OMEE, 1994).

Two policies are used to interpret the water management goal and application of the PWQOs to specific water bodies (OMEE, 1994).

> *"Policy 1: In areas which have water quality better than the Provincial Water Quality Objectives, water quality shall be maintained at or above the Objectives. Although some lowering of water quality is permissible in these areas, degradation below the Provincial Water Quality Objectives will not be allowed, ensuring continuing protection of aquatic communities and recreational uses.*
>
> *Policy 2: Water quality which presently does not meet the Provincial Water Quality Objectives shall not be further degraded and all practical measures shall be taken to upgrade the water quality to the Objectives."*

Municipal responsibilities for water quality in recreational lakes

The planning system in Ontario is established in the Planning Act, which provides the legislative jurisdiction for municipalities to monitor and regulate land use subject to policy provided in the 1997 Provincial Policy Statement. The Planning Act requires that in reviewing any planning application, a municipality must have regard to provincial policy. The provincial policy respecting water quality states:

> *"The quality and quantity of groundwater and surface water and the function of sensitive groundwater recharge/discharge areas, aquifers and headwaters will be protected or enhanced."*

In practice, municipal policy for water quality protection is interpreted against Ontario PWQOs, and against Policies 1 and 2 for protection of surface water quality, as outlined above. Municipalities are allowed to be flexible in their specific interpretations but must meet the intent of provincial policy. Municipalities can make use of trophic status models to develop shoreline development objectives, which can then be lodged in official plan documents.

Where a development objective exists, policy may then direct that new development shall not result in the predicted nutrient level exceeding the objective for that particular lake. Where a lake is at capacity, development or lot creation may only be permitted in limited circumstance, and in particular where:

1. There is an existing vacant lot of record.
2. The redevelopment of the property would not result in an increase of phosphorus loading to the water body.
3. The septic system and leaching bed can be set back from the waterbody by 300 m.
4. The septic system and leaching bed can be placed in another watershed that is not at capacity

More general policy should also encourage the maintenance of shoreline vegetation, the restoration and preservation of the waterfront shoreline where it has been artificially altered, and aesthetic factors such as setback, footprint or coverage of built structures. Site Plan Control is a planning instrument which can be used to implement site-specific development details.

Ontario's Planning Act requires that municipal official plans be reviewed from time to time to ensure that they continue to provide appropriate policy direction. Such review should include the need to update water quality objectives and lake models to ensure that the most recent scientific understanding of nutrient loading is properly incorporated.

Ontario's existing PWQO for total phosphorus

The existing PWQO for total phosphorus was developed in the late 1970s (OMEE, 1979). It drew on the trophic status classification scheme of Dillon and Rigler (1975) to protect against aesthetic deterioration and nuisance concentrations of algae in lakes, and excessive plant growth in rivers and streams. The rationale (OMEE, 1979) acknowledges that elemental phosphorus can be toxic but that it is rare in nature and so toxicity is rarely of concern. (In fact, there is only one documented case of elemental phosphorus poisoning an aquatic (marine) system in Canada.) Instead, the purpose of the objective was to protect the aquatic ecosystem from nontoxic forms of phosphorus: *"phosphorus must be controlled, however, to prevent any undesirable changes in the aquatic ecosystem due to increased algal growth ..."* (OMEE, 1979).

The 1979 PWQO was given the status of "guideline" to reflect the uncertainty about the effects of phosphorus and to acknowledge the difference between managing toxic and nontoxic pollutants.

> "Current scientific evidence is insufficient to develop a firm objective at this time. Accordingly, the following phosphorus concentrations should be considered as general guidelines which should be supplemented by site-specific studies:
>
> To avoid nuisance concentrations of algae in lakes, average total phosphorus concentrations for the ice-free period should not exceed 20 µg/L;
>
> A high level of protection against aesthetic deterioration will be provided by a total phosphorus concentration for the ice-free period of 10 µg/L or less. This should apply to all lakes naturally below this value;
>
> Excessive plant growth in rivers and streams should be eliminated at a total phosphorus concentration below 30 µg/L."

Total phosphorus and the PWQO development process

There are several shortcomings with Ontario's existing PWQO for total phosphorus and the province has been reviewing its approach to phosphorus management. The approach is derived from that first proposed in Hutchinson et al. (1991), and elements of it are summarized in this section. These can be considered in cases where a municipality wishes to derive its own phosphorus objectives to assist with managing shoreline development.

Phosphorus as a pollutant

Development of a water quality objective for total phosphorus is distinctly different from that for toxic substances. Most aquatic pollutants are directly toxic to some target tissue, such as the fish gill, even if some of them are required nutrients at trace amounts, i.e., copper or zinc. As a result, the health of aquatic organisms, and hence the ecosystem, declines rapidly at concentrations slightly above ambient levels (Figure II.17.9). Phosphorus, on the other hand, is a major nutrient. Concentrations can increase substantially with no direct toxic effects. In fact, the first response of the aquatic system is increased productivity and biomass. Beyond a certain point, however, indirect detrimental effects become apparent, which ultimately decrease system health.

The first detrimental responses of a lake to enrichment (i.e., water clarity, algal blooms) are aesthetic and of concern mostly to humans. Assessment of aesthetic impacts is highly subjective; perceived changes in water clarity are based largely on what one is used to (Heiskary and Walker, 1988). The development of a phosphorus objective must therefore acknowledge an element of subjectivity in dealing with human concerns. The objective-development process may also consider that aesthetic impacts begin where a change in water clarity is first noticeable to the human eye, or where the mean water clarity first exceeds natural variation. Unfortunately, human perception of water clarity has not been established. Existing guidelines are based on trophic status classification schemes. They do not consider other water clarity influences such as inorganic turbidity or dissolved organic carbon or how lake users perceive changes in water clarity.

Finally, trophic status indicators such as water clarity, chlorophyll "a," or dissolved oxygen cannot be managed directly, but only through management of phosphorus. In addition, there may be delays of up to decades between the addition of phosphorus sources

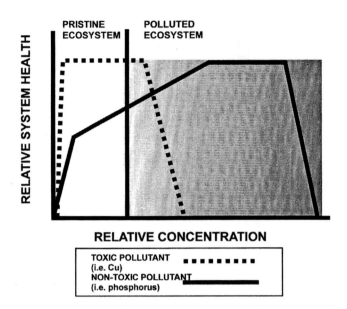

Figure II.17.9 Generalized responses of an ecosystem to toxic and nontoxic pollutants.

to a watershed (i.e., septic systems), its movement from the source to surface water (Robertson, 1998), and its expression as a change in trophic status. As a result, phosphorus management in Ontario requires the extensive use of models relating shoreline development to the trophic status of the receiving water. Phosphorus management may therefore be considered as a process of "predicting the predictor." Previous sections of this chapter have emphasized the importance of validating or accounting for these assumptions.

The shortcomings of a numeric phosphorus objective

Ontario's existing two-tiered numeric objectives for total phosphorus obscure fundamental differences between lake types and their nutrient status in the absence of human impact. Most Precambrian Shield lakes are characterized by excellent water quality as represented by low concentrations of total phosphorus. For example, the average of the mean annual measured concentration for 123 lakes in the Muskoka watershed was 9.3 ±4.2 µg/L and 62% of the lakes averaged 10 µg/L or less (Figure II.17.10). Within any set of lakes, however, there is still a large diversity of water clarity, controlled by both total phosphorus concentrations and dissolved organic carbon.

The provincial phosphorus objective allows lakes that currently contain less than 10 or 20 µg/L of phosphorus to increase to a maximum of 10 or 20 µg/L. The logical outcome of this two-tiered objective is that, over time, all recreational waters would converge on one or the other of the water quality objectives. This would produce a cluster of lakes slightly below 10 µg/L and another slightly below 20 µg/L, decreasing the existing diversity in water quality in lakes and, with it, the diversity of their associated aquatic communities.

The second shortcoming is that, over time, some lakes would sustain unacceptable changes in water quality while others would be unimpacted, producing both ecological and economic asymmetries as the resource was developed. A lake with a natural phosphorus concentration of 4 µg/L is a fundamentally different lake than one that exists at 9 µg/L. Both lakes, however, would be allowed to increase to 10 µg/L under the existing PWQO. One lake would experience no perceptible change (9 to 10 µg/L) and be overprotected, but the other (4 to 10 µg/L) would be underprotected and change dramatically. In both cases, human perceptions of aesthetics are ignored in the objective. Allocation of phosphorus loadings between these two lakes would be unfair as well. The higher-phosphorus lake could

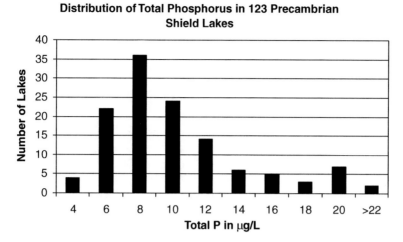

Figure II.17.10 Distribution of total phosphorus measurements in 123 lakes in the Muskoka River watershed, 1990–1998.

sustain a greater change than the low-phosphorus lake but would be restrained to a much lower load.

In summary, the existing two-tiered numeric objectives overprotect some lakes and do not protect others adequately. Allocation of phosphorus loadings is unnecessarily restricted in some lakes and overly generous in others. Neither biotic nor aesthetic attributes are adequately protected. Over time the diversity of trophic status presently represented in Ontario will decrease as lakes converge on one of two numeric objectives.

Environmental baselines and measured water quality

An emerging concern in environmental assessment is the need for a standard baseline for comparison against environmental change. Inland lakes respond quickly to point source phosphorus inputs. Detection of change is much more difficult, however, for nonpoint sources such as leachate from domestic septic systems.

The incremental nature of shoreline development (no lake is ever developed all at once) results in a slow and gradual increase in trophic status. The high degree of seasonal and annual variance in phosphorus levels in lakes (Clark and Hutchinson, 1992) means that changes may not be detectable without an intensive monitoring program, based on many samples and a precise and replicable analytical method.

Human observers may not notice a slow increase in trophic status over a generation. Environmental change that occurs over one generation becomes the status quo for the next. Over a long period, therefore, any assessment baseline based on measurements of total phosphorus will increase.

Any phosphorus objective that relies exclusively on measured water quality will therefore suffer from:

- Detection problems due to natural variance and analytical problems
- The lag time between addition of phosphorus to a watershed and its expression in a lake
- Failure to detect incremental changes in water quality
- Human perceptual conditioning, which reduces the apparent change in water quality over time

As a result, an increasing assessment baseline and incremental increases in nutrient loading may slowly degrade water quality past any objective. Impacts will accumulate by virtue of delay in their expression, repetition over time and space, extension of the impact boundary by downstream transport, or by triggering indirect changes in the system, such as anoxic sediment release. Nonpoint source phosphorus pollution, particularly from septic systems serving shoreline development, is thus an excellent example of a pollutant that may produce cumulative impacts to the aquatic environment.

An ecologically sound approach to objective setting

The emergence and validation of mass balance phosphorus models for lakes offers an opportunity to correct some of the disadvantages of water quality measurements and conventional assessment techniques. In the past, these models were used to establish the amounts of development that would maintain trophic status within the numeric objectives. They were used "after the fact," to implement a water quality objective. The recommended approach would be to use the water quality model itself to set lake-specific phosphorus objectives.

The merits of a modeled assessment baseline

The basis of the revised approach is increased reliance on water quality modeling in the objective setting process. Recent advances in trophic status models allow us to calculate the "predevelopment" phosphorus concentrations of inland lakes (Hutchinson et al., 1991). This is done by:

1. Modeling the total phosphorus budget for the lake
2. Comparing the predicted concentration to a reliable water quality measurement to validate the modeled result
3. Subtracting that portion of the budget attributable to human activities

The main advantage of the modeling approach is establishment of a constant assessment baseline. A modeled "predevelopment" baseline is based on an undeveloped watershed and so will not change over time. This serves as the starting point for all future assessments. Every generation of water quality managers will therefore have the same starting point for their decisions, instead of a steadily increasing baseline of phosphorus measurements.

The new approach therefore proposes phosphorus objectives based on modeled "predevelopment" phosphorus concentrations. This will provide water quality managers with:

1. A constant assessment baseline
2. A buffer against incremental loss of water quality
3. A buffer against variable water quality measurements

The "predevelopment" phosphorus concentration should not be interpreted as the objective itself. Pristine or "predevelopment" phosphorus levels have not existed in Ontario's Precambrian Shield lakes for over a century, and their attainment is not cost-effective on heavily developed lakes. The modeled "predevelopment" concentration only serves as the starting point for the objective and a reference point for future changes.

The degree of lake development can also guide the selection of a modeled baseline for use in objective derivation. The recreational settlement of lakes in Ontario began in the late 1800s and was very well advanced when trophic status models emerged 100 years later. There was no way to confirm how much of the phosphorus from a century of development was already being expressed in the lakes, and new development was added every year. Measured phosphorus concentrations were therefore higher than the natural baseline but potentially lower than their final, steady-state levels. The natural variability in phosphorus concentrations, and the lag time before the expression of septic-derived phosphorus (as discussed above), prevents the use of measured values as a planning baseline. Modeled, predevelopment baselines of phosphorus concentrations, validated against local lakes which are not yet developed, are therefore recommended as the starting point for objective development.

A model-based objective has two additional advantages. First, the modeled response of the watershed to future changes is instantaneous. It applies new development directly against capacity, without the intervening decades it takes for phosphorus to move to a lake and be expressed as a measured change in water quality (this approach, however, also requires assumptions on the ultimate mobility of phosphorus, which may not be valid (see Calibrating Anthropogenic Phosphorus Sources)). Second, the trophic status model is based on entire watersheds and so allows explicit consideration of downstream phosphorus transport in the assessment.

One disadvantage of the model-based baseline, however, lies with the inevitable changes in scientific understanding of lakes and watersheds. Any baseline derived from a water quality model is therefore subject to change as improved understanding or refinement produces changes in export coefficients, atmospheric deposition, or quantification of in-lake dynamics. For this reason, a modeling exercise must begin with locally validated coefficients and calibration of the water quality model. Model improvements should also be implemented within a defined schedule of Official Plan review, so that scientific understanding is incorporated at a schedule consistent with measured responses of lakes and the planning process.

The merits of a proportional increase

The second component of the objective is a proportional increase from the modeled predevelopment baseline. The proportional increase accommodates regional variation in natural or "background" water quality through the use of one numeric objective for all lakes. It is, in fact, a broader, yet simpler, application of the regionally specific, multitiered objectives proposed in other jurisdictions as a means of accommodating regional variation in background water quality (e.g., Heiskary and Walker, 1988).

One consideration is to adopt an allowable phosphorus increase of 50% above the modeled predevelopment level from anthropogenic phosphorus sources (Hutchinson et al. 1991). This approach is being considered by the Province of Ontario. Under this proposal, a lake modeled to a "predevelopment" phosphorus concentration of 4 µg/L would be allowed to increase to 6 µg/L from shoreline development or other human activities. Predevelopment concentrations of 6, 10, or 12 µg/L would increase to 9, 15, or 18 µg/L, respectively. A cap at 20 µg/L would still be maintained to protect against nuisance algal blooms.

There are numerous advantages to this approach.

1. Each waterbody would have its own water quality objective, but this could be described with one number (i.e., predevelopment plus 50%).
2. Development capacity and ultimate phosphorus status would be proportional to a lake's original trophic status.
3. As a result, each lake would maintain its original trophic status classification. A 4-µg/L lake would be developed to 6 µg/L and therefore maintain its distinction as oligotrophic. A 9-µg/L lake would be developed to 13.5 µg/L, would maintain its mesotrophic status, and development would not be unnecessarily constrained to 10 µg/L.

4. The existing diversity of trophic status in a region such as Ontario would therefore be maintained over the long term, instead of the ultimate outcome of a set of lakes at 10 µg/L and another at 20 µg/L.

Alternative water quality endpoints and objectives

Several other potential water quality objectives can also be considered in order to establish reasonable and defensible recommendations for development limits in recreational lakes. There is no one obvious "best" water quality objective. The uncertainty in objective setting for phosphorus and its status as a nontoxic pollutant (see Phosphorus as a Pollutant) mean that there will be an element of subjectivity in any figure that is ultimately used.

Background + 40%. One refinement could consider the "predevelopment" background phosphorus concentration plus an increase corresponding to the variance in phosphorus concentrations detected by a monitoring program. The average of the mean annual measured concentration for 123 lakes in the Muskoka River watershed was 9.3 ± 4.2 µg/L. Interannual variance in the mean total phosphorus concentration can be substantial, however, and ranged from 10 to 125% in individual lakes. The mean coefficient of variation for annual phosphorus measurements was ± 40% or 4 µg/L. A routine water quality monitoring program is, therefore, capable of determining the mean annual phosphorus concentrations in lakes to within 40% of the true value in any given year. More intensive, research-level programs, however, can detect changes of ± 20% between years (Hutchinson et al., 1991; Clark and Hutchinson, 1992), but this level of precision is not always attainable in routine programs.

This variance of ± 40% represents the ability of a routine monitoring program to detect changes and the natural variance in runoff or biotic factors from one year to the next. Any changes of 40% or less may not, therefore, be detectable over the long term, would be perceived as routine variance in water quality by lake users, and so may represent an objective based on the ability to detect change.

Background + 4 µg/L. A standard increase of 4 µg/L of phosphorus from background may also be considered and modeled as an objective. The figure of 4 µg/L represents the natural variance in phosphorus concentration determined by a routine monitoring program but is expressed as an absolute limit instead of a proportional change. This approach is not recommended, however, because the "proportional" increase (40 or 50%) provides a fairer allocation of development to a diversity of lakes. An absolute increase of 4-µg/L allows a doubling of phosphorus from development in nutrient-poor waters (i.e., 4–8 µg/L) but only a 25% increase for mesotrophic waters (i.e., 16–20 µg/L).

Existing + 10%. Reviews of the model accuracy showed that much of the septic phosphorus loading predicted by the model was not expressed as increases in the measured phosphorus concentration in lakes. A final approach would acknowledge this lack of response but still provide a margin of error for future changes. In this scenario, the existing measured phosphorus concentration would be used as a baseline and development allocated to a modeled increase of 10% higher than the measured mean. The small increase would acknowledge the uncertainty in the assumptions of phosphorus migration from septic fields so that development would proceed cautiously. Changes in the measured response would be reviewed against monitoring results every 10 years. If lakes began to respond to shoreline development, then the objectives could be reviewed against the nature and degree of response and development limits revised accordingly. Although this approach represents an adaptable response to uncertainty in model assumptions and maintains a proportional increase, it does not provide the desired attribute of a stable baseline over time.

Background + 50%. The Province of Ontario has considered a revised water quality objective for total phosphorus in surface waters which is based on a 50% increase in anthropogenic loadings above the modelled natural background. Although the figure of 50% can be debated, it does reflect the merits of a proportional increase and a modelled baseline, as discussed previously. Municipalities must consider Provincial Policy, and so an objective of "Background + 50%" is used here to illustrate the implications of a potential Provincial Water Quality Objective.

"Filters" and water quality objectives

A water quality objective is not the only determinant of development capacity for lakes. Other physical factors can be considered as "filters," additional constraints to development that will modify any numeric objective developed for water quality. They apply equally to any water quality objective. Review of water quality objectives against other development filters helps to determine which aspects of a lake are most limiting to development and to place the water quality objective in the context of other capacity determinants.

Perimeter filter. The first such filter is shoreline perimeter, or the availability of waterfront lots based on physical constraints to development. The amount of lakeshore is limiting for any lake; some Ontario municipalities, for example, require a minimum lot frontage of 200 ft for new lots. Other municipalities may adopt larger or smaller lot sizes as a response to narrow embayments or other biophysical limits. Sensitive wetland areas, steep topography, or lack of soil may impose additional physical constraints that exclude portions of a lakeshore from development or require larger lot frontages.

In many cases, particularly lakes at the low end of the watershed, perimeter may be a more restrictive development limit than water quality. Total shoreline perimeter for each lake in the model can be determined using a geographic information system (GIS). The perimeter is then divided into 200-ft lots to provide an estimate of maximum shoreline available for development. This presents an overestimate, however, as steep shorelines or other physical constraints to development may further reduce the number of developable lots. These must be assessed on a lake-by-lake basis.

Crown land filter. A second physical filter possibly reducing development potential is consideration of "Crown" land. Many lakes in Ontario are surrounded by "Crown" land, publicly owned land managed by the province in the name of Her Majesty, the Queen. These lands are not developed at present and cannot be subdivided by private interests unless their status is revised by the province. This has been an uncommon occurrence in southern Ontario and is unlikely to occur in the immediate planning horizon. Modeling of all lakes in a watershed to the water quality objective of "Background + 50%" (including Crown land lakes) thus overestimates their ultimate phosphorus loading. In a watershed-based water quality model, the phosphorus loading from these lakes is accumulated in the anthropogenic load for downstream lakes, thus restricting future development. Because Crown lands are unlikely to be developed (and any consideration will involve extensive consultation between the Crown and municipal governments), the potential future load from shoreline development on these lakes can be removed from the modeling exercise as an additional limit or "filter" on development.

Vacant lot filter. Many lakes contain vacant lots on their shorelines. These lots have been legally created but have not yet been developed. Owners of these lots retain the legal right to build on them at any time in the future and so their potential phosphorus load must be subtracted from the future development capacity to account for their ultimate development. The 123 monitored lakes that are used in this exercise contain 4,400 vacant shoreline lots in addition to over 18,000 developed shoreline lots. These vacant lots represent a limit to future development of new lots and should be removed from the future capacity of all lakes in the model as an additional filter.

Lake trout habitat filter. The final filter which can be considered is based on the presence of lake trout (*Salvelinus namaycush*) in lakes. The requirement of the species for cold, hypolimnetic waters and high-oxygen tensions makes it particularly vulnerable to phosphorus loading and associated oxygen demand (MacLean et al., 1990). Management of lake trout and their habitat remains a responsibility of the federal and provincial governments in Ontario and so lake-specific habitat requirements are not managed by municipalities. Nevertheless, any program must consider the reduced phosphorus contribution imposed by development restrictions on lake trout lakes to acknowledge their potential contribution to downstream capacity.

In practice, some development of these lakes may be allowed, subject to review by provincial agencies. The Ministry of Natural Resources (MNR) and the Ministry of the Environment (MOE) model those lakes that support lake trout populations. In application the MNR will allow development on lake trout lakes up to a limit based on hypolimnetic oxygen status. The MOE and the MNR determine this limit through a separate modeling exercise. When the MNR has identified a problem of depleted oxygen in a lake that supports lake trout, no further development is assumed for that lake.

Watershed-based development limits

Water quality models are set up to account for phosphorus movement from one lake to the next lake downstream, and hence throughout the watershed (Dillon and Rigler, 1975; Dillon et al., 1986). The need for such "watershed-based planning" has long been encouraged by provincial management agencies, but specific guidance has been limited. Watershed-based trophic status models include all hydrologic and phosphorus sources in a watershed and so meet the intent of watershed-based planning. The models add the total phosphorus load from one lake to the load for the next lake downstream, after accounting for in-lake retention. As a result, export of development-derived phosphorus from upstream lakes makes up part of the development loading for downstream lakes and must be included as part of the contribution from shoreline development. High levels of shoreline development on upstream lakes will therefore limit the development potential of downstream lakes. Sensitive downstream lakes will also limit development upstream.

The greatest problem with implementing development limits on a watershed basis is determining the extent of upstream influence of a lake that has reached capacity. Strict interpretation of watershed-based planning involves extending development limits to all lakes upstream of any lake that has reached its development capacity. This is logically consistent, as some of the phosphorus that enters the upper end of a watershed will ultimately reach the lower end of the watershed.

Although it is logically consistent to account for downstream transport of phosphorus in setting development capacities it is difficult to develop a spatial limit governing how far upstream of a capacity lake any limit on development should extend. The implications are considerable. If a lake in the lower watershed had no additional development capacity, then strict interpretation of a watershed-based planning policy would see future development restricted on all upstream lakes to prevent downstream transport of phosphorus.

In many cases, significant urban development and shoreline development potential may be located in the upper watershed. The trophic status of downstream lakes could, therefore, in the strictest sense, prevent any further development upstream.

In the practical sense, the watershed manager must determine the balance between protecting recreational water quality through watershed-based planning and the strictest approach, which would limit all development upstream of a capacity lake. Watershed-based planning remains an attractive concept, but quantitative advice on its implementation limits its utility.

Implications of watershed-based phosphorus limits

One approach to watershed-based planning is to first assess its implications:

1. Will the standards of water quality protection be reduced if development is not limited upstream of a lake that has reached capacity?
2. Does the added protection of a watershed-based approach translate into measurable or predictable improvements in water quality?

The implications can be assessed by running the watershed model in two scenarios:

1. In the first, future development is added to every lake in the watershed up to the limits prescribed by the "Background + 50%" or any other objective. These total future loads to upstream lakes are added to the downstream loading before downstream capacities are set. This approach represents true watershed-based planning.
2. In the second scenario, all lakes are developed to their individual limits, based on the difference between present day development and development to "Background + 50%" without accounting for the additional load from future upstream development. The phosphorus from this future development is then added to the downstream lakes, which are already at their own, independent capacity limits. This approach represents the implications of ignoring watershed-based planning.

The results of this exercise show that the implications of allocating development capacity independently of downstream transport are not significant for most lakes. Figure II.17.11 shows little overall deviation between watershed-based and independent allocations of development to the limits of the "Background + 50%" objective for lakes in the Muskoka River watershed. Lakes where the two development outcomes do not differ are shown as a straight 1:1 relationship. Departure from the 1:1 relationship, where the figure shows points above the 1:1 line, represent the degree to which the final water quality exceeds or "overshoots" the "background + 50%" objective because of upstream loading. The maximum deviation is 4.4 µg/L, in which a lake that should be at 11.1 µg/L ends up at 15.5 µg/L. (Figure II.17.11 shows one point with a very high deviation. This lake has already exceeded its objective as a result of upstream agricultural inputs.)

The ultimate, post-development phosphorus concentrations remained within 20% of the "Background + 50%" objective in 93% of the lakes modeled (Figure II.17.12). Water quality in 12 of the 376 modeled lakes could overshoot the objective by 40% or more as a result of downstream phosphorus transport. The water quality in these individual lakes would still be better, however, than that allowed if all were developed to Ontario's present-day Provincial Water Quality Objectives of 10 or 20 µg/L. The implications of exceeding the revised objective are therefore minor, and less than those of adhering to the present objective.

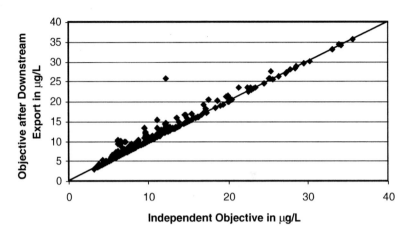

Figure II.17.11 Implications of independent versus watershed-based development allocation.

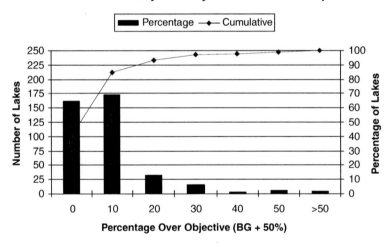

Figure II.17.12 Number of lakes in which water quality exceeds objective if downstream transport of phosphorus is not considered.

Although it is logically consistent to account for downstream transport of phosphorus in setting development capacities, it is difficult to develop a spatial limit governing how far upstream of a capacity lake any limit should extend. It is therefore clear that watershed-based planning should be interpreted and implemented in a reasonable fashion. The model already accounts for downstream transport of anthropogenic phosphorus from all existing development. The scenarios tested here reveal that the implications of not implementing a development freeze upstream of capacity lakes are minor and not necessary to ensure that water quality is protected. It is therefore evident that a watershed-based orientation of the trophic status model provides for sustainable levels of water quality. We note, however, that this conclusion is based on the relatively low loads associated with shoreline cottage development. Higher loadings, for example from point source, agricultural, or urban inputs, may have a greater impact on downstream water quality.

Setting development objectives

The objective-setting approach must protect water quality as well as assess the implications of model assumptions to policy for shoreline development. Development and calibration of the model for the Muskoka River watershed produced accurate estimates of water quality of the subject lakes, after accounting for reduced phosphorus loading to septic systems and the role of soils in attenuating phosphorus migration from shoreline septic systems to lakes. An objective of "Background + 50%" was tested to correspond to provincial initiatives and filters were developed to compare water quality and other determinants of development capacity. The resultant development capacities can then be summarized for each lake, each subwatershed, and the entire watershed. The capacities can then be compared with existing development density to provide insight into the degree of social stability and public expectations of policy.

Development objectives are best expressed as a phosphorus load in policy, so that the model can be used to compare resort, point source, or other development loadings against objectives. For ease of interpretation at the implementation stage that load can be converted to "Seasonal Residential" development using the occupancy figures for individual lakes (see "Validation of cottage usage" figures), or to any other type of development.

For this exercise, water quality objectives were stated as the total allowable anthrpogenic phosphorus load in kg. Objectives were set for each of the 376 lakes, bays, or rivers in the Muskoka River watershed as an index of potential development load, and for the 123 "monitored" lakes as an index of the loading of those lakes that are currently managed.

Although some jurisdictions may wish to optimize the allocation of future development to each lake, optimization was not attempted in this exercise. Certain popular lakes may well benefit from the adoption of stricter upstream controls on development, in order to maintain development opportunities in the popular lakes. Optimization would involve reducing the development capacity upstream of highly desired lakes, in order to maximize development of preferred locations. There is a near-infinite number of optimization strategies inherent in a large watershed, and optimization would involve a variety of stakeholders with different interests. Optimization could be considered on a case-by-case basis where there is a need to reallocate development opportunities.

Reconciliation of model accuracy and assumptions with objectives

Although the Muskoka River watershed model produced a very good correspondence between measured water quality and modeled estimates on average, there were many lakes in which a large discrepancy between measured and modeled phosphorus concentrations remained in the final model. The objective of a 50% increase in phosphorus against background was used here to illustrate a potential starting point for setting development limits. For this exercise, the 50% increase was modified, based on the agreement of the modeled estimate of phosphorus concentrations with phosphorus measurements for individual lakes, to accommodate the observed (versus the theoretical) expression of phosphorus in lakes. Where water quality measurements do not exist, the model should be assumed to be accurate.

The logic of reconciling objective development with model accuracy can be summarized as follows.

> *Criterion #1.* If the measured phosphorus concentration exceeds the modeled "Background + 50%" objective and the modeled total phosphorus exceeds "Background + 50%,"

Then no additional development is allocated (accurate model, lake at capacity).

Criterion #2. If the measured phosphorus concentration is 80% or more of the modeled total phosphorus concentration and the modeled total phosphorus exceeds "Background + 50%,"
Then no additional development is allocated (accurate model, lake at capacity); agreement of 80% or better is considered an acceptable indicator of model accuracy.

Criterion #3. If the measured phosphorus concentration is between 40% and 80% of the modeled total phosphorus concentration,
Then additional development is allocated up to the "Background + 50%" objective, but the objective is modified (increased) to account for the discrepancy between measurements and modeled estimates (inaccurate model, lake at capacity).
A conservative (protective) approach assumes that 80% of the modeled phosphorus would ultimately be expressed. The objective is therefore modified to Background + (1/0.8 * 50%) = Background + 62.5%.

Criterion #4. If the measured phosphorus concentration is less than 40% of the modeled total phosphorus concentration,
Then additional development is allocated up to the "Background + 50%" objective, but the objective is modified (increased) to account for the discrepancy between measurements and modeled estimates (inaccurate model, lake at capacity).
A conservative (protective) approach assumes that 40% of the modeled phosphorus would ultimately be expressed. The objective is therefore modified to Background + (1/0.4 * 50%) = Background + 125%.

Criterion #5. If no water quality measurements exist,
Then additional development is allocated up to the "Background + 50%" objective, under the assumption that the model is accurate.

Resultant phosphorus objectives for the Muskoka River watershed example

The total number of shoreline lots on the 123 "measured" lakes was approximately 14,000 in 1999. This corresponds to an anthropogenic phosphorus load of 9,204 kg. The "Background + 50%" objective allows an additional 3,403 kg of phosphorus from shoreline development, after consideration of the perimeter and Crown land filters. This is reduced to 3,010 kg when the vacant lot filter is added. Application of the filter restricting development on lake trout lakes reduces the allowable load to 2,638 kg. Development of all 123 lakes to the water quality objective of "Background + 50%" will thus allow a 37% increase in phosphorus loading from existing levels, or 29% when lake trout and vacant lots are considered.

Including all of the 376 modelled lakes in the exercise produces a corresponding increase in phosphorus loading. There are approximately 18,000 lots on these 376 lakes, corresponding to an anthropogenic phosphorus load of approximately 11,700 kg. The "Background + 50%" objective allows an additional 34,143 kg of phosphorus from shoreline development, or 9,451 kg after consideration of the perimeter and Crown land filters. This is reduced to 8,860 kg when the vacant lot filter is added. Application of the filter restricting development on lake trout lakes reduces the allowable load to 8,071 kg. Development of all 376 lakes to the water quality objective of "Background + 50%" will thus allow a 289% increase in phosphorus loading from existing levels, or 68% when all filters (perimeter, Crown land, vacant lots, and lake trout) are considered. The implications of each of the filters, particularly the perimeter and Crown land filters, are shown in Figure II.17.13. It is clear that water quality alone is not the most sensitive determinant of development capacity, and that physical restrictions are more limiting.

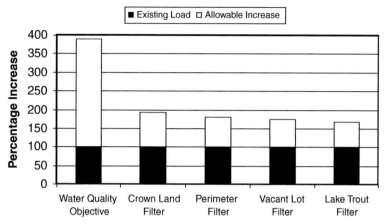

Figure II.17.13 Comparison of development constraints ("filters") in 525 lakes in the Muskoka River watershed.

Value of nutrient-based water quality objectives

Phosphorus is the nutrient limiting the growth of algae in the nutrient-poor lakes of the Precambrian Shield. When the phosphorus load to a lake increases because of anthropogenic sources and water quality declines, the recreational value of a lake will be diminished. In many municipalities on the southern Precambrian Shield in Ontario, lake-based recreation and tourism are the foundations of the local economy. A mechanism that allows local decision makers to define and understand the carrying capacity (whether based on water quality or otherwise) of the lakes within a municipality will ensure that further development does not unduly stress the natural resources upon which the area depends.

The intent of water quality-based development policies is to protect water quality from eutrophication induced by overdevelopment. It is therefore surprising that water quality is not always the strongest limitation on development capacity in lakes. The Muskoka River example was tested with a "Background + 50%" water quality objective and filters that limited development based on

1. Physical limits of available shoreline ("perimeter")
2. The presence of undevelopable Crown Lands
3. Vacant lots that were already committed to development
4. The presence of lake trout in lakes

Water quality alone did not represent the most significant restriction on shoreline development potential.

The exercise of modeling and monitoring an entire watershed is complex and costly. Water quality-based development limits are a worthy exercise in cases where the effort produces a substantial improvement in water quality protection; for example, if large point sources or urban areas are present. Biophysical and regulatory concerns may constrain development capacity far more, however, than water quality. Consideration of Crown Land and the physical shoreline limits may also reduce development capacity below that allowable under a very conservative phosphorus water quality objective of "Background + 50%." Lake trout habitat may be the most conservative filter and result in the lowest estimate of development capacity.

Simple consideration of physical development constraints may therefore, in a whole watershed analysis, provide sufficient protection of water quality in all but the most sensitive lake trout lakes. In all cases, however, the manager must review the large-scale findings at a finer resolution, as there will always be individual lakes to which the physical constraint does not apply and water quality is the most sensitive determinant. This conclusion applies to shoreline development in the form of recreational or residential development built on individual 200-ft lots. It must be reconsidered, however, for more intensive forms of phosphorus loading such as urban runoff, sewage treatment plants, agriculture, and high-density resort development.

Conclusions

Incorporation of the most up-to-date science on phosphorus loadings and dynamics in Precambrian Shield lakes produced very accurate estimates of phosphorus concentrations in the lakes of the Muskoka River watershed. The major conclusion was that previous modeling exercises, which assumed that 100% of the phosphorus in septic systems within 300 m of a lakeshore was mobile, could not be substantiated on an empirical or a mechanistic basis. Adoption of soil-based attenuation produced a substantial improvement in predictive capability and defensibility of the planning exercise. Although the lakes showed some response to shoreline development, the degree of response was much less than that originally predicted.

In the final analysis it is clear that water quality is not the most sensitive determinant of shoreline development and that adoption of simple standards such as a 200-ft minimum frontage on shoreline lots may achieve a high level of water quality protection without the need to rely on a whole watershed model of phosphorus dynamics and lake-specific water quality objectives. Some form of water quality assessment is essential, but it may not have to take the form of a complex predictive model — a sensitivity analysis may suffice. Water quality was also well protected in most cases without the need to "freeze" all development upstream of a lake which had reached capacity.

Social pressures and user conflicts are becoming increasingly important in cottage country. These may be partly managed if lake area ("recreational space") is used to help determine cottage density, independently of water quality. It is also clear that properly managed near-shore development, in which minimum lot sizes are coupled with: enhanced septic system setback, ensured naturalization of shorelines, protection of significant natural areas, wetlands, and scenic vistas, modern septic systems, mandatory septic inspection and a strong stewardship program, will likely be as successful in maintaining water quality as will development limits. Expansion of resorts and cluster or subdivision-type cottage development provide the potential to use package sewage plants that are capable of achieving very high effluent quality and that can be monitored to confirm their inputs, thus reducing the need to rely on assumptions regarding the effectiveness of septic systems. Finally, the development of lake-specific management plans by lake users or residents in cooperation with provincial or municipal authorities may bring all users together to draft a common vision of a lake's future and actions to achieve it based on site-specific local concerns and consensus.

In summary, lake management planning extends beyond consideration of plumbing. There will always be some phosphorus enrichment as lakes are developed, whether from land clearing, septic systems, urbanization, or expanded point sources. Recreational water quality must therefore remain a component of lake management. Development capacities based only on phosphorus, however, are costly, complex, and vulnerable to challenge.

Managers are encouraged to consider and implement a broader spectrum of management approaches, in addition to phosphorus-based development capacities.

Literature cited

Barnett, P.J., Cowan, W.R., and Henry, A.P., Quaternary Geology of Ontario, Southern Sheet; Ontario Geological Survey Map 2556, 1991.

Canada Department of Fisheries and Environment, *Hydrological Atlas of Canada*, Surveys and Mapping Branch, Dept. of Energy, Mines, and Resources, Ottawa, ON, 1978.

Clark, B.J. and Hutchinson, N.J., Measuring the trophic status of lakes: Sampling protocols, *Ont. Min. Envir. Tech. Report.*, 1992.

Cox, E.T., *Counts and Measurements of Ontario Lakes*, Fisheries Branch, Ontario Ministry of Natural Resources, 1978.

Dillon, P.J and Molot, L.A., Effect of landscape form on export of dissolved organic carbon, iron, and phosphorus from forested stream catchments, *Water Resour. Res.*, 33, 2591, 1997.

Dillon, P.J. and Rigler, F.H., A simple method for predicting the capacity of a lake for development based on lake trophic status, *J. Fish. Res. Board. Can.*, 32, 1519, 1975.

Dillon, P.J., Nicholls, K.H., Scheider, W.A., Yan, N.D., and Jeffries, D.S., Lakeshore capacity study, trophic status, Research and Special Projects Branch, Ontario Ministry of Municipal Affairs, 1986.

Dillon, P.J., Reid, R.A., and Evans, H.E., The relative magnitude of phosphorus sources for small, oligotrophic lakes in Ontario, Canada, *Verh. Internat. Verein. Limnol.*, 1992.

Dillon, P.J., Scheider, W.A., Reid, R.A., and Jeffries, D.S., Lakeshore capacity study: Part 1 — Test of effects of shoreline development on the trophic status of lakes, *Lake and Reserv. Manage.*, 8, 121, 1994.

District of Muskoka Planning and Economic Development Department, Water Quality Program, Lot Usage Study, 1995, unpublished.

Downing, J., Lakeshore capacity study committee report, Research and Special Projects Branch, Ontario Ministry of Municipal Affairs, 1986.

Gartner Lee Ltd., *Alternative Technologies for Removal of Phosphorus from Domestic Septic Effluent*, prepared for Science and Technology Branch, Ontario Ministry of the Environ. Dorset Environmental Science Centre, 2002 (in preparation).

Heiskary, S.A. and Walker, W.W., Developing phosphorus criteria for Minnesota lakes, *Lake Reserv. Manage.*, 4, 1, 1988.

Hutchinson, N.J., Dillon, P.J. and Neary, B.P., Validation and use of Ontario's trophic status model in establishing development guidelines, *Lake and Reservoir Manage.*, 7, 13, 1991.

Isenbeck-Schroter, M., Doring, U., Moller, A., Schroter, J., and Matthe, G., Experimental approach and simulation of the retention processes limiting orthophosphate transport in groundwater, *J. Contam. Hydrol.*, 14, 143, 1993.

Jenkins, D., Ferguson, J.F., and Menar, A.B., Chemical processes for phosphate removal, *Water Res.*, 5, 369, 1971.

MacLean, N.G., Gunn, J.M., Hicks, F.J., Ihssen, P.E., Malhiot, M., Mosindy, T.E., and Wilson, W., Environmental and genetic factors affecting the physiology and ecology of lake trout, in *Lake Trout Synthesis — Physiology and Ecology Working Group*, Ontario Ministry of Natural Resources, Toronto, 1990.

Michalski, M., Shoreline development, phosphorus loadings and lake trout — a scientific and bureaucratic boondoggle, *Conf. Proc. Wastewater Nutrient Removal Technol. and Onsite Manage. Dist.*, Univ. of Waterloo Centre for Groundwater Research, June 6, 1994.

Michalski, M., Nicholls, K.H., and Johnson, M.G., Phosphorus removal and water quality improvements in Gravenhurst Bay, Ontario, *Verh. Internat. Verein. Limnol.*, 19, 1871, 1975.

Mollard, D.G., Southern Ontario Engineering Geology Terrain Study. Database Map, Muskoka Area. Parry Sound and Muskoka District, Ontario Ministry of Natural Resources. Ontario Geological Survey Open File Report 5323, 1980.

Molot, L.A., Dillon, P.J., Clark, B.J., and Neary, B.P., Predicting end-of-summer oxygen profiles in stratified lakes, *Can. J. Fish. Aquat. Sci.*, 49, 2363, 1992.

OMEE, *Rationale for the Establishment of Ontario's Provincial Water Quality Objectives*, Ontario Ministry of the Environ., 1979.

OMEE, *Ontario's Water Quality Objective Development Process*, Ontario Ministry of the Environ., 1992.

OMEE, *Water Management: Policies, Guidelines, Provincial Water Quality Objectives of the Ministry of Environ. and Energy*, Ontario Ministry of Environ. and Energy, 1994.

Robertson, W.D., Schiff, S.L., and Ptacek, C.J., Review of phosphate mobility and persistence in 10 septic system plumes, *Ground Water*, 36, 1000, 1998.

Stumm, W. and Morgan, J.J., Aquatic Chemistry. An Introduction Emphasizing Chemical Equilibria in Natural Waters, Wiley Interscience, New York, 1970.

Winter, J.G. and Duthie, H.C., Export coefficient modelling to asses phosphorus loading in an urban watershed, *J. Am. Wat. Res. Assoc.*, 36, 1053–1061, 2000.

Wood, J.S.A., Migration of septic system contaminated groundwater to a lake in a Precambrian Shield setting: a case study. M.Sc. thesis, Dept. of Earth Sciences, University of Waterloo, Ontario, 1983.

Response

Land–lake linkages and land-use limits

This chapter by Neil Hutchinson represents what may very well be the start of an important paradigm shift in water sensitive planning for rural lakes. It is the contention of the authors that planning based on setting water quality goals may have been overemphasized in the past. Such approaches have attempted to implement development carrying capacities based on the use of water monitoring data such as total phosphorus models in relation to the "trophic status" of lakes. Hutchinson argues that the tens of thousands of calculations that some of these models require may not really be the most accurate barometer of shoreline development pressures. This occurs due to the inattention devoted to soil processes in nutrient transport and lake eutrophication (see Chapter II.13 by Margot Cantwell for another discussion of the role of soils in watershed planning).

Instead, this chapter informs us of the need for continual reexamination and reappraisal of existing models of land–lake linkages. Due to the important economic link between tourism and water quality, it is also critical to address such social determinants as noise, crowding, powerboats, beach use, etc. In other words, diversified planning should shift its focus from simply septic system plumbing to more comprehensive land-use pattern analyses. For example, Hutchinson advances the idea that adoption of a minimum standard for shoreline lot frontage may achieve a high level of water quality protection without the need to rely exclusively on watershed phosphorus modeling.

In certain situations then, it may be unnecessary to halt all development if a program of regulating lot size restrictions and locations could be invoked to keep lakes oligotrophic. This represents a stage in moving beyond water quality monitoring toward setting development objectives through build-out analyses and futures alternatives such as described in Chapter II.15 by Amir Mueller, Robert France, and Carl Steinitz.

Social and political issues in managing riparian buffers and corridors

Participants: Kathy Freemark, Stephen Hurst, Frank Mitchell, Arnold Valdez, Francois de Kock, Leslie Zucker, and Robert France

Past

- Prevalence of a fragmented rather than integrated approach, which fails to treat landscapes as mosaics; e.g., wetland protection separated from riparian buffer management
- Disciplinary instead of inter-/transdisciplinary focus of study
- Failure to include social components and understanding of processes
- Confounding by legal constraints and variability among jurisdictions with respect to the management of riparian corridors
- Traditional engineering solutions
- Regulatory approaches subject to political compromises that can result in a minimum amount of riparian protection
- Limited abilities to link across spatial scales; i.e., to ratchet up the implications of site-specific disturbances and interventions to the entire riverine corridor/network

Present

- Development of experiential learning to engender an environmental ethic among landowners
- Beginning emphasis on voluntary participation, community-based planning, collaborative approaches, and partnering incentives
- Understanding of, and integration among, ecological processes/patterns across spatio-temporal scales, including technical innovations and socioeconomic implications
- Cumulative impact assessment in relation to integrated land-use planning and management in concordance with the carrying capacity or limits to growth
- Emerging recognition of indigenous knowledge and its inclusion in modern land-use planning decisions
- Fledgling support for the need for a cultural shift toward humility in dealing with rivers and their sustaining landscapes
- Early moves from a reactive to a proactive mode in protecting water resources within a framework of open-space planning

Future

- Evolution toward regionally appropriate land-use management
- Social equity; i.e., elevation of the rights of the many in relation to the rights of the individual, or in other words, a move toward elements of land-use policy and governance as practiced in Canada

- Exploration of establishing a network of "learning regions" as currently being developed in Europe, facilitated by use of the Internet for bioregional ecological educational
- Acknowledgment of the full range of human characteristics in conscious evolution (i.e., intellectual, spiritual, emotional, physical) involved in respect for the land
- Encompassing such difficult-to-quantify aspects of land respect in decision making in addition to factual knowledge
- Recognition and embracing of landscape architecture, land art, and environmental ritual in fostering river stewardship
- Greater federal government-led aggressiveness in enforcing environmental regulation as imagined to be practiced in Canada
- Equitable distribution of resources on a global scale and acceptance of the working principles of sustainability

Multiple objectives in watershed management through use of GIS analysis

Participants: Jeffrey Schloss, Paul Zandbergen, Markley Bavinger, and Robert France

Past

- Lack of shared understanding about ecological strengths and development threats in a watershed context by both the public and regulatory agencies
- Absence of long-term vision; specifically the failure to comprehend the implications of alterations in development and development guidelines
- Piecemeal assessment of the cumulative effects of imperviousness
- Impression of drowning in a cacophony of Internet information with no clear understanding of what may or may not be important to the interests of any specific watershed group
- Ignorance of what has been previously accomplished in other locations so impression is left that the wheel must be invented anew every time for each watershed
- Only limited involvement of locals in land-use planning
- Limited use of GIS in only a top-down management approach; i.e., identification of locations for future development including consideration of zoning, carrying capacity, and build-out analyses

Present

- Recognition and circumvention of the gaps in knowledge through use of GIS modeling
- Education of communities to think in an entire watershed context through enabling visual comprehension of everyone's place in the landscape
- Establishment of voluntary controls to development and acknowledgment of the importance of land conservation
- Movement toward use of GIS in a bottom-up management approach; i.e., identification of resources to be protected from any future development including wetlands, groundwater recharge areas, habitat for threatened species, fisheries, etc.
- Fusion of the two GIS approaches through iterative feedback to provide the different interest groups with a logical strategy for spatial analysis; i.e., first, where are the critical areas that need to be protected?; second, where can new development easily occur due to available infrastructure?; third, where are the intersection points?
- Increased communication facilitation and community building by taking the message directly to the town council
- Understanding of the importance of proactive planning; i.e., need to act now to preserve special features of the watershed
- Embracing the guiding principles of low-impact development

Future

- Further transparency and access to information for the public to foster local empowerment in decision making
- Small decisions on the local scale have to be informed from a watershed perspective that must link immediate site-specific knowledge to long-term watershed inventories and future projections
- Increased ecological sophistication of the public fostered through computer technologies
- Moves toward technical standardization and national referencing of appropriate GIS approaches
- Development of clear visions of the future through alternate futures land-use planning
- Strengthen governmental management and enforcement
- Need to jolt people into thinking and planning on larger scales instead of being lost in the idiosyncrasies of individual sites

Postscript: Implementing water sensitive planning and design

The chapters selected for this volume represent the best models about how to engage in water sensitive planning and design as put forward by the most advanced thinkers and practitioners in the field. For many readers, though, a pressing and fair question may very well be "How can we get there from here?"

To this end, a companion volume is currently produced that addresses such concerns. Titled *Practical Support for Watershed Management: Environmental Communication, Demonstration Projects, and Education Outreach*, this work is designed to help provide a foundation from which to move toward effecting positive change in watersheds. Topics covered include ecosystem health, economic sustainability, innovative diffusion learning, dispute resolution, ecological integrity, environmental indices, alternate futures modeling, project scope generation, historical perspectives, combining art and science, buffer gardens, public art installations, design–build training workshops, education CD-ROMs, Internet watershed atlases, setting limits to growth, community participation, laboratory training, design workshops, production of educational publications, exploring evolutionary infrastructure, GIS decision-maker training, watershed street celebrations, and technology transfer through extension school outreach.

Understanding the wealth of opportunities that exist for promoting water sensitive planning and design in watersheds through adopting these nontraditional approaches will mean that the examples and precedents covered in the present book have a good likelihood of becoming realized in other locations by other participants.

Index

A

Abbotsford wetland park, British Columbia, 193–194
   amenities afforded by, 198–201
   design for, 196–197
   educational opportunities afforded by, 200
   Fishtrap Creek site for, 194–196
   habitat provided by, 201–202
   management of, 201
   public use of, 201–202
   vegetation and planting for, 197–198
Aquifer recharge management, 615–617
   model of, 615
   aquifer protection alternatives, 626, 629–632
   conventional alternatives, 622–628
   data required, 619, 621, 622
   data variability, 619
   digital elevation model, 619, 621–622
Areas of concern, Ontario, Canada
   Ontario Great Lakes Remedial Action Plan, 449–450
   timber management guidelines, GIS analysis, 602–611
Arizona, *see* Sweetwater Wetlands, Tucson (AZ)
Art Creek, 83
Atlanta (GA), *see* Peachtree Creek watershed
Australia, *see* Homebush Bay area (Sydney, Australia)

B

Basins, *see* Detention storage, basins/ponds; Infiltration
Beltway Plaza Expansion (MD), low-impact design in, 299–301
Bioretention, 125–126
   combined with wetland systems, *see* Wetland(s); Wetland ponds
   design manual, Prince George's County (MD), 298–299
   designs for, 131, 298
   eco-roofs, 137–142
   infiltration gardens, 131–134
   stormwater planters, 136–137
   swales, 128–131, 135–136, 150–151
   water garden/pond, 142–148
   in suburban watersheds, 49–64, *see also* Low-impact development (LID); Open space design
   in urban watersheds, 11–27, *see also* Low-impact development (LID); Urban watershed(s)
   long-term study, University of Virginia, 310
   swales for, design testing, 128–131
Boston area (MA)
   pond and water supply restoration in, 407–429, *see also* Fresh Pond Reservation
   wetland ponds restoration in, 215–216
   Chandler Pond, 317–338, *see also* Chandler Pond restoration
   funds needed for, 229
   Hall's Pond, 216–223
   Hill's Pond, 223–228
   professional consultation in, 229
   public educational process in, 231
   soils testing in, 228–229
   vegetation management plan required in, 231
   wetland expertise required for, 231
British Columbia, Canada, *see* Abbotsford wetland park, British Columbia
Buckman Heights (Portland, OR), watershed restoration projects, 131–134
Buckman Terrace (Portland, OR), watershed restoration projects, 134–141

C

California
   Rio Hondo Spreading Grounds, 21–23
   San Francisco Bay Area, low-impact design manual, 310
   San Francisco Bay area watershed management plans, 459–474
Cambridge (MA), Fresh Pond Reservation restoration, 407–429
Canada
   Abbotsford wetland park, British Columbia, 193–202
   Gander Lake, Newfoundland, land–water interaction modeling, 577–596
   lakeshore capacity planning, 647–679, *see also* Lakeshore capacity planning

689

timber management/land use, *see also* Ontario, Canada, timber management/land use
    erosion plumes study, 635–643
    GIS modeling, 601–611
    Ontario Great Lakes Remedial Action Plan, 445–455
Canada–United States Great Lakes Water Quality Agreement, 446
CeDES (Conservation Development Evaluation System), 62–63
Chandler Pond restoration, Boston (MA)
    biolog installation in, 332, 333
    community involvement in, 326
    condition prior to, 317–319
    demonstration project for, 324–326
    design planning for, 322–324
    management options for, 325
    planning for, and dredging, 319–324
    plantings and bank stabilization in, 326–329
    protection of, 330–332, 334
    plantings depth zones in, 331
    waterfowl fence for, 334, 335
    watershed management planning in, 332–338
Channel formation, river, 435–438
Chesapeake Bay Riparian Handbook: A Guide for Establishing and Maintaining Riparian Forest Buffers (Palone and Todd), 6
Comparative hydrology, *see* Subdivision design, residential
Connecticut
    Harbor Brook restoration, Meriden, 384, 390–393
    Merrick Brook restoration, Scotland, 349–350
    Naugatuck River restoration, 383
    river restoration, 379
    channel daylighting in, 386
        dam removal in, 382–383
    dechannelization in, 383–386
    flow management in, 387
    geomorphic considerations in, 388–390
    habitat improvements in, 387
    public participation in, 390
    river front access planning in, 386–387
    science of, evolution of, 388–390
    watersheds in, 380–382
Conservation Development Evaluation System (CeDES), 62–63
Conveyances, 13
Cornell University, Ithaca (NY), bioretention/wetlands system
    construction of, for parking lot runoff, 161–162
    educational opportunities afforded by, 171–172
    enhancement of, for parking lot runoff at, 159–161
    habitat improvements at, 172
    maintenance plan for, 171
    parking lot design for Ornithology Laboratory complex, 158–159
    plant selection for, 164–165
    soils installation in, 170–171
    soils profiles/creation for, 165–170
    swale construction in, for parking lot runoff, 162–164

Curve number (CN), 35, 109
    redevelopment, 111–112
Custer Park, Portland (OR), swale design, 151–152

D

Design
    interdisciplinary, 343–344
    earth science application in, 346–347
    ecologic assessment in, 344–346
    goals and objectives of, 344–346
    process of, 344
    sustainability and, 341–343, 347–349
    low-impact development, 97–122, *see also* Low-impact development (LID)
    open space, 49–50, 52, 56–58
    comparative analysis of, 49–60
        in low-density subdivision, 50–54
        in medium-density subdivision, 54–60
    other comparisons in, 60–61
    residential development, *see also* Subdivision design
    in Prince George's County (MD), 97–122
Designing Wetlands: Principles and Practices for Landscape Architects and Land-Use Planners (France), 4
Detention storage, 109, 113–114
    basins/ponds,13–14, 31–32
    critical data acquisition for, 34
    design goals for, 33–34
    ecologic effect of, 36–38
    flood control function of, 35
    functions of, 32–33
    habitat provided by, 36–38
        in limited space, 40
    mosquito breeding in, 37
    muddy discharge from, 42–44
    outflow control structures of, 37–38
    slope design for, 40–42
    water quality function of, 35–36
    Hall's pond restoration, Brookline (MA)
    approach to, and process of, 220–223
    environmental issues before, 216–220
    wetland plant list for, 233
Digital elevation model (DEM), 513
    creating, 520–522, 529, 606–608
Duck Crossing (MD), redesign comparisons, 50–54
DuPont Victoria Wetlands, Victoria (TX), 286–290

E

Eco-roofs, 137–142
Edgewood Crossroads, Pittsburgh (PA), watershed restoration, 78–82
Educational opportunities
    Low-impact development (LID), 120
    Sterrett school site, Pittsburgh (PA), 82
        Art Creek at, 83
    cisterns for roof runoff from, 82–84
    porous paving at, 85
    restorative design for, 82–85

swale at, 85
water garden at, 83, 85
Abbotsford wetland park, British Columbia, 201–202
Boston (MA), 231
Cornell University, Ithaca (NY), 171–172
DuPont Victoria Wetlands, Victoria (TX), 286, 288, 290
Exploration Place, Wichita (KS), 244
Farmington waterway (MN), 191
Greenbush Bay Natural Resource Area, Kent (WA), 281
Sweetwater Wetlands, Tucson (AZ), 284
Wakodahatchee Wetlands, Palm Beach County (FL), 293–294
Escher catchment (Germany), 118
Environmental quality indexing, 453–455
Evapotranspiration, 15–17
Experimental Sewer System (Japan), 118–119
Exploration Place, Wichita (KS), 235–236
  Educational opportunities afforded by, 244
  habitat provided by, 243–244
  maintenance of, 243
  park design concepts for, 237
  plantings unwieldiness of, 239–240
  second planting of, 243
  site descriptions, 236–237
  wetland construction for, 240–243
  planting process in, 241–243
  wetland design for, 238–239
  integrated with reflecting, 237–238
  plantings in, 239–240

F

Farmington waterway (MN), 175
  aesthetic unidirectional opportunities of, 191
  design of, 175,177–181
  educational opportunities provided by, 191
  habitats provided by, 190
  maintenance of, 190
  municipal approval of, 177
  outcomes of,183–187
  planting recommendations for, 188, 189–190
  site for, 176–177
  soils descriptions for, 189
  stability of,189–190
  stormwater treatment provided by, 190
  vegetation for wetlands and swales of, 188–190
  water systems of,189
    and water flow, 181–183
Filter strip, definition of, 361–362
Florida
  Florida Aquarium, Tampa Bay, low-impact design, 307
  Wakodahatchee Wetlands, Palm Beach County, 290–294
  Watershed Interactive Network Plan in, 545–552

Wetlands-Based Indirect Potable Reuse Project, 205
  advanced wastewater treatment system development for, 207
    design goals for, 208–211
  baseline monitoring of wetlands in, 207–208
  goals of, 205
  hydrologic features of West Palm Beach, 205–206
  hydrologic modeling in, 208
  Wetland Reuse Site in, 211
  wetlands demonstration project for, 207
Frederick County (MD), low-impact subdivision, 305–306
Fresh Pond Reservation restoration, Cambridge (MA), 407–408
  current and anticipated capital projects at, 414
  history of, 408
  landscape restoration and maintenance plan for, 413, 415
    bikeway corridor improvements, 425–427
    erosion control and slope stabilization, 419–420
    golf course drainage plans, 425
    northeast sector, 424–425
    perimeter fencing, 422–424
    road resurfacing, 422–424
    vegetation, 415–419
    water purification facility landscape improvements, 420, 422
  natural resource inventory of, 408
  vegetative cover types, 412, 413
  original plan for, 412
  stewardship plan for, 408–410, 428–429
  recommendations of, 410, 413

G

Galilee, Israel, aquifer recharge management, 615–632
Gander Lake, Newfoundland, land–water interaction models, 577–579
  comparisons of, 589, 592
  development of, 580–584
  GIS maps used, 582, 584
  land use changes, evaluating effect of, 592–595
  land use model, 579–580, 584
  predicted land use impact, 592, 593
  suspended solids contribution, 585–586
  spatially explicit model, 579–580, 584
  inherent sensitivity to soil loss, 590–592
  landscape features, 589
  predicted land use impact, 592, 594
  slope sensitivity, 589–590
  suspended solids contribution, 586–588
Geographic information systems (GIS)
  data commonly available from, 560–561
  data manipulation by, 559–560
  data obtained by,559, 560
  definition of, 559
  in aquifer recharge modeling, 615–632
  in land cover change modeling, 515–516, *see also* Land cover change, modeling

in land–water interaction, 580–592, *see also* Land–water interaction
in watershed natural resource inventory mapping, 557–558, *see also* Watershed Natural Resource Inventory mapping
repositories of data from, 561
schematic diagram of, 559
topology analysis by, 559–560
Global positioning system (GPS), 562–563
in watershed natural resource inventory mapping, 557
Great Lakes, *see* Ontario Great Lakes Remedial Action Plan; Rouge River National Wet Weather Demonstration Project
Green River Natural Resource Area, Kent (WA), 278–281
Greenways
as infrastructure, 395–396, 399, 404–405
environmental education and, 403
public participation in, 404
solid waste recycling and, 402
transportation design, 399–400

H

Hall's pond restoration, Brookline (MA)
approach to, and process of, 220–223
environmental issues before, 216–220
wetland plant list for, 233
Harbor Brook river restoration, Meriden (CT), 390–393
Hearthstone Quarry brook restoration, Chicopee (MA), 351
High Point (NC), watershed/water supply protection, low-impact design, 306–307
Hill's pond restoration, Arlington (MA)
approach to, and process of, 223–228
goals of, 223
Homebush Bay area, Sydney, Australia
design concepts for restoration of, 247–250
Northern Water Feature, 252–260
pollution control, 251–252
resource conservation, 251
species conservation, 250–251
water cycle strategy, 252
restoration of, 247
site description of, 247–250
Hunter Park, Pittsburgh (PA), watershed restoration, 74–78
Hydrologic Soil Group (HSG), 514
definitions, Natural Resources Conservation Service, 530
estimation, using field capacity measurements, 524–527
Hydrology, comparative, *see* Subdivision design, residential

I

Impervious cover
as indicator of watershed quality, 104–107
effects of, on watersheds, 11–12, 17–18, 49, 616–617

in suburban setting, 49–64, *see also* Open space design
in urban setting, 11–27, *see also* Urban watershed(s)
Infiltration, 17
as basis of stormwater management, 27
basins for, 18–19, *see also* Detention storage; Retention storage
designing, 19, 20–23
ecologic use of, 21, 22
examples of, 20–23
human use of, 21
during storm event, 26–27
effects of, 17–18
gardens for, 131–134
swales for, 20–21, *see also* Swale(s)
vegetation and, 125–153, *see also* Bioretention
Infrastructure
definitions of, 396–397
greenways as, 395–396, 399
in United States, 397–399
Integrated Watershed Management in the Global Ecosystem (Rattan, Ed.), 6
Integrating Stormwater into the Urban Fabric (American Association of Landscape Architects), 1
Interdisciplinary design, 343–344
Israel, *see* Galilee, Israel

J

Johnson County Wetlands Mitigation Bank project (KS), 479–481, 489
Blue River floodplain, typical cross section, 483
Blue River Watershed Map, 481
concept and plan development of, 481–484
construction phases of, 484–486
design of, 484
habitat provided by, 484–486
plantings for, 485
review and approval of, 486, 488
city and state exemptions in, 488–489
water budget calculations for, monthly, 487

K

Kansas
Exploration Place, Wichita (KS), 235–244
Johnson County Watershed Map, 478
Johnson County Wetlands Mitigation Bank project, 479–489
Kansas Urban Resource Assessment Project, 477–479
Wilderness Valley project, 479–489

L

Lake(s)
Gander Lake, Newfoundland, land–water interaction modeling, 577–596
Muskoka River watershed, Ontario, lakeshore capacity planning, 647–679

Lake Chocorua, 569–571
Squam Lake, 564–569
Ontario Great Lakes Remedial Action Plan, 445–455
Rouge River National Wet Weather Demonstration Project, 491–511
Lakeshore capacity planning, Ontario, 650
   background of, 649–650
   development capacity, in Muskoka River Watershed, 676–677
   municipal responsibilities in, 663–664
   need for, in Muskoka River Watershed, 647–648
   shoreline development policy in, 651
   objectives, 675–676
   trophic status modeling in, 647–662, *see also* Lake trophic status modeling
   water management and, 663
   water quality objectives in, 663
   alternative, 670–671
   baseline, modeled assessment, 668–669
   baselines, and measured water quality, 667–668
   filters in, 671–672
   phosphorus, total, 664–668
   proportional increases, 669–670
   watershed based phosphorus limits, implications, 673–674
Lake trophic status modeling, Muskoka River Watershed, 647, 651
   accuracy of, 660–662
   calibration in, 654
   anthropogenic phosphorus sources, 656, 659–660
   development distance from shoreline, 658–659
   natural phosphorus sources, 654–656
   phosphorous retention in soils, 656–658
   Dillon–Rigler model, 650
   Ontario Lakeshore Capacity Model, 650
   Ontario Trophic Status model, 652
   results of, 660, 676–677
   technical basis of, 651–653
   validation of, 653
   monitoring programs, 653–654
   value of, 677–679
Land cover change, modeling, 513–515
   algorithm in, 519–520
   data sources for, 517–519, 561–562
   digital elevation model creation in, 520–522
   geographic information systems in, 515–516
   hydrologic soil group estimation in, 515, 524–527
   land cover classification in, 522–523
   precipitation grid formation in, 524
   remote sensing technology in, 515–516
   soil moisture indices in, 516–517
   estimation process, 527–529
Land cover classification, data collection and modeling, 522–523, 529, 532, 534
LANDSAT multi-spectral scanner, and thematic mapper, 518, 561–562
Landscape(s), 15
Land–water interaction
   in urban watersheds, 15–18, *see also* Bioretention
   model comparisons in, 577–579, 584–592
   model development in, 580–584

delivery ratios, 583–584, 584, 586
   GIS map combinations used in spatially explicit model, 582, 584
   landscape features, 589
   slope sensitivity, 589–590
   soil loss values, 583
   Universal Soil Loss Equation, 580–581
Low-impact development (LID)
   aesthetic and educational aspects of, 120
   case studies of, 297–298
   bioretention demonstration project, University of Virginia, 310
   commercial applications, Tampa Bay (FL), 307, 309
   in Maryland, 298–306, *see also* Duck Crossing (MD); Frederick County (MD); Prince George's County (MD)
   infill development in Beltway Plaza Expansion, 299–301
   international, 311
   Jordan Cove (CT), 311
   Maryland State Highway Administration, 311
   national, 311
   parking lot retrofit, Prince George's County Office Complex, 301–302
   subdivision, Frederick County (MD), 305–306
   U.S. Environmental Protection Agency, 311
   watershed/water supply protection, High Point (NC), 306–307
   compared with conventional stormwater management, 302–304
   costs of, 121
   design approach in, 108–109
   conservation, 109
   detention storage, 113–114
   minimization, 109–110
   pollution prevention in, 112–113
   redevelopment curve number in, 111–112
   runoff volume maintenance in, 111–112
   time of concentration in, predevelopment, 110–111
   Maryland Department of Environment, 306
   national (EPA), 304–305
   Prince George's County, 298–299, 302
   San Francisco Bay Area Stormwater Management Agencies Association, 310
   ecosystem protection in, 103–104
   environmental impact of, 100–101
   examples of, national and international, 117–120
   habitat restoration in, 102–107
   implementation of, 121–122
   in current development trends, 101–102
   in Prince George's County (MD), 97–99, 297–298, *see also* Prince George's County (MD)
   in restorative development, 100
   in urban watershed restoration, 114, 116–117
   integrated management practices in, 114
   political pressures and, 101
   purpose of, 99–100
   watershed protection in, 102–103
Lyons residential infiltration strategies (France), 119–120

M

Maryland
  Department of Environment, design manual (2000), 306
  Duck Crossing, redesign comparisons, 50–54
  Frederick County, low-impact subdivision, 305–306
  Prince George's County, low-impact development in, 97–99, 297–298, *see also* Low-impact development (LID)
  Beltway Plaza Expansion, infill development, 299–301
  cost comparisons with conventional stormwater management, 302–304
  Count Office Complex, parking lot retrofit, 301–302
  design manuals published on, 298–299, 302
Massachusetts
  Chandler Pond, 317–338
  Hall's Pond and Hill's Pond, 215–231
  Cambridge, Fresh Pond Reservation restoration, 407–429
  Hearthstone Quarry Brook restoration, Chicopee, 351
  Salem Salt Marsh restoration, Salem, 351–353
  stormwater pond design in urban, 31–32
  critical data acquisition for, 34
  design goals for, 33–34
  ecologic effect of, 36–38
  flood control function of, 35
  functions of, 32–33
    in limited space, 40
  mosquito breeding in, 37
  muddy discharge from, 42–44
  outflow control structures of, 37–38
  slope design for, 40–42
  water quality function of, 35–36
Merrick Brook restoration, Scotland (CT), 349–350
Michigan, *see* Rouge River National Wet Weather Demonstration Project
Minnesota, *see* Farmington waterway (MN)
Modeling
  aquifer recharge management, Galilee, Israel, 615–632
  hydrographic, Northern Water Feature (Sydney, Australia), 257, 258, 259
  Rouge River National Wet Weather Demonstration Project, 496–497
  Wetlands-Based Indirect Potable Reuse Project (FL), 208
  hydrologic soils/land cover change, developing countries, 513–534
  lake shore capacity, Muskoka River Watershed, Ontario, Canada, 647–679
  Gander Lake, Newfoundland, Canada, 577–596
  San Francisco Bay area, 461–466
  SUNOM (Simplified Urban Nutrient Output Model), 50, 53–54, 59
Mosquito breeding, in detention basins/ponds, 37–38
Muskoka River Watershed, 647–648
  development capacity, based on modeling, 676–677
  phosphorus distribution in, 666–667
  quaternary geology of, 657
  subwatersheds of, 660

N

Natural resource inventory(ies), *see* Watershed natural resource inventory mapping
Natural Resources Conservation Service (NRCS), U. S., 514
  as GIS data repository, 561
  hydrologic soil group classifications, 530
  modeling from field measurements, 524–527, 527–529
Natural Step, 342–343
New Hampshire
  shoreland buffers, 361–362
  attitudes toward resource protection, 371–373
  attitudes toward value of surface waters, wetlands and shoreland property, 370–371
  concepts of, 374–375
  definition of, 361
  effectiveness of, 362–363
  establishing and maintaining, 368–369
  habitat provided, 363
  habitat provided by, 364–368
  historical interest in, 362
  hydrologic effects of, 364
  opinions of appropriate protective methods, 373–374
  public opinion survey results regarding, 370–374
  recreational opportunities provided by, 364
  water quality maintenance by, 364–365
  watershed natural resource inventory mapping, 557–558
  critical watershed analysis, 573–575
  information systems and data analysis, 557–563
  Lake Chocorua, 569–571
  Squam Lake, 564–569
  wetland/riparian buffer zones, Deerfield, 571–573
New York, *see* Cornell University, Ithaca (NY)
Nine Mile Run watershed, Pittsburgh (PA)
  Edgewood Crossroads site in, 77–78
  infiltration basins in, 78, 80, 81
  open space and catchment design for, 81
  porous pavement use in, 81
  redirection of roof runoff in, 81
  restoration of, 78–82
  habitat restoration in, 88
  Hunter Park site of, 74
  bioretention areas in, 74
  cover modification in, 77, 78
  restorative design for, 74–77
  stream restoration in, 74, 77
  swales in, 74, 77
  industrial use and restoration planning for, 69–72
  Regent Square Gateway site in, 85–86
  multiple use design for, 86–88
  restorative design for, 86–90
  sample designs for restoration of, 72–73
  status of restoration of, 91

Sterrett school site in, 82
  Art Creek at, 83
  cisterns for roof runoff from, 82–84
  porous paving at, 85
  restorative design for, 82–85
  swale at, 85
  water garden at, 83, 85
Northern Water Feature (Sydney, Australia), 248, 252–253, 254
  continuous deflection separation systems of, 253
  design team for, 256
  hydrographic model of, 257
  hydrographic models of, 257, 258, 259
  plantings for, 256, 257
  stormwater treatment in, 253–260
  wetland ponds of, 255, 258

O

Ontario, Canada, timber management/land use, 601–602
  areas of concern in, 602–603
  harvestable timber in, results, 609–611
  background information, 635–637
  buffer adequacy findings, 641–643
  characteristics of erosion plumes, 638–639
  data collection, 637–638
  factors influencing erosion, 639–641
  accuracy of, 608
  concept and approach, 603–604, 605
  data manipulation/GIS formatting, 606
  digital elevation model creation, 606–608
  digital topography data, 604–605
  forest cover data, 605, 608
  slope determination, in riparian zones, 608
  study area, 602–603, 604
  timber grid merged with areas of concern grid, 608, 609
Ontario Great Lakes Remedial Action Plan, 445
  criteria for delisting, 449
  status of, 449–450
  background of, 445–446
  environmental quality indexing in, 453–455
  funding concerns in, 452
  habitat rehabilitation in, 451–452
  progress tracking methods in, 446–449
  progress variability in, 450–452
  public involvement in, 451–452
Open space design, residential, 49–50
  advantages of, 61–62
  comparative analysis of, 49–60
  development costs, 54, 59–60, 61–62
  Duck Crossing (MD), 50–54
  hydrology, 52–53, 59
  nutrient output, 53–54, 59
  Stonehill Estates (Virginia), 54–60
  evaluation of, 62–63
  in watershed protection, 63–64
  low density, 50–54
  medium density, 54–60
  other comparative studies of, 60–61

Oregon, *see* Portland (OR)
Outflow control structures, of detention basins/ponds, 37–38, *see also* specific projects

P

Parking lot design
  bioretentive structures in, 131–153, *see also* Portland (OR)
  bioretentive-wetland system in, 155–172, *see also* Cornell University, Ithaca (NY)
  in parking lot expansion, Beltway Plaza, 301–302
  in suburban areas, *see* Low-impact development (LID); Open space design
  in urban areas, *see* Nine Mile Run watershed, Pittsburgh (PA)
  retrofit of, Prince George's County Office Complex, 301–302
Pavements, porous, 23–26, *see also* specific projects
Peachtree Creek watershed, Atlanta (GA), 15–17
Pembroke subdivision, Frederick County (MD), 305–306
Pittsburgh (PA), *see* Nine Mile Run watershed, Pittsburgh (PA)
Plant selection, in design, *see also* specific projects
  bioretentive structures, Portland (OR), 128–153
  wetlands, and bioretention swales, 271
  Abbotsford Park, British Columbia, 197–198
  Farmington waterway (MN), 188–190
  Ithaca (NY), 164–165
  Northern Water Feature (Sydney, Australia), 257
  Boston (MA), 326–329
  Brookline (MA), 233
Pollutants, in stormwater, 616
  removal efficiency for, 35–36
Ponds
  stormwater, 31–45, *see also* Detention storage, basins/ponds
  wetland, *see* Wetland(s); Wetland ponds
Porous pavements, 23–26
Portland (OR)
  bioretentive structures and design in, 125–126, 131
  Buckman Height watershed restoration projects in, 131–134
  Buckman Terrace watershed restoration projects in, 134–141
  Bureau of Environmental Services test swales, 128–131
  Bureau of Environmental Services Water Pollution Control Laboratory, 141–142
  retention pond at, 142–147
  vegetative filter at, 147–148
  Custer Park, 151–152
  Oregon Museum of Science and Industry, riparian zone management, 148–151
  rainfall analysis of, 126
  stormwater analysis, pre- and post-development, 126–128
  watershed restoration projects in, 131–141
Precipitation, 15
Precipitation grid formation, 524

Prince George's County (MD), low-impact development in, 97–99, 297–298, *see also* Low-impact development (LID)
    Beltway Plaza Expansion, infill development, 299–301
    cost comparisons with conventional stormwater management, 302–304
    Count Office Complex, parking lot retrofit, 301–302
    design manuals published on, 298–299, 302

Q

Qualitative habitat Index, 104

R

Rain gardens, *see* Bioretention
Redesign comparisons, residential subdivision, 49–64, *see also* Subdivision design, residential
Regent Square Gateway, Pittsburgh (PA), watershed restoration, 85–90
Remote sensing technology
    in land cover change modeling, 515–516, *see also* Land cover change, modeling
    in watershed natural resource inventory, 561–562, *see also* Watershed natural resource inventory
Retention storage, 109,111–112, *see also* Wetland(s); Wetland ponds
    basins/ponds,142–147
Rio Hondo Spreading Grounds, 21–23
Riparian Management in Forests of the Continental Eastern United States (Verry, Hornbeck, and Dolloff, Eds.), 5
Riparian zones, 5–6, *see also* Lake(s); Wetland(s); Wetland ponds
    Abbotsford wetland park, British Columbia, 193–202
    definition, 362
    New Hampshire, *see* under New Hampshire
    Newfoundland, Canada, *see* Gander Lake, Newfoundland
    Ontario, Canada, *see* Ontario, Canada, timber management/land use; Ontario, erosion plumes study
    Farmington waterway (MN), 175–191
    Green River Natural Resource Area, Kent (WA), 278–281
    Hearthstone Quarry Brook, Chicopee (MA), 351
    Johnson County Wetlands Mitigation Bank project (KS), 479–489
    Merrick Brook, Scotland (CT), 349–350
    natural resource inventory mapping, 571–573, *see also* Watershed natural resource inventory mapping
    Nine Mile Run watershed, Pittsburgh (PA), 49–64
    Oregon Museum of Science and Industry on Willamette River, 148–151
    restoration of, *see* River restoration, Connecticut
    river behavior, as system, 433–435
    channel evolution in, 435–438
    connectivity, 433
    energy dissipation and free flow, 435–438
    human activities impact, 380, 381
    management practices regarding, 431–433, 438–442
    natural processes of flow, 431
    nitrogen removal, 434
    periodic ecosystem disturbances, 434–435
    Rosgen stream classification system, 440, 441
    watershed management, 380–382
    Salem Salt Marsh restoration, Salem (MA), 351–353
    timber cutting/land use management, *see* Gander Lake, Newfoundland; Ontario, Canada, timber management/land use; Ontario, erosion plumes study
River restoration, Connecticut, 379
    channel daylighting in, 386
    dam removal in, 382–383
    dechannelization in, 383–386
    flow management in, 387
    geomorphic considerations in, 388–390
    habitat improvements in, 387
    Harbor Brook, 384, 390–393
    Naugatuck, 383
    public participation in, 390
    river front access planning in, 386–387
    science of, evolution of, 388–390
    watersheds in, 380–382
Roof design
    for redirection of runoff
        in Lyons, France, 82–84
        in Nine Mile Run (Pittsburgh) watershed, 81, 82–84
    for retention/detention, 38–39, *see also* Bioretention; Eco-roofs
Rosgen stream classification system, 440, 441
Rouge River National Wet Weather Demonstration Project, Detroit (MI), 491–493, 510–511
    combined sewer overflow management in, 494–495, 502–503
    computer modeling in, 496–497
    consensus building in, 496–497
    cost effectiveness of, 496
    data collection effort in, 495–496
    data management/information system in, 506
    financial arrangements in, 508–509
    greenway concepts for, 500
    initiation of, 493
    institutional arrangements in, 508–509
    lessons learned in, 498, 500–509
    luck and political advantages in, 509
    monitoring versus modeling in, 506–508
    public education and involvement in, 497, 498, 504–506
    regulatory framework formation for, 497–498
    Rouge Gateway Master Plan, 499
    Rouge River Watershed, 493
        as Area of Concern in Great Lakes system, 494
    characteristics of, 493–494
    stormwater management, 503–504
    water flow and water quality issue in, 509
    watershed approach to, 500–502

Index 697

Runoff
  parking lot, *see* Parking lot design
  roof, *see* Roof design
  volume maintenance,111–112
Runoff curve number (RCN), 514
  grid formation, 519–520, 527–529

S

Salem Salt Marsh restoration, Salem (MA), 351–353
San Francisco Bay area watershed management plans, 459–460
  baseline studies for, 463
    GIS database, 463–464
    land use/water quality impact models, 464–465
    map/tool kits, 465
    resource sensitivity zones, 465–466
    water quality vulnerability zones, 465
  land management challenges for, 470
  implementation and funding, 472, 474
  policy adjustment over time, 470–471
  short-term versus long-term planning, 471–472
  planning issues for, key, 466
    fire hazard management, 466–467
    golf course management, 468–470
    road management, 467
    trail management, 467–468
  planning process in
    database establishment, 461
    environmental impact reports, 462
    goal establishment, 461
    management plan development, 462
    newsletters, 463
    plan evaluation and selection, 461–462
    public involvement, 461, 462–463
    public opinion surveys, 463
    resource vulnerability maps, 461
      web page, 463
    San Francisco Public Utilities Commission vision of, 460–461
    upstream activities monitoring, 470
Science instruction, *see* Educational opportunities
Sediment transport, logging roads, 635–643
Setback, definition of, 361
Shoreland buffers, *see* Riparian zones, buffers
Simplified Urban Nutrient Output Model (SUNOM), *see* SUNOM (Simplified Urban Nutrient Output Model)
Soil moisture, modeling, 513–515
  data sources for, 517–519
  digital elevation model creation in, 520–522
  geographic information systems in, 515–516
  hydrologic soil group estimation in, 515, 524–527
  indices from, 516–517
    estimation process, 527–529
      flowchart, 526
  land cover classification in, 522–523
  modeling algorithm in, 519–520
  precipitation grid formation in, 524
  remote sensing technology in, 515–516

Squam Lake (NH), watershed natural resource inventory of, 564
  conventional land capability analysis, 564–565
  geographic data display, 565–566
  geographic information systems visualization, 565–566
  geographic referencing and spatial analysis, 566–567
  in-lake resource inventories, 567
  integration of data layers, 567
  results, and implementation, 569
  useful geographic information system products, 567–569
Start at the Source. Design Guidance Manual for Stormwater Quality Protection (Bay Area Stormwater Management Agencies Association), 2
Sterrett school, Pittsburgh (PA), watershed restoration, 82–85
Stonehill Estates (VA), redesign comparisons, 54–60
Stormwater discharge permits, 32
Stormwater management, 1–3,155–157
  bioretention in,125–153, *see also* Bioretention
  detention basins in, 13–14, 31–45, *see also* Detention storage
  development alternative, and impacts, 617–619, 620
  for flood control, *see* Abbotsford wetland park, British Columbia; Farmington waterway (MN)
  in suburban watersheds, 49–64, *see also* Low-impact development (LID); Open space design; Subdivision design
  in urban watersheds, 11–27, *see also* Urban watershed(s)
  infiltration basins in, 18–23, *see also* Infiltration
  low-impact development in, 97–122, *see also* Low-impact development (LID)
  porous pavements in, 23–26, *see also* Porous pavements
  roof detention in,38–39, *see also* Eco-roofs
  wetlands in, 3–4,14, *see also* Wetland(s)
Stormwater parks, *see* Wetland(s)
Stormwater planters, 136–137
Stormwater ponds, 31–45, *see also* Detention storage
Stormwater restoration, *see* Bioretention; Detention storage; Infiltration; Retention storage; Stormwater management; Wetland(s)
  in suburban watersheds, *see* Low-impact development (LID); Open space design; Subdivision design
  in urban watersheds, *see* Urban watershed(s)
Stream quality, studies of, 105
Stream restoration, 105, 385, *see also* River restoration
Subdivision design, residential, 49–50, *see also* Farmington waterway (MN); Frederick County (MD); Low-impact development (LID); Prince George's County (MD)
  comparative analysis of, 49–60
  development costs, 54, 59–60

Duck Crossing (MD), 50–54
   hydrology, 52–53, 59
   nutrient output, 53–54, 59
   Stonehill Estates (Virginia), 54–60
   low density, 50–54
   medium density, 54–60
   open space, 52, 56–58
   other comparisons in, 60–61
SUNOM (Simplified Urban Nutrient Output Model), 50
   in low-density residential modeling, 53–54
   in medium-density residential modeling, 59
Sustainability, watershed
   designing for, 341–343, 347–349
   earth science applications in, 346–347
   ecologic assessments in, 344–346
   examples of, 349–353
   goals and objectives of, 344–346
   interdisciplinary, 343–344
   process of, 344
   planning for, 541–542
      in Florida, 545–552
   process of, 548–549
   regional, 542–545
   Watershed Interactive Network Plan for, in Florida, 549–553
Swale(s), infiltration/bioretention, 20–21
   design tests, Portland (OR), 128–131
   for parking lot runoff, 157–158, 162–164
   in low-impact design, 108–114
   in urban watershed restoration, *see* Fresh Pond Reservation restoration, Cambridge (MA); Nine Mile Run watershed, Pittsburgh (PA)
   Portland (OR), 135–136, 150–151
Sweetwater Wetlands, Tucson (AZ), 281–286
Sydney (Australia), *see* Homebush Bay area (Sydney, Australia)

T

Texas, *see* DuPont Victoria Wetlands, Victoria (TX)
Thailand, Sisaket and Chachoengsao Provinces
   digital elevation models, 527, 528
   land cover change in, 519, 520
   classifications, 522–523, 529, 532, 534
   modeling, 513–529, 530–531, 533, *see also* Soil moisture, modeling
   study areas and data sources, 517–519, 522
   rainfall gauge stations in, 521
   relative soil moisture index grids, 533
   runoff curve number grids, 531
Thermal battery, 82–83
Timber management/land use, *see* Gander Lake, Newfoundland, land–water interaction models; Ontario, Canada, timber management/land use
Time of concentration (Tc), 109
   predevelopment, maintaining, 110–111
Top Ten Watershed Lessons Learned (EPA), 7
Trophic status modeling, *see* Lake trophic status modeling

U

Universal Soil Loss Equation, 580–581
Urban watershed(s), 11–12,107–108
   landscape dynamics in, 15–18
   management/restoration of
   Boston (MA), 332–338
   detention basins and,13–14, 31–45
   Galilee, Israel, 615–632
   High Point (NC), 306–307
   Homebush Bay area (Sydney, Australia), 247–260
   infiltration basins and, 18–23
      low-impact development in, 100, 114, 116–117
   Nine Mile Run, Pittsburgh (PA), 69–94
   Peachtree Creek, Atlanta (GA), 15–17
   policies for, 88–91
   porous pavements and, 23–26
   Portland (OR), 125–153
   Rouge River National Wet Weather Demonstration Project, Detroit (MI), 491–511
      San Francisco Bay area, 459–474
   wetland ponds and, 14
   stormwater discharge in, 12–13
   surface and subsurface processes of, 27

V

Vegetative filters, 147–148, *see also* Bioretention; Plant selection
Vernal pool hydrology, effects of bioretention on, 172
Virginia
   Stonehill Estates, redesign comparisons, 54–60
   University of, long-term bioretention study, 310

W

Wakodahatchee Wetlands, Palm Beach County (FL), 290–294
Washington, *see* Green River Natural Resource Area, Kent (WA)
Water design, 1
Water gardens, 142–148, *see also* Bioretention
Water planning, 1
Water sensitive design, 9–10, 359–360
Water wall, 82–84
Watershed(s), 6–7, 341
   assessment of, 477–489, *see also* Watershed(s), modeling of
   greenways and design of, 400–402
   impervious cover as quality indicator for, 104–107
   in river restoration, 380–382, *see also* Riparian zones
   in suburban areas,49–64, *see also* Farmington waterway (MN); Johnson County Wetlands Mitigation Bank project (KS); Open space design; Prince George's County (MD); Wetland(s); Wetland ponds
   in urban areas,11–27, 67–69, 107–108, *see also* Urban watershed(s)

Index 699

water quality improvement for water supply, *see* Homebush Bay area (Sydney, Australia); Sweetwater Wetlands, Tucson (AZ); Wakodahatchee Wetlands, Palm Beach County (FL); Wetlands-Based Indirect Potable Reuse Project
    lake shore capacity planning in, 647–679, *see also* Lake shore capacity planning
    low-impact development in, 100, 114, 116–117, *see also* Low-impact development (LID)
    modeling of aquifer recharge, 615–632
    hydrologic soils/land cover, 513–534
    land–water interaction, comparison of models, 577–596
    natural resource inventory mapping, 557–575
    natural resource inventory of, *see* Watershed natural resource inventory mapping
    restoration of, 67–69, 91–93, *see also* Riparian zones; Urban watershed(s); Wetland(s); Wetland ponds
        low-impact development in, 100, 114, 116–117, *see also* Low-impact development (LID)
    sustainability of, *see also* under Sustainability
    designing for, 341–349
    planning for, 541–553
Watershed natural resource inventory mapping, 557
    critical land and resources analysis, 573–575, *see also* Land cover change, modeling
    data collection and analysis in
    global information systems, 557–561
    global positioning systems, 562–563
    remote sensing technology, 561–562
    in New Hampshire, 558
    Lake Chocorua (NH), 569–571
    Squam Lake (NH), 564
    conventional land capability analysis, 564–565
    geographic data display, 565–566
    geographic information systems visualization, 565–566
    geographic referencing and spatial analysis, 566–567
    in-lake resource inventories, 567
    integration of data layers, 567
    results, and implementation, 569
    useful geographic information system products, 567–569
    wetland/riparian buffers, Deerfield (NH), 571–573
West Palm Beach (FL), Wetlands Based Potable Reuse Project, 205–211
Wetland(s), and wetland parks, 3–4, 263–264, 275–276, 294, *see also* Wetland ponds
    buffers of, 361–375, *see also* Riparian zones, buffers in
    construction of, 271–272, *see also* specific wetland project
    plans typically required in, 273–275
    design criteria for, typical, 266–267
    for science instruction, 235–244, *see also* Exploration Place, Wichita (KS)
    goals and objectives in creation of, 277–278, *see also* specific wetland project
    maintenance of, 272, 275, *see also* specific wetland project
    of Cornell University, Ithaca (NY), 155–172, *see also* Cornell University, Ithaca (NY)
    of DuPont Victoria Wetlands, Victoria (TX), 286–290
    of Farmington (MN) waterway, 175–191, *see also* Farmington waterway (MN)
    of Fishtrap Creek, Abbotsford, British Columbia, 193–202, *see also* Abbotsford wetland park, British Columbia
    of Green River Natural Resource Area (Kent, WA), 278–281
    of Johnson County Wetlands Mitigation Bank project (KS), 479–489
    of Sweetwater Wetlands, Tucson (AZ), 281–286
    of Wakodahatchee Wetlands, Palm Beach County (FL), 290–294
    planning and feasibility assessment for, 264–271
    data requirements, 269
    plant selection for, 271, *see also* specific wetland project
    site selection criteria for, 268–269
    water balance design for, 270
    water quality improvement for urban water supply, *see* Homebush Bay area (Sydney, Australia); Sweetwater Wetlands, Tucson (AZ); Wakodahatchee Wetlands, Palm Beach County (FL); Wetlands-Based Indirect Potable Reuse Project
Wetland ponds, 14, *see also* projects under Wetland(s)
    buffers of, 361–375, *see also* Riparian zones, buffers in
    of Northern Water Feature, Sydney, Australia, 255
    restoration of, in Boston (MA) area, 215–216
    Chandler Pond, 317–338, *see also* Chandler Pond restoration
    funds needed for, 229
    Hall's Pond, 216–223
    Hill's Pond, 223–228
    professional consultation in, 229
    public educational process in, 231
    soils testing in, 228–229
    vegetation management plan required in, 231
    wetland expertise required for, 231
Wetland treatment system, 205–206
Wetlands and Urbanization (Azous and Horner, Eds.), 3
Wetlands-Based Indirect Potable Reuse Project, 205
    advanced wastewater treatment system development for, 207
    design goals for, 208–211
    baseline monitoring of wetlands in, 207–208
    goals of, 205
    hydrologic features of West Palm Beach (FL), 205–206
    hydrologic modeling in, 208
    Wetland Reuse Site in, 211
    wetlands demonstration project for, 207
Wilderness Valley project, *see* Johnson County Wetlands Mitigation Bank project (KS)